T0348662

Optimization in Chemical Engineering

Also of interest

Process Intensification.
Design Methodologies
Gómez-Castro, Segovia-Hernández (Eds.), 2019
ISBN 978-3-11-059607-6, e-ISBN 978-3-11-059612-0

Process Optimization.
Decision-Making Tools for Chemical Engineers
Zondervan, 2025
ISBN 978-3-11-134208-5, e-ISBN 978-3-11-134228-3

Sustainable Process Integration and Intensification.
Saving Energy, Water and Resources
Klemeš, Varbanov, Wan Alwi, Manan, Fan, Chin, 2023
ISBN 978-3-11-078283-7, e-ISBN 978-3-11-078298-1

Industrial Process Plants.
Global Optimization of Utility Systems
Nath, 2024
ISBN 978-3-11-101531-6, e-ISBN 978-3-11-102067-9

Product and Process Design.
Driving Sustainable Innovation
Harmsen, de Haan, Swinkels, 2024
ISBN 978-3-11-078206-6, e-ISBN 978-3-11-078212-7

Optimization in Chemical Engineering

Deterministic, Meta-Heuristic and Data-Driven Techniques

Edited by
Fernando Israel Gómez-Castro and Vicente Rico-Ramírez

DE GRUYTER

Authors
Fernando Israel Gómez-Castro
Departamento de Ingeniería Química
Universidad de Guanajuato
Noria Alta S/N
36050 Guanajuato
Mexico

Vicente Rico-Ramírez
Departamento de Ingeniería Química
Tecnológico Nacional de México en Celaya
Celaya
38010 Guanajuato
Mexico

ISBN 978-3-11-138338-5
e-ISBN (PDF) 978-3-11-138343-9
e-ISBN (EPUB) 978-3-11-138362-0

Library of Congress Control Number: 2024951445

Bibliographic information published by the Deutsche Nationalbibliothek
The Deutsche Nationalbibliothek lists this publication in the Deutsche Nationalbibliografie;
detailed bibliographic data are available on the Internet at http://dnb.dnb.de.

© 2025 Walter de Gruyter GmbH, Berlin/Boston, Genthiner Straße 13, 10785 Berlin
Cover image: metamorworks/iStock/Getty Images Plus
Typesetting: Integra Software Services Pvt. Ltd.

www.degruyter.com
Questions about General Product Safety Regulation:
productsafety@degruyterbrill.com

Contents

List of contributing authors

Juan Gabriel Segovia-Hernández
División of Natural and Exact Sciences
Departament of Chemical Engineering
University of Guanajuato
Campus Guanajuato
Noria Alta S/N Col. Noria Alta
36050Guanajuato
Mexico
gsegovia@ugto.mx

Maricruz Juárez-García
División of Natural and Exact Sciences
Departament of Chemical Engineering
University of Guanajuato
Campus Guanajuato
Noria Alta S/N Col. Noria Alta
36050 Guanajuato
Mexico

Eduardo Sánchez-Ramírez
División of Natural and Exact Sciences
Departament of Chemical Engineering
University of Guanajuato
Campus Guanajuato
Noria Alta S/N Col. Noria Alta
36050 Guanajuato
Mexico

José A. Caballero
Institute of Chemical Process Engineering
University of Alicante
Carretera de S. Vicente s.n.
03690 Alicante
Spain

Luis Fernando Lira-Barragán
Chemical Engineering Department
Universidad Michoacana de San Nicolás de
Hidalgo
Morelia, Michoacán
México

Fabricio Nápoles-Rivera
Chemical Engineering Department
Universidad Michoacana de San Nicolás de
Hidalgo
Morelia, Michoacán
México

José María Ponce-Ortega
Chemical Engineering Department
Universidad Michoacana de San Nicolás de
Hidalgo
Morelia, Michoacán
México
jose.ponce@umich.mx

Oscar Daniel Lara-Montaño
Facultad de Ingeniería
Universidad Autónoma de Querétaro
Cerro de las Campanas S/N Col. Las Campanas
76010 Querétaro
Mexico

Manuel Toledano-Ayala
Facultad de Ingeniería
Universidad Autónoma de Querétaro
Cerro de las Campanas S/N Col. Las Campanas
76010 Querétaro
Mexico
toledano@uaq.mx

Claudia Gutiérrez-Antonio
Facultad de Ingeniería
Universidad Autónoma de Querétaro
Cerro de las Campanas S/N Col. Las Campanas
76010 Querétaro
Mexico

Fernando Israel Gómez-Castro
División de Ciencias Naturales y Exactas
Departamento de Ingeniería Química
Universidad de Guanajuato
Campus Guanajuato
Noria Alta S/N Col. Noria Alta
36050 Guanajuato
Mexico

https://doi.org/10.1515/9783111383439-203

Elena Niculina Dragoi
Cristofor Simionescu Faculty of Faculty of
Automatic Control and Computer Engineering
Gheorghe Asachi Technical University of Iasi
Str. Prof. Dr. Doc. Dimitrie Mangeron, nr. 27
Iași, 700050
Romania

Salvador Hernández
División de Ciencias Naturales y Exactas
Departamento de Ingeniería Química
University of Guanajuato
Campus Guanajuato
Noria Alta S/N Col. Noria Alta
36050 Guanajuato
Mexico

Mathias Neufang
Department of Chemical Engineering
Imperial College London
Exhibition Rd
South Kensington
London SW7 2AZ
United Kingdom
mathias.neufang22@imperial.ac.uk

Emma Pajak
Department of Chemical Engineering
Imperial College London
United Kingdom

Damien van de Berg
Department of Chemical Engineering
Imperial College London
United Kingdom

Ye Seol Lee
Department of Chemical Engineering
Imperial College London
United Kingdom

Ehecatl Antonio del Rio Chanona
Department of Chemical Engineering
Sargent Centre for Process Systems Engineering
Imperial College London
London SW72AZ
United Kingdom
a.del-rio-chanona@imperial.ac.uk

Daniel Rangel-Martínez
Department of Chemical Engineering
University of Waterloo
200 University Ave W
Waterloo
Ontario, Canada N2L 3G1

Luis A. Ricardez-Sandoval
Department of Chemical Engineering
University of Waterloo
200 University Ave W
Waterloo
Ontario, Canada N2L 3G1
laricard@uwaterloo.ca

Antonio Flores-Tlacuahuac
Institute of Advanced Materials for Sustainable
Manufacturing
Instituto Tecnológico y de Estudios Superiores
de Monterrey
Av. Eugenio Garza Sada 2501
Monterrey, Nuevo León 64849
México
antonio.flores.t@tec.mx

Seyed Reza Nabavi
Department of Applied Chemistry
Faculty of Chemistry
University of Mazandaran
Babolsar
Iran

Zhiyuan Wang
Department of Computer Science
DigiPen Institute of Technology Singapore
Singapore 139660
Singapore

Gade Pandu Rangaiah
Department of Chemical and Biomolecular
Engineering
National University of Singapore
Singapore 117585
Singapore
And
School of Chemical Engineering
Vellore Institute of Technology
Vellore 632014, Tamil Nadu
India
chegpr@nus.edu.sg

Kai Kruber
Institute of Process Systems Engineering
Hamburg University of Technology
Am Schwarzenberg-Campus 4
Hamburg 21073
Germany

Siv Kinau
Institute of Process Systems Engineering
Hamburg University of Technology
Am Schwarzenberg-Campus 4
Hamburg 21073
Germany

Mirko Skiborowski
Institute of Process Systems Engineering
Hamburg University of Technology
Am Schwarzenberg-Campus 4
Hamburg 21073
Germany
mirko.skiborowski@tuhh.de

Urmila Diwekar
Vishwamitra Research Institute
Clarendon Hills, IL 60514
United States of America
and
Richard and Loan Hill Department of Biomedical
Engineering
University of Illinois at Chicago
Crystal Lake, IL 60012
United States of America
urmila@vri-custom.org

Riju De
Department of Chemical Engineering
Birla Institute of Technology and Science Pilani
K. K. Birla Goa Campus
Zuarinagar 403726, Goa
India

Yogendra Shastri
Department of Chemical Engineering
Indian Institute of Technology
Mumbai 400076, Maharashtra
India
yshastri@che.iitb.ac.in

Yulissa Mercedes Espinoza-Vázquez
División de Ciencias Naturales y Exactas
Departamento de Ingeniería Química
Universidad de Guanajuato
Campus Guanajuato
Noria Alta S/N Col. Noria Alta
36050 Guanajuato
Mexico

Nereyda Vanessa Hernández-Camacho
División de Ciencias Naturales y Exactas
Departamento de Ingeniería Química
Universidad de Guanajuato
Campus Guanajuato
Noria Alta S/N Col. Noria Alta
36050 Guanajuato
Mexico

Jahaziel Alberto Sánchez-Gómez
División de Ciencias Naturales y Exactas
Departamento de Ingeniería Química
Universidad de Guanajuato
Campus Guanajuato
Noria Alta S/N Col. Noria Alta
36050 Guanajuato
Mexico

José Ezequiel Santibañez-Aguilar
Escuela de Ingeniería y Ciencias
Instituto Tecnológico y de Estudios Superiores
de Monterrey
Av. Eugenio Garza Sada 2501
Monterrey, 64849 Nuevo León
México

David Esteban Bernal Neira
Davidson School of Chemical Engineering
Purdue University
West Lafayette, Indiana
United States of America
And
Research Institute of Advanced Computer
Science
Universities Space Research Association
Mountain View, California
United States of America
And
Quantum Artificial Intelligence Laboratory
NASA Ames Research Center
Moffett Field, California
United States of America

Vicente Rico-Ramírez
Departamento de Ingeniería Química
Tecnológico Nacional de México en Celaya
Av. Tecnológico y García Cubas S/N
Celaya, 38010 Guanajuato
Mexico

Juan Gabriel Segovia-Hernández*, Maricruz Juárez-García, and Eduardo Sánchez-Ramírez

Chapter 1
Optimization and its importance for chemical engineers: challenges, opportunities, and innovations

Abstract: This chapter delves into the critical role of mathematical optimization in chemical engineering, exploring the challenges and opportunities that shape process design, operation, and control. The intricate complexity of chemical processes, characterized by numerous variables, nonlinearity, high dimensionality, and uncertainty, poses significant challenges for optimization. Advanced optimization techniques, including mixed-integer nonlinear programming, dynamic optimization, and stochastic optimization, are essential for addressing these complexities.

This chapter presents substantial opportunities for improving optimization efficiency, promoting sustainability, fostering innovation, and providing robust decision support. By optimizing resource allocation, production scheduling, and energy utilization, chemical engineers can achieve cost savings and operational improvements. Incorporating sustainability metrics into optimization models aids in minimizing environmental impact and enhancing resource efficiency. Furthermore, optimization fosters innovation by enabling novel process configurations and advanced control strategies, driving technological advancements in the field.

The chapter also explores future directions in optimization within the context of circular economy, artificial intelligence (AI), and Industry 4.0. Integration of circular economy principles, advancements in AI and machine learning, and digitalization are revolutionizing chemical engineering processes. Multi-objective and multi-scale optimization approaches are increasingly crucial for addressing the complexity of modern chemical engineering systems. Collaborative and interdisciplinary research is emphasized as a key driver of innovation, enabling the development of cutting-edge optimization techniques and tools.

Thus, this chapter highlights how optimization in chemical engineering is evolving to meet the demands of efficiency, sustainability, and innovation, paving the way for transformative changes in the industry.

*Corresponding author: Juan Gabriel Segovia-Hernández, Department of Chemical Engineering, University of Guanajuato, Campus Guanajuato, División of Natural and Exact Sciences, Noria Alta S/N Col. Noria Alta, Guanajuato 36050, Mexico, e-mail: gsegovia@ugto.mx
Maricruz Juárez-García, Eduardo Sánchez-Ramírez, Department of Chemical Engineering, University of Guanajuato, Campus Guanajuato, División of Natural and Exact Sciences, Noria Alta S/N Col. Noria Alta, Guanajuato 36050, Mexico

https://doi.org/10.1515/9783111383439-001

Keywords: optimization, deterministic methods, stochastic methods, linear problems, nonlinear problems

1.1 Introduction

Since the beginnings of chemical engineering, there has been a need to develop processes that meet certain established objectives under certain conditions. This problem of attaining an objective (minimizing or maximizing an objective function), subject to a set of conditions (constraints), is what we know as optimization problem. Initially, the objectives to meet were primarily economic, but this has evolved over time. Nowadays, numerous aspects, encompassing efficiency and sustainability, must be taken into account in the design of a chemical process. In this regard, optimization has been a fundamental tool for solving countless problems within the field of chemical engineering. From process design, plant location, and material selection to maximizing production, minimizing waste and resource utilization, and even in process control, optimization is a major tool in the field, and this chapter delves into the fundamental role that optimization plays in the realm of chemical engineering and how its application influences every aspect of this discipline. However, it is also important to note that while optimization has helped to innovate the industry, as we know it today, the emergence of increasingly complex optimization problems within the field has also driven the development of various optimization techniques, leading to very sophisticated tools that are used today [1].

With the advent of Industry 4.0 and the integration of digital technologies, optimization has taken on an even greater significance. Through the utilization of advanced algorithms, machine learning, and real-time data analytics, chemical engineers can fine-tune processes, optimize resource utilization, and enhance overall operational performance like never before. Furthermore, optimization plays a crucial role in promoting sustainability within the chemical industry. By optimizing processes to minimize energy consumption, reduce emissions, and maximize resource efficiency, chemical engineers contribute to the global effort toward a more environmentally friendly and sustainable future. Whether it is through the development of greener production methods, the implementation of circular economy principles or the design of eco-friendly products, optimization serves as a driving force in fostering a more sustainable chemical industry.

In this chapter, we will discuss some of the historic facts that allowed the development of the optimization as a field. Then, some of the commonly used optimization techniques in the field of chemical engineering will be reviewed and some concrete examples of application in the area will be given. The importance of optimization in driving innovation, enhancing competitiveness, and addressing the complex challenges of the modern chemical industry will be discussed as well as its role in shaping the future of chemical engineering in the era of Industry 4.0 and sustainability. Additionally, we will discuss emerging trends such as digital twins, predictive maintenance, and autonomous systems, which further underscore the vital role of optimization in optimizing processes and decision-making in the dynamic landscape of the chemical industry. By embracing optimization as a strategic imperative, chemical engineers can unlock new opportunities, mitigate risks, and lead the way toward a more prosperous and sustainable future for the industry and society as a whole.

1.1.1 History of optimization

The history of mathematical optimization is a journey marked by the relentless pursuit of efficiency, precision, and optimal solutions across diverse domains of science, engineering, and industry.

The origins of mathematical optimization date back to ancient civilizations, where early mathematicians grappled with problems of allocation, logistics, and resource management. From the ancient Greeks' study of geometric optimization to the optimization challenges faced by Renaissance-era architects and engineers, the quest for optimal solutions has been a recurring theme throughout history. However, it was not until the late nineteenth and the early twentieth century that mathematical optimization began to emerge as a distinct field of study. With the advent of linear programming (LP) in the mid-twentieth century, spearheaded by luminaries such as George Dantzig and John von Neumann, optimization took a giant leap forward, offering powerful tools for solving complex optimization problems with linear constraints. The subsequent decades witnessed an explosion of research and innovation in mathematical optimization, fueled by advancements in computer science, algorithms, and numerical methods. Today, mathematical optimization stands at the forefront of scientific and technological progress, playing a critical role in fields as varied as operations research, finance, engineering design, logistics, and data science [2]. Next, we embark on a journey through the rich tapestry of the history of mathematical optimization, tracing its evolution from ancient origins to contemporary applications.

1.1.1.1 Pre-Christian times

Although everyday problems inherent to preservation and development of civilization were not explicitly formulated as optimization problems, such problems have always been solved by considering an objective or multiple objectives as guideline. Therefore, the origins of mathematical optimization can be traced back to antiquity, where early civilizations faced practical challenges that require efficient allocation of resources and optimal decision-making. In the pre-Christian era, diverse civilizations across Mesopotamia, China, the Indus Valley, Mesoamerica, and Persia demonstrated a remarkable utilization of optimization principles in various domains despite the absence of formalized mathematical frameworks [3].

In ancient Egypt, architects and engineers employed geometric techniques to optimize the construction of pyramids and other monumental structures. Mathematicians such as Euclid and Archimedes made significant contributions to geometry and mechanics, and Euclidean geometry provided a rigorous framework for studying geometric optimization problems, including the determination of optimal shapes and configurations for various engineering and architectural applications. In the Indus Valley civilization, urban planners showcased a sophisticated understanding of optimization by meticulously organizing cities like Mohenjo Daro and Harappa to maximize efficiency in transportation, sanitation, and trade.

The Babylonians were renowned for their advancements in mathematics, notably applied optimization techniques to enhance agricultural practices. Similarly, the Qin dynasty in ancient China employed optimization concepts in urban planning endeavors, strategically designing city layouts to optimize defense, commerce, and social cohesion [4–5]. Meanwhile, the Maya civilization in Mesoamerica harnessed advanced mathematical concepts to optimize agricultural production, leveraging intricate irrigation systems and celestial-based agricultural calendars [4].

1.1.1.2 Middle Ages

The Middle Ages witnessed significant developments in the field of mathematical optimization. In medieval Europe, the legacy of ancient Greek and Roman knowledge was preserved and enriched by scholars such as Boethius, Alcuin of York, and Gerbert of Aurillac, who played key roles in transmitting mathematical and scientific texts from antiquity to the medieval era. While mathematical optimization per se was not a dominant focus of medieval scholarship, the principles of geometry, arithmetic, and algebra laid the groundwork for practical applications in architecture, engineering, and commerce [6].

One notable area of optimization during the Middle Ages was the design and construction of Gothic cathedrals, architectural marvels characterized by their soaring spires, intricate vaulted ceilings, and a vast number of stained glass windows. Master builders and craftsmen employed geometric and structural optimization techniques to achieve architectural grandeur while ensuring structural stability and load-bearing capacity. In the realm of agriculture, medieval farmers employed heuristic techniques to maximize crop yields and manage land efficiently, despite the limitations of medieval agricultural technology [7]. Additionally, medieval guilds and merchant associations applied optimization principles in commerce and trade. By coordinating production, distribution, and pricing strategies, these organizations sought to maximize profits and minimize risks, contributing to the economic prosperity of medieval cities [8].

In the Islamic world, mathematicians advanced the study of algebra, geometry, and trigonometry. One of the most enduring legacies of medieval Islamic mathematics is the development of algebra, the systematic approach to problem-solving, advocated by scholars like Al-Khwarizmi, who in his seminal work "Al-Kitab al-Mukhtasar fi Hisab al-Jabr wal-Muqabala" (*The Compendious Book on Calculation by Completion and Balancing*) paved the way for the emergence of calculus and algebraic optimization methods in subsequent centuries [6].

1.1.1.3 The Renaissance period

Renaissance scholars such as Leonardo da Vinci, Luca Pacioli, and Niccolò Tartaglia embraced the principles of mathematical rigor and empirical observation, applying them to a wide range of disciplines, including art, architecture, mechanics, and commerce. Renaissance geometers such as Johannes Kepler and René Descartes developed new geometric methods and techniques, laying the groundwork for the modern field of analytic geometry.

One of the most significant advancements during the Renaissance period was the development of techniques for solving algebraic equations and systems of equations. Mathematicians such as François Viète, Rafael Bombelli, and Girolamo Cardano introduced new methods for solving polynomial equations, including the famous cubic and quartic equations. These algebraic techniques provided powerful tools for solving optimization problems in fields such as commerce, engineering, and astronomy, paving the way for the emergence of calculus and modern optimization theory [9].

Navigators and cartographers utilized mathematical algorithms and celestial observations to optimize sea routes, reducing travel time and enhancing maritime commerce [10]. Renaissance-era military commanders, inspired by mathematical theories of geometry and calculus, devised strategic formations and maneuvers to optimize troop deployment and battlefield tactics. Treatises on military engineering and fortification, authored by scholars like Niccolò Machiavelli, provided insights into the strategic application of optimization principles in warfare, influencing military tactics for centuries to come [11].

1.1.1.4 Seventeenth–nineteenth centuries

During the seventeenth century, mathematicians such as Pierre de Fermat, Blaise Pascal, and John Wallis developed new methods for finding maxima and minima of functions, laying the groundwork for the calculus of variations and optimization theory. The development of differential calculus allowed for the rigorous analysis of optimization problems involving continuously varying quantities, opening new avenues for mathematical exploration and practical applications.

The eighteenth century saw the consolidation of optimization principles and techniques into a coherent mathematical framework, driven by the work of mathematicians such as Leonhard Euler, Joseph-Louis Lagrange, and Carl Friedrich Gauss. Lagrange formalized the study of constrained optimization with the introduction of Lagrange multipliers, providing a powerful tool for solving optimization problems subject to equality constraints. Meanwhile, Euler and Gauss made significant contributions to the calculus of variations, extending the principles of differential calculus to the optimization of functionals and integral quantities. These developments laid the foundation for the modern theory of optimization and its applications in physics, engineering, and economics [12].

The nineteenth century witnessed the confluence of mathematical optimization with the burgeoning fields of industrialization, transportation, and urban planning. Engineers and economists applied optimization principles to solve practical problems such as optimal resource allocation, production planning, and transportation logistics. The development of numerical analysis, probability theory, and statistical methods paved the way for the development of optimization algorithms such as gradient descent, genetic algorithms (GAs), and simulated annealing, further expanding the scope and applicability of optimization techniques.

The centuries spanning from the seventeenth to the nineteenth witnessed remarkable advancements in the field of mathematical optimization, driven by the convergence of mathematical theory, technological innovation, and practical application [3]. Through the pioneering efforts of mathematicians, engineers, and economists, optimization principles became essential tools for solving complex problems in science, engineering, and industry, laying the foundation for the modern era of computational optimization.

1.1.1.5 Twentieth century

One of the most significant milestones in optimization in the twentieth century was the development of LP, which revolutionized decision-making processes and resource allocation in diverse fields. In 1947, George Dantzig introduced the simplex method, a powerful algorithm for solving LP problems, laying the foundation for the widespread adoption of optimization techniques in industry, logistics, and economics. LP enabled businesses to optimize production processes, transportation networks, and supply chains, leading to significant improvements in efficiency, productivity, and profitability.

The field of operations research emerged as a distinct discipline in the mid-twentieth century, driven by the need to apply mathematical and scientific methods to solve complex problems in military planning, logistics, and management [13]. During World War II, mathematicians and engineers such as John von Neumann, Norbert Wiener, and George B. Dantzig pioneered operations research techniques to optimize military operations, including troop movements, supply distribution, and strategic planning. After the war, operations research found applications in civilian sectors such as manufacturing, telecommunications, and healthcare, becoming an essential tool for optimizing business processes and decision-making.

The advent of digital computers in the mid-twentieth century revolutionized the field of optimization, enabling the development of powerful algorithms for solving complex optimization problems with unprecedented speed and accuracy. In the 1950s and 1960s, researchers such as George Dantzig, John Kemeny, and Harold Kuhn made significant advancements in computational optimization techniques, including the revised simplex method. Their work paved the way for the automation of optimization processes and the solution of large-scale optimization problems. It is important to note, however, that the interior-point method was later introduced by Narendra Karmarkar, and the branch-and-bound algorithm was developed by Alisa Land and Alison Doig [14].

The latter half of the twentieth century witnessed further advancements in optimization theory and algorithms, driven by the rapid progress in computer science, numerical analysis, and mathematical modeling. Nonlinear programming (NLP), integer programming, dynamic programming (DP), and stochastic optimization emerged as prominent subfields of optimization, addressing a wide range of optimization problems with nonlinear, discrete, and uncertain characteristics. The development of optimization software packages such as AMPL, GAMS, and MATLAB provided researchers and practitioners with powerful tools for modeling, solving, and analyzing optimization problems in diverse domains [15].

The twentieth century was a transformative period for mathematical optimization, characterized by the development of novel theories, algorithms, and applications that revolutionized science, engineering, economics, and industry. From the emergence of LP and operations research to the advent of modern computational optimization techniques, the achievements of the twentieth century laid the foundation for the continued evolution of optimization in the twenty-first century and beyond [14].

1.1.1.6 Twenty-first century

The twenty-first century has witnessed unprecedented advancements in the field of mathematical optimization, driven by the convergence of computational power, algorithmic innovation, and interdisciplinary collaboration. From the optimization of complex systems in transportation and logistics to the design of efficient algorithms for machine learning and data analysis, optimization techniques have become indispensable tools for addressing the challenges of the modern world [14].

One of the defining features of optimization in the twenty-first century is the proliferation of large-scale optimization problems with diverse constraints and objectives. With the exponential growth of data and the complexity of real-world systems, researchers and practitioners are confronted with optimization challenges of unprecedented scale and complexity. Optimization techniques such as mixed-integer programming, metaheuristic algorithms, and stochastic optimization have emerged as essential tools for tackling these challenges, enabling the solution of large-scale optimization problems in diverse domains [16].

The rise of machine learning and data-driven decision-making has fueled the demand for optimization techniques that can handle uncertainty, nonlinearity, and high-dimensional data. In recent years, researchers have developed novel optimization algorithms, tailored to the unique characteristics of machine learning models such as deep learning networks, reinforcement learning algorithms, and probabilistic graphical models. These advancements have enabled the optimization of machine learning models for tasks such as classification, regression, clustering, and reinforcement learning, leading to significant improvements in predictive accuracy, scalability, and efficiency.

The integration of optimization techniques with emerging technologies such as cloud computing, big data analytics, and Internet of things (IoT) has further expanded the scope and applicability of optimization in the twenty-first century. By harnessing the power of distributed computing platforms and real-time data streams, researchers and practitioners can solve optimization problems in dynamic and uncertain environments, optimizing resource allocation, scheduling, and decision-making in real time [17]. Applications of optimization in areas such as smart cities, renewable energy systems, healthcare delivery, and financial services are transforming industries and improving the quality of life for people around the world.

In addition to its applications in science, engineering, and industry, optimization has also found new frontiers in interdisciplinary research and education. Collaborations between mathematicians, computer scientists, engineers, and domain experts are leading to groundbreaking advancements in optimization theory, algorithms, and applications. Interdisciplinary programs and initiatives are fostering the development of the next generation of optimization researchers and practitioners, equipping them with the knowledge and skills needed to address the complex challenges of the twenty-first century.

The twenty-first century has seen remarkable advancements in the field of mathematical optimization, driven by the convergence of computational power, algorithmic innovation, and interdisciplinary collaboration. From the optimization of complex systems in transportation and logistics to the design of efficient algorithms for machine learning and data analysis, optimization techniques are playing an increasingly important role in addressing the challenges of the modern world. By harnessing the power of optimization, researchers and practitioners are shaping the future of science, engineering, and society in the twenty-first century and beyond (Figure 1.1).

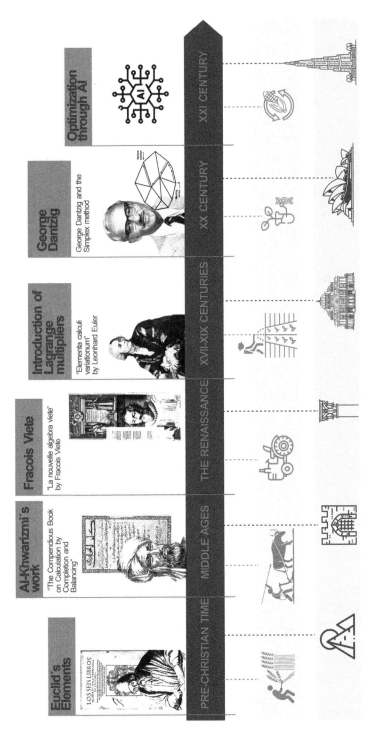

Figure 1.1: Timeline of advances in mathematical optimization in history.

1.2 Optimization techniques in chemical engineering

In chemical engineering, the application of mathematical optimization techniques encompasses a broad spectrum of challenges encountered in the design, operation, and optimization of chemical processes. Each type of optimization technique offers unique strengths and capabilities, making them suitable for different types of problems encountered in the field, such as linear or nonlinear problems, discrete or continuous, constrained or unconstrained, static or dynamic, single objective or multi-objective, and deterministic or stochastic. By leveraging these diverse capabilities of the optimization techniques, chemical engineers can address a wide range of complex challenges and drive innovation in process design, operation, and control [18].

1.2.1 Deterministic optimization techniques

Mathematical programming was defined by Dantzig and Thapa (1997) as the branch of mathematics that deals with techniques for maximizing or minimizing an objective function, subject to linear, nonlinear, and integer constraints on the variables. Mathematical programming involves the study of the mathematical structure of optimization problems and proposes methods to solve these problems.

LP, for instance, provides a powerful framework for optimizing processes with linear relationships between variables. Its ability to efficiently handle large-scale problems makes it well-suited for tasks such as production planning, supply chain optimization, and resource allocation. LP techniques are particularly valuable in situations where decision variables can be represented as continuous quantities and constraints can be expressed as linear equations or inequalities [19].

NLP extends the principles of LP to problems with nonlinear objective functions or the feasible region is bounded by nonlinear constraints. This allows NLP techniques to address a wider range of optimization problems encountered in chemical engineering, including those involving nonlinearity in reaction kinetics, thermodynamics, or material balances. Applications of NLP in chemical engineering span reactor design, process optimization under nonlinear constraints, and parameter estimation for mathematical models [20].

Mixed-integer linear programming and mixed-integer nonlinear programming (MINLP) techniques are essential for handling optimization problems with discrete decision variables. These techniques are commonly employed in chemical engineering for tasks such as equipment sizing, production scheduling, and process synthesis, where decisions must be made on the selection and configuration of discrete components or operating modes.

DP is particularly well-suited for optimizing processes over multiple time periods or stages. In chemical engineering, DP finds applications in dynamic optimization problems involving time-varying constraints, such as the optimization of batch processes, continuous processes with time-varying parameters, and the design of optimal control strategies for chemical reactors and distillation columns [19].

1.2.2 Metaheuristic optimization techniques

Metaheuristic optimization is a family of approximate optimization and general-purpose search algorithms. These methods are based on iterative procedures where the search is guided under a subordinate heuristic by intelligently combining different concepts to explore and properly exploit the search space. These methods are useful when an exact method requires a considerable amount of time and resources, and when a global optimum solution is not required and an approximate solution of satisfactory quality is sufficient. In general, metaheuristic methods seem to be a generic algorithm framework that can be applied to almost all optimization problems. Metaheuristic algorithms have two main components: intensification or exploitation and diversification or exploration [21]. Diversifying the objective is to generate several solutions or explore the search space in distant regions, while intensifying the objective is to focus the search effort on a local region (exploiting the space), knowing that a current good solution is found in this region. A reliable metaheuristic algorithm will have an appropriate combination of these two components, thus guaranteeing acceptable solutions close to the global optimum.

Some ways to classify metaheuristic methods are: a) constructive methods, in which an initial solution is started and components are added until a solution is built, b) path-based methods, in which an initial solution is started and a local search algorithm is used to iteratively replace it with a better solution, and c) population-based methods, in which multiple starting points in the search space are used and evolve in parallel. Metaheuristic optimization techniques, including simulated annealing, particle swarm optimization, and ant colony optimization, offer versatile tools for tackling combinatorial optimization problems encountered in chemical engineering. These methods are often used for optimizing network configurations, facility layouts, and process synthesis problems, where the search space is large and discrete [22].

GAs and other evolutionary optimization techniques offer a robust approach to solving optimization problems where traditional methods may struggle due to nonconvexity, discontinuity or high dimensionality. GA techniques are widely used in chemical engineering for global optimization problems, such as process synthesis, optimization of complex reaction networks, and design of experiments, to explore large design spaces [23].

Stochastic optimization techniques are essential for addressing optimization problems under uncertainty, which are prevalent in chemical engineering due to variability in raw material properties, market conditions, and process disturbances. Stochastic optimization methods consider probabilistic models of uncertain parameters and optimize decision-making strategies to minimize risks and maximize expected performance under uncertainty.

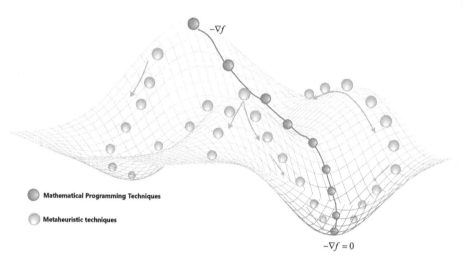

Figure 1.2: Graphical interpretation of search methodology for metaheuristic and deterministic optimization.

In mathematical optimization, deterministic and metaheuristic methods represent two distinct approaches for finding optimal solutions (Figure 1.2). Deterministic methods follow a predefined sequence of operations, ensuring reproducibility and guaranteeing the optimal solution, if one exists. These methods, such as LP and branch and bound, offer algorithmic precision and proofs of convergence, but can be computationally expensive for large or complex problems, e.g., highly non-convex or non-differentiable functions. These algorithms are best suited for problems with well-defined and continuous objective functions and constraints [18]. In contrast, metaheuristic methods employ flexible, adaptive search strategies using probabilistic rules to explore the solution space and escape local optima. Examples include GAs, simulated annealing, and particle swarm optimization. While they do not guarantee finding the exact optimal solution, they aim to achieve near-optimal solutions efficiently, making them scalable and suitable for large-scale or real-time problems. Metaheuristics are particularly useful for complex, discontinuous or poorly defined problems. However, some metaheuristics algorithms can also be computationally expensive due to the need to evaluate several potential solutions. In summary, deterministic methods provide precision and reliability for exact optimization, whereas metaheuristic methods offer flexibility and efficiency for tackling complex and large-scale problems [20].

The selection of the most appropriate optimization technique depends on factors such as the problem structure, computational requirements, and the presence of uncertainty, highlighting the importance of understanding and applying a diverse set of optimization tools in the field of chemical engineering. Mathematical programming algorithms, which mainly constitute classical optimization, are based on rigorous mathematical theory, derived from calculus, linear algebra, and other mathematical fields. They can be classified as gradient-based or gradient-free, and work in a mechanical way without any random nature (very often called deterministic). These algorithms tend to provide exact optimal solution, local or global, and will depend heavily on the initialization of the problem as it can lead to a region of local optima in which the algorithm will be unable to leave this region [18]. On the other hand, metaheuristic techniques involve randomness (stochastic), both in the randomness of choosing starting points for the search and in the orientation of the search. Most modern metaheuristic methods are intended for global optimization, but unlike classical optimization algorithms, these algorithms do not guarantee optimality. In turn, they make a general exploration of the search space by using many initial points with reasonable resource consumption, and generally offer a good quality solution that, depending on the nature of the problem, could not be achieved with a classical approach.

1.2.3 Problem definition and formulation

The formulation of the optimization problem is a crucial task, and this will influence whether the problem is tractable and solvable, how difficult it will be to solve and the proper method of solution. Some considerations should be taken into account when carrying out this task. The model equations, typically mass and energy balance, must be based on the physical, thermodynamic and transport laws that govern the study of chemical engineering, but it can include rules of thumb and other empirical relations. The number of independent and dependent variables must be carefully chosen to get the number of degrees of freedom, and the relationships between them should be reflected in the model. The nature of the variables and their interactions can give rise to the initial nonlinearities and lead to mixed-integer problems as well as contribute to the size of the problem. The definition of the constraints allows defining the feasible region and affects the choice of the optimal solution, but since these constraints are design parameters and operating conditions in many of the problems, they also ensure that the solution to the problem is technically feasible. The objective function involves the optimization criteria. In chemical engineering, cost minimization or profit maximization is typically used, and it is stated explicitly or implicitly as a function of the decision variables. It should be considered that the nonlinearity of the ob-

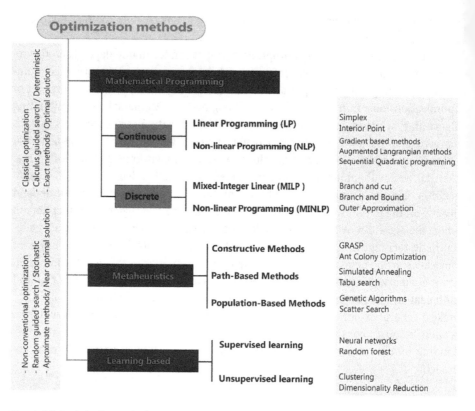

Figure 1.3: Optimization methods.

jective function increases the complexity of the model [20]. In addition, if multiple optimization criteria are to be satisfied, the problem statement will end up in a multi-objective optimization problem; environmental objectives are usually used in addition to economic objectives. The nature of the process or phenomenon being modeled will also affect the type of the resulting problem. Continuous processes, whose parameters vary over time, are best modeled with multi-period or dynamic optimization [24].

It is clear that, depending on these considerations and the nature of the problem, the optimization algorithm should be selected for the problem posed, and although there are currently many methods that allow solving complex and large problems, it is advisable to simplify the problem as much as possible without losing the essence of the modeled process or phenomenon.

1.3 Relevance of optimization in chemical engineering

Typically, the applications of optimization in chemical engineering address three sectors: a) synthesis and design, which includes plant design, equipment design, material design, heat exchange networks and material exchange networks, flow sheeting, etc.; b) operation planning, which includes supply chains, scheduling, and real-time optimization; and c) process control [25].

The classical optimization problem in chemical engineering is minimizing the energy consumption in chemical processes through energy integration and the design of heat exchange networks (HENs). This problem has been researched for more than 50 years; the first proposed approach was the pinch point, which is based on thermodynamics insights and heuristics. The second approach for HENs design is to formulate the synthesis as an optimization problem; the formulation presented in 1990 by Yee and Grossman has been a benchmark that has led to multiple approaches to solving the problem of HEN design. Generally, this problem has been formulated as a MINLP problem to model the existence or absence of an exchanger in the network. These problems have been proposed to be solved sequentially, where the main problem is decomposed into several subproblems that are solved hierarchically or simultaneously; the advantage of this approach is the relatively easy solution and the disadvantage is that the solutions will be suboptimal [24]. Another method is to solve the problem simultaneously where the general problem is solved in a single step. Although this method leads to an optimal solution, the disadvantage is the resources expended for its solution.

Another important optimization problem in the field of chemical engineering is reactor optimization, where the objective is to achieve the optimal performance for a given chemical reaction in determining the type of reactor, operating conditions and control [11, 26]. In the first instance, a heuristic approach was used to determine the type and size of a reactor or a reactor network, while operating parameters and conditions are optimized afterwards. But this approach is very limited due the countless variables associated with the reaction systems. In the 1990s, the attainable region method gained popularity. In this method, first an optimum or objective in the attainable region is identified and then the route to reach this point is determined. However, these methods yield solutions that are limited by the set of predefined reactors, and are not good options for complex reaction networks. With the introduction of the concept of superstructure, this optimization problem was proposed as an MINLP problem, whose main drawback is that it is not easy to solve and often yields local optima [13]. In this context, dynamic optimization has gained popularity, when combined with other approaches such as the optimal reaction route and the electuary process function.

Other important examples of optimization applications in chemical engineering are the fluid flow system design, which involves sizing and layout of pipelines, storage tanks, and pump configuration – determining the best site for plant location, subject to geographic and climatic conditions, communication and transportation infrastructure, transportation cost, and location of suppliers and buyers.

Planning production and scheduling are crucial tasks to cost reduction and are interrelated, the result of the planning being the scheduling objective. The scheduling problem has been addressed mainly through probabilistic models and stochastic optimization techniques. These techniques, in combination with MINLP models, result in robust approaches to accomplish this task. Probabilistic models use the probability distribution to model uncertain parameters, but this is only possible and realistic when historical data are available to propose the probabilistic models [13].

Process control has also been posed as an optimization problem for many decades [27], which essentially consists of proposing the dynamic control structure between manipulated and control variables in order to achieve certain objectives, which can be economics, operational feasibility, safety, and product quality. In this case, the optimization problem may be intractable without the use of decomposition methods.

An optimization problem that may be more challenging is the integration of process design and process control, in which the most promising approach seems to be mixed-integer dynamic optimization.

The relevance of optimization in the twenty-first century chemical engineering extends far beyond these technical tasks. Optimization in the field of chemical engineering has been extended to accomplish these essential tasks and also from a holistic point of view, for different aspects such as efficiency, sustainability, safety, and innovation.

Efficiency is intricately intertwined with optimization principles in chemical engineering. In today's global context, where resource conservation and energy efficiency are paramount, optimization techniques serve as indispensable tools. Through the meticulous application of advanced optimization methodologies, chemical engineers can systematically enhance process efficiency [18]. This results in tangible benefits such as reduced operational costs, minimized waste of raw materials, and streamlined energy consumption, all contributing to a more sustainable and competitive operational model.

Whether conceptualizing novel facilities or modernizing existing ones, optimization techniques guide engineers in designing the most efficient plant layouts, operational flowsheets, equipment, etc. By leveraging advanced optimization tools, engineers can meticulously optimize equipment arrangement, select the most suitable technologies, and orchestrate process flows to maximize productivity, while minimizing environmental footprints. This holistic approach to plant design not only enhances operational efficiency but also ensures adherence to stringent environmental regulations, reinforcing the industry's commitment to sustainability.

In sustainability, optimization emerges as a strategic imperative for the chemical industry. With mounting pressure to minimize environmental impact and maximize resource efficiency, optimization techniques offer a pathway to more sustainable operations. By optimizing processes to curtail environmental burdens, such as minimizing the use of nonrenewable resources and maximizing the recyclability of products and by-products, chemical engineers play a pivotal role in steering the industry toward a more sustainable future.

Safety and regulatory compliance remain paramount considerations in chemical engineering operations. Here, optimization serves as a critical tool to bolster safety protocols and ensure adherence to stringent regulatory standards. By optimizing processes to minimize operational risks and hazards, while concurrently guaranteeing compliance with safety regulations, chemical engineers safeguard the well-being of workers and local communities, fostering an environment of operational reliability and regulatory compliance [25].

Innovation and competitiveness represent the lifeblood of success in today's dynamic business landscape. Optimization strategies empower chemical companies to enhance operational efficiency, trim production costs, and deliver superior-quality products and services at competitive prices. By optimizing supply chain logistics and operational workflows, companies bolster their resilience, adaptability, and responsiveness to market dynamics, ensuring sustained competitiveness in an ever-evolving marketplace.

Encompassing multifaceted approaches that address the critical challenges and opportunities shaping the modern chemical industry, from process design to supply chain management, chemical engineers navigate the complex challenges of the contemporary industrial landscape.

1.4 Challenges and opportunities

In the field of chemical engineering, mathematical optimization presents both challenges and opportunities that shape the way processes are designed, operated, and optimized. Understanding these dynamics is crucial for leveraging optimization techniques effectively and addressing the complex demands of the modern chemical industry [2].

1.4.1 Challenges

– Complexity of models: Chemical processes often involve intricate interactions between numerous variables, parameters, and constraints. Modeling these systems accurately can be challenging, leading to complex mathematical formulations that are computationally demanding to solve. Moreover, the integration of various physical, chemical, and thermodynamic phenomena further increases model complexity, making it difficult to capture all relevant aspects accurately.

– Nonlinearity: Many chemical processes exhibit nonlinear behavior due to phenomena such as reaction kinetics, phase equilibria, and heat transfer. Nonlinear optimization problems can be more difficult to solve than their linear counterparts, requiring specialized algorithms and computational techniques. Dealing with nonlinearity introduces challenges such as multiple local optima, discontinuities, and sensitivity to initial conditions, which can complicate the optimization process and necessitate robust solution methods.

– High dimensionality: Optimization problems in chemical engineering often involve a large number of decision variables, making the search space exponentially large. Exploring this high-dimensional space efficiently presents significant computational challenges, particularly for global optimization problems. The curse of dimensionality exacerbates computational complexity as the number of possible solutions grows exponentially with the number of decision variables, leading to increased computational time and memory requirements.

– Uncertainty: Chemical processes are subject to various sources of uncertainty, including fluctuations in raw material properties, market conditions, and process disturbances. Accounting for uncertainty in optimization models adds another layer of complexity and requires advanced stochastic optimization techniques. Robust optimization approaches, scenario-based optimization, and probabilistic programming are among the methods used to address uncertainty, but they often entail increased computational overhead and complexity.

1.4.2 Opportunities

– Improved efficiency: Mathematical optimization offers the potential to enhance the efficiency of chemical processes by optimizing resource allocation, production scheduling, and energy utilization. By minimizing waste and maximizing yield, optimization techniques can lead to significant cost savings and operational improvements. Advanced optimization algorithms, such as MINLP and dynamic optimization, enable the identification of optimal process configurations and operating conditions that maximize productivity and profitability.

– Sustainability: Optimization can play a crucial role in promoting sustainability in the chemical industry by optimizing processes to minimize environmental impact, reduce emissions, and conserve resources. Sustainable process design and operation are increasingly important considerations for chemical engineers, and optimization techniques can help achieve these objectives. By incorporating sustainability metrics into optimization models such as carbon footprint, water

usage, and waste generation, engineers can identify environmentally friendly solutions that meet both economic and environmental objectives.
– Innovation: Optimization fosters innovation in chemical engineering by enabling the exploration of new process configurations, the design of novel materials, and the development of advanced control strategies. By pushing the boundaries of what is possible, optimization techniques drive technological advancements and facilitate the development of more efficient and sustainable processes. Optimization-driven innovation often involves interdisciplinary collaboration between chemical engineers, mathematicians, computer scientists, and domain experts to tackle complex problems and develop cutting-edge solutions.
– Decision support: Optimization provides valuable decision support tools for chemical engineers, helping them make informed decisions in complex and uncertain environments. By quantifying trade-offs and analyzing alternative scenarios, optimization models can guide decision-making and facilitate the identification of optimal solutions. Decision support systems based on optimization techniques enable engineers to evaluate the potential impact of different decisions on key performance indicators, risk factors, and sustainability metrics, empowering them to make data-driven decisions that align with organizational goals and objectives.

While mathematical optimization poses challenges in the field of chemical engineering, it also offers significant opportunities for improving efficiency, promoting sustainability, fostering innovation, and providing decision support. By addressing these challenges and leveraging the opportunities presented by optimization, chemical engineers can drive progress and innovation in the industry while meeting the evolving needs of society.

1.5 Where are we going?

In the era of the circular economy, artificial intelligence (AI), and Industry 4.0, the trajectory of mathematical optimization in chemical engineering is undergoing profound transformations, and is poised to revolutionize processes toward sustainability, efficiency, and intelligence [1]. Several key trends and directions shape the evolution of optimization in this context, each contributing to a holistic approach to process design, operation, and control:
– Integration of circular economy principles: Mathematical optimization is increasingly playing a pivotal role in facilitating the transition toward a circular economy within chemical engineering. Optimization models are being meticulously crafted to optimize resource utilization, minimize waste generation, and maximize the reuse, recycling, and recovery of materials and energy. Incorporating

constraints related to material flows, recycling technologies, and lifecycle assessments, these frameworks aim to design processes that are not only environmentally sustainable but also economically viable in the long term.

– Advancements in AI and machine learning: The fusion of AI and machine learning techniques with optimization methodologies is unlocking unprecedented possibilities for process optimization and control. AI-driven optimization algorithms, including reinforcement learning and deep reinforcement learning, empower autonomous decision-making and adaptive control strategies in chemical processes. By harnessing data-driven insights, these techniques optimize process performance, predict system behavior, and dynamically adapt to evolving operating conditions in real time, fostering flexibility, responsiveness, and efficiency across operations.

– Industry 4.0 and digitalization: The convergence of optimization with Industry 4.0 technologies such as the IoT, cyber-physical systems, and big data analytics is reshaping the landscape of chemical engineering processes. Optimization models are seamlessly integrated with digital twins and real-time process monitoring systems to enable predictive maintenance, condition-based monitoring, and dynamic optimization of production processes. This integration facilitates proactive decision-making, reduces downtime, and enhances overall process efficiency and reliability.

– Multi-objective and multi-scale optimization: In response to the increasing complexity and interconnectedness of chemical engineering systems, there is a growing emphasis on multi-objective and multi-scale optimization approaches. Optimization models are being refined to concurrently optimize multiple conflicting objectives, such as cost, energy efficiency, environmental impact, and product quality. Additionally, multi-scale optimization techniques are employed to harmonize optimization across different levels of process hierarchy, from molecular-level reactions to plant-wide operations, fostering holistic optimization strategies that account for interactions and trade-offs at various scales.

Collaborative and interdisciplinary research: The future of optimization in chemical engineering hinges on collaborative and interdisciplinary research endeavors that amalgamate expertise from diverse fields, including mathematics, computer science, chemistry, and economics. Cross-disciplinary collaborations engender the development of innovative optimization techniques, algorithms, and tools tailored to the unique challenges and opportunities of the circular economy, AI-driven process optimization, and Industry 4.0 applications. By nurturing synergies between disparate disciplines, collaborative research initiatives accelerate the advancement of optimization in chemical engineering, propelling transformative changes in the industry landscape (Figure 1.4).

In summary, the trajectory of mathematical optimization in chemical engineering unfolds amidst the integration of circular economy principles, advancements in AI

Figure 1.4: Role of AI in chemical engineering optimization.

and machine learning, digitalization of processes, multi-objective and multi-scale optimization approaches, and collaborative interdisciplinary research endeavors. These trends collectively redefine the paradigm of optimization in chemical engineering, fostering the emergence of sustainable, efficient, and intelligent processes that resonate with the demands of the contemporary economy and society.

1.6 Conclusions

This chapter has shown the transformative trajectory of mathematical optimization within the realm of chemical engineering, emphasizing its key role in driving the industry toward sustainability, efficiency, and intelligence. The integration of circular economy principles, advancements in AI and machine learning, digitalization through Industry 4.0 technologies, and the adoption of multi-objective and multi-scale optimization approaches collectively redefine the landscape of process optimization. Furthermore, the importance of collaborative and interdisciplinary research has been underscored as a catalyst for developing innovative optimization techniques tailored to contemporary challenges.

As we look to the future, several predictions emerge for the evolution of optimization in chemical engineering. Optimization models will continue to evolve, integrating more sophisticated constraints and variables to maximize resource efficiency and minimize waste, ultimately leading to closed-loop systems that are both economically and environmentally sustainable. The synergy between AI and optimization will usher in an era of autonomous, self-optimizing chemical processes capable of real-time adaptation to changing conditions, thereby enhancing process efficiency, reliability, and responsiveness.

The integration of digital twins and real-time data analytics with optimization frameworks will become ubiquitous, enabling predictive maintenance, dynamic process control, and continuous improvement in production environments. The adoption of multi-objective and multi-scale optimization will grow, facilitating comprehensive optimization strategies that consider diverse factors such as cost, energy use, environmental impact, and product quality across various levels of process hierarchy. Continued emphasis on collaborative research across disciplines will lead to groundbreaking optimization methodologies and tools, addressing the complex and multifaceted challenges of modern chemical engineering.

Thus, the future of mathematical optimization in chemical engineering is poised to be shaped by these emerging trends and innovations, driving the industry toward a more sustainable, efficient, and intelligent future.

References

[1]   He, C., Zhang, C., Bian, T., Jiao, K., Su, W., Wu, K. J., Su, A. A review on artificial intelligence enabled design, synthesis, and process optimization of chemical products for industry 4.0. *Processes*, 2023, 11(2), 330.

[2]   Sadat Lavasani, M., Raeisi Ardali, N., Sotudeh-Gharebagh, R., Zarghami, R., Abonyi, J., Mostoufi, N. Big data analytics opportunities for applications in process engineering. *Reviews in Chemical Engineering*, 2023, 39(3), 479–511.

[3]   Mala-Jetmarova, H., Barton, A., Bagirov, A. A history of water distribution systems and their optimisation. *Water Science and Technology: Water Supply*, 2015, 15(2), 224–235.

[4]   Zhang, Y. *Insights into Chinese Agriculture*. Springer, 2018.

[5]   Shao, Z. *The New Urban Area Development: A Case Study in China*. Springer, 2015.

[6]   Simon, D. *Evolutionary Optimization Algorithms*. John Wiley & Sons, 2013.

[7]   Mueller, L., Eulenstein, F., Dronin, N. M., Mirschel, W., McKenzie, B. M., Antrop, M., . . . Poulton, P. Agricultural landscapes: History, status and challenges. *Exploring and Optimizing Agricultural Landscapes*, 2021, 3–54.

[8]   Gelderblom, O., Grafe, R. The rise and fall of the merchant guilds: Re-thinking the comparative study of commercial institutions in premodern Europe. *Journal of Interdisciplinary History*, 2010, 40(4), 477–511.

[9]   Koetsier, T., Bergmans, L. (Eds.). *Mathematics and the Divine: A Historical Study*. Elsevier, 2004.

[10]   Song, D. P. *Container Logistics and Maritime Transport*. Routledge, 2021.

[11]   Hanska, J. *War of Time: Managing Time and Temporality in Operational Art*. Springer Nature, 2020.

[12]   Chondros, T. G. Archimedes Influence in Science and Engineering. In *The Genius of Archimedes–23 Centuries of Influence on Mathematics, Science and Engineering: Proceedings of an International Conference Held at Syracuse*. Springer, 2010, pp. 411–425.

[13]   Padberg, M. *Linear Optimization and Extensions*. Vol. 12, Springer Science & Business Media, 2013.

[14] Diwekar, U. M. *Introduction to Applied Optimization*. Vol. 22, Springer Nature, 2020.

[15] Ekeland, I. *The Best of All Possible Worlds: Mathematics and Destiny*. University of Chicago Press, 2006.

[16] Fakhouri, H. N., Alawadi, S., Awaysheh, F. M., Hamad, F. Novel hybrid success history intelligent optimizer with Gaussian transformation: Application in CNN hyperparameter tuning. *Cluster Computing*, 2024, 27(3), 3717–3739.

[17] Meng, Z., Yıldız, B. S., Li, G., Zhong, C., Mirjalili, S., Yildiz, A. R. Application of state-of-the-art multiobjective metaheuristic algorithms in reliability-based design optimization: A comparative study. *Structural and Multidisciplinary Optimization*, 2023, 66(8), 191.

[18] Al Ani, Z., Gujarathi, A. M., Al-Muhtaseb, A. A. H. A state of art review on applications of multi-objective evolutionary algorithms in chemicals production reactors. *Artificial Intelligence Review*, 2023, 56(3), 2435–2496.

[19] Bishnu, S. K., Alnouri, S. Y., Al-Mohannadi, D. M. Computational applications using data-driven modeling in process Systems: A review. *Digital Chemical Engineering*, 2023, 8, 100111.

[20] Franzoi, R. E., Menezes, B. C., Kelly, J. D., Gut, J. A., Grossmann, I. E. Large-scale optimization of nonconvex MINLP refinery scheduling. *Computers & Chemical Engineering*, 2024, 186, 108678.

[21] Cui, E. H., Zhang, Z., Chen, C. J., Wong, W. K. Applications of nature-inspired metaheuristic algorithms for tackling optimization problems across disciplines. *Scientific Reports*, 2024, 14(1), 9403.

[22] Selvarajan, S. A comprehensive study on modern optimization techniques for engineering applications. *Artificial Intelligence Review*, 2024, 57(8), 194.

[23] Rajwar, K., Deep, K., Das, S. An exhaustive review of the metaheuristic algorithms for search and optimization: Taxonomy, applications, and open challenges. *Artificial Intelligence Review*, 2023, 56(11), 13187–13257.

[24] Turgut, O. E., Turgut, M. S., Kırtepe, E. A systematic review of the emerging metaheuristic algorithms on solving complex optimization problems. *Neural Computing and Applications*, 2023, 35(19), 14275–14378.

[25] Matoušová, I., Trojovský, P., Dehghani, M., Trojovská, E., Kostra, J. Mother optimization algorithm: A new human-based metaheuristic approach for solving engineering optimization. *Scientific Reports*, 2023, 13(1), 10312.

[26] Horn, F., Klein, J. Optimization theory and reactor performance. 1972.

[27] Morari, M., Arkun, Y., Stephanopoulos, G. Studies in the synthesis of control structures for chemical processes: Part I: Formulation of the problem. Process decomposition and the classification of the control tasks. Analysis of the optimizing control structures. *AIChE Journal*, 1980, 26(2), 220–232.

José A. Caballero

Chapter 2
Deterministic optimization of distillation processes

Abstract: In this chapter, we present an overview of deterministic optimization methods for the design of column sequences for separating zeotropic mixtures. First, we focus on models developed in the last years of the twentieth century that deal with sequences formed by conventional columns and the sharp separation of consecutive key components. The extension to models for the sharp separation of components while allowing sloppy separations (nonconsecutive key components) is untimely related to thermally coupled distillation; therefore then we do a discussion on the structural characteristics that characterize thermally coupled distillation and present deterministic models for optimizing these sequences.

Column sequencing is usually based on shortcut models, which are usually good enough, especially if we are interested in sequence comparison. However, in some situations rigorous models are mandatory; therefore we introduce the main models developed for the optimization of distillation columns. Using generalized disjunctive programming as a modeling framework, we show that it is possible to formulate different models with varying levels of complexity (equilibrium or transport-based, or even including reactions) without altering the model structure. This enables, for instance, the use of advanced commercial simulators within hybrid optimization models.

Keywords: distillation, superstructure representation, process intensification, modelling and optimization

2.1 Introduction

The industrial sector accounts for around a third of the global energy consumption, which in 2020 reached up to 156 EJ ($1.56 \cdot 10^{20}$ J) and a prevision of 207 EJ in 2050 [1]. Chemical and petrochemical industries account for around 20–30% of the energy consumption of the industrial sector, e.g., in the European Union, it was 21.5% in 2021, that is, around 2.159 EJ [2].

Separation processes are everywhere in chemical, petrochemical, and biochemical industries. They are crucial for ensuring product purity, reducing pollutants to acceptable levels, and sorting out unreacted substances for recycling. Various methods,

José A. Caballero, Institute of Chemical Process Engineering, University of Alicante, Carretera de S. Vicente s.n., 03690 Alicante, Spain

https://doi.org/10.1515/9783111383439-002

like distillation, membrane permeation, absorption, adsorption, and liquid-liquid extraction among others, are used for separating mixtures. However, in the chemical and petrochemical sectors, distillation is the predominant method used for approximately 90–95% of all separation and purification processes. This scenario is expected to persist in the foreseeable future and contributes to around 40–60% of the energy consumption [3, 4], which is equivalent to 2.5–5.6 PJ/year or around 47–105 millions of tons of oil per year.

Considering its widespread use and the need to reduce carbon footprints, it would be expected that distillation would be a major focus of research. However, according to Agrawal and Tumbalam Gooty [5], academic interest in distillation has decreased since the late 1980s, and only recently it is acquiring renewed interest due to the necessity of reducing energy consumption. Agrawal and Tumbalam Gooty [5] identified two main reasons for this decline in research: 1. the belief that distillation is the most energy-intensive process among all the separations, and thus alternative processes were needed, and 2. distillation technology is mature and there was nothing worthwhile to research. Energy is indeed introduced in the reboiler of the distillation columns, and approximately the same amount of energy is removed in the condensers at a lower temperature, which renders an inefficient process. But it is also true that it is one of the most effective alternatives for the separation of mixtures [6]. For example, Agrawal and Tumbalam Gooty [5] showed that in an industrial process, distillation needs only a fraction of the energy requirement for membranes.

The origins of distillation can be traced to around 3500 BC when the Sumerians were the first to apply evaporation and condensation of a liquid to refine a substance by extracting oils from herbs. From that time, distillation has been rediscovered and/or perfectionated by different civilizations (China, Greek, Romans, Arabs, etc.) [7] and continued evolving during the Middle Ages. Around the year 1800, the designs for distillation closely resembled modern distillation columns. However, the first calculations for distillation dated from the late nineteenth century and the first years of the twentieth century. Eugen Hausbrand was likely one of the first to publish quantitative calculations for rectification [7]. In 1902, Rayleigh performed the first calculation for batch distillation, followed by contributions by Robinson in 1922 [8], and McCabe in 1925 [9], among many others. An excellent review of the history of distillation can be found in [7].

We can differentiate two kinds of problems depending on whether the initial mixture contains azeotropes or not. While the azeotropic problem is complex, the number of alternatives is considerably smaller than the zeotropic one. The techniques for generating different sequences in azeotropic systems are mainly based on conceptual-graphical approaches. However, the number of structurally different new sequences in zeotropic systems can be enormous. Therefore, in this chapter, we will focus on zeotropic systems.

Early works using optimization techniques for distillation can be dated around the 1960s and focused on how to separate an N-component mixture into a set of specified products. At that time, the researchers realized that when attempting to separate multiple components from a mixture using distillation, selecting the correct sequence for the separation has a much greater impact on energy and, consequently more economical than the detailed optimization of a particular column.

Therefore, in this chapter, we first address the design of distillation sequences, and later we will focus on the rigorous design of a particular distillation column and the integration of both approaches.

2.2 Superstructures

In the application of mathematical programming techniques to design and synthesis problems it is always necessary to postulate a superstructure of alternatives [10, 11]. This is true independently of the complexity of the model used to represent the system from the shortcut, high-level aggregation, or detailed model. While in some cases generating a superstructure is rather straightforward, this is not true in the general case. Even more, except maybe in some simple cases, the representation of alternatives is not unique. Two major issues arise in postulating a superstructure. The first is, given a set of alternatives that are to be analyzed, what are the major types of representations that can be used? And what are the implications for the modeling? The second is, for a given representation that is selected, what are all the feasible alternatives that must be included to guarantee that the global optimum is not overlooked?

Yeomans and Grosmann [12] have characterized two major types of representation using the concepts of tasks, states, and equipment. A state is the minimum set of physical and chemical properties needed to characterize a stream in a given context. They can be quantitative like pressure or temperature, or qualitative, i.e., mixture of CDE indicating that we have a stream formed by the compounds C, D, E (each capital letter refers to a component in the mixture) in some specifications that do not exclude the presence or other compounds. A task is the chemical or physical transformation that relates two or more states. The equipment is the physical device in which a task is performed.

The first major representation is the state-task-network (STN) which is motivated by the work in scheduling by Kondili et al. [13]. The basic idea here is to use a representation that has only two types of nodes: states and tasks. (As an example, Figure 2.1 shows the superstructure for separating a four-component mixture using conventional columns and separations between consecutive key components – it coincides with the superstructure previously proposed by Andrecovich and Westerberg [14].) The assignment of equipment is dealt with implicitly through the model. Both the cases of one-task one-equipment (OTOE) in which a given task is assigned a single piece of equipment and the variable task equipment (VTE), in which a given task can be performed by different equipment, were considered. The second representation is the state equipment network (SEN) which was motivated by the work of Smith [15]. In this case, the superstructure uses two nodes: states and equipment, which assume an a priori assignment of the different tasks to equipment based on the knowledge of the designer about the process. (As an example, Figure 2.2 shows the SEN superstructure for the same case as in Figure 2.1; it coincides with the superstructure previously proposed by Novak et al. [16].

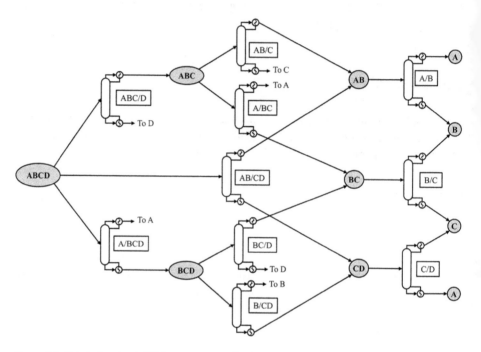

Figure 2.1: State task network (STN) superstructure for the separation of a four-component mixture. Note that there is a one-to-one relationship between the separation tasks and the equipment assigned (STN-OTOE).

It is worth noting that the STN and SEN are not the only options for superstructure optimization. First, between these two options, there are a large number of superstructures with intermediate characteristics that can be generated by aggregation of some tasks, or by an initial partial assignment of equipment [17]. Other superstructures include the process graph (P-graph) [18], which is very similar to STN formalism but provides a systematic algorithm for generating the P-graph. An accelerated branch-and-bound strategy was also developed to exploit the P-graph representation by restricting the search to feasible structures. In the state space representation (SSR) the design problem is posed as the identification of the properties of the input-state-output relations between input and output variables. In the SSR a distillation network is introduced as the integration of heat and mass exchanger operations [19]. Other alternatives include the R-graph representation [20, 21], the generalized modular framework [22], or the unit port

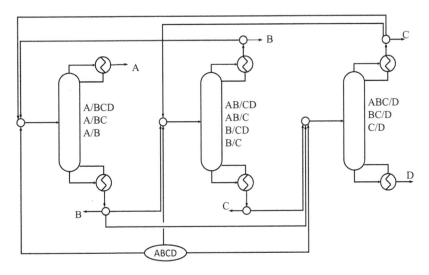

Figure 2.2: State equipment network (SEN) superstructure for the separation of a four-component mixture. Note that the assignment of tasks to equipment is not unique.

conditioning stream [23] among others. An excellent review of superstructures in optimization has been recently published by Mencarelli et al. [24]

2.3 Deterministic optimization models for the design of distillation sequences

The general separation problem was defined in 1968 by Rudd and Watson [25] as the separation of several source mixtures into several product mixtures with specific characteristics. However, in 1983 Westerberg [26] stated that the general separation problem was still essentially unsolved, and nowadays, more than 50 years later, the general separation problem is still not completely solved. This is not surprising because the definition of Rudd and Watson is too general for at least two reasons. First, it includes all possible separation technologies, most of them in continuous development both for single and hybrid equipment. Second, defining the final products in terms of "specific characteristics" includes complex problems of mixtures, typical, for example, in the refining processes. Instead, the first works focused on a simpler problem: the separation of a single source mixture into several of its components using distillation columns. If the sequence of distillation columns separates the feed into products with no overlap between them (or do it inside some tolerance), then the sequence is carrying out a "sharp separation."

The first optimization models developed for the distillation sequence problem assumed that the sequence was formed by "conventional" or simple columns. A conventional column is a distillation column having one feed and producing two products – distillate and bottoms – with one condenser and one reboiler.

The sharp separation of a mixture using distillation does not necessarily require that each distillation column performs a sharp separation; for example, the sharp separation of a zeotropic mixture ABC, where the components are sorted by decreasing volatilities can be done by first separating AB/BC where the component B is optimally distributed between condenser and reboiler and then A is separated from B (separation A/B) and B from C (separation B/C). However, it has been shown that this separation sequence using conventional columns is suboptimal, at least in terms of total cost (e.g., it requires three conventional columns vs. two distillation columns using non-conventional columns with the same energy consumption) [27, 28]. Although most of the theoretical aspects of distillation sequencing including sloppy separations started in the 1960s [29] and extended for all of the second half of the twentieth century, it was not until the first decade of the 2000s that the foundations for generating and optimizing distillation sequences with "sloppy separations" were established. It is worth noting that this development is closely linked to thermally coupled distillation (TCD), which will be addressed in more detail later in this chapter.

2.3.1 Distillation sequences with conventional columns

Considering only conventional columns, the earliest effort focused on the development of heuristics to aid in selecting the preferred structure. Lockhart [30] examined the separation of a three-component mixture. Harbert in 1957 [31] clearly stated the problem in a paper with the suggestive title of "Which column goes where." Heaven in 1969 [32] and Rudd et al. in 1973 [33] proposed a reasonable set of heuristics, extended by Seader and Westerberg [34], which allowed generating quickly reasonable sequences that would prove to be close to the best sequence [26]. The main principles behind those rules are:

Heuristic 1: Remove dangerous and/or corrosive species first.

Heuristic 2: Do not use distillation when the relative volatility between the key components is less than 1.05.

Heuristic 3: Do the easy splits (i.e., those having the largest relative volatilities) first in the sequence.

Heuristic 4: Remove the most abundant components first.

Heuristic 5: Remove the most volatile component first.

Heuristic 6: The species leading to desired products should appear in a distillate product somewhere in the sequence if possible.

Heuristic 7: These heuristics are listed in order of importance.

Some of these heuristic rules may be contradictory to each other, and although they are listed in order of importance, it is common for the "weight" of a rule lower down on the list to outweigh another listed earlier. In these cases, it is up to the designer to make a decision. However, in any case, it allows for obtaining a valid and probably "good" sequence very quickly. This is especially suitable for preliminary estimates."

If we focus on models based on deterministic optimization, the earlier approaches take advantage of the fact that generating all the feasible separation sequences is straightforward. However, the number of alternatives can be very large if the number of components to be separated is large. These early models were based on algorithms for searching for the best sequence using the "tree of alternatives" (see Figure 2.3).

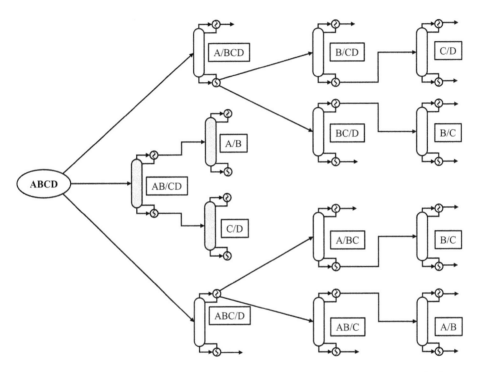

Figure 2.3: Separation tree of alternatives for a five-component mixture. Components are named with capital letters and sorted by decreasing volatility.

Thompson and King in 1972 [35] used a pseudo algorithm that resembled a branch-and-bound search. It did not work always, but when it worked, it was fast. Hendry and Hughes [36] used dynamic programming, and among the earliest works using branch-and-bound search are the models presented by Westerberg and Stephanopoulos [37], Rodrigo and Seader [38], or Gomez and Seader [39].

Perhaps the most influential model for addressing the sequencing problem was proposed by Andrecovich and Westerberg [14]. Instead of employing a "tree" to search for the optimal column sequence, they introduced a "network" that coincides with the STN-OTOE superstructure shown in Figure 2.1. The network offers the advantage of eliminating redundant "separations," resulting in a more concise representation that, simultaneously, enhances the model's writability.

The model proposed by Andrecovich and Westerberg takes the form of an MILP (mixed-integer linear programming) model. It incorporates many features of actual models and has the advantage of being linear. In the following section, we will introduce linear models inspired by these early works using the formalism of generalized disjunctive programming (GDP), which allows for a clearer and more compact visualization. However, some algebraic modeling languages like GAMS [40] or Pyomo.GDP [41] can perform this transformation automatically.

2.3.1.1 Distillation sequences with conventional columns: linear models

As commented earlier, the sharp separation of a zeotropic mixture using conventional columns requires that each column performs a sharp separation between two components adjacent in volatilities. In that case, it is possible to sequentially calculate each one of the possible columns either using a shortcut, typically the Fenske [42] Underwood [43] Gilliland [44] (FUG) equations, aggregated [17], or even rigorous models. For example, from Figure 2.1 if we optimize separation ABC/D, we know the mixture ABC that can be used to optimize separations AB/C and A/BC, and so forth. Besides, for each possible distillation column, we obtain all the relevant parameters to calculate its cost (i.e., number of trays, internal flows, condenser and reboiler duties, etc.). It is even possible to assume a 100% recovery of key components, without significantly increasing the error. It is important to note that the number of possible sequences increases with the number of components to be separated, much more rapidly than the number of potential separation tasks (columns). The concept is to make a preliminary effort by optimizing each of the potential columns instead of optimizing both the columns and the sequence simultaneously. Moreover, the use of shortcut models that capture the main features of the columns enables the execution of this initial stage very efficiently.

In those conditions, the compositions of the streams connecting two columns, as well as all the parameters of each column are known parameters, and the objective of the model consists of selecting the best sequence. The disjunctive model can be written as follows:

Index Sets:

COL	$\{j \mid j$ is a potential column$\}$
	e.g. {A/BCD, AB/CD, ABC/, A/BC, AB/C, B/CD, BC/D, A/B, B/C, C/D}
STATE	$\{s \mid s$ is a stream, it includes the feed, internal sub-mixtures, and final products$\}$
	e.g. {ABCD, ABC, BCD, AB; BC, CD, A, B, C, D}
INI_s	Initial State (Feed stream)
	e.g. ABCD

$PROD_s$	Final Products
	e.g. {A, B, C, D}
$INTER_s$	Intermediate sub-mixtures.
	e.g. {ABC, BCD, AB, BC, CD}
$CS_{j,s}$	Connectivity relationship. The column j can generate the state s
	e.g.{ ABCD.(A/BCD, AB/CD, ABC/D), ABC.(A/BC, AB/C), BCD.(B/CD, BC/D), AB.(A/B),
	BC.(B/C), CD./C/D) }
$SC_{s,j}$	Connectivity relationship. The state s can generate the column j

Data

$Tcost_j$	Cost of Column j. It includes the operating and investment costs.

Variables

Total cost	Total Cost. (objective variable)
$Cost_j$	Cost of Column j
Y_j	Boolean Variable. Takes the value "True" if the column j is selected and "False" otherwise.

$$\min \sum_{j \in COL} Cost_j$$

s.t.

$$\begin{bmatrix} Y_j \\ Cost_j = Tcost_j \end{bmatrix} \underline{\vee} \begin{bmatrix} \neg Y_j \\ Cost_j = 0 \end{bmatrix} \quad \forall j \in COL \tag{M.2.1}$$

$$\Omega(Y) = True$$

$$Y \in \{True, False\}^{|COL|}$$

The expression $\Omega(Y) = True$ refers to a set of logical relationships that allows only feasible separation sequences. These logical relationships can be obtained directly from connectivity relations and the knowledge of the characteristics of the separation sequence. A valid set is the next one:

The feed must be entered in exactly one column

$$\underset{(s,j) \in SC_{s,j} \cap INI_s}{\underline{\vee}} Y_j \tag{2.1}$$

A given state, except the initial one, must be produced by at most one column:

$$\underset{(i,s) \in CS_{j,s} \setminus s \notin INI_s}{\underline{\vee}} Y_j \vee Z \tag{2.2}$$

where Z is a "dummy" variable that takes the value of True if all the other Boolean variables are False.

Forward connectivity: If there is a column i generating state s, it must lead to one of the columns generated by that state

$$Y_j \Rightarrow \bigvee_{i \in SC_{s,i}} Y_i \quad \forall j \in CS_{j,s} \tag{2.3}$$

Backward connectivity: If column j exists, generated by state s, then there must be at least one of the columns generating that state

$$Y_j \Rightarrow \bigvee_{i \in SC_{i,s}} Y_i \quad \forall j \in SC_{s,j} \tag{2.4}$$

Final products must be generated exactly once

$$\bigvee_{(i,s) \in CS_{j,s} \cap FIN_S} Y_j \tag{2.5}$$

The final model is formed by eqs. in (M.2.1) and (2.1)–(2.5).

The previous model can be written as an MILP. In this model, the Boolean variable Y is substituted by the binary variable y that takes the value of 1 when the Boolean takes the value of "True" and zero otherwise:

$$\min \sum_{j \in COL} Tcost_j \, y_j$$

s.t.

$$
\begin{aligned}
& \sum_{j \in SC_{s,j}} y_j = 1 \quad s \in INI_s \\
& \sum_{j \in CS_{j,s}} y_j \leq 1 \quad s \notin INI_s \\
& 1 - y_j + \sum_{i \in SC_{s,j}} y_i \geq 1 \quad \forall j \in CS_{s,j} \\
& 1 - y_j + \sum_{i \in CS_{j,s}} y_i \geq 1 \quad \forall j \in SC_{s,j} \\
& y \in \{0,1\}^{|COL|}
\end{aligned}
\tag{M.2.2}
$$

In previous models we assumed that all the utilities are provided by external sources, however, it is often desirable to perform heat integration in distillation sequences because energy tends to be the dominant cost. Andrecovich and Westerberg [14] presented a model based on discretizing temperatures, which together with the direct heat integration between condensers and reboilers facilitates the consideration of multi-effect heat integration. The main idea consists of aggregating the heat through the "**heat transfer transshipment model**" [45]. Here we present a version of such a model.

The first step consists of developing a superstructure. In this case, we only need to modify the network of Figure 2.1 by postulating several candidate columns for each separation task (Figure 2.4). This number is either the maximum number of columns we are willing to have in multi-effect separation or the maximum number of columns we can stack inside the lowest cold utility temperature and highest hot utility temperature.

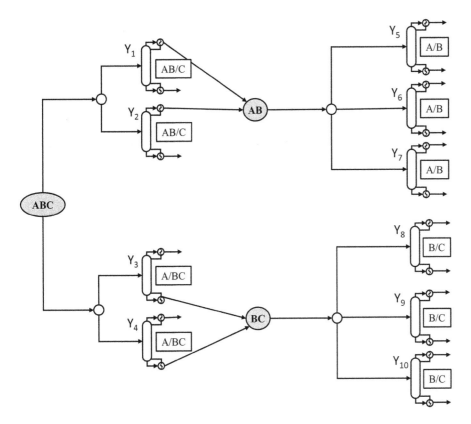

Figure 2.4: Superstructure for a three-component mixture with two columns for three-component separation and three columns for two-component separation.

Unlike the previous model, we do not know the feed flow entering each column. However, this is not a problem in estimating the heat duties because, for a fixed pressure, the heat duties change linearly with the feed flow. To capture the capital cost, a linear relation using a fixed and variable cost in terms of total feed flow entering a column produces good results.

Due to the assumption of sharp separation in each conventional column, independently of whether we assume a 100% recovery or not, we can know a priori the composition of each one of the states (feed to each separation task).

Now we can embed the heat integration into the model through a heat cascade. Treating the condensers as hot streams, and the reboilers as cold streams, it is possible to construct a heat cascade based on the temperatures of these streams. In the original work those streams were assumed to be isothermal; however, except for final pure products, they are not. From a practical point of view, the only difference is that with non-isothermal streams the number of temperature intervals can be large. Although in linear problems this is not a major problem nowadays, it could have been so in the 1980s, which justifies the assumption.

The disjunctive model is as follows:

Index Sets:

Task	$\{t \mid t$ is a separation task$\}$
COL	$\{c \mid c$ is a potential column$\}$
$CT_{c,t}$	Columns c performing the same separation task t
STATE	$\{s \mid s$ is a stream, it includes the feed, internal sub-mixtures, and final products$\}$
$SC_{s,c}$	Connectivity relationship. The state s can generate the column c.
TI	$\{k \mid k$ is a temperature interval$\}$
HU	$\{m \mid m$ is a hot utility$\}$
CU	$\{n \mid n$ is a cold utility$\}$
$C_{c,k}$	$\{c \mid c$ is the reboiler – cold stream – of column c that demands heat from interval $k\}$
$H_{c,k}$	$\{c \mid c$ is the condenser -hot stream- that supplies heat to interval $k\}$
$S_{m,k}$	$\{m \mid m$ is a hot utility able to supply heat to interval $k\}$
$W_{m,k}$	$\{n \mid n$ is a cold utility able to remove heat from interval $k\}$

Known Data

a_c	Fixed Cost of Column c
β_c	Variable cost of column c
F_t^{TOT}	Total flow entering a set of separation tasks t
$CostHU_m$	Cost of Hot utility m
$CostCU_n$	Cost of Cold utility n
$\overline{Q}_c^{COND}$	Condenser heat duty if all the flow from state s that generates column c goes to that column
$\overline{Q}_c^{REB}$	Reboiler heat duty if all the flow from state s that generates column c goes to that column

Variables

Y_c	Boolean Variable. It takes the value True if column c is selected and False otherwise
Z_t	Boolean Variable. It takes the value True if separation task t is selected and False otherwise
Q_c^{Cond}	Heat supplied by the condenser (hot stream) of column c
Q_c^{Reb}	Heat demanded by reboiler (cold stream) of column c
$Feed_c$	Feed flow to column c
Q_m^{HU}	Heat provided by hot utility m
Q_n^{CU}	Heat provided by cold utility n
R_k	Heat residual exiting from interval k

$$\min: \sum_{c \in COL} CostC_c + \sum_{m \in HU} CostHU_m Q_m^{HU} + \sum_{n \in CU} CostCU_n Q_n^{CU}$$

s.t.

$$\begin{bmatrix} Y_c \\ CostC_c = \alpha_c + \beta_c Feed_c \\ Q_c^{Reb} = f_c \overline{Q}_c^{REB} \\ Q_c^{Cond} = f_c \overline{Q}_c^{COND} \end{bmatrix} \forall f_c \in CT_{c,t} \end{bmatrix} \lor \begin{bmatrix} \neg Y_c \\ CostC_c = 0 \\ Feed_c = 0 \\ Q_c^{Reb} = 0 \\ Q_c^{Cond} = 0 \end{bmatrix} \forall c \in COL$$

$$Feed_c = \sum_{c \in CT_{c,t}} f_c F_t^{TOT} \quad \forall t \in TASKS \tag{M.2.3}$$

$$R_k - R_{k-1} - \sum_{m \in S_{m,k}} Q_m^{HU} + \sum_{n \in W_{n,k}} Q_n^{CU} - \sum_{c \in H_{c,k}} Q_c^{Cond} + \sum_{c \in C_{c,k}} Q_c^{Reb} = 0 \quad \forall k \in TI$$

$$\Omega(Z,Y) = True$$

$$0 \le f_c \le 1$$

$$\left. \begin{array}{l} Feed_c \ge 0, \quad Q_c^{Cond} \ge 0, \quad Q_c^{Reb} \ge 0 \\ Y_c \in \{True, False\} \end{array} \right\} \forall c \in COL$$

$$R_k \ge 0 \ \forall k \in TI \setminus k \ne |TI|, \quad R_k = 0 \ k = |TI|$$

The set of logical relationships $\Omega(Z,Y) = True$ is the same as in the previous model (M.2.1), but instead of referring to a particular column Y, the logical relationships must be in terms of separation task Z (remember that as a consequence of the multi-effect heat integration, the same separation task can be performed by more than a column). And we must ensure that if a given task Z is selected then at least one of the columns that can perform that separation task must be selected:

$$Z_t \Rightarrow \underset{c \in CT_{c,t}}{\lor} Y_c \quad \forall t \in Task \tag{2.6}$$

Besides, if we are not interested in multi-effect integration but still in energy integration we can ensure that at most one column among all that can perform a given separation task is selected:

$$\underset{c \in CT_{c,t}}{\lor} Y_c \lor D \quad \forall t \in Task \tag{2.7}$$

In this case, D is a dummy Boolean variable that takes the value of True if all the rest takes a value of False.

2.3.1.2 Distillation sequences with conventional columns: nonlinear models

If the assumption that the recovery of each component can be calculated a priori is not met, for example, because small variations in the concentration of a component in the feed have a significant effect on its behavior, or "excessive" recovery of a component may result in an exponential increase in energy consumption, then it is necessary to simultaneously determine the optimal sequence and operating conditions of the column.

Due to the complexity of the mathematical model associated with the optimization of a distillation column, the resulting model is nonlinear and non-convex. Besides, there is an intrinsic relationship between the superstructure and the numerical performance of the final model. Although, as commented above there are other alternatives, we will show the disjunctive models of the two superstructures presented in Section 2.2: STN-OTOE and SEN. One nice thing about the disjunctive representation is that the structure of the model is the same independently of the model used in the column; then it is possible to move from shortcut to rigorous models by only changing the corresponding equations.

The following models are the disjunctive models for the separation of an N-component mixture using conventional columns without specifying any particular model for the column.

For the STN-OTOE model, we will use the same index sets as in problem (M.2.1) plus the set of components. We repeated them here for the sake of clarity:

$COMP$	$\{i \mid i$ is a component to be separated$\}$
COL	$\{c \mid c$ is a potential column or a separation task$\}$
$STATE$	$\{s \mid s$ is a stream, it includes the feed, internal states, and final products$\}$
INI_s	Initial State (Feed stream)
$PROD_s$	Final Products
$INTER_s$	Intermediate sub-mixtures.
$CS_{c,s}$	Connectivity relationship. Column c can generate the state s
$SC_{s,c}$	Connectivity relationship. The state s can generate the column c

The known data are the component molar flow in the feed to the system (F_i^0) and the final recovery (rec_i) of each component, all the thermodynamics and cost correlations needed to estimate the cost (or any other performance metric).

The variables are:

$Cost_c$	Cost of column j
Y_c	Boolean variable. Takes the value "true" if the column j is selected and "false" otherwise.
$F_{i,c}$	Component molar flow i, entering the column c
$D_{i,c}$	Distillate molar flows in column c
$D_{i,c}$	Bottoms molar flows in column c

$$\min: \sum_{c \in COL} Cost_c$$

s.t.

$$\sum_{c \in SC_{s,c} \cap INI_s} F_{i,c} = F_i^0 \quad \forall i \in COMP$$

$$\sum_{c \in CS_{c,s}} D_{i,c} + \sum_{c \in CS_{c,s}} B_{i,c} = \sum_{c \in SC_{s,c}} F_{i,c} \quad \forall i \in COMP, \ s \in INTER_s$$

$$\sum_{c \in CS_{c,s} \cap PROD_s} D_{i,c} + \sum_{c \in CS_{c,s} \cap PROD_s} B_{i,c} \geq rec_i F_i^0 \quad \forall i \in COMP$$

(M.2.4)

$$\begin{bmatrix} Y_c \\ [D_{i,c}, B_{i,c}, Cost_c] = f_c(column \ equations) \end{bmatrix} \underline{\vee} \begin{bmatrix} \neg Y_c \\ D_{i,c} = 0 \\ B_{i,c} = 0 \\ Cost_c = 0 \end{bmatrix} \quad \forall c \in COL$$

$$\Omega(Y) = True$$

$$Y \in \{True, False\}^{|COL|}$$

$$D_{i,c}, B_{i,c}, F_{i,c} \geq 0$$

The three first equations are mass balances in the feed state, intermediate states, and final products respectively. The logical relationships are the same as those presented in model (M.2.2) and in eqs. (2.1)–(2.7).

In the SEN model, the columns are fixed a priori, and the combinatorial part is related to the assignment of a separation task to a given column. The sets of components and states are the same as in the STN-OTOE model, but now there is not a one-to-one relationship between columns and tasks. Besides, the connectivity relations must be adapted to the SEN superstructure.

The model can be written as follows:

The sets of components (*COMP*), states (*STATES*), feed (*INIs*), intermediate states (*INTER$_s$*), and final products (*PROD$_s$*) are the same as in the STN model. The rest must be modified:

COL	{c | c is a column}
TASK	{t | t is a separation task}
$CT_{c,t}$	{Column c can perform the separation task t}
ST	{k | k is an external stream connecting two mixers/splitters in the superstructure}
STF_k	{Stream k that exits from the feed splitter}
$STC_{k,c}$	{Stream k enters to the mixer that feeds the column c}
$STD_{k,c}$	{Stream k exits from the splitter after the distillate in column c}
$STB_{k,c}$	{Stream k exits from the splitter after the bottoms in column c}
$STP_{k,i}$	{Streams k that can produce the final product i}
$STT_{k,t}$	{Stream k that must be different from zero if task t is selected}

The known data is the same as in the STN model.

In the set of variables, we now must add the component molar flow of external streams $R_{i,k}$

$$\min: \sum_{c \in COL} Cost_c$$

s.t.

$$[D_{i,c}, B_{i,c}, Cost_c] = f_c(\text{column equations}) \quad \forall c \in COL$$

$$\sum_{k \in STF_k} R_{i,k} = F_i^0 \quad \forall i \in COMP$$

$$\sum_{k \in STC_{k,c}} R_{i,k} = F_{i,c} \quad \forall i \in COMP, \forall c \in COL$$

$$D_{i,c} = \sum_{k \in STD_{k,c}} R_{i,k} \quad \forall i \in COMP, \forall c \in COL$$

$$B_{i,c} = \sum_{k \in STB_{k,c}} R_{i,k} \quad \forall i \in COMP, \forall c \in COL \qquad \text{(M.2.5)}$$

$$\sum_{k \in STP_{k,i}} R_{i,k} \geq rec_i F_i^0 \quad \forall i \in COMP$$

$$\bigvee_{t \in CT_{c,t}} \begin{bmatrix} Y_{c,t} \\ R_{i,k} = 0 \quad \forall k \notin STT_{k,t}, \forall i \in COMP \end{bmatrix} \quad \forall c \in COL$$

$$\Omega(Y) = True$$

$$Y \in \{True, False\}^{|TASK|}$$

$$D_{i,c}, B_{i,c}, F_{i,c}, R_{i,k} \geq 0$$

In the model SEN the first equation refers to the set of equations needed to calculate each one of the columns. The second is a mass balance in the mixer previous to each column. The third and fourth are mass balances in the splitters after the distillate and bottoms of each column, respectively. The fifth equation forces the model to get a minimum recovery of each final product. Note that the disjunction may force a set of the streams to be zero, depending on which separation task is performed in each column. The set of logical relationships is the same as in previous models. The subscript "c" in the Boolean variable does not introduce any problem because in a column only a single separation task can be selected, and then eqs. (2.1)–(2.7) can be used by merely dragging the "c" index.

Upon reformulating both models as mixed-integer nonlinear programming (MINLP) problems, it is observed that the SEN generates models with a reduced number of equations than the STN approach. However, the numerical performance depends on the model used in the columns. In models using shortcuts or aggregated equations, the SEN models, overall, demonstrate greater numerical robustness. However, with rigorous models, the same variables can take very different values depending on the task assigned to each column, which can eventually hinder the convergence of subproblems (both in decomposition algorithms and in branch-and-bound methods) while in the STN approach, each column can be efficiently initialized. In any case, the reformulation process into MINLP is typically simpler when utilizing an STN model.

A comprehensive description of the STN/SEN disjunctive models and their reformulation to MINLP using the FUG equations can be found in the work by Yeomans and Grossmann [46]. A comparison of MINLP models for SEN and STN can be found in the pioneering work by Novak et al. [16]. A discussion of SEN/STN together with other superstructures with intermediate characteristics using an aggregated model was presented by Caballero and Grossmann [17]. Yeomans and Grossmann also addressed the design of a column sequence using the SEN superstructure together with a disjunctive model for the rigorous optimization of each one of the columns [47].

2.3.2 Distillation sequences with nonconventional columns: thermally coupled distillation

As discussed in the introduction, distillation is a highly energy-intensive separation method, making any alternative that enhances energy efficiency of significant interest to the chemical industry. In pursuit of this goal, TCD has garnered considerable attention. TCD eliminates the need for some condensers and reboilers by directly transferring heat between columns through two streams (liquid and vapor). Reported cost savings range from 10–50% when compared to scenarios involving the use of simple columns arranged in series to achieve the desired product purity levels [48–50].

Due to the potential energy and investment savings in TCD, an important effort has been dedicated during the last decades to develop TCD models. In the 1980s and the early 1990s, the main focus was on developing methods for the optimization, design, and control of specific configurations, usually constrained to three-component mixtures. See, for example, the works by Fidkowski and Krolikowski [51, 52], Carlberg and Westerberg [53, 54], Alatiqi and Luyben [55], Glinos et al. [56], Glinos and Malone [57], Nikolaides and Malone [58], or Tryantafyllou and Smith [59] among others. Most of the theoretical results and modeling approaches can be generalized from three-component mixtures to N-component mixtures; therefore, we first introduce the most important results for three-component mixtures, and later we extend to general N-component systems.

2.3.3 Three-component systems

Although a growing interest in TCD appeared in the 1980s and 1990s of the last century, the first apparatus that used TCD concepts was due to Brugma in a US patent from 1942 [60] and Wright in a patent from 1949 [61]. The theoretical basis of TCD was developed more than 10 years later in the 1960s by Petlyuk and coworkers [29]. Petlyuk noticed that separation sequences using conventional columns (a single feed with two product streams, a condenser, and a reboiler) suffer from an inherent inefficiency produced by the thermodynamic irreversibility during the mixing of streams at the feed, top and bottom of the column. This remixing is inherent to any separation that involves an intermediate boiling component and can be generalized to an N-component mixture. The theoretical studies developed by Petluyk and coworkers showed that this inefficiency can be improved by removing some heat exchangers and introducing thermal coupling between columns. If a heat exchanger is removed, the liquid reflux (or vapor load) is provided by a new stream that is withdrawn from another column. In this way, it is possible to reduce energy consumption, and under some circumstances, the capital costs.

A fully thermally coupled (FTC) configuration is reached when the entire vapor load is provided by a single reboiler and all the reflux by a single condenser (Figure 2.5). Fidkowski and Krolokowski [52] proved that the FTC, in the case of three-component system known as Petlyuk configuration in honor to F. Petluyk, is the one with lowest energy consumption. Later this result was generalized for an ideal N-component mixture by Halvorsen and Skogestad [62].

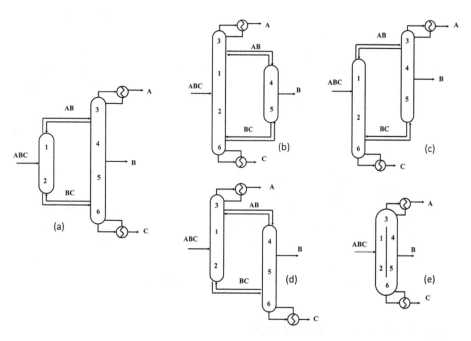

Figure 2.5: (a) Fully thermally coupled configuration for the separation of a three-component mixture, also known as Petlyuk configuration. (b–d) Thermodynamically equivalent configurations to the Petlyuk one. A divided wall column (e) can be considered thermodynamically equivalent to the Petlyuk arrangement.

When a thermal couple appears in a given design, a thermodynamically equivalent configuration (TEC) can be derived [63–67]. Two distillation sequences are thermodynamically equivalent if it is possible to go from one to another by moving column sections using the two streams of the thermal couple. Divided wall columns (traditional or with vertical partitions) can be generated by placing in the same shell columns connected by thermal couples and therefore they are also thermodynamically equivalent to the set of columns from which were generated (see Figure 2.5).

Between sequences formed by conventional columns and the Petlyuk arrangement, there is a set of alternatives that should be considered. Figure 2.6 shows the set alternatives for separating a single-feed three-component mixture including TECs using one or two columns. In Figure 2.6, we have included configurations with vertical partitions (internal walls) and multiple condensers and/or reboilers [68–71] that usually are also known as divided wall columns. If we can neglect the effects of heat transfer across the walls and the different pressure drops inherent to different locations of column sections, all TECs will have the same flows, temperatures, compositions and pressures. Thus, the energy consumption is the same in all TECs – and consequently also the operating costs-. Differences appear in their operability characteristics and the investment costs. Among all the structural different alternatives, the optimal one depends mainly on relative volatilities and initial concentration even though other factors like relative costs of energy and investment, operational pressure, and the complexity of the control system could eventually be important.

In three-component systems, it is possible to evaluate all the alternatives (as we will see latter this is not possible – or at least practical – for a general N-component separation). But even in the case of three-component systems a rigorous simulation and/or optimization of all the alternatives could be very time-consuming. Instead, a typical approach consists of using shortcut models. In the case of three-component mixtures FUG equations have been used by different researchers to get explicit formulations for minimum energy consumption (equivalent to minimum vapor load) [51–54, 72, 73]. Even though it is possible to use an explicit equation for most of the structurally different alternatives shown in Figure 2.6, with modern computers it is not necessary, and for the case of three component systems the optimization using FUG equations can be efficiently done with a simple spreadsheet. An interesting and detailed example of the application of FUG equations for the optimization and design of Petluyk and DWCs is the paper proposed by Ramirez-Corona et al. [74].

2.3.3.1 General thermally coupled sequences

The benefits in the reduction of energy consumption that can be obtained in three-component systems can be extended to N-component systems, or in general to systems in which we want to separate N key components from a very complex mixture. However, when the number of components increases the number of sequences also

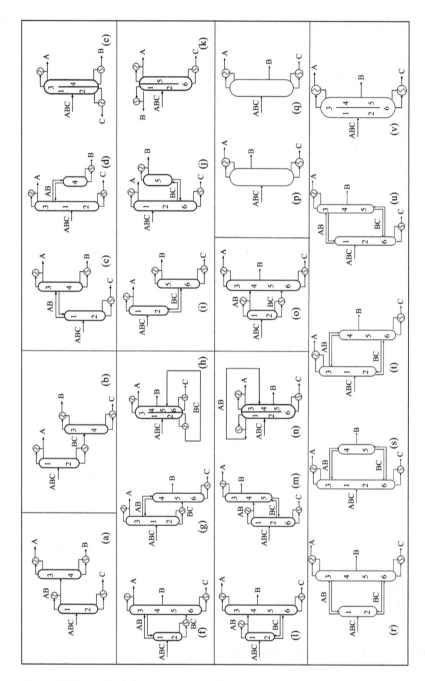

Figure 2.6: Alternatives for separating a single-feed three-component mixture including thermodynamically equivalent configurations using one or two columns. Columns with vertical partitions and DWCs have also been included. Configurations (p) and (q) are only of interest if the purity of B is not important.

increases and it is no longer possible (or at least practical) to perform a complete eval-uation of the search space. It is necessary to use mathematical programming-based approaches that allow obtaining the best configuration from the complete set of alter-natives without either explicitly generating all the alternatives or optimizing each al-ternative.

Maybe the first attempt to generalize the separation using TCD to N components was due also to Petlyuk et al. [29]. These authors proposed that a FTC sequence must obey the following rules:

1. The total number of column sections must be equal to $N(N–1)$.
2. The sequence must include only one condenser and one reboiler.
3. The key components in each column are those components with extreme (highest and lowest) relative volatilities.
4. From the final column (product column), N streams are withdrawn, each one con-taining a pure product.

In 1976 Sargent and Gaminibandara [75] used the resulting sequence – see Figure 2.7 – as the base of a nonlinear programming model (NLP) to optimize the separation of four- and five-components systems.

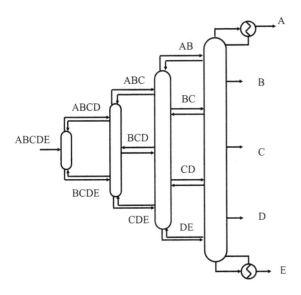

Figure 2.7: Superstructure proposed by Sargent and Gaminibandara. It is also the only alternative when Petlyuk rules for FTC sequencing are applied.

However, the generalization proposed by Petlyuk et al. is just one of a very large num-ber of alternatives that appear in TCD. The first rule avoids a large number of alterna-tives even if we consider only FTC configurations. As shown by Agrawal [76] the number of sections in FTC distillation sequences ranges between $4(N-6)$ and $N(N-1)$. The second one removes a large number of configurations including from conventional columns to partially thermally coupled configurations with more than two heat exchangers. The third constrains the large number of alternatives (even considering only FTC sequences) to a single one. And finally, the fourth does not consider the possibility of thermodynamically equivalent alternatives.

Following a conceptual design approach, Rong and coworkers [77–79] presented different alternatives for sequences with four and five components. Kim and coworkers [80–83], Hernández and Jiménez [84], Hernández et al [85], Blancarte-Palacios et al [86], and Calzon-McConville et al [87] also presented a set of interesting papers with the focus on the rigorous design of some specific configurations for three-, four-, or five-component mixtures.

The first interesting attempt to solve the problem of TCD was presented by Shah and Kokossis [88, 89] who introduced the concept of "supertask." A supertask groups some known complex structures like side columns, Petlyuk arrangements, DWCs and side-streams. In that way it is possible to calculate a priori all the possible individual tasks or supertasks, and determine the optimal configuration through an MILP model. Therefore, it is possible to generate complex arrangements, where each task or supertask can be optimized using shortcuts (used in the original paper) or rigorous models. The major drawback is that there are a large number of feasible configurations that are not considered (e.g., all those that include thermal couples between more than two consecutive columns).

However, it was Agrawal in 1996 [76] who established the structural considerations to separate an N-component mixture. He showed that it is always possible to separate an N-component mixture using $N-1$ columns. ***The same as in "classical column sequencing"***! These results were later used to systematically generate all the feasible sequences. Agrawal's results can be summarized in three basic rules:

1. Given a sequence of conventional columns, it is possible to remove the heat exchangers (condensers and/or reboilers) associated with intermediate streams (streams connecting two columns) without changing the structure or the sequence of separation tasks (see Figure 2.8).

2. It is possible to remove heat exchangers (reboilers and/or condensers) associated with final products of intermediate volatility. But in this case, it is necessary to add two column sections for each heat exchanger removed. It is worth noting that the total number of columns does not increase, but the number of column sections does. The set of separation tasks performed by the sequence increases by one for each heat exchanger removed (Figure 2.9). Following this approach, and taking into account that there are $N-2$ products of intermediate volatility, it is possible to get an FTC system with the minimum number of column sections (MNCS):

$$\text{MNCS} = 2(N-1) + 2(N-2) = 4N - 6 \tag{2.8}$$

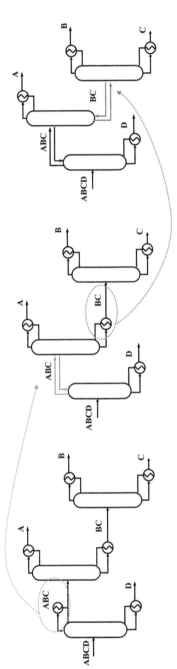

Figure 2.8: Substituting internal heat exchangers by thermal couples does not modify either the sequence of separation tasks (ABC/D – A/BC – B/C) or the number of column sections (six column sections).

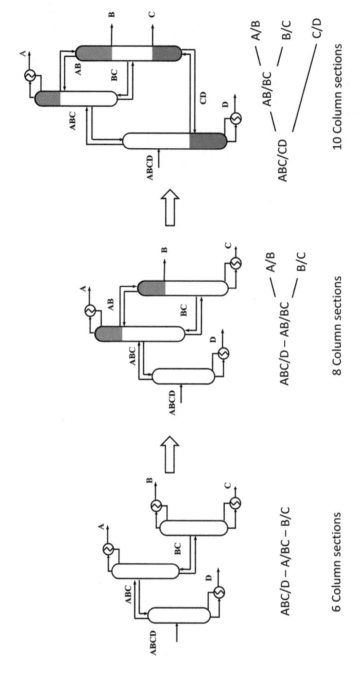

Figure 2.9: It is possible to remove heat exchangers associated with intermediate volatility final products. However, we must add two column sections for each heat exchanger removed. The sequence of separation tasks also changes. The three alternatives in the figure perform a different sequence of separation tasks.

3. From an FTC system with the MNCSs (4 N–6), it is possible to reduce energy consumption by substituting those separation tasks that do not involve key components with extreme volatilities by a sequence of separations in which the key components are the lightest and the heaviest ones. Again, we have to add two more column sections. Note that the number of total columns continues to be equal to N–1 (Figure 2.10).

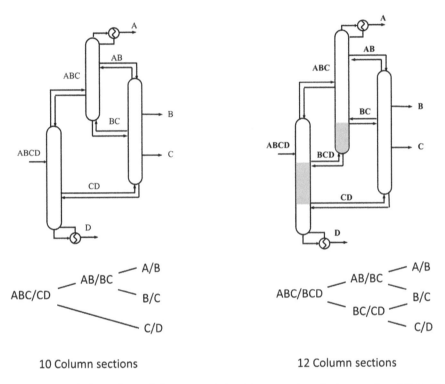

10 Column sections   12 Column sections

Figure 2.10: Fully thermally coupled sequences. Left, minimum number of column sections 4 N–6 = 10. Right maximum number of column sections $N(N$–1) = 12.

A detailed discussion on the number of column sections needed for a given separation can be found in the mentioned paper by Agrawal [76] and in Caballero and Grossmann [27, 90, 91].

In light of prior structural considerations, distillation sequences can be categorized based on the number of distillation columns employed to separate an N-component mixture into N-product streams. These classifications include sequences utilizing more than N–1 columns, sequences employing precisely N–1 columns, and sequences requiring fewer than N–1 columns.

In the case of zeotropic mixtures, sequences with exactly N–1 columns, named **regular configurations** by Agrawal [92], are characterized by the following three features:

1) Mixtures (or states) with the same components are transferred only once from one distillation column to another.
2) A final product is obtained in a single location of the sequence.
3) The feed stream and all the intermediate mixtures are split into exactly two product streams by two column sections that form a separation task.

Configurations that violate the first two features and obey the third produce sequences with more than N–1 columns. These configurations, also referred to as **nonbasic configurations**, have higher operating costs than the best basic configuration [27, 28, 93]. Nonbasic configurations also tend to have higher capital costs due to the additional distillation columns, and therefore nonbasic configurations can be removed from the search space.

Configurations that violate the third feature have higher operating costs than the best basic configuration due to increased heat duty, especially for getting high-purity products. However, the reduced number of columns could compensate for the extra energy consumption. In the literature, some of these cases can be found, for example, those of Brugma [60], Kaibel [94], Kim et al [95], or Errico and Rong [96]. These alternatives can be generated from some regular configurations by removing some column sections or merging columns (task) in a single shell. The inclusion of these alternatives from the beginning in the search space considerably complicates the problem. On the other hand, it is not yet clear how to systematically generate all these "intensified" sequences. As an alternative, it has been proposed to use a sequential approach in which the first stage is constrained to regular configurations [97–100]. An excellent review of the advances in intensification in distillation can be found in [101].

It is important to make some remarks on previous features and classification:

– Sequences obeying the three distinguishing features can always be arranged in N–1 columns although the total number of separation tasks can be larger. Consider the example in Figure 2.8, left side. We want to separate a four-component mixture ABCD where the components are sorted by decreasing volatilities. We perform the following separation tasks (ABC/CD, which means separating AB from D letting the C component be optimally distributed between distillate and bottoms; AB/BC; A/B, B/C and C/D). In this example, we can identify five separation tasks, and this sequence can be performed using five distillation columns, but this sequence can be easily arranged in three distillation columns. It can be rearranged in 16 TECs using three distillation columns [64].
– In the presence of a thermal couple, the arrangement of separation tasks in the actual columns is not unique. Utilizing the two flows of the thermal couple, it is possible to relocate a column section from one actual column to another, resulting in different arrangements of tasks in the actual columns. Figure 2.11 illustrates this concept. All the configurations obtained by moving column sections using a thermal couple are considered thermodynamically equivalent. From a practical standpoint, there are some differences due to varying pressure losses and practical considerations in trans-

ferring vapor streams. However, at the preliminary design stage, these configurations can be treated as equivalent. A detailed discussion on TECs can be found in [63, 64, 102, 103].

- The total number of TECs can be very large. For example, in a five-component mixture, there are 203 **basic configurations** (a basic configuration is a regular one in which each column has a reboiler and a condenser); 6,128 regular configurations, and more than 10^5 arrangements in actual columns – most of them with very similar performance in terms of total cost. To avoid this degeneracy, it is convenient to represent a sequence in terms of the separation tasks involved instead of the particular equipment used to perform the separation. In other words, the search space must be formed only by the set of regular configurations [28].

The first rule-based algorithm for generating the full set of basic configurations was proposed by Agrawal [92, 103]. Following this line, Ivakpour and Kasiri [104] proposed a formulation in which distillation configurations are represented mathematically as upper triangular matrices.

While those approaches provide a straightforward method to generate all basic or regular configurations, they are not readily adaptable for inclusion in a mathematical programming model. Caballero and Grossmann [90, 91, 105, 106] employed a distinct approach by developing a set of logical rules, expressed in terms of Boolean or binary variables, to guarantee basic or regular configurations. These rules were incorporated into an optimization-based framework for identifying the optimal column sequence, encompassing conventional to FTC distillation sequences. The primary objective in these works was not to explicitly generate all basic configurations but rather to ensure a robust relaxation during the solution of the resulting MINLPs, which incorporate all distillation column performance equations. This approach aims to extract the optimal configuration without an exhaustive enumeration of all alternatives. It is noteworthy that Shah and Agrawal [107] presented a viable alternative set of equations expressed in terms of binary variables that could also be integrated into a mathematical programming environment. However, their focus was on rapidly verifying whether a given alternative constitutes a basic one (with superior performance) rather than on performance when those equations are combined with the column model in a mathematical programming environment. Some of their equations can be obtained by aggregating some of the logical relations presented by Caballero and Grossmann [27, 91], and therefore a worse relaxation can be expected. Giridhar and Agrawal [93] proposed alternative methods for generating the complete space of basic alternatives.

An overview of a disjunctive model for identifying the optimal separation sequence for separating N key components of a mixture using distillation follows. The set of alternatives includes sequences formed only by conventional columns to FTC configurations going through all intermediate alternatives. The model is based on separation tasks and the optimal solution of tasks and heat exchangers (condensers and reboilers) must be rearranged in $N-1$ actual columns. Even though no particular con-

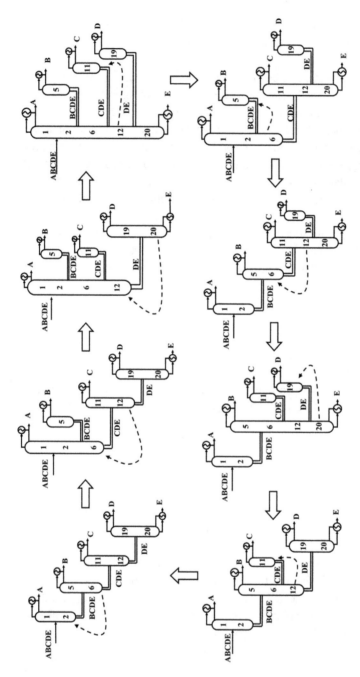

Figure 2.11: The eight thermodynamically equivalent configurations for the separation A/BCDE – B/CDE – C/DE – D/E, with only heat exchangers in final products.

figuration in actual columns is assumed, a given separation task is formed by two column sections: a rectifying and a stripping section (even though, in the final arrangement of tasks in actual columns, these two sections may be located in different columns). This perspective allows treating a separation task as a pseudo-column, conceptually simplifying the modeling process. This approach also enables the calculation of the cost of each column section.

Moreover, considering that the total number of actual columns is always N–1 and that operating costs are independent of the particular arrangement of tasks within columns, the model facilitates an accurate estimation of the total annual cost or, at the very least, a tight lower bound to the final cost.

We use an STN superstructure (Figure 2.12). But at the difference of sequences formed by conventional columns, there is no one-to-one relationship between separation tasks and columns, and the final assignment of tasks to actual columns should be based on other criteria (operability, building constraints, etc.) [63, 102].

For modeling the logical relationships, it is necessary to define the following index sets (we include an example of the members of the set for a four-component separation):

TASK	$\{t \mid t$ is a separation task$\}$
	e.g., *TASK* = [(ABC/BCD), (AB/BCD), (ABC/CD), (AB/BC), (AB/C), (B/CD), (BC/CD), (A/B), (B/C), (C/D)]
STATES	$\{s \mid s$ is a state$\}$
	e.g., *STATES* = [(ABCD), (ABC), (BCD), (AB), (BC), (CD), (A), (B), (C), (D)]
IS_s	$\{s \mid s$ is an intermediate state. All but initial and final products$\}$
	e.g., IM_S = [(ABC), (BCD), (AB), (BC), (CD)]
COMP	$\{i \mid i$ is a component to be separated in the mixture$\}$
	e.g., *COMP* = [A, B, C, D]
FS_t	$\{t \mid t$ is a possible initial task. The task that receives the external feed$\}$
	e.g., FS_t = [(A/BCD), (AB/BCD), (AB/CD), (ABC/BCD), (ABC/CD), (ABC/D)]
$TS_{s,t}$	$\{$tasks t that the state s can produce$\}$
	e.g., $TS_{ABCD.}$ = [(A/BCD), (AB/BCD), (AB/CD), (ABC/BCD),(ABC/CD), (ABC/D)]
	TS_{ABC} = [(A/BC), (AB/BC), (AB/C)]
	TS_{BCD} = [(B/CD), (BC/CD), (BC/D)]
	TS_{AB} = [(A/B)]; TS_{BC} = [(B/C)]; TS_{CD} = [(C/D)];
$RECT_{s,t}$	$\{$task t that produces state s by a rectifying section$\}$
	e.g., $RECT_{ABC}$ = [(ABC/CD), (ABC/BCD)]
	$RECT_{AB}$ = [(AB/BCD), (AB/BC), (AB/C)]
	$RECT_{BC}$ = [(BC/CD)]
$STRIP_{s,t}$	$\{$task t that produces state s by a stripping section$\}$
	e.g., $STRIP_{BCD}$ = [(AB/BCD), (ABC/BCD)]
	$STRIP_{BC}$ = [(AB/BC)]
	$STRIP_{CD}$ = [(ABC/CD), (B/CD), (BC/CD)]
$ST_{s,t}$	$\{$tasks t that can produce state $s\}$ $ST_{s,t} = RECT_s \cup STRIP_s$
FPs	$\{s \mid s$ is a final state (pure products) $\}$
	e.g., *FP* = [(A), (B), (C), (D)] do not confuse with components, although the name is the same

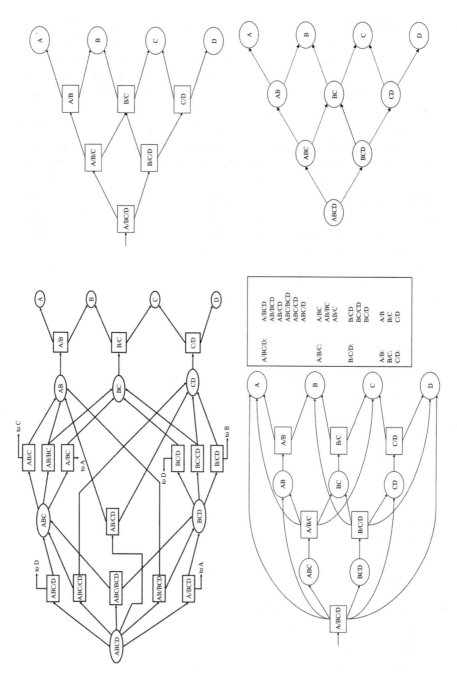

Figure 2.12: STN superstructure for separating a four-component mixture. (a) STN explicitly includes all states and tasks. (b) STN in which some tasks have been grouped. (c) STN in which only groups of tasks – and final states – are represented. If a given group of tasks does not appear a bypass must be included. (d) STN using only states. If a given state does not appear a bypass must be used instead.

$PREC_{s,t}$ {tasks t that produce final product s through a rectifying section}

         e.g., $PREC_A = [(A/B)]$

         $PREC_B = [(B/CD), (B/C)]$

         $PREC_C = [(C/D)]$

$PSTR_{s,t}$ {tasks t that produce final product s through a stripping section}

         e.g., $PST_B = [(A/B)]$

         $PST_C = [(AB/C), (B/C)]$

         $PST_D = [(C/D)]$

and the following Boolean variables:

Y_t	True if the separation task t exists. False otherwise.
Z_s	True if the state s exists. False, otherwise.
W_s	True if the heat exchanger associated with the state s exists. False, otherwise.
W_s^{Cond}	True if the heat exchanger in state s exists and it is a condenser.
W_s^{Reb}	True if the heat exchanger in state s exists and it is a reboiler.

1. *A given state s can give rise to at most one task:*

$$\underset{t \in TS_t}{\vee} Y_t \vee R; \ s \in STATES \tag{2.9}$$

where R is a dummy Boolean variable that means "do not choose any of the previous options."

2. *A given state can be produced at most by two tasks: one must come from the rectifying section of a task and the other from the stripping section of a task:*

$$\left. \begin{array}{c} \underset{t \in RECT_{s,t}}{\vee} Y_t \vee R \\ \underset{t \in STRIP_{s,t}}{\vee} Y_t \vee R \end{array} \right\} \ \forall s \in STATES \tag{2.10}$$

where R has the same meaning as in eq. (2.9). Note that if we want only systems with the MNCSs, a given state, except products, must be produced at most by one contribution. Note also that when at least a state is produced by two contributions, the number of separation tasks is not the minimum (see Figure 2.13).

3. *All the products must be produced at least by one task:*

$$\underset{t \in \left(PREC_{s,t} \cup PSTR_{s,t} \right)}{\vee} Y_t; \ \forall s \in FP_s \tag{2.11}$$

4. *If a given final product stream is produced only by one task, the heat exchanger associated with this state (product stream) must be selected. A given final product must always exist, produced by a rectifying section, by a stripping section, or by both. Therefore, an equivalent form to express this logical relationship is that if a final product is not produced by any rectifying (stripping) section the heat exchanger related to that product must exist:*

Valid alternatives

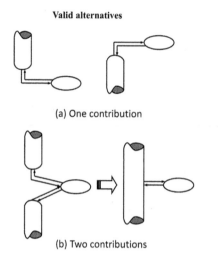

(a) One contribution

(b) Two contributions

No-valid alternative
Increases the number of columns, the number of heat exchangers and it is sub-optimal from an energetic point of view.

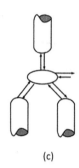

(c)

Figure 2.13: Graphical illustration of logical relationship 2. The state on the right side (c) is produced by two rectifying sections simultaneously. The number of columns increases and cannot be rearranged into just N–1 columns.

$$\left. \begin{array}{l} \neg \left(\bigvee_{t \in PREC_{s,t}} Y_t \right) \Rightarrow W_s \\[2ex] \neg \left(\bigvee_{t \in PSTR_{s,t}} Y_t \right) \Rightarrow W_s \end{array} \right\} \quad \forall s \in FP_s \tag{2.12}$$

5. *If a given state is produced by two tasks (a contribution coming from a rectifying section and the other from a stripping section of a task) then there is no heat exchanger associated with that state (stream):*

$$(Y_t \wedge Y_k) \Rightarrow \neg W_s \quad \begin{cases} t \in RECT_{s,t} \\ k \in STRIP_{s,k} \\ s \in STATES \end{cases} \tag{2.13}$$

6. *Connectivity relationships between tasks in the superstructure:*

$$\left. \begin{array}{l} Y_t \Rightarrow \bigvee_{k \in TS_{s,t}} Y_k; \quad t \in ST_{s,t} \\[2ex] Y_t \Rightarrow \bigvee_{k \in ST_{s,t}} Y_k; \quad t \in TS_{s,t} \end{array} \right\} \quad s \in STATES \tag{2.14}$$

7. *If a heat exchanger associated with any state is selected then a task that generates that state must also be selected:*

$$W_s \Rightarrow \bigvee_{t \in ST_{s,t}} Y_t; \quad s \in STATES \tag{2.15}$$

8. *If a separation task t produces a state s by a rectifying section, and that state has a heat exchanger associated, then it must be a condenser. If the state is produced by a stripping section then it must be a reboiler:*

$$Y_t \wedge W_s \Rightarrow W_s^{Cond} \quad (s,t) \in RECT_{s,t}$$
$$Y_t \wedge W_s \Rightarrow W_s^{Reb} \quad (s,t) \in STRIP_{s,t}$$
$$(2.16)$$

It is convenient to complete the previous rule by adding that:

9. *If a given state does not have a heat exchanger, then both W_C and W_R associated with that state must be false:*

$$\neg W_s \Rightarrow \neg W_s^{Cond} \wedge \neg W_s^{Reb} \quad s \in STATES \qquad (2.17)$$

If the problem is solved as an MI(N)LP the binary variables w_s^{Cond}, w_s^{Reb} (related to the Booleans with the same name in capital letters) need not be declared as binary and they can be considered as continuous with values between 0 and 1. Equations (2.16) and (2.17) ensure that w_s^{Cond} and w_s^{Reb} take integer values when y and w are integers. Therefore, the variables $w_s^{Cond} w_s^{Reb}$ do not increase the combinatorial complexity of the problem.

10. *The set of logical relationships previously presented in terms of separation tasks can be easily rewritten in terms only of states: "There is a one-to-one correspondence between the sequence of tasks and the sequence of states and vice versa." The relationship between tasks and states is as follows:*

$$Y_t \Rightarrow Z_s; \quad (t,s) \in ST_{s,t} \qquad (2.18)$$

$$Z_s \Rightarrow \bigvee_{t \in TS_{s,t}} Y_t \quad \forall s \in STATES \qquad (2.19)$$

Equation (2.18) can be read as: "if the task t, which belongs to the set of tasks produced by the state s exists then the state s must exist." Equation (2.19) as: "If the state s exists at least one of the tasks that the state s can produce must exist."

We should note that if the problem is solved as an MI(N)LP, it is only necessary to declare as binary either y_t or z_s, but not both. Whether y_t is declared as binary z_s can be declared as continuous between 0 and 1 and vice versa.

The previous equations ensure that any sequence of tasks and the selected heat exchangers are feasible separations that can be arranged in N–1 distillation columns.

The rest of the model is formed by the equations associated with each Boolean variable. The particular equations of the model depend on the approximation used (shortcut, aggregated, rigorous, etc.). Instead, we present in Figure 2.14 a **disjunctive conceptual representation** that is independent of the model used.

If a given state exists (i.e., Y_t = True), then the equations of the separation task t must be calculated. Otherwise, all the variables related to that state must be fixed to zero. See the first disjunction in Figure 2.14. As previously commented, the separation task is considered as a pseudo-column formed by two sections (rectifying and stripping) by similarity with conventional columns. Note also that, in this disjunction, we are not considering the existence or nonexistence of condensers and reboilers. The rectifying section needs a reflux stream and the stripping section a vapor load, but they could come from a condenser/reboiler or directly from a stream withdrawn from another column (thermal couple).

$$\min : Total\ Cost$$

$$s.t.$$

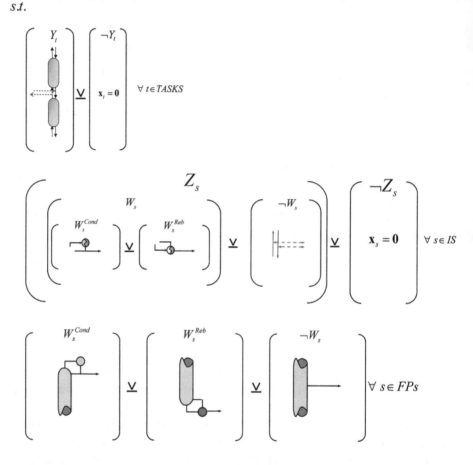

$$\Omega(Y_t, W_s, Z_s, W_s^{Cond}, W_s^{Reb}) = True \quad Eqs.\ \text{(L-1) to (L-10)}$$

Figure 2.14: Conceptual representation of the disjunctive model for the N components separation using thermally coupled distillation.

If a given intermediate state exists ($Z_s = True$) – second disjunction in Figure 2.14 – then either there is a heat exchanger associated with that state or there is a thermal couple. Otherwise, all the variables associated with that state must be zero. If there is a heat exchanger associated with the state, then it can be either a condenser ($W_s^{Cond} = True$) or a reboiler ($W_s^{Reb} = True$).

The thermal couple that appears in any intermediate state without a heat exchanger can be modeled by a vapor a liquid mass balances in the state s (Figure 2.15).

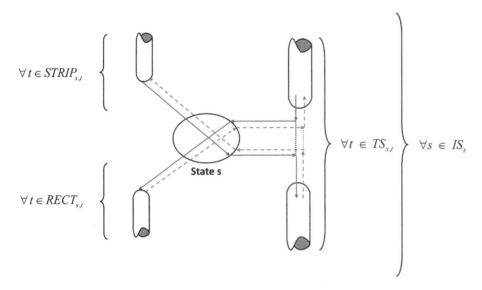

Figure 2.15: Mass balance in an intermediate state s when there is a thermal couple in that state.

The latest disjunction in Figure 2.14 indicates that a final product may require a condenser, a reboiler or neither of the two. In the latter case, the product is extracted at an intermediate point in a column corresponding to the ends of a task that produces it through an enrichment section and another task that produces it through a stripping section.

The previous model can be expanded to explicitly consider DWCs. Due to the widespread adoption of DWCs as a standard technology and the associated significant savings, Caballero and Grossmann [108] extended the previous model to include columns with a single wall. As the degree of thermal coupling increases, the potential for heat integration diminishes (i.e., the number of heat exchangers decreases).

The TCD is not always the optimal solution: for instance, in FTC systems, energy must be supplied at the highest system temperature and removed at the lowest, potentially limiting the utilization of less expensive utilities. Additionally, thermal couples could result in excessively high columns. Instead, heat integration between condensers and reboilers (or with other system components) could be a more viable approach.

Caballero et al. [27, 109] proposed an extension of the previous model that simultaneously considers the design of both heat-integrated and TCD sequences. Rong and colleagues [110–112] likewise presented an alternative procedure based on conceptual design for heat-integrated and TCD sequences. Agrawal [113] proposed a methodology for integrating TCD with multi-effect distillation, although it is a conceptual procedure not based on mathematical programming.

2.4 Models for the rigorous optimization of distillation columns

A significant challenge in the design of a distillation column is the rigorous tray-by-tray optimization of the column assuming phase equilibrium. The models include both continuous and discrete variables. Continuous variables are the operating conditions (pressure, temperatures, flows, heat loads, etc.) and the discrete variables are the total number of equilibrium trays as well as feed(s) and product(s) locations.

The first approximation to solve this problem using deterministic optimization was due to Sargent and Gaminibandara [75], who proposed a superstructure to determine the optimal feed location in a column with a fixed number of trays. (Figure 2.16). The model uses the MESH equations: mass and enthalpy balances, phase equilibrium equations and that molar fraction summation in each phase is equal to one.

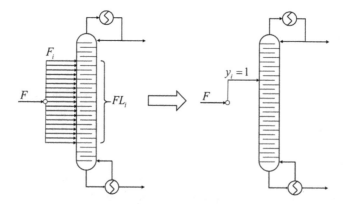

Figure 2.16: Sargent and Gaminibandara superstructure for the feed tray location problem.

Besides the MESH equations, it is necessary to add a set of equations to determine the optimal feed location. To this end, we need two index sets:

Tray	$\{i \mid i$ is a tray in the column$\}$
FL$_i$	Possible feed locations for the feed.

and a binary variable y_i that takes the value of 1 if the feed is added in tray i, and 0 otherwise.

The model can be written as follows:

$$\min \ Column \ Total \ Cost$$

$$s.t.$$

$$MESH_i \ \forall i \in Tray$$

$$Cost \ estimation \ equations$$

$$\sum_{i \in FL_i} F_i = F \qquad\qquad\qquad [\text{M.4.1}]$$

$$\sum_{i \in FL_i} y_i = 1$$

$$F_i \leq Fy_i \ \forall i \in FL_i$$

$$y \in \{0,1\}; \ F_i \geq 0$$

In the previous model, the last three equations ensure that the feed is entering a single tray. An interesting property is that, frequently, the problem is solved as a relaxed NLP model.

Notwithstanding, the first model, which optimizes not only the feed location but also the total number of trays, is due to Viswanathan and Grossmann [114]. They proposed a model in which both, the reflux and the vapor load can return at different trays: reflux above and vapor load below the feed that is fed to a fixed tray. If the postulated total number of trays is large enough, the optimal reflux/vapor load return creates some "deactivated trays": trays without vapor load or liquid reflux. A scheme of the Viswanathan and Grossmann superstructure is depicted in Figure 2.17.

Besides the MESH equations for each tray, cost estimation, recovery constraints, the variable reflux/vapor load part of the model and the rest of the equations of a column can be written as described below.

We need the following index sets:

Tray	$\{i \mid i$ is a tray in the column$\}$
RC_i	$\{i \mid i$ is a candidate tray for the reflux return$\}$
VC_i	$\{i \mid i$ is a candidate tray for the vapor from reboiler return$\}$

where L_i, V_i, the reflux and vapor load return to tray i in the column, respectively, LT and VT the total liquid and vapor exiting from condenser and reboiler, and y_i^R, y_i^V the binaries associated with reflux and vapor load return:

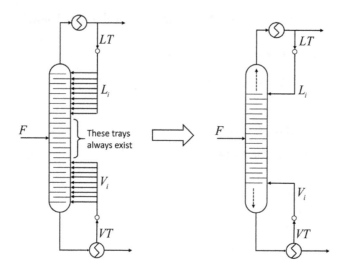

Figure 2.17: Superstructure proposed by Viswanathan and Grossmann for the rigorous optimization of a single column.

$$LT = \sum_{i \in RC_i} L_i$$

$$L_i \le LT^{up} y_i^R \quad \forall i \in RC_i$$

$$\sum_{i \in RC_i} y_i^R = 1$$

$$VT = \sum_{i \in VC_i} V_i \qquad \text{[M.4.2]}$$

$$V_i \le VT^{up} y_i^V \quad \forall i \in VC_i$$

$$\sum_{i \in VC_i} y_i^V = 1$$

$$L_i \ge 0, \ y_i^R \in \{0,1\}$$

$$V_i \ge 0, \ y_i^V \in \{0,1\}$$

The major drawback of this model is that for nonexisting trays there is a zero-liquid flow (rectifying section) and a zero-vapor flow (stripping section) which can produce important numerical problems due to equilibrium equations with zero flow in one of the phases, i.e., the vapor-liquid equilibrium must be satisfied in trays without mass transfer. Despite all the convergence problems different researchers have successfully used the previous model to optimize complex columns and column sequences. For example, Ciric and Gu [115] solved the MINLP model for synthesizing reactive distillation columns when chemical reaction equilibrium cannot be ensured and illustrated it with the ethylene glycol synthesis. Bauer and Stichlmair [116] used the model with zeotropic and azeotropic sequences of up to four components. These authors also studied other configurations for feed and product location: e.g., the feed entering in several stages, product removal in several stages or the inclusion of intermediate heat exchangers. Dunnebier and Pantelides [117] applied the model for the optimal design of a thermally coupled sequence.

In the line initiated by Bauer and Stichlmair [116], Barttfeld et al. [118] studied the impact of different superstructures that can be used to optimize a single distillation column. Figure 2.18 illustrates three representations that differ from the original one developed by Viswanathan and Grossmann but achieve the same objective. First, in Figure 2.18(a), condensers and reboilers are positioned on all candidate trays for energy exchange. This means that a variable reflux/vapor load stream is considered by moving the condenser/reboiler. In contrast, in the representation of the variable reflux location shown in Figure 2.17, the condenser and reboiler are fixed equipment. These two alternatives are equivalent if one piece of fixed equipment is considered at each column end. However, when variable heat exchange locations are incorporated into the optimization procedure, some distinctions arise. In one case, the challenge is to find the ideal location for energy exchange, whereas in the other, the optimal location for a secondary feed stream (reflux/vapor load) is taken into account. Figure 2.18 (b) and (c) shows the cases with variable feed and variable reboiler location and variable feed and variable condenser location, respectively.

Variable heat exchange offers a significant advantage by allowing energy to be exchanged at intermediate tray temperatures, potentially leading to more energy-efficient designs [119]. In the cases studied by Barttfeld et al., the results indicate that the most energy-efficient MINLP representation entails the variable reboiler and feed tray location. However, this decision is likely to be specific to the situation, for instance, in a column with the condenser operating at a low temperature (requiring expensive cold utility), the variable condenser location, enabling condensers to be placed at intermediate heat exchangers, could be the preferred superstructure.

To avoid numerical issues arising in the optimization of a column using MINLP, Yeomans and Grossmann [120] proposed a model based on GDP. The basic idea is to categorize the trays of a distillation column into permanent trays (that must always exist) and conditional trays (that may or may not exist depending on the optimal solution). For the existing trays, material and energy transfer are modeled using MESH equations. For the trays that do not exist, the model considers a simple liquid and vapor bypass along the tray without material and energy transfer. In this case, the MESH equations are trivially satisfied, preventing issues associated with flows being equal to zero in MINLP models. Figure 2.19 illustrates the GDP superstructure.

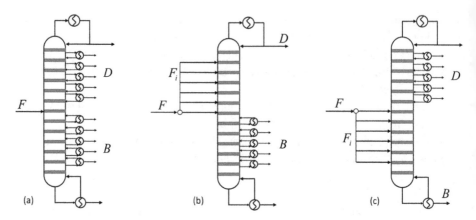

Figure 2.18: Representations for the MINLP optimization of a single column: (a) variable reboiler and condenser location; (b) variable feed and reboiler location; and (c) variable feed and condenser location.

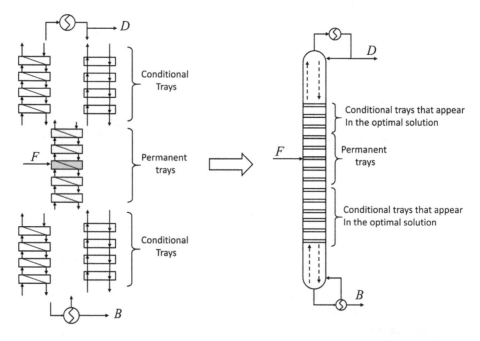

Figure 2.19: Superstructure for the optimal design of a distillation column using a GDP optimization.

Conceptually, the GDP model for the rigorous optimization of a single distillation column can be written as follows:

min *Total Cost*

s.t.

MESH equations for permanent trays

Mass/Energy Balances for conditional trays

[M.4.3]

$$\begin{bmatrix} Y_t \\ MESH\ equations \end{bmatrix} \underline{\vee} \begin{bmatrix} \neg Y_t \\ Bypass\ equations \end{bmatrix} t \in Conditional\ Trays$$

$$Y_t \Rightarrow Y_{t+1}\ t \in RECT$$

$$Y_t \Rightarrow Y_{t-1}\ t \in STRP$$

In the previous model, the last two equations force all existing trays to be consecutive in the final solution. This is necessary to avoid the possibility of degenerate solutions (e.g., existing trays interspersed with trays that do not exist). REC and STR refer to the set of conditional trays in the rectifying/stripping section and it is assumed that trays are numbered from the top to the bottom of the column.

In general, the GDP model avoids potential numerical issues of MINLP models. However, since both models are highly nonlinear and non-convex, they require good initial values to converge. Obtaining a good set of initial values may not be straightforward. Barttfeld and Aguirre [121] proposed using a preprocessing first phase to achieve a good estimate. Initially, it is assumed that all trays will be active (providing an upper limit to the total number of trays) and that the operation will be under conditions of minimum reflux and minimum entropy (reversible separation). With this strategy, the authors achieved good estimates to converge the rigorous tray-by-tray model.

Another option is to start with a simplified representation of the column, for example, using shortcut methods, and then gradually increase complexity until reaching the rigorous model [122]. For instance, Harwardt and Marquardt [123] followed this approach for the design of heat-integrated distillation columns (HIDiC) and columns with vapor recompression cycles (VRC). In this case, they began with a shortcut method based on pinch analysis to estimate concentration profiles and minimum energy requirements, which were used to initialize the optimization. Next, using the results of this initial step, they developed a model that considered only material balances and equilibrium but not enthalpy balances. Finally, energy balances were included to solve the final model.

Two additional interesting modifications were introduced. The first involved separating the equations for calculating liquid-vapor equilibrium from the rest of the model, using an external model for this calculation. In other words, liquid-vapor equilibrium is not directly accessible to the solver, which treats these equations as an implicit model. This approach avoids explicitly writing all thermodynamic equations, eliminating potential errors and making such calculations more robust. The second modification involved using the Successive Relaxed MINLP (SR-MINLP) proposed by Kraemer et al. [124]. In SR-MINLP, whether the model is MINLP or GDP, it is reformulated as a problem with only continuous variables with Big-M constraints. Discrete decisions are enforced through non-convex constraints that compel binary variables to take discrete values.

Even with all the difficulties and numerical convergence problems associated with these models, various groups have successfully solved complex problems, including reactive distillation models [115, 125], azeotropic sequences [116, 126, 127], hybrid distillation-membrane systems [128] and Kaibel and Petlyuk thermally coupled configurations [129, 130] among others.

One of the challenges associated with the rigorous simulation of distillation columns lies in the precise determination of liquid-vapor equilibrium, and the estimation of thermodynamic and transport properties, for several reasons. The first reason is that explicitly writing all equations is prone to error (and difficult to debug) due to the large number of involved parameters. Additionally, the equations are highly nonconvex, further complicating the convergence of an already complex model. Finally, if there is a need to change the thermodynamic model (or any correlation), it is necessary to rewrite those equations from scratch.

On the other hand, there are commonly used thermodynamic packages that enable not only the estimation of properties for pure components and mixtures but also efficient calculations of liquid-vapor (L-V) or liquid-liquid-vapor (L-L-V) equilibrium, as well as the determination of dew and bubble temperatures and pressures. Building on this idea, some researchers have introduced a decoupling approach by separating the estimation of liquid-vapor equilibrium (and, in general, all property estimation calculations) from the main model equations. This is achieved by treating thermodynamic packages as implicit models that the solver does not directly access. These models must provide not only the value of the property (such as bubble temperature, enthalpy, or entropy) but also accurate derivatives with respect to the independent variables used in the model.

For example, several authors [122, 123, 131] have shown how to interface GAMS with thermodynamic models via CAPE-OPEN by using external equations, or extrinsic functions [132]. However, the implementation depends on additional C code to create a Dynamic Link Library (DLL) that must be manually generated. Recently, Krone et al. [133] addressed this issue by automatically generating all the code for interfacing GAMS with an external CAPE-OPEN thermodynamic property package. In those cases, the optimization of a distillation column or a sequence of distillation columns is treated as a modular combination of general vapor-liquid equilibrium stage models.

On the other hand, since the early days of process engineering, rigorous simulation models for distillation columns have been developed and implemented in chemical process simulators. These models incorporate the latest advancements in both thermodynamics and numerical implementation to achieve robust, reliable and numerically efficient simulations. It would be desirable to integrate such modules into optimization models, similar to how it has been done with thermodynamic packages. However, two significant problems arise:

- Due to the black-box architecture, derivatives can only be calculated by perturbing the independent variables, which considerably increases the number of simulations. Even though the computational time increases, convergence from a perturbed variable is usually very fast. However, some unit operations inherently include numerical noise [134] (and distillation is the typical one). The noise can be insignificant from the simulation or design perspective, but it may lead to flawed estimations of the derivatives. Erratic solver behavior can be caused by inaccurate derivatives, and the KKT (Karush-Kuhn-Tucker) conditions may not even be met at an optimal solution.
- The number of trays and feed tray position are discrete variables that cannot be relaxed (like in branch-and-bound algorithms, for example) and must be fixed before each simulation.

Despite all these challenges, Caballero et al. [135] and Navarro et al. [136, 137] have developed a modeling framework that enables the use of rigorous process simulator models and incorporates them as a part of optimization models. In the case of distillation, Caballero et al. [138] adapted the disjunctive model of Yeomans and Grossmann [120] to seamlessly integrate process simulators for its calculations. To achieve accurate derivative estimates, they meticulously investigated and established the most appropriate perturbation values for each variable, in conjunction with adjusting convergence tolerances in material and energy balances within the columns. This approach enabled them to obtain sufficiently accurate gradient values. Additionally, these authors modified the "Master" problem by incorporating a term into the linearizations that consider the influence of the presence or absence of each tray within the column. These contributions were derived from simulations by systematically removing or adding trays from a column. The algorithm was successfully applied to the optimal design for the separation of hydrocarbon mixtures; alcohols with and without azeotropes, and the extractive separation of water and methanol using ethylene glycol as an extractant or the membrane distillation to separate propylene and propane [139]. Alternatively, Javaloyes-Anton et al. [140] effectively adapted the extended cutting-plane algorithm for solving MINLP problems, to optimize distillation columns. They successfully optimized Petlyuk configurations and extractive distillation schemes. By comparing their results with popular stochastic optimization approaches like Particle Swarm Optimization and Genetic Algorithms, they demonstrated superior solutions in all test problems compared to these two widely used stochastic global search algorithms.

An intermediate level of aggregation between implicit thermodynamics and the entire column involves optimizing a conventional distillation column as a combination of possible column sections, each differing from the other in the number of trays. The problem then entails selecting a column section with the appropriate number of trays from all the candidates for each section comprising the column to be optimized [141]. While this problem is challenging to solve as an MINLP, its disjunctive reformulation makes it highly efficient. This is because only small NLP problems need to be solved for columns with a fixed number of trays, and no heuristics are required to update the Master problem with contributions from each tray. Using this method, sequences of columns with thermal coupling, hydrocarbon separations or extractive distillation were optimized within computation times that, in the worst case, did not exceed 5 minutes.

In all previous models, it has been assumed that there is vapor-liquid equilibrium at each of the trays of the column. However, the actual operating conditions of the column prevent this assumption from being achieved. In these cases, correlations are usually employed to estimate the overall efficiency of the column (or efficiency per theoretical tray). With this data, the number of actual trays in the column can be recalculated. However, a more rigorous approach to addressing the problem involves using rate-based models. One of the major advantages of GDP is that the structure of the model remains unchanged, and the only difference is the need to replace the MESH equations with the set of equations that form the rate-based model. These equations include correlations for estimating individual coefficients, which require column sizing and hydraulic characteristics. This makes the problem numerically much more challenging to converge, although formally, it only requires exchanging one set of equations for another within the disjunctions.

2.5 Overview and final considerations

In this chapter, an overview of deterministic optimization methods for the design of distillation columns or sequences of columns without azeotropes has been provided. The initial models, developed in the last decades of the twentieth century focused primarily on optimizing sequences formed by conventional columns and the sharp separation of consecutive key components. The extension to sequences of columns maintaining complete separation but of nonconsecutive components also began to be developed during those years. In fact, the theoretical foundations were laid by Petlyuk and collaborators in the 1960s. However, the systematization for the complete development of the alternative space was primarily achieved during the first two decades of the twenty-first century and it is untimely related to TCD.

Although in this chapter we have focused on the separation of zeotropic mixtures, many of its results can be extended to extractive [142, 143], azeotropic or even reactive distillation [125]. The main difference lies in that although in these models the number of sequences is considerably smaller and concepts of states and tasks can still be used, generating valid separation sequences requires tools such as residue curves. For mixtures of three or four components, using residue curves, valid alternatives can be generated more or less systematically, employing concepts such as separation regions, types of nodes (stable or unstable), the curvature of distillation boundaries [144, 145] or the sensitivity of azeotropes to pressure – pressure swing distillation [146]. Extending to systems with more than four components also requires topological tools or graph theory to generate the complete space of alternatives [147]. All these considerations would require more than one chapter each, which is why it has been decided not to include them in this chapter. However, it is important to note that the rigorous models for a column presented in the final part are valid whether the separation presents azeotropes or not, or even for reactive distillation (in fact, different works considering all these alternatives are mentioned in that section).

Finally, it should be emphasized that the models presented in this chapter are by no means the end, far from it. Currently, there are many groups actively working on different intensification alternatives (the divided wall column being the most paradigmatic example, but not the only one). Likewise, there are significant advances in integration with the rest of the process, either through direct or indirect energy integration via thermodynamic cycles (vapor recompression; Bottom Flashing, or internally integrated columns), which are becoming increasingly important due to the need for decarbonization (e.g., through electrification) of the chemical industry (including distillation) where thermal energy based on fossil fuels can be replaced by mechanical (electric) energy based on renewable sources [148]. Furthermore, hybrid systems beyond reactive distillation, such as membrane distillation [149–151] or combinations of extractive, conventional and reactive distillation, are gaining increasing importance.

Distillation remains a very active area of research where there is still ample room for improvement.

References

[1]   International Energy Agency. World Energy Outlook 2021. IEA Publ [Internet]. 2021; Available from: www.iea.org/weo

[2]   Eurostat. Final energy consumption in industry. Detailed statistics [Internet]. Available from: https://ec.europa.eu/eurostat

[3]   Sholl, D. S., Lively, R. P. Seven chemical separations to change the world. *Nature*, 2016, 532, 435–437.

[4]   Humphrey, J., Siebert, L. Separation technologies; An opportunity for energy savings. *Chemical Engineering Progress*, 1992, 88, 32–41.

[5]   Agrawal, R., Tumbalam Gooty, R. Misconceptions about efficiency and maturity of distillation. *AIChE Journal [Internet]*, 2020 Aug, 66(8), 2–4. Available from: https://onlinelibrary.wiley.com/doi/10.1002/aic.16294.

[6]   Caballero, J. A., Grossmann, I. E. Optimization of Distillation Processes. In Gorak, A., Sorensen, E. (Eds.). *Distillation: Fundamentals and Principles*. Amsterdam: Academic Press, 2014, pp. 437–496.

[7]   Kockmann, N. History of Distillation. In Górak, A., Sorensen, E. (Eds.). *Distillation Fundamentals and Principles*. Amsterdam: Academic Press, 2014.

[8]   Mccabe, W. L., Thiele, E. W. Graphical Design of Fractionating Columns. *Industrial & Engineering Chemistry [Internet]*, 1925 Jun 1, 17(6), 605–611. Available from: https://pubs.acs.org/doi/abs/10.1021/ie50186a023.

[9]   Mccabe, W. L., Thiele, E. W. Graphical design of fractionating columns. *Industrial & Engineering Chemistry [Internet]*, 1925 Jun 1, 17(6), 605–611. Available from: https://pubs.acs.org/doi/abs/10.1021/ie50186a023.

[10]   Grossmann, I. E., Caballero, J. A., Yeomans, H. Mathematical programming approaches to the synthesis of chemical process systems. *Korean Journal of Chemical Engineering*, 1999, 16(4).

[11]   Grossmann, I. E., Caballero, J. A., Yeomans, H. Advances in mathematical programming for the synthesis of process systems. *Latin American Applied Research [Internet]*, 2000, 30(4). 263–284. Available from: http://www.scopus.com/inward/record.url?eid=2-s2.0-0034288326&partnerID=40&md5=33984e1df71ba2769a7328c17b1e9dce.

[12]   Yeomans, H., Grossmann, I. E. A systematic modeling framework of superstructure optimization in process synthesis. *Computers & Chemical Engineering*, 1999, 23(6), 709–731.

[13]   Kondili, E., Pantelides, C. C., Sargent, R. W. H. A general algorithm for short-term scheduling of batch operations – I. MILP formulation. *Computer Chemical Engineering [Internet]*, 1993 Feb, 17(2), 211–227. Available from: https://linkinghub.elsevier.com/retrieve/pii/009813549380015F.

[14]   Andrecovich, M. J., Westerberg, A. W. An MILP formulation for heat-integrated distillation sequence synthesis. *AIChE Journal [Internet]*, 1985 Sep, 31(9), 1461–1474. Available from: http://doi.wiley.com/10.1002/aic.690310908.

[15]   Smith, E. M. B., Pantelides, C. C. Design of reaction/separation networks using detailed models. *Computer Chemical Engineering [Internet]*, 1995 Jun, 19(95), 83–88. Available from: https://linkinghub.elsevier.com/retrieve/pii/0098135495870199.

[16]   Novak, Z., Kravanja, Z., Grossmann, I. E. Simultaneous synthesis of distillation sequences in overall process schemes using an improved MINLP approach. *Computer Chemical Engineering [Internet]*, 1996 Jan, 20(12), 1425–1440. Available from: https://linkinghub.elsevier.com/retrieve/pii/0098135495002405.

[17]   Caballero, J. A., Grossmann, I. E. Aggregated models for integrated distillation systems. *Industrial & Engineering Chemistry Research*, 1999, 38(6).

[18]   Friedler, F., Tarjan, K., Huang, Y. W., Fan, L. T. Graph-theoretic approach to process synthesis: Polynomial algorithm for maximal structure generation. *Computers & Chemical Engineering*, 1993, 17(9), 929–942.

[19]   Bagajewicz, M. J., Manousiouthakis, V. Mass/heat-exchange network representation of distillation networks. *AIChE Journal [Internet]*, 1992 Nov, 38(11), 1769–1800. Available from: http://doi.wiley.com/10.1002/aic.690381110.

[20]   Farkas, T., Rev, E., Lelkes, Z. Process flowsheet superstructures: Structural multiplicity and redundancy Part I: Basic GDP and MINLP representations. *Computers & Chemical Engineering*, 2005, 29(10), 2180–2197.

[21]   Farkas, T., Rev, E., Lelkes, Z. Process flowsheet superstructures: Structural multiplicity and redundancy Part II: Ideal and binarily minimal MINLP representations. *Computers & Chemical Engineering*, 2005, 29(10), 2198–2214.

[22]   Papalexandri, K. P., Pistikopoulos, E. N. Generalized modular representation framework for process synthesis. *AIChE Journal [Internet]*, 1996 Apr, 42(4), 1010–1032. Available from: http://doi.wiley.com/10.1002/aic.690420413.

[23]   Wu, W., Henao, C. A., Maravelias, C. T. A superstructure representation, generation, and modeling framework for chemical process synthesis. *AIChE Journal [Internet]*, 2016 Sep 16, 62(9), 3199–3214. Available from: https://aiche.onlinelibrary.wiley.com/doi/10.1002/aic.15300.

[24] Mencarelli, L., Chen, Q., Pagot, A., Grossmann, I. E. A review on superstructure optimization approaches in process system engineering. *Computer Chemical Engineering [Internet]*, 2020 May, 136, 106808. Available from: https://doi.org/10.1016/j.compchemeng.2020.106808.

[25] Rudd, D. F., Watson, C. C. *Strategy of Process Engineering*. New York: John Wiley & Sons Inc editor., 1968.

[26] Westerberg, A. W. A review of the synthesis of distillation based separation systems. *Dep Electr Comput Eng Pap*, 100. http://repository.cmu.edu/ece/100.1983.

[27] Caballero, J. A., Grossmann, I. E. Structural considerations and modeling in the synthesis of heat-integrated–thermally coupled distillation sequences. *Industrial & Engineering Chemistry Research [Internet]*, 2006, 45(25). 8454–8474. Available from: http://www.scopus.com/inward/record.url?eid= 2-s2.0-33846120154&partnerID=tZOtx3y1.

[28] Giridhar, A., Agrawal, R. Synthesis of distillation configurations: I. Characteristics of a good search space. *Computer Chemical Engineering [Internet]*, 2010, 34(1). 73. Available from: http://www.science direct.com/science/article/B6TFT-4W99W07-4/2/ba909c2f1e3b773800edbc2452492a75.

[29] Petlyuk, F. B., Platonov, V. M., Slavinsk, D. M. Thermodynamically optimal method for separating multicomponent mixtures. *International Chemical Engineering*, 1965, 5(3), 555–561.

[30] Lockhart, F. J. Multi-column distillation of natural gasoline. *Pet Refine*, 1947, 2JS, 104.

[31] Harbert, V. D. Which tower goes where? *Pet Refine*, 1957, 36(3), 169–174.

[32] Heaven, D. *Optimum Sequencing of Distillation Columns in Multicomponent Fractionation*. M.S. Thesis. Berkeley: University of California, 1969.

[33] Rudd, D. F., Powers, G. J., Siirola, J. J. *Process Synthesis*. New York: Prentice Hall, 1973.

[34] Seader, J. D., Westerberg, A. W. A combined heuristic and evolutionary strategy for synthesis of simple separation sequences. *AIChE Journal [Internet]*, 1977 Nov, 23(6), 951–954. Available from: http://doi.wiley.com/10.1002/aic.690230628.

[35] Thompson, R. W., King, C. J. Systematic synthesis of separation schemes. *AIChE Journal [Internet]*, 1972 Sep, 18(5), 941–948. Available from: http://doi.wiley.com/10.1002/aic.690180510.

[36] Hendry, J. E., Hughes, R. E. Generating process separation flowsheets. *Chemical Engineering Progress*, 1972, 68, 69.

[37] Westerberg, A. W., Stephanopoulos, G. Studies in process synthesis – I. *Chemical Engineering Science [Internet]*, 1975 Aug, 30(8), 963–972. Available from: https://linkinghub.elsevier.com/retrieve/pii/ 0009250975800637.

[38] Rodrigo, B. F. R., Seader, J. D. Synthesis of separation sequences by ordered branch search. *AIChE Journal [Internet]*, 1975 Sep, 21(5), 885–894. Available from: http://doi.wiley.com/10.1002/aic. 690210509.

[39] Gomez, M. A., Seader, J. D. Separation sequence synthesis by a predictor based ordered search. *AIChE Journal [Internet]*, 1976 Nov, 22(6), 970–979. Available from: http://doi.wiley.com/10.1002/aic. 690220604.

[40] GAMS Development Corporation. *General Algebraic Modeling System (GAMS) Release 44.3.0*. Fairfax, VA. USA, 2023.

[41] Chen, Q., Johnson, E. S., Bernal, D. E., Valentin, R., Kale, S., Bates, J. et al. Pyomo.GDP: An ecosystem for logic based modeling and optimization development. *Optimization and Engineering [Internet]*, 2022, 23(1). 607–642. Available from: https://doi.org/10.1007/s11081-021-09601-7.

[42] Fenske, M. R. Fractionation of straight-run Pennsylvania gasoline. *Industrial & Engineering Chemistry [Internet]*, 1932 May, 24(5), 482–485. Available from: http://pubs.acs.org/doi/abs/10.1021/ ie50269a003.

[43] Underwood, A. J. V. Fractional Distillation of Multicomponent Mixtures. *Industrial & Engineering Chemistry*, 1949, 41(12), 2844–2847.

[44] Gilliland, E. R. Multicomponent rectification: Estimation of the Number of Theoretical Plates as a Function of the Reflux Ratio. *Industrial & Engineering Chemistry*, 1940, 32(9), 1220–1223.

[45] Papoulias, S. A., Grossmann, I. E. A structural optimization approach in process synthesis-II. Heat recovery networks. *Computers & Chemical Engineering*, 1983, 7(6), 707–721.

[46] Yeomans, H., Grossmann, I. E. Nonlinear disjunctive programming models for the synthesis of heat integrated distillation sequences. *Computers & Chemical Engineering*, 1999, 23(9), 1135–1151.

[47] Yeomans, H., Grossmann, I. E. Disjunctive programming models for the optimal design of distillation columns and separation sequences. *Industrial & Engineering Chemistry Research [Internet]*, 2000, 39(6). 1637–1648. Available from: http://www.scopus.com/inward/record.url?eid=2-s2.0-0034099586&partnerID=tZOtx3y1.

[48] Wolff, E. A., Skogestad, S. Operation of Integrated 3-Product (Petlyuk) Distillation-Columns. *Industrial & Engineering Chemistry Research*, 1995, 34(6), 2094–2103.

[49] Fidkowski, Z. T., Agrawal, R. Multicomponent thermally coupled systems of distillation columns at minimum reflux. *AIChE Journal*, 2001, 47(12), 2713–2724.

[50] Halvorsen, I. J., Skogestad, S. Minimum energy consumption in multicomponent distillation. 2. Three-product Petlyuk arrangements. *Industrial & Engineering Chemistry Research [Internet]*, 2003, 42(3). 605–615. Available from: http://www.scopus.com/inward/record.url?eid=2-s2.0-0037419639&partnerID=tZOtx3y1.

[51] Fidkowski, Z., Krolikowski, L. Thermally coupled systems of distillation columns: optimization procedure. 1986, 32(4).

[52] Fidkowski, Z., Królikowski, L. Minimum energy requirements of thermally coupled distillation systems. *AIChE Journal [Internet]*, 1987 Apr, 33(4), 643–653. Available from: http://doi.wiley.com/10.1002/aic.690330412.

[53] Carlberg, N. A., Westerberg, A. W. Temperature-heat diagrams for complex columns: 2: Underwood'S method for side strippers and enrichers. *Industrial & Engineering Chemistry Research*, 1989, 28(9), 1379–1386.

[54] Carlberg, N. A., Westerberg, A. W. Temperature-heat diagrams for complex columns. 3. underwood's method for the Petlyuk configuration. *Industrial & Engineering Chemistry Research*, 1989, 28(9), 1386–1397.

[55] Alatiqi, I. M., Luyben, W. L. Control of a complex sidestream column/stripper distillation configuration. *Industrial & Engineering Chemistry Process Design and Development [Internet]*, 1986 Jul 1, 25(3), 762–767. Available from: https://pubs.acs.org/doi/abs/10.1021/i200034a028.

[56] Glinos, K. N., Nikolaides, I. P., Malone, M. F. New complex column arrangements for ideal distillation. *Industrial & Engineering Chemistry Process Design and Development [Internet]*, 1986 Jul 1, 25(3), 694–699. Available from: https://pubs.acs.org/doi/abs/10.1021/i200034a016.

[57] Glinos, K., Malone, M. F. Optimality regions for complex column alternatives in distillation systems. *Chemical Engineering Research and Design*, 1988, 66(3), 229–240.

[58] Nikolaides, I. P., Malone, M. F. Approximate design of multiple-feed/side-stream distillation systems. *Industrial & Engineering Chemistry Research*, 1987, 26(9), 1839–1845.

[59] Triantafyllou, C., Smith, R. The design and optimization of fully thermally coupled distillation columns. *Chemical Engineering Research and Design*, 1992, 70(2), 118–132.

[60] Brugma, A. J. Process and device for fractional distillation of liquid mixtures, more particularly petroleum. U.S. Patent 2.295.256, 1942.

[61] Wright, R. O. Fractionation Apparatus, United States. Vol. No. 247113, United States Patent Office. US2471134A, 1949. p. 4–6.

[62] Halvorsen, I. J., Skogestad, S. Minimum energy consumption in multicomponent distillation. 3. More than three products and generalized Petlyuk arrangements. *Industrial & Engineering Chemistry Research*, 2003, 42(3), 616–629.

[63] Agrawal, R., Fidkowski, Z. T. More operable arrangements of fully thermally coupled distillation columns. *AIChE Journal*, 1998, 44(11), 2565–2568.

[64] Caballero, J. A., Grossmann, I. E. Thermodynamically Equivalent Configurations for Thermally Coupled Distillation. *AIChE Journal*, 2003, 49(11).

[65] Hernández, S., Gabriel Segovia-Hernández, J., Rico-Ramírez, V. Thermodynamically equivalent distillation schemes to the Petlyuk column for ternary mixtures. *Energy*, 2006, 31(12), 2176–2183.

[66] Rong, B. G., Kraslawski, A., Turunen, I. Synthesis and optimal design of thermodynamically equivalent thermally coupled distillation systems. *Industrial & Engineering Chemistry Research*, 2004, 43(18), 5904–5915.

[67] Rong, B. G., Turunen, I. A new method for synthesis of thermodynamically equivalent structures for Petlyuk arrangements. *Chemical Engineering Research and Design*, 2006, 84(A12).

[68] Yildirim, Ö., Kiss, A. A., Kenig, E. Y. Dividing wall columns in chemical process industry: A review on current activities. *Separation and Purification Technology [Internet]*, 2011, 80(3). 403–417. Available from: http://www.sciencedirect.com/science/article/pii/S1383586611002978.

[69] Asprion, N., Kaibel, G. Dividing wall columns: Fundamentals and recent advances. *Chemical Engineering and Processing: Process Intensification [Internet]*, 2010, 49(2). 139–146. Available from: http://www.sciencedirect.com/science/article/pii/S025527011000022X.

[70] Agrawal, R. Multicomponent distillation columns with partitions and multiple reboilers and condensers. *Industrial & Engineering Chemistry Research*, 2001, 40(20), 4258–4266.

[71] Madenoor Ramapriya, G., Tawarmalani, M., Agrawal, R. Thermal coupling links to liquid-only transfer streams: A path for new dividing wall columns. *AIChE Journal [Internet]*, 2014, 60(8). 2949–2961. Available from: http://dx.doi.org/10.1002/aic.14468.

[72] Glinos, K., Malone, M. F. Minimum reflux, product distribution, and lumping rules for multicomponent distillation. *Industrial & Engineering Chemistry Process Design and Development*, 1984, 23(4), 764–768.

[73] Fidkowski, Z. T., Doherty, M. F., Malone, M. F. Feasibility of separations for distillation of nonideal ternary mixtures. *AIChE Journal*, 1993, 39(8), 1303–1321.

[74] Ramírez-Corona, N., Jiménez-Gutiérrez, A., Castro-Agüero, A., Rico-Ramírez, V. Optimum design of Petlyuk and divided-wall distillation systems using a shortcut model. *Chemical Engineering Research and Design [Internet]*, 2010, 88(10). 1405–1418. Available from: http://www.sciencedirect.com/science/article/pii/S0263876210000882.

[75] Sargent, R. W. H., Gaminibandara, K. Optimal design of plate distillation columns. *Optim Action*, 1976, 267–314.

[76] Agrawal, R. Synthesis of distillation column configurations for a multicomponent separation. *Industrial & Engineering Chemistry Research [Internet]*, 1996 Jan, 35(4), 1059–1071. Available from: https://pubs.acs.org/doi/10.1021/ie950323h.

[77] Rong, B. G., Kraslawski, A. Partially thermally coupled distillation systems for multicomponent separations. *AIChE Journal*, 2003, 49(5), 1340–1347.

[78] Rong, B. G., Kraslawski, A., Nystrom, L. The synthesis of thermally coupled distillation flowsheets for separations of five-component mixtures. *Computers & Chemical Engineering*, 2000, 24(2–7).

[79] Rong, B. G., Kraslawski, A., Nystrom, L. Design and synthesis of multicomponent thermally coupled distillation flowsheets. *Computers & Chemical Engineering*, 2001, 25(4–6).

[80] Kim, Y. H. Rigorous design of extended fully thermally coupled distillation columns. *Chemical Engineering Journal [Internet]*, 2002, 89(1–3). 89. Available from: http://www.sciencedirect.com/science/article/B6TFJ-4625TDV-4/2/855b1b9b9fb54d32d42882b3469e57c9.

[81] Kim, Y. H. Structural design and operation of a fully thermally coupled distillation column. *Chemical Engineering Journal*, 2002, 85(2–3), 289–301.

[82] Kim, Y. H. Structural design of fully thermally coupled distillation columns using a semi-rigorous model. *Computer Chemical Engineering [Internet]*, 2005, 29(7). 1555. Available from: http://www.sciencedirect.com/science/article/B6TFT-4F7VFJB-1/2/83ed85b36fbc7186f6521dc86dbfeadc.

[83] Kim, Y. H. Evaluation of three-column distillation system for ternary separation. *Chemical Engineering and Processing: Process Intensification [Internet]*, 2005 Oct 1 [cited 2019 Jul 2], 44(10), 1108–1116. Available from: https://www.sciencedirect.com/science/article/pii/S0255270105000735.

[84] Hernandez, S., Jimenez, A. Design of energy-efficient Petlyuk systems. *Computer Chemical Engineering [Internet]*, 1999, 23(8). 1005. Available from: http://www.sciencedirect.com/science/article/B6TFT-3X70SJM-3/2/05ec3ad89370f293e4b4bbe5dae95560.

[85] Hernandez, S., Pereira-Pech, S., Jimenez, A., Rico-Ramirez, V. Energy efficiency of an indirect thermally coupled distillation sequence. *The Canadian Journal of Chemical Engineering*, 2003, 81(5), 1087–1091.

[86] Blancarte-Palacios, J. L., Bautista-Valdes, M. N., Hernandez, S., Rico-Ramirez, V., Jimenez, A. Energy-efficient designs of thermally coupled distillation sequences for four-component mixtures. *Industrial & Engineering Chemistry Research*, 2003, 42(21), 5157–5164.

[87] Calzon-McConville, C. J., Rosales-Zamora, M. B., Segovia-Hernández, J. G., Hernández, S., Rico-Ramírez, V. Design and Optimization of Thermally Coupled Distillation Schemes for the Separation of Multicomponent Mixtures. *Industrial & Engineering Chemistry Research [Internet]*, 2005, 45(2). 724–732. Available from: http://dx.doi.org/10.1021/ie050961s.

[88] Shah, P. B., Kokossis, A. C. Knowledge based models for the analysis of complex separation processes. *Computer Chemical Engineering [Internet]*, 2001 May 1 [cited 2019 Jul 3], 25(4–6), 867–878. Available from: https://www.sciencedirect.com/science/article/pii/S0098135401006615.

[89] Shah, P. B., Kokossis, A. C. New synthesis framework for the optimization of complex distillation systems. *AIChE Journal [Internet]*, 2002, 48(3). 527–550. Available from: http://dx.doi.org/10.1002/aic.690480311.

[90] Caballero, J. A., Grossmann, I. E. Generalized disjunctive programming model for the synthesis of thermally linked distillation systems. *Computer Aided Chemical Engineering*, 2001, 9.

[91] Caballero, J. A., Grossmann, I. E. Design of distillation sequences: From conventional to fully thermally coupled distillation systems. *Computers & Chemical Engineering*, 2004, 28(11), 2307–2329.

[92] Agrawal, R. Synthesis of multicomponent distillation column configurations. *AIChE Journal*, 2003, 49(2), 379–401.

[93] Giridhar, A., Agrawal, R. Synthesis of distillation configurations. II: A search formulation for basic configurations. *Computer Chemical Engineering [Internet]*, 2010, 34(1). 84. Available from: http://www.sciencedirect.com/science/article/B6TFT-4W99W07-2/2/b64f7e47d987dbc6f14eb52bcfe73250.

[94] Kaibel, G. Distillation columns with vertical partitions. *Chemical Engineering & Technology*, 1987, 10(1), 92–98.

[95] Kim, J. K., Wankat, P. C. Quaternary Distillation Systems with Less than N – 1 Columns. *Industrial & Engineering Chemistry Research [Internet]*, 2004 Jul, 43(14), 3838–3846. Available from: https://pubs.acs.org/doi/10.1021/ie030640l.

[96] Errico, M., Rong, B.-G. Modified simple column configurations for quaternary distillations. *Computer Chemical Engineering [Internet]*, 2012 Jan 10 [cited 2019 Jul 3], 36, 160–173. Available from: https://www.sciencedirect.com/science/article/pii/S0098135411002031.

[97] Errico, M., Pirellas, P., Torres-Ortega, C. E., Rong, B.-G., Segovia-Hernandez, J. G. A combined method for the design and optimization of intensified distillation systems. *Chemical Engineering and Processing: Process Intensification [Internet]*, 2014, 85(0). 69–76. Available from: http://www.sciencedirect.com/science/article/pii/S0255270114001585.

[98] Errico, M., Rong, B.-G., Torres-Ortega, C. E., Segovia-Hernandez, J. G. The importance of the sequential synthesis methodology in the optimal distillation sequences design. *Computer Chemical Engineering [Internet]*, 2014, 62(0). 1–9. Available from: http://www.sciencedirect.com/science/article/pii/S009813541300361X.

[99] Rong, B.-G., Errico, M. Synthesis of intensified simple column configurations for multicomponent distillations. *Chemical Engineering and Processing: Process Intensification [Internet]*, 2012 Dec 1 [cited

2019 Jul 3], 62, 1–17. Available from: https://www.sciencedirect.com/science/article/pii/S0255270112001936.

[100] Caballero, J. A., Reyes-Labarta, J. A., Grossmann, I. E. A sequential algorithm for the rigorous design of thermally coupled distillation sequences. *Computer Aided Chemical Engineering*, 2015, 37.

[101] Jiang, Z., Agrawal, R. Process intensification in multicomponent distillation: A review of recent advancements. *Chemical Engineering Research and Design [Internet]*, 2019 Jul 1 *[cited 2019* Jul 3], 147, 122–145. Available from: https://www.sciencedirect.com/science/article/pii/S0263876219301819?via%3Dihub.

[102] Agrawal, R. More operable fully thermally coupled distillation column configurations for multicomponent distillation. *Chemical Engineering Research and Design [Internet]*, 1999 Sep 1 [cited 2019 Jul 3], 77(6), 543–553. Available from: https://www.sciencedirect.com/science/article/pii/S026387629971823X.

[103] Agrawal, R. A method to draw fully thermally coupled distillation column configurations for multicomponent distillation. *Chemical Engineering Research and Design*, 2000, 78(A3), 454–464.

[104] Ivakpour, J., Kasiri, N. Synthesis of distillation column sequences for nonsharp separations. *Industrial & Engineering Chemistry Research*, 2009, 48(18), 8635–8649.

[105] Caballero, J. A., Grossmann, I. E. Logic-based methods for generating and optimizing thermally coupled distillation systems. *Computer Aided Chemical Engineering*, 2002, 10.

[106] Caballero, J. A., Reyes-Labarta, J. A., Grossmann, I. E. Synthesis of integrated distillation systems. *Computer Aided Chemical Engineering*, 2003, 14.

[107] Shah, V. H., Agrawal, R. A matrix method for multicomponent distillation sequences. *AIChE Journal [Internet]*, 2010, 56(7). 1759–1775. Available from: http://dx.doi.org/10.1002/aic.12118.

[108] Caballero, J. A., Grossmann, I. E. Synthesis of complex thermally coupled distillation systems including divided wall columns. *AIChE Journal*, 2013.

[109] Caballero, J. A., Reyes-Labarta, J. A., Grossmann, I. E. Simultaneous design of heat integrated and thermally coupled distillation systems. *Computer Aided Chemical Engineering*, 2004, 18.

[110] Rong, B. G., Kraslawski, A., Turunen, I. Synthesis of heat-integrated thermally coupled distillation systems for multicomponent separations. *Industrial & Engineering Chemistry Research*, 2003, 42(19), 4329–4339.

[111] Rong, B. G., Turunen, I. New heat-integrated distillation configurations for Petlyuk arrangements. *Chemical Engineering Research and Design*, 2006, 84(A12).

[112] Rong, B. G., Turunen, I. Synthesis of new distillation systems by simultaneous thermal coupling and heat integration. *Industrial & Engineering Chemistry Research*, 2006, 45(11), 3830–3842.

[113] Agrawal, R. Multieffect distillation for thermally coupled configurations. *AIChE Journal*, 2000, 46(11), 2211–2224.

[114] Viswanathan, J., Grossmann, I. E. Optimal feed locations and number of trays for distillation columns with multiple feeds. *Industrial & Engineering Chemistry Research*, 1993, 32(11), 2942–2949.

[115] Ciric, A. R., Gu, D. Synthesis of nonequilibrium reactive distillation processes by MINLP optimization. *AIChE Journal*, 1994, 40(9), 1479–1487.

[116] Bauer, M. H., Stichlmair, J. Design and economic optimization of azeotropic distillation processes using mixed-integer nonlinear programming. *Computer Chemical Engineering [Internet]*, 1998 Aug, 22(9), 1271–1286. Available from: https://linkinghub.elsevier.com/retrieve/pii/S0098135498000118.

[117] Dünnebier, G., Pantelides, C. C. Optimal design of thermally coupled distillation columns. *Industrial & Engineering Chemistry Research [Internet]*, 1999 Jan [cited 2018 Apr 30], 38(1), 162–176. Available from: https://pubs.acs.org/doi/pdf/10.1021/ie9802919.

[118] Barttfeld, M., Aguirre, P. A., Grossmann, I. E. Alternative representations and formulations for the economic optimization of multicomponent distillation columns. *Computers & Chemical Engineering*, 2003, 27(3), 363–383.

[119] Grossmann, I. E., Aguirre, P. A., Barttfeld, M. Optimal synthesis of complex distillation columns using rigorous models. *Computers & Chemical Engineering*, 2005, 29(6 SPEC. ISS.), 1203–1215.

[120] Yeomans, H., Grossmann, I. E. Optimal design of complex distillation columns using rigorous tray-by-tray disjunctive programming models. *Industrial & Engineering Chemistry Research [Internet]*, 2000, 39(11). 4326–4335. Available from: http://pubs.acs.org/doi/abs/10.1021/ie0001974.

[121] Barttfeld, M., Aguirre, P. A. Optimal synthesis of multicomponent zeotropic distillation processes. 2. Preprocessing phase and rigorous optimization of efficient sequences. *Industrial & Engineering Chemistry Research*, 2003, 42(14), 3441–3457.

[122] Kossack, S., Kraemer, K., Marquardt, W. Efficient Optimization-Based Design of Distillation Columns for Homogenous Azeotropic Mixtures. 2006, 8492–8502.

[123] Harwardt, A., Marquardt, W. Heat-integrated distillation columns: Vapor recompression or internal heat integration? *AIChE Journal [Internet]*, 2012 Dec, 58(12), 3740–3750. Available from: https://onlinelibrary.wiley.com/doi/10.1002/aic.13775.

[124] Kraemer, K., Kossack, S., Marquardt, W. An efficient solution method for the MINLP optimization of chemical processes. *Computer Aided Chemical Engineering*, 2007, 24, 105–110.

[125] Jackson, J. R., Grossmann, I. E. A disjunctive programming approach for the optimal design of reactive distillation columns. *Computers & Chemical Engineering*, 2001, 25(11–12).

[126] Barttfeld, M., Aguirre, P. A., Grossmann, I. E. A decomposition method for synthesizing complex column configurations using tray-by-tray GDP models. *Computer Chemical Engineering [Internet]*, 2004 Oct, 28(11), 2165–2188. Available from: https://linkinghub.elsevier.com/retrieve/pii/S0098135404000699.

[127] Bauer, M. H., Stichlmair, J. Superstructures for the mixed integer optimization of nonideal and azeotropic distillation processes. *Computers & Chemical Engineering*, 1996, 20(SUPPL.1), 25–30.

[128] Kookos, I. K. Optimal design of membrane/distillation column hybrid processes. *Industrial & Engineering Chemistry Research*, 2003, 42(8), 1731–1738.

[129] Soraya Rawlings, E., Chen, Q., Grossmann, I. E., Caballero, J. A. Kaibel column: Modeling, optimization, and conceptual design of multi-product dividing wall columns. *Computer Chemical Engineering [Internet]*, 2019 Jun 9 [cited 2019 May 14], 125, 31–39. Available from: https://www.sciencedirect.com/science/article/pii/S0098135419300250.

[130] Waltermann, T., Sibbing, S., Skiborowski, M. Optimization-based design of dividing wall columns with extended and multiple dividing walls for three- and four-product separations. *Chemical Engineering and Processing – Process Intensification*, 2019, 146(September).

[131] Skiborowski, M., Harwardt, A., Marquardt, W. Efficient optimization-based design for the separation of heterogeneous azeotropic mixtures. *Computer Chemical Engineering [Internet]*, 2015, 72, 34–51. Available from: http://dx.doi.org/10.1016/j.compchemeng.2014.03.012.

[132] Manassaldi, J. I., Mussati, M. C., Scenna, N. J., Mussati, S. F. Development of extrinsic functions for optimal synthesis and design – Application to distillation-based separation processes. *Computer Chemical Engineering [Internet]*, 2019, 125, 532–544. Available from: https://doi.org/10.1016/j.compchemeng.2019.03.028.

[133] Krone, D., Esche, E., Asprion, N., Skiborowski, M., Repke, J. U. Enabling optimization of complex distillation configurations in GAMS with CAPE-OPEN thermodynamic models. *Computer Chemical Engineering [Internet]*, 2022, 157, 107626. Available from: https://doi.org/10.1016/j.compchemeng.2021.107626.

[134] Caballero, J. A., Grossmann, I. E. An algorithm for the use of surrogate models in modular flowsheet optimization. *AIChE Journal [Internet]*, 2008 Oct, 54(10), 2633–2650. Available from: http://doi.wiley.com/10.1002/aic.11579.

[135] Caballero, J. A., Odjo, A., Grossmann, I. E. Flowsheet optimization with complex cost and size functions using process simulators. *AIChE Journal*, 2007, 53(9).

[136] Caballero, J. A., Navarro, M. A., Ruiz-Femenia, R., Grossmann, I. E. Integration of different models in the design of chemical processes: Application to the design of a power plant. *Applied Energy*, 2014, 124.

[137] Navarro-Amorós, M. A., Ruiz-Femenia, R., Caballero, J. A. Integration of modular process simulators under the Generalized Disjunctive Programming framework for the structural flowsheet optimization. *Computers & Chemical Engineering*, 2014, 67.

[138] Caballero, J. A., Milán-Yañez, D., Grossmann, I. E. Rigorous design of distillation columns: Integration of disjunctive programming and process simulators. *Industrial & Engineering Chemistry Research*, 2005, 44(17), 6760–6775.

[139] Caballero, J. A., Grossmann, I. E., Keyvani, M., Lenz, E. S. Design of hybrid distillation-vapor membrane separation systems. *Industrial & Engineering Chemistry Research*, 2009, 48(20).

[140] Javaloyes-Antón, J., Kronqvist, J., Caballero, J. A. Simulation-based optimization of chemical processes using the extended cutting plane algorithm. *Computer Aided Chemical Engineering [Internet]*, 2018, 463–469. Available from: https://linkinghub.elsevier.com/retrieve/pii/B9780444642356500838.

[141] Caballero, J. A. Logic hybrid simulation-optimization algorithm for distillation design. *Computer Chemical Engineering [Internet]*, 2015, 72, 284–299. Available from: http://www.sciencedirect.com/science/article/pii/S0098135414000970.

[142] Gerbaud, V., Rodriguez-Donis, I., Hegely, L., Lang, P., Denes, F., You, X. Q. Review of extractive distillation. Process design, operation, optimization and control. *Chemical Engineering Research and Design [Internet]*, 2019, 141, 229–271. Available from: https://doi.org/10.1016/j.cherd.2018.09.020.

[143] Sun, S., Lü, L., Yang, A., Wei, S., Shen, W. Extractive distillation: Advances in conceptual design, solvent selection, and separation strategies. *Chinese Journal of Chemical Engineering [Internet]*, 2019, 27(6). 1247–1256. Available from: https://doi.org/10.1016/j.cjche.2018.08.018.

[144] Shen, W. F., Benyounes, H., Song, J. A review of ternary azeotropic mixtures advanced separation strategies. *Theoretical Foundations of Chemical Engineering*, 2016, 50(1), 28–40.

[145] Lucia, A., Taylor, R. The geometry of separation boundaries: I. *Basic Theory and Numerical Support. AIChE Journal*, 2006, 52(2), 582–594.

[146] Gu, J., Lu, S., Shi, F., Wang, X., You, X. Economic and environmental evaluation of heat-integrated pressure-swing distillation by multiobjective optimization. *Industrial & Engineering Chemistry Research [Internet]*, 2022 Jun 29, 61(25), 9004–9014. Available from: https://pubs.acs.org/doi/10.1021/acs.iecr.2c01043.

[147] Marquardt, W., Kossack, S., Kraemer, K. A framework for the systematic design of hybrid separation processes. *Chinese Journal of Chemical Engineering*, 2008, 16(3), 333–342.

[148] Kiss, A. A., Smith, R. Rethinking energy use in distillation processes for a more sustainable chemical industry. *Energy [Internet]*, 2020, 203, 117788. Available from. https://doi.org/10.1016/j.energy.2020.117788.

[149] Stephan, W., Noble, R. D., Koval, C. A. Design methodology for a membrane/distillation column hybrid process. *Journal of Membrane Science*, 1995, 99(3), 259–272.

[150] Eliceche, A. M., Carolina Daviou, M., Hoch, P. M., Uribe, I. O. Optimization of azeotropic distillation columns combined with pervaporation membranes. *Computers & Chemical Engineering*, 2002, 26 (4–5).

[151] Alkhudhiri, A., Darwish, N., Hilal, N. Membrane distillation: A comprehensive review. *Desalination*, 2012, 287, 2–18.

Luis Fernando Lira-Barragán, Fabricio Nápoles-Rivera,
and José María Ponce-Ortega*

Chapter 3
Optimal design of process energy systems integrating sustainable considerations

Abstract: This chapter presents a novel approach for designing sustainable trigeneration systems (i.e., heating, cooling, and power generation cycles) integrated with heat exchanger networks and accounting simultaneously for economic, environmental, and social objectives. The trigeneration system comprises steam and organic Rankine cycles and an absorption refrigeration cycle. Multiple sustainable energy sources such as solar energy, biofuels, and fossil fuels are considered to drive the steam Rankine cycle. The model aims to select the optimal working fluid to operate the organic Rankine cycle and to determine the optimal system to drive the absorption refrigeration cycle. The residual energy available in the steam Rankine cycle and/or the process excess heat can be employed to run both the organic Rankine cycle and the absorption refrigeration cycle to produce electricity and refrigeration below the ambient temperature, respectively. Two example problems are presented to show the applicability of the proposed methodology.

Keywords: sustainability, trigeneration, organic Rankine cycle, absorption refrigeration, multi-objective optimization

3.1 Introduction

Nowadays, energy is one of the most important resources and at the same time one of the most relevant concerns around the world, owing to the fast depletion of non-renewable fuels, global warming, and climate change. For these reasons, several governments have promoted the use of cleaner energies through tax credits and significant economic resources have been invested in searching for alternative energies to mitigate the environmental issues. In this sense, power plants and industry consume an enormous amount of fossil fuels to satisfy the electricity and utilities demands. A few decades ago, several researchers prioritized this topic and focused their investigations on the maximization of recovery process heat through the minimization of the

---

*__Corresponding author: José María Ponce-Ortega__, Chemical Engineering Department, Universidad Michoacana de San Nicolás de Hidalgo, Morelia, Michoacán, México, e-mail: jose.ponce@umich.mx
__Luis Fernando Lira-Barragán, Fabricio Nápoles-Rivera__, Chemical Engineering Department, Universidad Michoacana de San Nicolás de Hidalgo, Morelia, Michoacán, México

https://doi.org/10.1515/9783111383439-003

external utilities for the heat exchanger networks (HENs), where there are some streams requiring cooling and others need heating (see [1–3]). In this regard, the methodologies for synthesizing HENs have been classified into the following categories: sequential approaches using pinch analysis [4–6], stochastic methods [7–11], and mathematical programming-based techniques [12–16].

Nonetheless, previous approaches have concentrated on synthesizing HENs without considering the energy interactions among HENs and the associated utility systems. Other methodologies have considered the energy integration among power cycles and HEN as well as with a utility system, which has been based on thermodynamic principles and heuristic rules [17–19]. In this sense, Townsend and Linnhoff [17] presented rules for the appropriate placement of heat pumps and engines in process networks relative to the heat recovery pinch; Maréchal and Kalitventzeff [18] proposed a procedure for computing the integrated combined heat and power target of industrial processes based on the analysis of the shape of the balanced grand composite curves and improves the qualitative guidelines for positioning combined heat and power engines; and Desai and Bandyopadhyay [19] developed a sequential method based on pinch analysis for integrating organic Rankine cycles (ORCs) with processes to generate power and, at the same time, reduce the overall consumption of cold utility. Additionally, other methodologies have employed optimization techniques for the problem of heat and power integration with the process and utility systems. Thus, Papoulias and Grossmann [20] presented a mixed-integer linear programming (MILP) formulation for the optimal synthesis of total processing systems consisting of a chemical plant, with its HEN and utility system including a novel superstructure-based method for the synthesis of heat and power integration in process networks. Colmenares and Seider [21] developed a nonlinear programming (NLP) strategy for integrating heat engines and heat pumps with the process heat cascade. Swaney [22] proposed an extended transportation array formulation for the design of process heat recovery networks incorporating Rankine cycles and heat pumps. Holiastos and Manousiouthakis [23] introduced a mathematical formulation for the optimal integration of heat exchangers, heat engines, and heat pumps in HEN. Chen and Lin [24] formulated a mixed-integer nonlinear programming (MINLP) model for synthesizing an entire energy system, which includes the interactions between the steam network and the HEN of process plants. Hipólito-Valencia et al. [25] designed an MINLP formulation based on an integrated stagewise superstructure to simultaneously determine the optimal configuration, design parameters and operation conditions of integrated energy systems that consist of an HEN and an ORC. In this context, ORC is a novel strategy to use residual energy to produce electricity [26–28]. A significant number of methodologies have made their efforts to find the optimal organic working fluid to operate properly an ORC [29–33]. Nevertheless, these approaches have not considered the optimal selection of the working fluid for an ORC interacting energetically at the same time with an HEN and other thermodynamic cycles.

Afterward, a graphical method based on the total site profiles, which is an extension of pinch analysis, has played an important role in solving the problem of heat and power integration in a set of processes served by a central utility system. This method is called "the total site analysis" and was introduced by Dhole and Linnhoff [34], while Raissi [35] used this methodology for targeting energy recovery, heat and power cogeneration, and emissions from utility systems. Later, this approach was widely employed by other authors (see [36–39]). Furthermore, Bagajewicz and Barbaro presented nonlinear targeting models to get maximum cost savings and optimum integration of heat pumps in total sites [40], Varbanov et al. [41] applied the total site analysis to address the steam system design with the total site integration considering the reduction of the greenhouse gas emissions, whereas Perry et al. [42] used the method to reduce carbon footprint through the integration of renewable energy sources, and Varbanov and Klemeš [43] considered that the variations in the energy supplies and demands are necessary.

On the other hand, there are several industrial processes requiring cooling above ambient temperature; in this regard, regularly recirculating cooling water systems are used to provide the required cold utility. Several methodologies have been published [44–48] considering the interactions between cooling systems and processes, in other words, the cooling tower is designed together with the synthesis of the associated HEN or cooling water network. Additionally, some strategies have been reported for processes that require cooling below ambient temperature. In this way, Ponce-Ortega et al. [49] and Lira-Barragán et al. [50] reported mathematical programming-based approaches to address the problem of multi-objective optimization of absorption refrigeration (AR) cycles that are integrated with HEN. In these integrated energy systems, the AR cycle uses excess process heat to drive the cycle generator to simultaneously decrease the associated costs and the use of fossil fuels that negatively impact the environment through GHGE. These works show how the optimal integration of AR cycles with processes can meet economic, environmental, and social goals.

According to the environmental concerns previously commented such as depletion of natural resources and concerns over socio-economic development and global climate change due to GHGE, especially CO_2 emissions associated with the use of fossil fuels, in recent times, different methodologies have included economic and environmental aspects to develop more sustainable solutions to engineering problems of different processes and scales [51–57]. However, researchers must address the three aspects of sustainability for the long-term development of industries and effective environmental protection. Thus, some works have balanced economic and social dimensions of sustainability [58–61] in industrial systems' sustainable development (see [62, 63]).

In this context, significant socio-economic and environmental benefits can be achieved through the optimal synthesis of trigeneration systems (i.e., cogeneration systems for combined heat, cooling, and power production) that are integrated with HEN and use renewable energy resources as primary heat sources to reduce fossil fuel consumption and GHGE. To properly illustrate the general scheme for the problem considered by this work, it is incorporated in Figure 3.1. In this scheme, the primary energy sources (i.e., solar collectors, biofuels, and fossil fuels) only are provided to the SRC, which produces electricity and low-pressure steam (LPS). This steam can be divided to be supplied to the evaporator that belongs to the ORC, as well as it can be provided to the AR cycle to satisfy the refrigeration demands of hot process streams in the HEN, and finally it can be reused as a hot utility in the HEN. The hot process streams (HPS) inside the HEN can transfer their excess heat to the AR cycle and the ORC. It also allows the heat exchange between the ORC and the CPS. It should be noted that the model formulation must determine the optimal working fluid to operate the ORC. This decision depends on the operational temperatures (see Figure 3.1) and the thermal efficiency; also, the model selects the optimal system to run the AR cycle.

Therefore, the objective of this chapter is to present a mathematical formulation for the simultaneous synthesis of sustainable trigeneration systems and heat exchanger networks, where the optimal working fluid to operate the ORC is determined as well as the optimal system to run the AR cycle, considering the energy connections among the different subsystems accounted. In this regard, the problem takes into account a set of HPS and a set of cold process streams (CPSs), as well as three thermodynamic cycles: single-effect AR to supply below ambient cooling, a steam Rankine cycle (SRC) and an ORC cycle to produce electricity. The proposed mathematical model is based on a superstructure that includes all feasible heat integration options and connections between the system components. The synthesis problem is formulated as a multi-objective mixed integer nonlinear programming problem with economic, environmental, and social concerns of sustainability. The environmental objective is evaluated by the overall GHGE, and the social objective is quantified by the number of generated jobs. The multi-objective optimization model is solved with the ε-constraint method. The optimal solutions lead to the Pareto set for the problem that shows the tradeoffs between the total annual profit, the greenhouse gas emissions, and the number of generated jobs in the entire life cycle of the integrated energy system. The capabilities of the proposed approach are illustrated through its application to two example problems. The results indicate that the integration of HEN and trigeneration schemes yields energy savings and more environmentally benign solutions.

3.2 Model approach

The proposed formulation has been subdivided into the following sections:

3.2.1 Heat exchanger network

The proposed superstructure for the HEN considers the following:

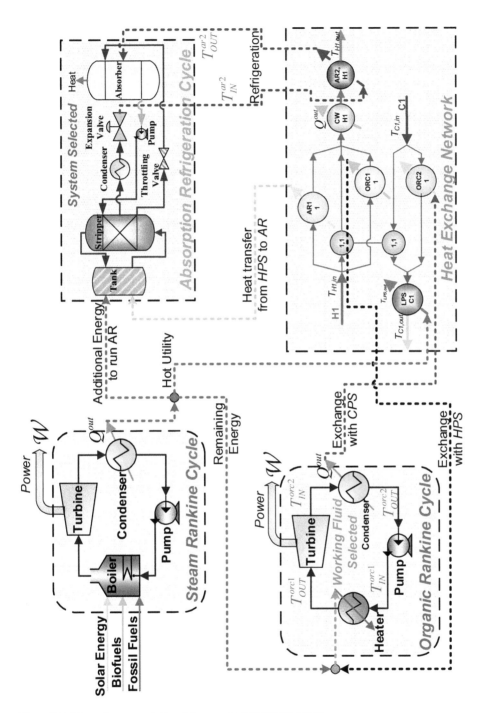

Figure 3.1: Schematic representation of the proposed integrated system.

- Each HPS entering the inner stages (in Figure 3.2 these stages are represented by stages 1 and 2; however for each specific case the required number of stages must be determined according to the maximum number of HPS or CPS) can have any match with any CPS, as well as can transfer their heat to the AR system (AR1 units) and the ORC (ORC1 units).
- Once the HPS have left the inner stages, they can get the first cooling using cold water (CW units).
- For the HPS requiring refrigeration below room temperature, the heat transfer units located at the cold end (AR2 units) of the superstructure are installed to complete the cooling requirements.
- Each CPS can start its heating in the internal stages through heat exchange with HPS and using the energy available in the condenser of the ORC (ORC2 units).
- To complete the heating requirements, each CPS can use external hot utilities (LPS units).

Then, the following relationships model all the abovementioned exchangers.

Total energy balances for the process streams:

$$(T_{IN_i} - T_{OUT_i})FCp_i = \sum_{k \in ST}\left[\sum_{j \in CPS} q_{i,j,k} + q_{i,k}^{ar1} + q_{i,k}^{orc1}\right] + q_i^{cw} + q_i^{ar2}, \quad \forall i \in HPS \tag{3.1}$$

$$(T_{OUT_j} - T_{IN_j})FCp_j = \sum_{k \in ST}\left[\sum_{i \in HPS} q_{i,j,k} + q_{j,k}^{orc2}\right] + q_j^{lps}, \quad \forall j \in CPS \tag{3.2}$$

Energy balance for internal stages:

$$(t_{i,k} - t_{i,k+1})FCp_i = \sum_{j \in CPS} q_{i,j,k} + q_{i,k}^{lar1} + q_{i,k}^{orc1}, \quad \forall i \in HPS, k \in ST \tag{3.3}$$

$$(t_{j,k} - t_{j,k+1})FCp_j = \sum_{i \in HPS} q_{i,j,k} + q_{j,k}^{orc2}, \quad \forall j \in CPS, k \in ST \tag{3.4}$$

Energy balance for external stages:

$$(t_{i,NOK+1} - t_i^{ar2})FCp_i = q_i^{cw}, \quad \forall i \in HPS \tag{3.5}$$

$$(t_i^{ar2} - T_{OUT_i})FCp_i = q_i^{ar2}, \quad \forall i \in HPS \tag{3.6}$$

$$(T_{OUT_j} - t_{j,1})FCp_j = q_j^{lps}, \quad \forall j \in CPS \tag{3.7}$$

Assignment of temperature for the extreme borders of the inner stages:

$$T_{IN_i} - t_{i,1}, \quad \forall i \in HPS \tag{3.8}$$

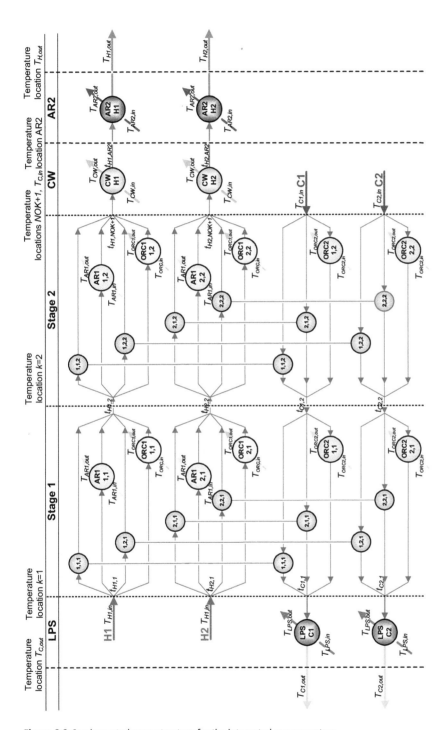

Figure 3.2: Implemented superstructure for the integrated energy system.

$$T_{IN_j} - t_{j,NOK+1}, \quad \forall j \in CPS \tag{3.9}$$

Feasibility constraints:

$$t_{i,k} \geq t_{i,k+1}, \quad \forall i \in HPS, k \in ST \tag{3.10}$$

$$t_{i,NOK+1} \geq t_i^{ar2}, \quad \forall i \in HPS \tag{3.11}$$

$$t_i^{ar2} \geq T_{OUT_i}, \quad \forall i \in HPS \tag{3.12}$$

$$T_{OUT_j} \geq t_{j,1}, \quad \forall j \in CPS \tag{3.13}$$

$$t_{j,k} \geq t_{j,k+1}, \quad \forall j \in CPS, k \in ST \tag{3.14}$$

Existence of the heat exchangers:

$$q_{i,j,k} - Q_{i,j}^{\max} z_{i,j,k} \leq 0, \quad \forall i \in HPS, j \in CPS, k \in ST \tag{3.15}$$

$$q_{i,k}^{ar1} - Q_{i,k}^{\max} z_{i,k}^{ar1} \leq 0, \quad \forall i \in HPS, k \in ST \tag{3.16}$$

$$q_{i,k}^{orc1} - Q_{i,k}^{\max} z_{i,k}^{orc1} \leq 0, \quad \forall i \in HPS, k \in ST \tag{3.17}$$

$$q_i^{cw} - Q_i^{\max} z_i^{cw} \leq 0, \quad \forall i \in HPS \tag{3.18}$$

$$q_i^{ar2} - Q_i^{\max} z_i^{ar2} \leq 0, \quad \forall i \in HPS \tag{3.19}$$

$$q_{j,k}^{orc2} - Q_{j,k}^{\max} z_{j,k}^{orc2} \leq 0, \quad \forall j \in CPS, k \in ST \tag{3.20}$$

$$q_j^{lps} - Q_j^{\max} z_j^{orc2} \leq 0, \quad \forall j \in CPS \tag{3.21}$$

Feasibilities for the temperature differences:

$$dt_{i,j,k} \leq t_{i,k} - t_{j,k} + \Delta T_{i,j}^{\max}(1 - z_{i,j,k}), \quad \forall i \in HPS, j \in CPS, k \in ST \tag{3.22}$$

$$dt_{i,j,k+1} \leq t_{i,k+1} - t_{j,k+1} + \Delta T_{i,j}^{\max}(1 - z_{i,j,k}), \quad \forall i \in HPS, j \in CPS, k \in ST \tag{3.23}$$

$$dt_{i,k}^{ar1} \leq t_{i,k} - T_{OUT}^{ar1} + \Delta T_i^{ar1\,\max}(1 - z_{i,k}^{ar1}), \quad \forall i \in HPS, k \in ST \tag{3.24}$$

$$dt_{i,k+1}^{ar1} \leq t_{i,k+1} - T_{IN}^{ar1} + \Delta T_i^{ar1\,\max}(1 - z_{i,k}^{ar1}), \quad \forall i \in HPS, k \in ST \tag{3.25}$$

$$dt_{i,k}^{orc1} \leq t_{i,k} - T_{OUT}^{orc1} + \Delta T_i^{orc1\,\max}(1 - z_{i,k}^{orc1}), \quad \forall i \in HPS, k \in ST \tag{3.26}$$

$$dt_{i,k+1}^{orc1} \leq t_{i,k+1} - T_{IN}^{orc1} + \Delta T_i^{orc1\,\max}(1 - z_{i,k}^{orc1}), \quad \forall i \in HPS, k \in ST \tag{3.27}$$

$$dt_i^{cw-1} \leq t_{i,NOK+1} - T_{OUT}^{cw} + \Delta T_i^{cw\,\max}(1 - z_i^{cw}), \quad \forall i \in HPS \tag{3.28}$$

$$dt_i^{cw-2} \leq t_i^{ar2} - T_{IN}^{cw} + \Delta T_i^{cw\,\max}(1 - z_i^{cw}), \quad \forall i \in HPS \tag{3.29}$$

$$dt_i^{ar2-1} \leq t_i^{ar2} - T_{OUT}^{ar2} + \Delta T_i^{ar2\,\max}(1 - z_i^{ar2}), \quad \forall i \in HPS \tag{3.30}$$

$$dt_{j,k}^{orc2} \leq T_{IN}^{orc2} - t_{j,k} + \Delta T_j^{orc2\,max}\left(1 - z_{j,k}^{orc2}\right), \quad \forall i \in CPS, k \in ST \tag{3.31}$$

$$dt_{j,k+1}^{orc2} \leq T_{OUT}^{orc2} - t_{j,k+1} + \Delta T_j^{orc2\,max}\left(1 - z_{j,k}^{orc2}\right), \quad \forall j \in CPS, k \in ST \tag{3.32}$$

$$dt_j^{lps-2} \leq T_{OUT}^{lps} - t_{j,1} + \Delta T_j^{lps\,max}\left(1 - z_j^{lps}\right), \quad \forall j \in CPS \tag{3.33}$$

Temperature differences:

$$\Delta T_{min} \leq dt_{i,j,k}, \quad \forall i \in HPS, j \in CPS, k \in ST \tag{3.34}$$

$$\Delta T_{min} \leq dt_{i,k}^{ar1}, \quad \forall i \in HPS, k \in ST \tag{3.35}$$

$$\Delta T_{min} \leq dt_{i,k}^{orc1}, \quad \forall i \in HPS, k \in ST \tag{3.36}$$

$$\Delta T_{min} \leq dt_i^{cw-1}, \quad \forall i \in HPS \tag{3.37}$$

$$\Delta T_{min} \leq dt_i^{cw-2}, \quad \forall i \in HPS \tag{3.38}$$

$$\Delta T_{min} \leq dt_i^{ar2-1}, \quad \forall i \in HPS \tag{3.39}$$

$$\Delta T_{min} \leq dt_{j,k}^{orc2}, \quad \forall j \in CPS, k \in ST \tag{3.40}$$

$$\Delta T_{min} \leq dt_j^{lps-2}, \quad \forall j \in CPS \tag{3.41}$$

3.2.2 Optimal selection of working fluids

This work considers selecting the optimal working fluid to operate the ORC between a set of available working fluids and a disjunctive model to determine the optimal system to run the AR cycle. It is important to remark that in each optimal solution, it is determined if the ORC is required or not; whereas the AR cycle is always needed because the proposed methodology considers the existence of cooling requirements.

Total energy balances for the process streams

Optimal selection of the working fluid for ORC: The following disjunctive model establishes that if the ORC is required, then it will generate a value for power production greater than zero. Besides, there is a set of available working fluids to be used by the ORC; thus, when the ORC exists, the optimal working fluid to run it must be selected. In this regard, each working fluid has a characteristic value for the efficiency factor, as well as specific operational temperatures in each step of the ORC. Then, considering these variations between the available working fluids, the disjunctive model will select the optimal working fluid. Finally, it is accounted for the scenario where the ORC does not exist and all the abovementioned parameters are set as zero.

$$
\begin{bmatrix}
Y^{orc} \\
Power^{orc} \geq 0 \\
\begin{bmatrix}
W_1^{orc} \\
\mu^{orc} = \mu_1^{orc} \\
T_{IN}^{orc1} = T_{IN,1}^{orc1} \\
T_{OUT}^{orc1} = T_{OUT,1}^{orc1} \\
T_{IN}^{orc2} = T_{IN,1}^{orc2} \\
T_{OUT}^{orc2} = T_{OUT,1}^{orc2}
\end{bmatrix}
\vee
\begin{bmatrix}
W_1^{orc} \\
\mu^{orc} = \mu_2^{orc} \\
T_{IN}^{orc1} = T_{IN,2}^{orc1} \\
T_{OUT}^{orc1} = T_{OUT,2}^{orc1} \\
T_{IN}^{orc2} = T_{IN,2}^{orc2} \\
T_{OUT}^{orc2} = T_{OUT,2}^{orc2}
\end{bmatrix}
\vee \ldots \vee
\begin{bmatrix}
W_g^{orc} \\
\mu^{orc} = \mu_g^{orc} \\
T_{IN}^{orc1} = T_{IN,g}^{orc1} \\
T_{OUT}^{orc1} = T_{OUT,g}^{orc1} \\
T_{IN}^{orc2} = T_{IN,g}^{orc2} \\
T_{OUT}^{orc2} = T_{OUT,g}^{orc2}
\end{bmatrix}
\end{bmatrix}
\vee
\begin{bmatrix}
\neg Y^{orc} \\
Power^{orc} = 0 \\
\mu^{orc} = 0 \\
T_{IN}^{orc1} = 0 \\
T_{OUT}^{orc1} = 0 \\
T_{IN}^{orc2} = 0 \\
T_{OUT}^{orc2} = 0
\end{bmatrix}
$$

On the other hand, the previous disjunction is transformed into a set of algebraic relationships through the convex hull technique [64, 65]. In this sense, each Boolean variable is associated with a binary variable to carry out the proper reformulation. Thus, when the Boolean variables are true the associated binary variables are set as one; otherwise if the Boolean variables are false, the binary variables must be set as zero. Then, the following logical relationship states that when the ORC is required, one working fluid must be selected; otherwise, if the solar collector is not needed, there is no activated working fluid:

$$
y^{orc} = \sum_{g \in G} w_g^{orc} \tag{3.42}
$$

The next relationship states that if the ORC exists, the amount of electricity produced must be lower than an upper limit ($Power^{orc-Max}$),

$$
Power^{orc} \leq Power^{orc-Max} y^{orc} \tag{3.43}
$$

Vélez et al. [26] reported that ORC can produce up to 2 MW. Moreover, the chosen efficiency factor is obtained through the following relationship:

$$
\mu^{orc} = \sum_{g \in G} \mu_g^{orc} w_g^{orc} \tag{3.44}
$$

It should be noted that only one working fluid can be selected. The same type of relationships is applied to select the four temperatures involved in the ORC:

$$
T_{IN}^{orc1} = \sum_{g \in G} T_{IN,g}^{orc1} w_g^{orc} \tag{3.45}
$$

$$
T_{OUT}^{orc1} = \sum_{g \in G} T_{OUT,g}^{orc1} w_g^{orc} \tag{3.46}
$$

$$T_{IN}^{orc2} = \sum_{g \in G} T_{INg}^{orc2} w_g^{orc} \tag{3.47}$$

$$T_{OUT}^{orc2} = \sum_{g \in G} T_{OUT,g}^{orc2} w_g^{orc} \tag{3.48}$$

Additionally, the following relationships establish that if the *ORC* is required, then it can be the heat exchange between the *ORC* with *HPS* and *CPS*; in the other case, if the *ORC* does not exist, then the possibility of the heat exchange with *ORC* is eliminated:

$$y^{orc} \geq z_{i,k}^{orc1}, \quad \forall i \in HPS, k \in ST \tag{3.49}$$

$$y^{orc} \geq z_{j,k}^{orc2}, \quad \forall j \in CPS, k \in ST \tag{3.50}$$

Optimal selection of the system for AR cycle: This work considers the optimal selection of the system to operate the *AR* cycle. To carry out this task, the following disjunctive model (where Y_h^{ar} represents the Boolean variable associated with the selection of the system h to run the *AR* system) is employed:

$$
\begin{bmatrix} Y_1^{ar} \\ COP^{ar} = COP_1^{ar} \\ T_{IN}^{ar2} = T_{IN,1}^{ar2} \\ T_{OUT}^{ar2} = T_{OUT,1}^{ar2} \end{bmatrix}
\vee
\begin{bmatrix} Y_1^{ar} \\ COP^{ar} = COP_2^{ar} \\ T_{IN}^{ar2} = T_{IN,2}^{ar2} \\ T_{OUT}^{ar2} = T_{OUT,2}^{ar2} \end{bmatrix}
\vee \ldots \vee
\begin{bmatrix} Y_h^{ar} \\ COP^{ar} = COP_h^{ar} \\ T_{IN}^{ar2} = T_{IN,h}^{ar2} \\ T_{OUT}^{ar2} = T_{OUT,h}^{ar2} \end{bmatrix}
$$

The previous disjunction selects the optimal system between a set of available systems. Each available system has a characteristic value COP_h^{ar} also, according to the system, it changes the values for the inlet and outlet temperatures for the *AR2* units ($T_{IN,h}^{ar2}$ and $T_{OUT,h}^{ar2}$, respectively). Previous temperatures affect drastically the lowest temperature that can be achieved by the *HPS*.

When the previous disjunctive model is reformulated into a set of algebraic equations, it obtains the next logical relationships. Firstly, one system must be chosen:

$$\sum_{h \in H} y_h^{ar} = 1 \tag{3.51}$$

Then, the *COP*, inlet, and outlet temperatures for the operation of the *AR* cycle are selected according to the selected system:

$$COP^{ar} = \sum_{h \in H} COP_h^{ar} y_h^{ar} \tag{3.52}$$

$$T_{IN}^{ar2} = \sum_{h \in H} T_{IN,h}^{ar2} y_h^{ar} \tag{3.53}$$

$$T_{OUT}^{ar2} = \sum_{h \in H} T_{OUT,h}^{ar2} y_h^{ar} \tag{3.54}$$

3.2.3 Thermodynamic cycles and their interactions

To model the operation of the *AR* cycle, a coefficient of performance (*COP*) is considered; whereas the modeling for thermodynamic cycles is carried out through efficiency factors (μ). Additionally, in this section it is established that the available heat in the condenser in the *SRC* (residual energy) can be taken to be used in the *ORC*, *AR* cycle, and the *HEN* (as hot utility).

Modeling for the SRC: The energy balance for all the external energy sources supplied to the boiler placed in the *SRC* is stated as follows:

$$Q^{External} = Q_t^{Solar} + \sum_{b \in B} Q_{b,t}^{Biofuel} + \sum_{f \in F} Q_{f,t}^{Fossil}, \qquad \forall t \in T \tag{3.55}$$

where $Q^{External}$ represents the total energy provided to the *SRC*.

Later, the power produced by the *SRC* is equal to the total energy fed to the cycle multiplied by an efficiency factor that represents the thermal efficiency to convert the external energy provided to the cycle into electricity:

$$Power^{src} = Q^{External} \mu^{src} \tag{3.56}$$

Then, the remaining residual energy (the heat that cannot be converted into power) is sent to the *ORC* (to produce electricity), *AR* system (to yield refrigeration), and the *HEN* (as hot utility):

$$Q^{src-mps} = Q^{ar-mps} + Q^{orc-mps} + \sum_{j \in CPS} q_j^{lps} \tag{3.57}$$

The overall energy balance for the *SRC* establishes that the total energy supplied to the cycle ($Q^{External}$) is converted into electricity ($Power^{src}$) or it can be sent to other subsystems (Q^{src_mps}) as well as it is considered the possible heat transfer to cooling water (Q^{src_cw}):

$$Q^{External} = Power^{src} + Q^{src-mps} + Q^{src-sw} \tag{3.58}$$

The worst scenario for the residual heat is the heat transfer with cold water, owing to this option generating an operational cost (cooling water cost) and it is not harnessed for relevant applications.

Modeling for the ORC: According to the proposed scheme, the *ORC* can receive heat coming from the *SRC* (Q^{orc_mps}), and it also takes into account the possible heat transfer from *HPS* to the *ORC* ($q_{i,k}^{orc1}$). Then to calculate the power produced by the *ORC* ($Power^{orc}$), it considers the efficiency factor (μ^{orc}) as follows:

$$Power^{orc} = \left(Q^{orc-mps} + \sum_{i \in HPS} \sum_{k \in ST} q_{i,k}^{orc1} \right) \mu^{orc} \tag{3.59}$$

Additionally, the residual energy (available in the condense of the ORC) can be used by the cold process steams ($q_{j,k}^{orc2}$) or can be transferred to cold water (Q^{orc_cw}):

$$Q^{orc\_mps} + \sum_{i \in HPS} \sum_{k \in ST} q_{i,k}^{orc1} = Power^{orc} + \sum_{j \in CPS} \sum_{k \in ST} q_{j,k}^{orc2} + Q^{orc\_cw} \tag{3.60}$$

It is important to highlight that the efficiency factor and operational temperatures for the ORC depend on the selected working fluid to run the cycle. In this regard, the value for the operational temperatures is an important factor for the feasibility of the heat transfer among the HPS and the ORC, as well as for the heat exchange from the ORC to the CPS, which can decrease the use of external energy sources.

Modeling for the AR cycle: The balance for the AR system establishes that the cooling load below the ambient temperature required by hot process streams (q_i^{ar2}) must be supplied by the excess heat of the HPS ($q_{i,k}^{ar1}$) as well as by the energy available in the condenser of the SRC (Q^{ar-mps}). This energy balance also considers a COP (which depends on the chosen system) as a factor to describe the energy conversion between the heat provided and the cooling obtained; therefore, it is stated as follows:

$$\frac{\sum_{i \in HPS} q_i^{ar2}}{COP^{ar}} = \sum_{i \in HPS} \sum_{k \in ST} q_{i,k}^{orc1} + Q_{ar}^{mps} \tag{3.61}$$

Maximum production of power: Before the optimization process the maximum desired power generation (i.e., the maximum electricity that will be produced by the SRC and the ORC) must be analyzed and determined and must be obtained through a study considering the electricity requirements of the project and the maximum amount of power that can be sold to other industries or local governments. This constraint is modeled as follows:

$$Power^{src} + Power^{orc} \leq Power^{Max} \tag{3.62}$$

where $Power^{Max}$ represents the maximum desired power production.

The optimal size for the solar collector: The solar collector can exist or not in the optimal solution (owing to it representing the most expensive energy source; but at the same time is the cleanest energy); then, if the solar collector is required, there is need to determine its optimal area. First, the following relationship determines the total solar energy provided to the system (Q_t^{Solar}):

$$Q_t^{Solar} \leq Q_t^{Useful\_Solar} A_c^{Solar} \frac{1}{D_t}, \quad \forall t \in T \tag{3.63}$$

$Q_t^{Useful\_Solar}$ represents the useful solar energy for the available solar radiation in the specific location where the solar collector can be installed (in this case, this value includes the efficiency associated with the equipment), A_c^{Solar} is the optimal area of the solar collector and D_t is a conversion factor to change the units of time. Then, once the solar energy requirements and the optimal size have been determined, the capital (C_{cap}^{Solar}) and operational (C_{op}^{Solar}) costs for the solar collector are calculated :

$$C_{cap}^{Solar} = FC^{Solar} y^{Solar} + VC^{Solar} \left(A_c^{Solar} \right)^{a^{Solar}} \tag{3.64}$$

$$C_{cap}^{Solar} = Cu^{Solar} \sum_{t \in T} \left(Q_t^{Solar} D_t \right) \tag{3.65}$$

FC^{Solar} and VC^{Solar} represent the fixed and variable unit costs for the capital costs of the solar collector, while y^{Solar} is a binary variable used to model the existence of the solar collector, a^{Solar} is an exponent for the area to consider the economies of scale in the capital cost function for the solar collector, and finally, Cu^{Solar} is the unitary operating cost.

Maximum availability for biofuels:. On the other hand, since the availability of biofuels changes drastically through the year, the following constraint must be included in the model:

$$Q_{bt}^{Biofuel} \le \frac{Heating_b^{Power} Avail_{bt}^{Max}}{D_t}, \quad \forall b \in B, t \in T \tag{3.66}$$

$Heating_b^{Power}$ is the heating power for biofuel b and $Avail_{bt}^{Max}$ denotes the maximum amount of biofuel b available in period t.

3.2.4 Objective functions

The addressed problem is a multi-objective MINLP model that simultaneously considers economic, environmental, and social issues (which are important criteria included in sustainability). Thus, the economic goal consists in the maximization of the total annual profit (*TAP*), whereas the environmental target accounted for in the proposed methodology is the minimization of the net *GHGE*, and finally, the social objective is to maximize the jobs generated by the project:

$$OF = \{ Max\ TAP;\ Min\ NGHGE^{overall};\ MaxJobs \} \tag{3.67}$$

Since the first two objective functions (*Max TAP, Min NGHGE^{Overall}*) contradict each other, the number of jobs that can be created by the project plays an important role for the decision-makers and local governments involved in the project.

Economic objective function: The economic objective function is aimed at maximizing the profit, which includes the selling of power (*SP*) and tax credit reduction (*TCR*), minus the capital cost (*CaC*), fixed cost (*FiC*), operating cost (*OC*), and energy sources cost (*ESC*):

$$Max \; Profit = SP + TCR - CaC - FiC - OC - ESC \tag{3.68}$$

These terms are explained as follows: The main economic benefits are obtained through the selling of the power generated by the *SRC* and the *ORC*:

$$SP = H_Y D^{sh} \left(GaPow^{src} Power^{src} + GaPow^{orc} Power^{orc} \right) \tag{3.69}$$

H_Y denotes the hours of operation per year of the plant, D^{sh} is a conversion factor that transforms seconds into hours, and $GaPow^{src}$ and $GaPow^{orc}$ represent the unitary gains for the selling of power produced in the *SRC* and *ORC*. These unitary gains are determined considering the unitary price for the power, as well as the power production costs for the *SRC* ($PPCost^{src}$) and the *ORC* ($PPCost^{orc}$), respectively:

$$GaPow^{ran} = SuP^{Power} - PPCost^{src}$$

$$GaPow^{orc} = SuP^{Power} - PPCost^{orc}$$

Several governments have promoted the use of cleaner energy through tax credits given to the users of this type of energy. Based on this, the project can obtain tax credits for the use of solar energy, biofuels, and even some fossil fuels:

$$TCR = H_Y \left\{ R^{Solar} \left[\sum_{t \in T} \left(Q_t^{solar} D_t \right) \right] + \sum_{t \in T} \sum_{b \in B} \left[R_b^{Biofuel} Q_{bt}^{Biofuel} D_t \right] + \sum_{t \in T} \sum_{f \in F} \left[R_f^{Fossil} Q_{f,t}^{Fossil} D_t \right] \right\}$$

$$\tag{3.70}$$

R^{Solar}, $R_b^{Biofuel}$, and R_f^{Fossil} are the unitary tax credits for solar energy, biofuels, and fossil fuels, respectively. The capital costs included are the capital cost for the *AR* cycle and the capital costs for all the heat exchanger units shown in Figure 3.2. In this order, the proposed methodology uses Chen's approximation [66] to avoid logarithmic terms in the optimization model to determine *LMTD* for the heat exchange units:

$$CaC = k_f \left[C_{cap}^{ar} \sum_{i \in HPS} q_i^{ar2} + \sum_{i \in HPS} \sum_{j \in CPS} \sum_{k \in ST} C_{ij}^{exc} \left\{ \frac{q_{i,j,k} \left(\frac{1}{h_i} + \frac{1}{h_j} \right)}{\left[\left(dt_{i,j,k} \right) \left(dt_{i,j,k+1} \right) \left(\frac{dt_{i,j,k} + dt_{i,j,k+1}}{2} \right) + \delta \right]^{1/3}} \right\}^{\beta} + \right.$$

$$\sum_{i \in HPS} \sum_{k \in ST} C_i^{ar1} \left\{ \frac{q_{i,k}^{ar1} \left(\frac{1}{h_i} + \frac{1}{h_{ar1}} \right)}{\left[\left(dt_{i,k}^{ar1} \right) \left(dt_{i,k+1}^{ar1} \right) \left(\frac{dt_{i,k}^{ar1} + dt_{i,k+1}^{ar1}}{2} \right) + \delta \right]^{1/3}} \right\}^{\beta}$$

$$+ \sum_{i \in HPS} \sum_{k \in ST} C_i^{orc1} \left\{ \frac{q_{i,k}^{orc1} \left(\frac{1}{h_i} + \frac{1}{h_{orc1}} \right)}{\left[\left(dt_{i,k}^{orc1} \right) \left(dt_{i,k+1}^{orc1} \right) \left(\frac{dt_{i,k}^{orc1} + dt_{i,k+1}^{orc1}}{2} \right) + \delta \right]^{1/3}} \right\}^{\beta} +$$

$$\sum_{i \in HPS} C_i^{cw} \left\{ \frac{q_i^{cw} \left(\frac{1}{h_i} + \frac{1}{h_{cw}} \right)}{\left[\left(dt_i^{cw-1} \right) \left(dt_i^{cw-2} \right) \left(\frac{dt_i^{cw-1} + dt_i^{cw-2}}{2} \right) + \delta \right]^{1/3}} \right\}^{\beta}$$

$$+ \sum_{i \in HPS} C_i^{ar2} \left\{ \frac{q_i^{ar2} \left(\frac{1}{h_i} + \frac{1}{h_{ar2}} \right)}{\left[\left(dt_i^{ar2-1} \right) \left(T_{OUT_i} - T_{IN}^{ar2} \right) \left(\frac{dt_i^{ar2-1} + \left(T_{OUT_i} - T_{IN}^{ar2} \right)}{2} \right) + \delta \right]^{1/3}} \right\}^{\beta} +$$

$$\sum_{j \in CPS} \sum_{k \in ST} C_j^{orc2} \left\{ \frac{q_{j,k}^{orc2} \left(\frac{1}{h_j} + \frac{1}{h_{orc2}} \right)}{\left[\left(dt_{j,k}^{orc2} \right) \left(dt_{j,k+1}^{orc2} \right) \left(\frac{dt_{j,k}^{orc2} + dt_{j,k}^{orc2}}{2} \right) + \delta \right]^{1/3}} \right\}^{\beta}$$

$$\left. + \sum_{j \in CPS} C_j^{lps} \left\{ \frac{q_j^{lps} \left(\frac{1}{h_j} + \frac{1}{h_{lps}} \right)}{\left[\left(T_{IN}^{lps} - T_{OUT_j} \right) \left(dt_j^{lps-2} \right) \left(\frac{\left(T_{IN}^{lps} - T_{OUT_j} \right) + dt_j^{lps-2}}{2} \right) + \delta \right]^{1/3}} \right\} \right]$$

(3.71)

k_f is a factor used to annualize the inversion, C_{cap}^{ar} represents the unitary capital cost for the *AR* cycle, δ is a small parameter used to avoid infeasibilities in the optimization process, while C_{ij}^{exc}, C_i^{ar1}, C_i^{orc1}, C_i^{cw}, C_i^{ar2}, C_j^{orc2} and C_j^{lps} are the area cost coefficients for exchangers between process streams, *AR1* exchangers, *ORC1* coolers, units that exchange heat between *HPS* with cooling water, *AR2* coolers, exchangers transferring heat from the *ORC* to the *CPS* and heaters using *LPS* as hot utility, respectively; and h represents the film heat transfer coefficients.

The fixed costs accounted by this work are the fixed costs for all the exchangers required in the optimal solution:

$$FiC = k_f \left[\begin{array}{l} \sum\limits_{i \in HPS} \sum\limits_{j \in CPS} \sum\limits_{k \in ST} C_{F_{ij}}^{exc} Z_{i,j,k} + \sum\limits_{i \in HPS} \sum\limits_{k \in ST} C_{F_i}^{ar1} Z_{i,k}^{ar1} + \sum\limits_{i \in HPS} \sum\limits_{k \in ST} C_{F_i}^{orc1} Z_{i,k}^{orc1} + \\ \sum\limits_{i \in HPS} C_{F_i}^{cw} Z_i^{cw} + \sum\limits_{i \in HPS} C_{F_i}^{ar2} Z_i^{ar2} + \sum\limits_{j \in CPS} \sum\limits_{k \in ST} C_{Fj}^{orc2} Z_{j,k}^{orc2} + \sum\limits_{j \in CPS} C_{Fj}^{lps} Z_j^{lps} \end{array} \right] \quad (3.72)$$

$C_{F_{ij}}^{exc}$, $C_{F_i}^{ar1}$, $C_{F_i}^{orc1}$, $C_{F_i}^{cw}$, $C_{F_i}^{ar2}$, C_{Fj}^{orc2}, and C_{Fj}^{lps} are the fixed costs for exchangers, $AR1$ coolers, $ORC1$ coolers, CW coolers, $AR2$ coolers, $ORC2$ heaters, and LPS heaters, respectively.

The operating costs taken into account are the cooling water costs to cool HPS (q_i^{cw}), as well as in the condenser of the SRC (Q^{src_cw}) and for the ORC (Q^{orc_cw}):

$$OC = C_{cw} \left[\sum_{i \in HPS} q_i^{cw} + Q^{src-cw} + Q^{orc-cw} \right] \quad (3.73)$$

The economic function includes the costs related to the primary energy sources (fossil fuels, biofuels, and solar collectors):

$$ESC = H_Y \left\{ \sum_{f \in z} \left[C_f^{Fossil} \sum_{t \in T} \left(Q_{f,t}^{Fossil} D_t \right) \right] + \sum_{b \in B} \left[C_b^{Biofuel} \sum_{t \in T} \left(Q_{b,t}^{Biofuel} D_t \right) \right] \right\}$$

$$+ H_Y C_{op}^{Solar} + k_f C_{cap}^{Solar} \quad (3.74)$$

C_f^{Fossil} and $C_b^{Biofuel}$ are the unitary costs for fossil fuel f and biofuel b.

It should be noted that depending on the value of the counterweights among the revenues and costs, for some cases the profit can be positive, while for other cases it can have a negative value.

Environmental objective function: This chapter considers that the environmental impact assessment is carried out through the overall quantification of the $GHGE$ because fossil fuels and biofuels release carbon dioxide when they are burned:

$$Min\ NGHGE^{Overall} = \sum_{t \in T} \sum_{f \in z} \left[GHGE_f^{Fossil} Q_{f,t}^{Fossil} D_t \right] + \sum_{t \in T} \sum_{b \in B} \left[GHGE_b^{Biofuel} Q_{b,t}^{Biofuel} D_t \right]$$

$$(3.75)$$

$NGHGE^{Overall}$ is the overall $GHGE$ released to the environment, while $GHGE_b^{Biofuel}$ and $GHGE_f^{Fossil}$ are the individual $GHGE$ for fossil fuel f and biofuel b. It is important to remark that the individual $GHGE$ are determined through the life cycle analysis (the $GREET$ software can be used for this purpose [67]) given in units of ton of CO_{2eq} reduction per kJ generated.

Social objective function:. The model is aimed at considering sustainable aspects. In this order, a sustainable process includes economic, environmental, and social criteria [68–71]. In this context, a very important social issue is the lack of employment, especially in rural areas; to mitigate this problem the proposed methodology accounts for maximizing the number of jobs that can be created by the implementation of the project. Then, these jobs are generated by the production of fossil fuels, biofuels, and for the operation of the solar collector to provide the energy requirements of the project. However, most of the jobs are attributed to the production of biofuels, which usually are produced in rural zones.

On the other hand, to quantify the number of jobs created the JEDI (jobs and economic development impact) model is employed. In this sense, Miller and Blair [72] have used the JEDI model for economic and social sciences, which is based on an input-output analysis. The input-output analysis is based on the use of multipliers, where a multiplier is a simple ratio of total systemic change over the initial change resulting from a given economic activity. This provides estimates of the total impact resulting from an initial change in economic output (e.g., employment) through the implementation or termination of a project. To determine the total effect of yielding biofuels, fossil fuels, and solar collectors, three separate impacts are examined: direct, indirect, and induced.

Direct effect: the immediate (or on-site) effect created by an expenditure. For example, in constructing a plant, direct effects include the on-site contractors and crews hired to construct the plant. Direct effects also include the jobs at the plant that build the process equipment.

Indirect effect: the increase in economic activity that occurs when contractors, vendors, or manufacturers receive payment for goods or services and in turn can pay others who support their business. For instance, indirect effects include the banker who finances the contractor, the accountant who keeps the contractor's books, and the steel mills and electrical manufacturers and other suppliers that provide the required materials.

Induced effect: The change in wealth that occurs or is induced by the spending of those persons directly and indirectly employed by the project.

The total effect from a single expenditure can be calculated by summing all three effects, using region-specific multipliers and personal expenditure patterns [62, 63, 73]. Hence, the social function consists of maximizing the number of jobs created by the project for the production of fossil fuels, biofuels, and solar collectors:

$$Max\ NJOBS^{Overall} = \sum_{t \in T} \sum_{f \in B} \left[NJOB_f^{Fossil} Q_{f,t}^{Fossil} D_t \right] + \sum_{t \in T} \sum_{b \in B} \left[NJOB_b^{Biofuel} Q_{b,t}^{Biofuel} D_t \right]$$

$$+ \sum_{t \in T} \left[NJOB^{Solar} Q_t^{Solar} D_t \right] \tag{3.76}$$

$NJOB_f^{Fossil}$, $NJOB_b^{Biofuel}$, and $NJOB^{Solar}$ are the number of jobs generated per kilo joule provided by fossil, biofuels, and solar collectors, respectively. The social concern included in this chapter represents an invaluable tool for governments, decision-makers, and investors because it is an instrument for the social and economic development of marginalized zones.

Finally, it is important to establish that the proposed methodology corresponds to a preliminary design and a detailed design that incorporates other aspects such as pressure drops or the multi-period operation should be incorporated in a future contribution.

3.3 Results and discussion

The applicability of the proposed mathematical programming model is shown in the following example. In this order, the solver DICOPT and the solvers CONOPT and CPLEX were implemented in the general algebraic modeling system (GAMS [75]). Additionally, this case study considers the following assumptions:

The project will be placed at Morelia, México, which has the coordinates N 19° 42′ 08″ and W 101° 11′ 08″. Table 3.1 shows the useful energy that can be collected per month by the solar collector in that place. To determine these values the solar radiation available in the specific location is computed by the efficiency to catch the solar energy of the solar collector (parabolic trough solar collector – $PTSC$).

The hours of operation of H_Y are 8,760 h/year.

The unitary price of the power SuP^{Power} is \$0.14/kWh; whereas the power production costs are $PPCost^{src}$ = \$0.10/kWh and $PPCost^{orc}$ = \$0.12/kWh.

The fixed and variable capital costs for the solar collector are \$75,526.61 and \$40.78/ m², respectively, while the unitary operating cost is \$0.012/kWh, and finally a^{Solar} is 1.

The inlet and outlet temperatures for some external services for the HEN are as follows: $T_{IN}^{ar1} = 313K$, $T_{OUT}^{ar1} = 353K$, $T_{IN}^{cw} = 303K$, $T_{OUT}^{cw} = 313K$, $T_{IN}^{lps} = 623K$, and $T_{OUT}^{lps} = 622K$.

The unitary cost for cooling water C_{cw} is \$20/(year K).

The unit number of jobs generated by the solar collector is 9.95459 × 10⁻¹⁰ Jobs/kJ.

The minimum temperature difference ΔT_{min} is 10 K.

Table 3.1: Useful collected energy per month for the solar collector.

Month/type of solar collector	PTSC (kJ/(m² month))
January	409,293
February	443,016
March	577,530
April	571,860
May	555,768
June	454,410
July	443,610
August	439,425
September	394,470
October	410,967
November	407,430
December	522,288

Moreover, Table 3.2 presents the fossil fuels and biofuels considered in this example, as well as the values for the parameters associated with them. Notice that only four biofuels are considered in the example.

Table 3.2: Data for the fossil fuels and biofuels considered in the examples presented.

No.	Fuel	Heating power (kJ/kg)	Overall *GHGE* (ton CO$_2$-eq/kJ)	Cost ($/mm kJ)	Generation of jobs (Jobs/kJ)
Fossil fuels					
1.	Coal	35,000	2.21357×10^{-7}	1.5559	1.06281×10^{-11}
2.	Oil	45,200	8.05408×10^{-8}	18.2447	1.81677×10^{-11}
3.	Natural gas	54,000	7.90892×10^{-8}	5.8349	5.25431×10^{-11}
Biofuels					
1.	Biomass	17,200	2.44307×10^{-8}	2.0303	6.6964×10^{-8}
2.	Biogas	52,000	2.68216×10^{-8}	8.5388	5.25431×10^{-7}
3.	Softwood	20,400	3.3482×10^{-8}	2.5332	1.46691×10^{-8}
4.	Hardwood	18,400	3.3482×10^{-8}	2.8975	5.43641×10^{-8}

3.3.1 Case study

A new industrial project is planning to construct auxiliary units to provide hot and cold utilities, refrigeration, and electricity; then the optimal integrated design must be determined. In this context, according to the given flowsheet, there are six *HPS* and five *CPS* available to carry out the energy integration (see Table 3.3).

Table 3.3: Stream data for the example.

Stream	Inlet temperature (K)	Outlet temperature (K)	FCp (kW/K)
H1	345	278	22.40
H2	448	342	25.84
H3	300	273	77.80
H4	288	271	226.50
H5	488	383	27.85
H6	543	395	9.40
C1	358	453	19.50
C2	370	418	55.30
C3	283	315	135.45
C4	278	298	215.65
C5	338	416	62.15

Also, three potential candidates of organic fluids are considered to run an *ORC* (only if it is required) and the two most common systems to operate the *AR* cycle. The required information is shown in Table 3.4 (as can be seen, when the working fluid is more efficient, it requires higher turbine inlet temperatures). It has been estimated that an *SRC* can perform with an efficiency of 0.28 and it is desired to achieve an electricity production up to 10 MW.

Table 3.4: Working fluids for ORC and systems for AR available in the example [30, 32, 33].

		Organic Rankine cycle					Absorption refrigeration cycle			
No.	Working fluid	μ_g^{orc}	$T_{IN,g}^{orc1}(K)$	$T_{OUT,g}^{orc1}(K)$	$T_{IN,g}^{orc2}(K)$	$T_{OUT,g}^{orc2}(K)$	No.	System	COP	$T_{IN,h}^{ar2}(K)$	$T_{OUT,h}^{ar2}(K)$
1.	R113	0.19	294	458	298	295	1.	H$_2$O–LiBr	1.2	268	268
2.	R245ca	0.17	305	443	308	306	2.	NH$_4$–H$_2$O	0.7	243	243
3.	Isobutene	0.145	304	398	306	305	–	–	–	–	–

Then, the proposed methodology is applied to this case considering a tax credit of $5/ton CO$_2$-eq and the obtained results are described in detail. Firstly, in Figure 3.3 the continuous Pareto curve illustrates the optimal solutions for a power production equal to or lower than 10 MW; whereas the suboptimal solutions drawn through the dashed line represent the best scenarios when the power production is fixed to 10 MW.

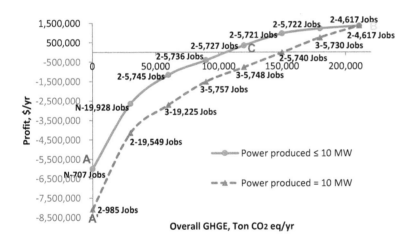

Figure 3.3: Pareto solutions for the case study.

For both Pareto curves, the zone operating with economic losses is bigger than the area with positive gains. Also, the organic fluid R113 (working fluid with the highest efficiency) is not found in any optimal point reported in Figure 3.3, and consequently is more favorable to a fluid that allows the heat exchange between the *HEN* and the *ORC* than a fluid with higher efficiencies to reduce the consumption of external energies. In addition, for all the solutions found in this example, the chosen system to run the *AR* is NH$_4$-H$_2$O. Nonetheless, the model can include a greater set of systems to run the *AR* cycle, and thus it is possible to find an interesting variation for the chosen system. In this figure, the solutions A, A', B, and C belong to both Pareto curves (it should be noted that both lines end at Point B) to discuss the most relevant differences among them. Points A and A' represent the optimal solutions with the lowest overall *GHGE* (zero), and at the same time, the worst profit (these solutions have been associated with an important number of economic losses) and the number of jobs that can be created by the project is the lowest with respect to all the solutions presented. While Point B represents the best possible economic solution, however, also is the worst environmental solution and the corresponding number of generated jobs is relatively high; but other solutions offer better values for the creation of jobs.

To overcome the last two aspects, Point C has been selected. This solution operates in positive gains for the profit ($349,474.42/year), releases 120,000 tons CO_2-eq/year to the environment (90,889.31 tons CO_2-eq/year lower than point B), and creates 5,727 jobs (1,110 jobs more than point B). Table 3.5 contains a detailed description of these points. Making a comparison between points A and A', it can be seen that solution A produces approximately 6.3 MW and the value for the profit is almost –$6 MM/year; then if the produced power is strictly equal to 10 MW (solution A') the solar collector cost augments approximately from $8.65 MM/year to $12 MM/year (i.e., it requires a bigger solar collector) producing an increase in the economic losses over $2 MM.

Table 3.5: Details for solutions to the case study.

Concept/case	$Power^{Max} \leq 10$ MW		$Power^{Max} = 10$ MW	
	Point A (min *profit*, min *GHGE*)	Point C (selected solution)	Point B (max *profit*, max GHGE)	Point A' (min *profit*, min GHGE)
Fossil fuels cost, $/year	0	1,630,270.99	1,470,509.05	0
Biofuels cost, $/year	0	229,033	139,684.64	0
Solar collector cost, $/year	8,652,546.38	0	0	12,049,060
Power produced, kW	6,309.71	6,907.84	10,000	10,000
Selling of power, $/year	2,210,922.19	2,345,056.25	3,329,205.73	3,291,404.33
Tax credits reduction, $/year	786,540.70	207,417.55	67,742.65	1,095,293.32
Capital cost for exchangers, $/year	284,443.56	289,276.52	362,748.60	315,057.35
Fixed cost for exchangers, $/year	54,418	54,418	46,046	50,232
Operational cost, $/year	0	0	11,160.69	57,765.40
Energy sources cost, $/year	8,652,546.38	1,859,304.85	1,610,193.69	12,049,060
Profit, $/year	–5,993,945.04	349,474.42	1,366,799.16	–8,085,417.33
Overall GHGE, ton CO_2-eq/year	0	120,000	210,889.31	0
Jobs	707	5,727	4,617	985
Working fluid selected for ORC	No ORC	R245ca	R245ca	R245ca

On the other hand, Figure 3.4 shows the optimal solution, where the *ORC* is not required. The solar collector needed to provide the total energy has an area of 154,080 m²; notice also that the total heat available in the condenser of the *SRC* is employed as a hot utility in the *HEN* and to run the *AR* cycle (typically this residual energy is not used as it is proposed by this work).

Figure 3.5 requires an *ORC* running with the organic fluid R245ca. The operational temperatures related to this fluid contribute to allowing the heat exchange between the *ORC* and the *HEN*; specifically 1,328.89 kW are transferred from the stream H5 to the heater inside the *ORC* to produce electricity and the condenser belonging to the *ORC* sent 4,313 kW to completely heat the CPS C4 and thus only 558.05 kW are wasted and transferred to cooling water.

This type of configuration simultaneously maximizes the reuse of heat and minimizes the external utilities, which means a minimum consumption of external energies; for this reason, the levels for operational temperatures for the *ORC* (which depends on the selected fluid) can be more advantageous than a more efficient fluid. Also, Figure 3.6 (this figure illustrates the optimal configuration for Point C) shows the following advantages:

All the energy available in the condenser of the *SRC* is reused to run the AR cycle, to run an *ORC* as well, and to heat the *CPS* in the *HEN* (remember that a cooler operating with cooling water, which represents the typical performance for *SRC* has been considered).

For this case, the levels for operational temperatures for the *ORC* (that the R245ca involves for its operation) also is more beneficial than a more efficient fluid (i.e., R113), to allow the heat transfer between the stream H5 and the heater of the *ORC* as well as among the condensers corresponding to the *ORC* and the cold stream C4 (notice that for this case all the heat available in the condenser of the ORC is sent to the stream C4).

Finally, this type of design simultaneously maximizes the reuse of heat and minimizes the utility requirements; this represents a minimum consumption of primary energy sources.

Afterward, Figure 3.7 shows the type of primary energy source required monthly for solution C and Figure 3.8 illustrates a comparison with the case when the tax credit increases to $20/ton CO_2-eq, here again the gap between these two scenarios appears.

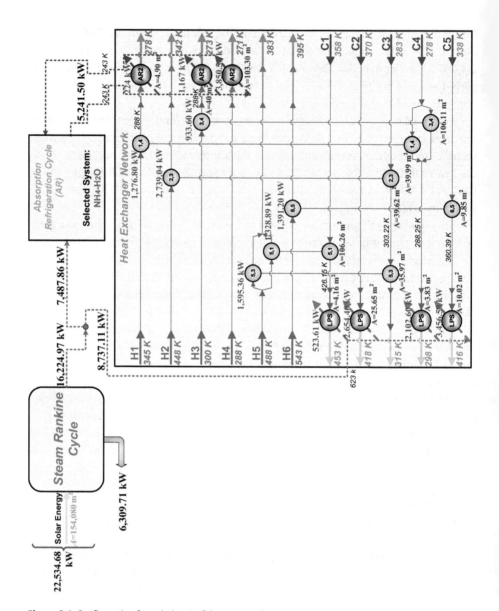

Figure 3.4: Configuration for solution A of the case study.

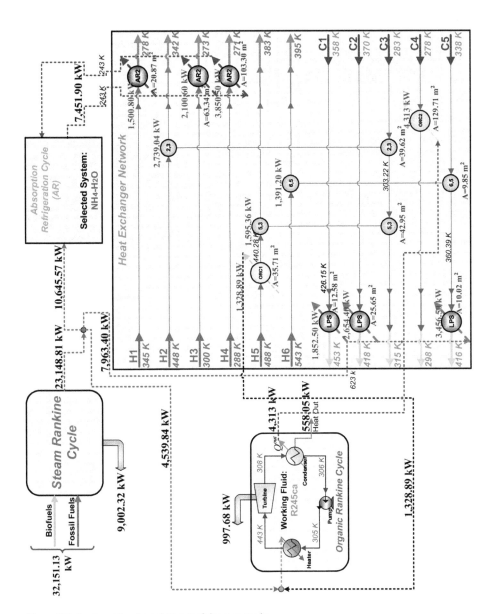

Figure 3.5: Configuration for solution B of the case study.

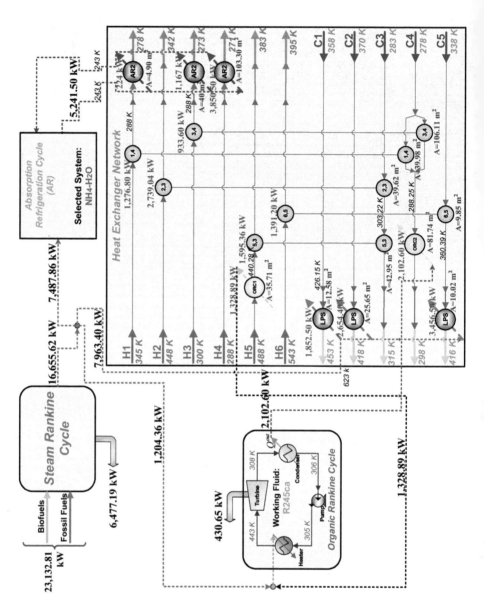

Figure 3.6: Configuration for solution C of the case study.

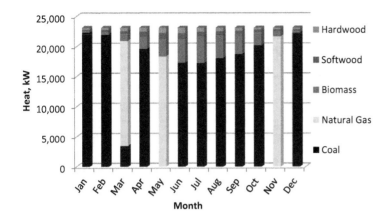

Figure 3.7: Energy sources required for each month for solution C of the case study.

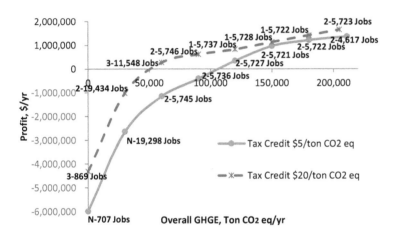

Figure 3.8: Sensitivity analysis for different values of the tax credit for the case study.

3.4 Conclusions

This chapter has presented a mathematical programming model for the integrated optimal design of power cycles, absorption refrigeration cycles, and HENs taking into
account economic, environmental, and social aspects (which are the most important
criteria in sustainable processes). The economic objective consists of the maximization of the total annual profit; the environmental objective is aimed at maximizing
the overall greenhouse gas emissions, while the social objective maximizes the number of jobs that can be created by the use of different types of primary energy sources.
The proposed methodology is followed to determine the optimal use and distribution

of sustainable energy sources (solar and biofuels) to obtain tax credits. The results are shown through Pareto curves to discuss the tradeoffs among the economic, environmental, and social objectives of sustainability.

Results highlight the benefits of considering the energy integration of the available heat in the condenser of the SRC. In this order, the residual energy available in the SRC can be reused to run the absorption refrigeration cycle (satisfying the cooling requirements), in addition to running an ORC (generating electricity) as it can also be used as a hot utility in the HEN (reducing the external hot utilities). Typically, this heat is wasted without being used for relevant applications as it is proposed by the present methodology. Additionally, another relevant point considered by this work consists in the optimal selection of the fluid to operate the ORC; in fact, the results emphasize the importance of including the optimal selection for the working fluid in this type of methodology because it can be more useful to have a set of operational temperatures for the ORC that allows the heat exchange among the HEN and the ORC to maximize the reuse of heat and reduce the consumption of primary energy sources than a more efficient fluid.

Since Pareto curves show the tradeoffs among the economic and environmental goals, the social function can be the decisive criteria to decide the best feasible solution to be implemented to improve the social and economic conditions for rural areas; however, the final decision will be taken by the decision-makers and local governments. Nevertheless, the proposed methodology represents a valuable tool for investors to visualize all the possible scenarios.

As can be seen in the case study, solution A corresponds to the best environmental configurations as primary energy only is provided by the solar collector; however, this is the most expensive energy and as a consequence it is economically unfeasible. Otherwise, points B represent the best economic solutions where the profits are considerable; however, they employ mainly fossil fuels; for this reason, the GHGE exceeds the 200,00 ton CO_2-eq/year. C points are addressed to balance both aspects, and even the number of jobs generated by the project is adequate. Finally, the proposed approach is general and it was formulated in a way that does not represent significant numerical problems in its solution.

Nomenclature

Avail	Maximum availability, kg/month
C	Cost, $
Cp	Specific heat capacity, kJ/kg K
COP	Coefficient of performance
Cu	Unitary operational cost, $/kJ
Dsh	Time conversion factor, s/h

Dt	Time conversion factor, s/month
FC	Fixed charge, $/year
FCp	Heat capacity flowrate, kJ/s K
Ga	Pow unitary gains obtained by selling power, $/kW
GHGE	Unitary greenhouse gas emissions, ton CO_{2eq}/kJ
h	Film heat transfer coefficient, kW/m^2 K
Heating power	Heating power, kJ/kg
HY	Hours of operation per year, h/year
kf	Factor used to annualize the capital costs
NJOBS	Number of generated jobs, jobs
NOK	Total number of stages
Q^{max}	Upper bound for heat exchange, kJ/s
Q^{Useful_Solar}	Usable solar radiation in the specific location, kJ/m^2 month
R	Tax credit for the reduction of GHGE, $/kJ
Sup	Unitary price for the power, $/kW
T_{IN}	Inlet temperature, K
T_{OUT}	Outlet temperature, K
VC	Unit variable charge, $/m^3 year

Greek symbols

α	Exponent for area cost of the solar collector
β	Exponent for area cost of exchangers
δ	Small number (i.e., 1×10^{-5})
μ	Efficiency factor
ΔT^{max}	Upper bound for temperature difference, K
ΔTmin	Minimum approach temperature difference, K
ε_a	Parameter of the interval between minimum GHGE and maximum GHGE for the constraint method, ton $CO_{2eq/kJ}$

Variables

Ac	Area, m^2
CaC	Capital cost for exchangers, $/year
dt	Temperature approach difference, K
ESC	Energy sources cost, $/year
FiC	Fixed cost for exchangers, $/year
NGHGE	Greenhouse gas emissions, ton CO_{2eq}/year
OC	Operational cost, $/year
Power	Power output, kW
PPCost	Power production cost, $/kW

q	Heat transfer in the heat exchanger units, kJ/s
Q	Heat exchanged between the energy subsystems, kJ/s
$Q^{External}$	Total energy supplied to the Rankine cycle, kJ/s
Q^{mps}	Available heat in the condenser of the Rankine cycle, kJ/s
SP	Selling of power, $/year
t	Internal temperature, K
TAP	Total annual profit, $/year
TCR	Tax credit reduction, $/year
y	Binary variable used to model the existence of the solar collector
z	Binary variables used to model the existence of heat exchange units

Sets

B	{b | b is a biofuel}
CPS	{j | j is a cold process stream}
F	{f | f is a fossil fuel}
G	{g | g is a working fluid for the organic Rankine cycle}
H	{h | h is a system for the absorption refrigeration system}
HPS	{i | i is a hot process stream}
ST	{k | k is a stage in the superstructure, k = 1, . . ., NOK}
T	{t | t is a period }

Subscripts and superscripts

Ar	Absorption refrigeration
ar1	Heat exchange unit where the heat excess from hot process streams is removed
ar2	Stage of the superstructure where the hot process streams are cooled
b	Biofuel
cap	Capital
cw	Cold water
f	Fossil fuel
i	Hot process stream
j	Cold process stream
k	Index for stage (1, . . ., NOK) and temperature location (1, . . ., NOK+1)
lps	Stage of the superstructure where the cold process streams exchange heat with low-pressure steam
Max	Maximum
op	Operational
orc	Organic Rankine cycle
orc1	Heat exchanger where hot process streams transfer energy to organic Rankine cycle
orc2	Heat exchanger where the organic Rankine cycle transfer energy to cold process streams
src	Steam Rankine cycle
t	Period

References

[1]   Gundersen, T., Naess, L. The synthesis of cost optimal heat exchanger network-an industrial review of the state-of-the art. *Computers and Chemical Engineering*, 1988, 12(6), 503–530.

[2]   Furman, K. C., Sahinidis, N. V. A critical review and annotated bibliography for heat exchanger network synthesis in the 20th century. *Industrial and Engineering Chemistry Research*, 2002, 41(10), 2335–2370.

[3]   Morar, M., Agachi, P. S. Review: Important contributions in development and improvement of the heat integration techniques. *Computers and Chemical Engineering*, 2010, 34(8), 1171–1179.

[4]   Linnhoff, B., Hindmarsh, E. The pinch design method for heat exchanger networks. *Chemical Engineering Science*, 1983, 38(5), 745–763.

[5]   Linnhoff, B., Ahmad, S. Cost optimum heat exchanger networks 1. Minimum energy and capital using simple-models for capital costs. *Computers and Chemical Engineering*, 1990, 14(7), 729–750.

[6]   Serna-González, M., Ponce-Ortega, J. M. Total cost target for heat exchanger networks considering simultaneously pumping power and area effects. *Applied Thermal Engineering*, 2011, 31(11–12).

[7]   Lewin, D. R., Wang, H., Shalev, O. A generalized method for HEN synthesis using stochastic optimization e I. General framework and MER optimal synthesis. *Computers and Chemical Engineering*, 1998, 22(10), 1503–1513.

[8]   Athier, G., Floquet, P., Pibouleau, L., Domenech, S. Synthesis of heat exchanger networks by simulated annealing and NLP procedures. *AIChE Journal*, 1997, 43(11), 3007–3019.

[9]   Lotfi, R., Boozarjomehry, R. B. Superstructure optimization in heat exchanger network (HEN) synthesis using modular simulators and a genetic algorithm. *Industrial and Engineering Chemistry Research*, 2010, 49(10), 4731–4737.

[10]   Ponce-Ortega, J. M., Serna-González, M., Jiménez-Gutiérrez, A. Heat exchanger networks synthesis including detailed heat exchanger design using genetic algorithms. *Industrial and Engineering Chemistry Research*, 2007, 46(25), 8767–8780.

[11]   Ponce-Ortega, J. M., Serna-González, M., Jiménez-Gutiérrez, A. Synthesis of multipass heat exchanger networks using genetic algorithms. *Computers and Chemical Engineering*, 2008, 32(10), 2320–2332.

[12]   Floudas, C. A., Ciric, A. R., Grossmann, I. E. Automatic synthesis of optimum heat exchanger network configurations. *AIChE Journal*, 1986, 32(2), 276–290.

[13]   Yee, T. F., Grossmann, I. E. Simultaneous optimization models for heat integration-II. Heat exchanger network synthesis. *Computers and Chemical Engineering*, 1990, 14(10), 1165–1184.

[14]   Ponce-Ortega, J. M., Jimenez-Gutierrez, A., Grossmann, I. E. Simultaneous retrofit and heat integration of chemical processes. *Industrial and Engineering Chemistry Research*, 2008, 47(15), 5512–5528.

[15]   Ponce-Ortega, J. M., Serna-González, M., Jimenez-Gutierrez, A. Synthesis of heat exchanger networks with optimal placement of multiple utilities. *Industrial Engineering Chemistry Research*, 2010, 49(6), 2849–2856.

[16]   Huang, K. F., Al-mutairi, E. M., Karimi, I. A. Heat exchanger network synthesis using a stagewise superstructure with non-isothermal mixing. *Chemical Engineering Science*, 2012, 73, 30–43.

[17]   Townsend, D. W., Linnhoff, B. Heat and power networks in process design. Part 1: Criteria for placement of heat engines and heat pumps in process networks. *AIChE Journal*, 1983, 29(5), 742–748.

[18]   Maréchal, F., Kalitventzeff, B. Identification of the optimal pressure levels in steam networks using integrated combined heat and power method. *Chemical Engineering Science*, 1997, 52(17), 2977–2989.

[19]   Desai, N. B., Bandyopadhyay, S. Process integration of organic Rankine cycle. *Energy*, 2009, 34(10), 1674–1686.

[20]   Papoulias, S. A., Grossmann, I. E. A structural optimization approach in process synthesis utility systems – III. Total processing systems. *Computers and Chemical Engineering*, 1983, 7(6), 723–734.

[21] Colmenares, T. R., Seider, W. D. Heat and power integration of chemical processes. *AIChE Journal*, 1987, 33(6), 898–915.

[22] Swaney, R. E. Thermal integration of processes with heat engines and heat pumps. *AIChE Journal*, 1989, 35(6), 1003–1016.

[23] Holiastos, K., Manousiouthakis, V. Minimum hot/cold/electric utility cost for heat exchange networks. *Computers and Chemical Engineering*, 2002, 26(1), 3–16.

[24] Chen, C. L., Lin, C. Y. Design of entire energy systems for chemical plants. *Industrial and Engineering Chemistry Research*, 2012, 51(30), 9980–9996.

[25] Hipólito-Valencia, B. J., Rubio-Castro, E., Ponce-Ortega, J. M., Serna-González, M., Nápoles-Rivera, F., El-Halwagi, M. M. Optimal integration of organic Rankine cycles with industrial processes. *Energy Conversion and Management*, 2013, 53(1), 285–302.

[26] Vélez, F., Segovia, J. J., Martín, M. C., Antolín, G., Chejne, F., Quijano, A. A technical, economical and market review of organic Rankine cycles for the conversion of low-grade heat for power generation. *Renewable and Sustainable Energy Reviews*, 2012, 16(6), 4175–4189.

[27] Pan, L., Wang, H., Shi, W. Performance analysis in near-critical conditions of organic Rankine cycle. *Energy*, 2012, 37, 281–286.

[28] Schuster, A., Karellas, S., Aumann, R. Efficiency optimization potential in supercritical Organic Rankine Cycles. *Energy*, 2010, 35, 1033–1039.

[29] Dongxiang, W., Ling, X., Peng, H., Liu, L., Tao, L. Efficiency and optimal performance evaluation of organic Rankine cycle for low grade waste heat power generation. *Energy*, 2013, 50, 343–352.

[30] Wang, E. H., Zhang, H. G., Fan, B. Y., Ouyang, M. G., Zhao, Y., Mu, Q. H. Study of working fluid selection of organic Rankine cycle for engine waste heat recovery. *Energy*, 2011, 36, 3406–3418.

[31] Lai, N. A., Wendland, M., Fischer, J. Working fluids for high-temperature organic Rankine cycles. *Energy*, 2011, 36, 199–211.

[32] Mago, P. J., Chamra, L. M., Somayaji, C. Performance analysis of different working fluids for use in organic Rankine cycles. *Journal of Power and Energy*, 2007, 221, 255–264.

[33] Bao, J., Zhao, L. A review of working fluid and expander selections for organic Rankine cycle. *Renewable and Sustainable Energy Reviews*, 2013, 24, 325–342.

[34] Dhole, V. R., Linnhoff, B. Total site targets for fuel co-generation, emissions, and cooling. *Computers and Chemical Engineering*, 1993, 17(Suppl), S101–S109.

[35] Raissi, K. Total site integration; Ph.D. thesis. Dept of Process Integration, UMIST, Manchester, UK, 1994.

[36] Klemeš, J., Dhole, V. R., Raissi, K., Perry, S. J., Puigjaner, L. Targeting and design methodology for reduction of fuel, power and CO_2 on total sites. *Applied Thermal Engineering*, 1997, 17(8–10).

[37] Maréchal, F., Kalitventzeff, B. Energy integration of industrial sites: Tools, methodology and application. *Applied Thermal Engineering*, 1998, 18(11), 921–933.

[38] Mavromatic, S. P., Kokossis, A. C. Conceptual optimisation of utility networks for operational variations. 1. Targets and level optimization. *Chemical Engineering Science*, 1998, 53(8), 1585–1608.

[39] Mohan, T., El-Halwagi, M. M. An algebraic targeting approach for effective utilization of biomass in combined heat and power systems through process integration. *Clean Technology and Environmental Policy*, 2007, 9(1), 13–25.

[40] Bagajewicz, M. J., Barbaro, A. F. On the use of heat pumps in total site integration. *Computers and Chemical Engineering*, 2003, 27(11), 1707–1719.

[41] Varbanov, P., Perry, S., Klemeš, J., Smith, R. Synthesis of industrial utility systems: Cost-effective de-carbonisation. *Applied Thermal Engineering*, 2005, 25(7), 985–1001.

[42] Perry, S., Klemeš, J., Bulatov, I. Integrating waste and renewable energy to reduce the carbon footprint of locally integrated energy sectors. *Energy*, 2008, 33(10), 1489–1497.

[43] Varbanov, P. S., Klemeš, J. J. Integration and management of renewables into total sites with variable supply and demand. *Computers and Chemical Engineering*, 2011, 3(9), 1815–1826.

[44] Kim, J. K., Smith, R. Cooling water system design. *Chemical Engineering Science*, 2001, 56(12), 3641–3658.
[45] Ponce-Ortega, J. M., Serna-González, M., Jiménez-Gutiérrez, A. Optimization model for re-circulating cooling water systems. *Computers and Chemical Engineering*, 2010, 34(2), 177–195.
[46] Gololo, K. V., Majozi, T. On synthesis and optimization of cooling water systems with multiple cooling towers. *Industrial and Engineering Chemistry Research*, 2011, 50(7), 3775–3787.
[47] Picón-Núñez, M., Polley, G. T., Canizales-Dávalos, L., Medina-Flores, J. M. Short cut performance method for the design of flexible cooling systems. *Energy*, 2011, 36(8), 4646–4653.
[48] Rubio-Castro, E., Serna-González, M., Ponce-Ortega, J. M., El-Halwagi, M. M. Synthesis of cooling water systems with multiple cooling towers. *Applied Thermal Engineering*, 2013, 50(1), 957–974.
[49] Ponce-Ortega, J. M., Tora, E. A., González-Campo, J. B., El-Halwagi, M. M. Integration of renewable energy with industrial absorption refrigeration systems: Systematic design and operation with technical, economic and environmental objectives. *Industrial and Engineering Chemistry Research*, 2011, 50(16), 9667–9684.
[50] Lira-Barragán, L. F., Ponce-Ortega, J. M., Serna-González, M., El-Halwagi, M. M. Synthesis of integrated absorption refrigeration systems involving economic and environmental objectives and quantifying social benefits. *Applied Thermal Engineering*, 2013, 52(2), 402–419.
[51] Yue, D., Kim, M. A., You, F. Design of sustainable product systems and supply chains with life cycle optimization based on functional unit: General modeling framework, MINLP algorithms and case study on hydrocarbon biofuels. *ACS Sustainable Chemistry and Engineering*, 2013, 1(8), 1003–1014.
[52] Murillo-Alvarado, E., Ponce-Ortega, J. M., Serna-González, M., Castro-Montoya, A. J., El-Halwagi, M. M. Optimization of pathways for biorefineries involving the selection of feedstocks, products and processing steps. *Industrial and Engineering Chemistry Research*, 2013, 52(14), 5177–5190.
[53] Yue, D., You, F. Sustainable scheduling of batch processes under economic and environmental criteria with MINLP models and algorithms. *Computers and Chemical Engineering*, 2013, 54, 44–59.
[54] Carvalho, A., Matos, H. A., Gani, R. SustainPro – A tool for systematic process analysis, generation and evaluation of sustainable design alternatives. *Computers and Chemical Engineering*, 2013, 50, 8–27.
[55] Čuček, L., Varbanov, P. S., Klemeš, J. J., Kravanja, Z. Total footprints-based multi-criteria optimisation of regional biomass energy supply chains. *Energy*, 2012, 44(1), 135–145.
[56] Grossmann, I. E., Guillén-Gosálbez, G. Scope for the application of mathematical programming techniques in the synthesis and planning of sustainable processes. *Computers and Chemical Engineering*, 2010, 34(9), 1365–1376.
[57] Guillén-Gosálbez, G., Caballero, J., Jiménez, L. Application of life cycle assessment to the structural optimization of process flowsheets. *Industrial and Engineering Chemistry Research*, 2008, 47(3), 777–789.
[58] Azapagic, A., Perdan, S. Indicator of sustainable development for industry: A general framework. *Process Safety and Environmental Protection*, 2000, 78(4), 243–261.
[59] Sikdar, S. K. Sustainable development and sustainability metrics. *AIChE Journal*, 2003, 49(8), 1928–1932.
[60] Lozano, R. Envisioning sustainability three-dimensionally. *Journal of Cleaner Production*, 2008, 16(17), 1838–1846.
[61] Čuček, L., Klemeš, J. J., Kravanja, Z. A review of footprint analysis for monitoring impacts on sustainability. *Journal of Cleaner Production*, 2012, 34, 9–20.
[62] Bamufleh, H. S., Ponce-Ortega, J. M., El-Halwagi, M. M. Multi-objective optimization of process cogeneration systems with economic, environmental, and social tradeoffs. *Clean Technologies and Environmental Policy*, 2013, 15(1), 185–197.

[63] You, F., Tao, L., Graziano, D. J., Snyder, S. W. Optimal design of sustainable cellulosic biofuel supply chains: Multiobjective optimization coupled with life cycle assessment and input-output analysis. *AIChE Journal*, 2012, 58(4), 1157–1180.

[64] Raman, R., Grossmann, I. E. Modeling and computational techniques for logic based integer programming. *Computers and Chemical Engineering*, 1994, 18(7), 563–578.

[65] Ponce-Ortega, J. M., Jimenez-Gutierrez, A., Grossmann, I. E. Optimal synthesis of heat exchanger networks involving isothermal process streams. *Computers and Chemical Engineering*, 2008, 32(8), 1918–1942.

[66] Chen, J. J. J. Letter to the editor: Comments on improvement on a replacement for the logarithmic mean. *Chemical Engineering Science*, 1987, 42(10), 2488–2489.

[67] GREET Version 2.7, Copyright _ 2007 UChicago Argonne, LLC. 2007.

[68] Phalan, B. The social and environmental impacts of biofuels in Asia: An overview. *Applied Energy*, 2009, 86(1), S21–9.

[69] Buchholz, T., Luzadis, V. A., Volk, T. A. Sustainability criteria for bioenergy systems: Results from an expert survey. *Journal of Cleaner Production*, 2009, 17(1).

[70] Lehtonen, M. Social sustainability of the Brazilian bioethanol: Power relations in a centre-periphery perspective. *Biomass and Bioenergy*, 2011, 35(6), 2425–2434.

[71] Diaz-Chavez, R. A. Assessing biofuels: Aiming for sustainable development or complying with the market. *Energy Policy*, 2011, 39(10), 5763–5769.

[72] Miller, R. E., Blair, P. D. *Input-Output Analysis: Foundations and Extensions*. Cambridge, UK: Cambridge University Press, 2009.

[73] Goldberg, M., Sinclair, K., Milligan, M. Job and economic development impact (JEDI) model: A user-friendly tool to calculate economic impacts from wind projects. In 2004 Global Windpower Conference. 2004, NREL/CP-500-35953.

[74] Diwekar, U. M. *Introduction to Applied Optimization and Modeling*. Dordrecht, The Netherlands: Kluwer Academic Publishers, 2003.

[75] Brooke, A., Kendrick, D., Meeruas, A., Raman, R. *GAMS-Language Guide*. Washington, D.C.: GAMS Development Corporation, 2012.

Oscar Daniel Lara-Montaño, Manuel Toledano-Ayala*,
Claudia Gutiérrez-Antonio, Fernando Israel Gómez-Castro,
Elena Niculina Dragoi, and Salvador Hernández

Chapter 4
Metaheuristics for the optimization
of chemical processes

Abstract: Metaheuristic optimization algorithms are effective for exploring complex multidimensional search spaces and avoiding convergence on suboptimal local solutions. They are used across a range of chemical engineering applications, from process design and reactor configuration to environmental engineering and energy management. These algorithms are used for adjusting operational parameters and identifying solutions that might not be detected by conventional methods. This chapter reviews the use of metaheuristic algorithms in these areas, referencing recent published works to illustrate their practical benefits. The discussion aims to provide a concise overview of how these algorithms enhance chemical engineering practices, contributing to increased efficiency, adaptability, and sustainability in solutions.

Keywords: metaheuristic optimization, process optimization, nonlinear optimization

Acknowledgments: The authors express their sincere gratitude to the Consejo Nacional de Humanidades, Ciencias y Tecnologías (CONAHCYT), for the support provided through the postdoctoral fellowship awarded to Oscar Daniel Lara-Montaño. This support has been essential for the development of the present research.

*Corresponding author: **Manuel Toledano-Ayala**, Universidad Autónoma de Querétaro, Facultad de Ingeniería, Cerro de las Campanas S/N Col. Las Campanas, Querétaro 76010, Mexico,
e-mail: toledano@uaq.mx
Oscar Daniel Lara-Montaño, Claudia Gutiérrez-Antonio, Universidad Autónoma de Querétaro, Facultad de Ingeniería, Cerro de las Campanas S/N Col. Las Campanas, Querétaro 76010, Mexico
Fernando Israel Gómez-Castro, Salvador Hernández, Universidad de Guanajuato, Campus Guanajuato, División de Ciencias Naturales y Exactas, Departamento de Ingeniería Química, Noria Alta S/N Col. Noria Alta, Guanajuato 36050, Mexico
Elena Niculina Dragoi, Gheorghe Asachi Technical University of Iasi, Cristofor Simionescu Faculty of Faculty of Automatic Control and Computer Engineering, Str. Prof. Dr. Doc. Dimitrie Mangeron, nr. 27, Iași 700050, Romania

https://doi.org/10.1515/9783111383439-004

4.1 Metaheuristic optimization of chemical processes

The field of chemical engineering is inherently complex, involving processes and systems that are primarily nonlinear, dependent on multiple design variables, and frequently subject to multiple constraints, in most cases derived from physical limitations. For instance, time cannot be negative, the diameter of a reactor is usually limited to meters due to space and construction restrictions, and pressure and temperature must remain within safe operational limits; moreover, flow rates are often constrained by pump capacities, and material properties such as viscosity and density can impose restrictions on process design. Optimization plays a crucial role in this field; applying optimization strategies can reduce costs or environmental impact, maximize profit or social impact, or even optimize according to more than one objective. The traditional optimization algorithms are suitable for some chemical engineering cases; nevertheless, sometimes, these approaches may not fully succeed because of the complex search space inherent in the systems under study (determined by the interaction between the process parameters and conditions interactions) and the dynamic nature of their behavior.

In this context, alternative strategies that aim to solve or reduce the inherent issues of classical optimization strategies are sought after. Metaheuristic algorithms represent such alternatives as their behavior (in determining the optimal or near-optimal solutions) is based on exploring the search space through a series of approaches, which are different from those used in classical mathematical programming. Metaheuristic optimization algorithms are a class of optimization strategies that take inspiration from natural phenomena and problem-solving behaviors observed in animals or humans. These methods have become widely used in solving engineering problems due to their design for versatility and robustness in addressing complex challenges. In contrast to deterministic optimization algorithms, metaheuristic algorithms often do not require an explicit model or calculation of derivatives of the model equations. Although they do not guarantee optimal solutions, they provide near-optimal solutions in reasonable time [1].

The interest in metaheuristic algorithms lies in their capacity to navigate non-convex search spaces, avoid local optima, and adapt to changing conditions. Also, they are highly used with black-box models from process simulators, where the models employed to perform rigorous calculations are not explicitly available. This makes them particularly suitable for chemical engineering problems, where complex search space and uncertainties are common. This chapter explores the field of metaheuristic algorithms, focusing on principles, methodologies, and their application in chemical engineering-related topics. We will examine some of the most known metaheuristic algorithms and how they have been used to address various chemical engineering problems.

While this chapter does not focus extensively on specific metaheuristic software, it is important noting that a variety of programming environments are commonly used to implement these algorithms. Tools such as MATLAB, Python, and MS Excel are frequently employed due to their flexibility and the extensive libraries available that support metaheuristic programming. Python, in particular, offers several free libraries such as DEAP (Distributed Evolutionary Algorithms in Python) and PySwarms, which are well-suited for researchers and practitioners looking for freely available resources to implement metaheuristic algorithms. These tools facilitate the experimentation and development of metaheuristic solutions across various applications.

4.1.1 Optimization in chemical engineering: an overview

Optimization is a fundamental step in chemical process design, analysis, and operation. The primary objective of optimization is to maximize or minimize a specific criterion. In chemical engineering, optimization is applied to various areas, such as process design and control [2–4], product formulation [5, 6], and resource management [7].

The goal of optimization in this field is to improve process outcomes, increase efficiency, minimize costs, and satisfy environmental regulations. Chemical engineering optimization faces challenges due to nonlinearities, interdependent variables, continuous and discrete design variables, and problem-related constraints. Additionally, some studies consider parameters' uncertainties in real-world situations due to operation or environmental factors. Traditional optimization methods, such as linear, nonlinear, or dynamic programming, have been widely used to optimize chemical engineering systems. Although sometimes adequate, these methods may not always be pertinent.

Optimization approaches can be categorized into deterministic and stochastic methods, each suited to different types of problems based on the certainty and complexity of the process parameters. Deterministic methods assume that all parameters are precisely known and are particularly effective for convex problems, where they can guarantee the identification of a global optimum. For non-convex problems, however, these methods may only ensure local optima, but still provide consistent results for the same input under unchanged conditions. Stochastic methods, on the other hand, utilize probability distributions to address uncertainty in the system, making them capable of handling a wide range of problems, including those with complex and unpredictable behaviors. These methods do not guarantee an optimal solution and can produce different outcomes with each run due to their inherent randomness. They are often favored for large-scale and complex problems where traditional deterministic approaches may falter due to the complex nature of the optimization space.

Metaheuristic algorithms, a subset of stochastic methods, while requiring some basic information such as objective functions, constraints, decision variables, and bounds similar to deterministic methods, excel in environments where complete and precise model understanding is lacking. This is because metaheuristics are designed to explore and exploit search spaces based on performance feedback rather than requiring exact mathematical formulations. These algorithms are robust and adaptable, making them suitable for both well-defined problems and black-box scenarios, where internal system mechanisms are not clear. By effectively navigating through complex, nonlinear spaces, metaheuristics identify sub-spaces that are likely to contain high-quality solutions. The performance of these algorithms heavily depends on the selection and tuning of parameters that balance the exploration of new possibilities and exploitation of known good solutions. Although a thorough understanding of the system's behavior enhances the selection process, metaheuristics can still operate effectively by adjusting their strategies based on the outcomes they generate, which makes them invaluable for tackling problems where traditional models do not suffice. Table 4.1 compiles characteristics of metaheuristic and deterministic algorithms to better visualize the particularities they exhibit [8–10].

Table 4.1: Comparison of metaheuristic and deterministic optimization methods.

Feature	Metaheuristic methods	Deterministic methods
Algorithmic basis	Employ nature-inspired algorithms that do not require derivative information to explore solution spaces.	Utilize algorithms that require information about derivatives of the function, or assume specific properties about the function's landscape.
Solution precision	Typically provide near-optimal solutions with variability in outcomes due to stochastic processes, making them suitable for complex or noisy environments.	Aim to achieve the exact global optimum under conditions of continuity, differentiability and convexity.
Gradient dependency	Operate without the need for gradient information, making them applicable to problems where such information is difficult to obtain or the function is non-differentiable.	Dependent on gradient information to navigate the search space, which restricts their application to problems where gradients can be calculated reliably.
Sensitivity to noise	Demonstrates robustness against noisy function evaluations due to inherent randomness.	Sensitive to inaccuracies in function evaluations, which can significantly affect the search trajectory and potentially lead to suboptimal solutions.
Convergence characteristics	Stochastic convergence mechanisms can lead to fast discovery of good-enough solutions but may require parameter tuning to balance exploration and exploitation effectively.	Deterministic convergence ensures that a local or global optimum is reached, depending on the algorithm and the structure of the problem, and often requires additional iterations for more complex cases.

Table 4.1 (continued)

Feature	Metaheuristic methods	Deterministic methods
Adaptability and flexibility	Exhibits high adaptability to various types of optimization problems, particularly in undefined or dynamic landscapes.	Generally less flexible, with effectiveness closely tied to the assumptions about the problem's mathematical properties and the landscape's characteristics.
Parameter tuning	Sensitive to parameter settings, often requiring extensive experimentation or adaptive mechanisms to optimize performance across different problems.	Typically involves fewer parameters, but the parameters that are used (like step size or tolerance) can significantly influence the efficiency and success of the search.
Handling of nonlinearity and discontinuities	Adequated for nonlinear, discontinuous, or non-differentiable objective functions due to their non-reliance on gradient calculations.	Performance may be limited by nonlinearity and discontinuities, as traditional methods rely on smoothness and continuity for effective search.
Exploration versus exploitation	Designed with mechanisms to switch between exploring new regions and exploiting known promising regions, though the effectiveness of this balance is often dependent on the specific metaheuristic used.	Primarily focus on exploiting the current region of the search space methodically, which can sometimes neglect potentially better regions elsewhere.
Computational cost considerations	Preferable when the cost of function evaluations is low, as they can perform a large number of evaluations quickly to explore the search space extensively.	More suitable when function evaluations are costly.

After detailing the principles and general applications of metaheuristic algorithms, it is important to highlight recent advancements that further enhance these strategies. Hybrid metaheuristic algorithms represent a significant breakthrough in optimization methods, blending elements from various established algorithms to overcome their individual shortcomings. These hybrids are especially adept at navigating the complex, multidimensional search spaces typical in chemical process optimization. By combining local search techniques from one algorithm with the global search strengths of another, these hybrids not only stabilize the optimization process but also accelerate convergence. This fusion results in superior solution quality and greater operational efficiency, which are crucial for effectively addressing the multifaceted challenges encountered in large-scale industrial chemical processes. Hybrid metaheuristic algorithms can be categorized into high-level and low-level combinations. High-level combinations, also known as "weak coupling," keep the original algorithms separate while cooperating through a well-defined interface. In contrast, low-level combinations, or "strong coupling," involve the integration of individual components or functions from each algorithm, making them more interdependent [11].

4.1.2 Metaheuristic optimization: concepts and relevance

Metaheuristic optimization algorithms are relevant in chemical engineering. These algorithms, inspired by diverse natural phenomena such as evolution, animal behavior, and physical processes, excel at managing complex, nonlinear optimization challenges. They effectively search and identify sub-spaces likely containing near-optimal solutions, making them valuable for optimizing chemical processes.

The primary strengths of metaheuristic optimization algorithms include their adaptability to various problems, their effectiveness in finding near-optimal solutions, and their ability to avoid local optimal locations. These characteristics are crucial in a field where the specific optimization problems of problems can vary greatly. However, the performance of these algorithms can differ depending on the problem at hand. Consequently, selecting the right algorithm for a specific optimization challenge is critical, as no single algorithm can efficiently solve all optimization problems universally [12].

Choosing the right metaheuristic algorithm for a specific application is crucial and often requires a consideration of the problem's characteristics, such as the complexity of the solution space and the nature of the constraints involved. It is important to note that more than one metaheuristic algorithm, such as genetic algorithms (GAs) and particle swarm optimization (PSO), can be effectively applied to solve the same application. The selection should be based on the algorithm's ability to handle the specific dimensions and constraints of the problem. Therefore, engineers and researchers are encouraged to evaluate multiple algorithms to determine which provides the most effective solution.

This chapter explores the principles and methodologies of various metaheuristic algorithms. It will cover GAs, which simulate natural selection and genetic inheritance; as well as PSO, inspired by the social behavior of birds and fish. In addition, ant colony optimization (ACO) is included, which mimics the path-finding behavior of ants. Each algorithm employs a unique method to explore the search space and converge on high-quality solutions.

4.2 Theoretical foundations of metaheuristic optimization

Glover coined the term metaheuristic in 1986 [13]. It is "a high-level problem-independent algorithmic framework that provides a set of guidelines or strategies to develop heuristic optimization algorithms" [14]. Unfortunately, in the computer science domain, the terms "heuristics" and "metaheuristics" are used interchangeably, which can lead to confusion as they refer to distinct algorithms [15]. Heuristics are problem-dependent (tailored to take advantage of the problem structure), and are usually based on the framework defined by the metaheuristic. Metaheuristic algorithms are broadly defined and are adaptable to a range of real-world optimization problems, both simple and complex. These algorithms are capable of handling nonlinear objectives and constraints, which differentiates them from some traditional optimization methods such as those used in Mixed Integer Nonlinear Programming (MINLP). It is relevant to note, however, that while MINLP techniques can manage nonlinearities associated with the continuous variables, metaheuristics may allow for more flexible problem formulations.

The effectiveness of metaheuristics can vary depending on the specific characteristics of each problem. Influential factors include the dimensionality of the problem, the presence of multiple objectives, and the complexity of the solution space. To optimize the performance of metaheuristic algorithms, careful consideration of how problems are formulated is essential. Effective problem formulation involves defining objective functions and constraints that capture the essence of the problem without overly restricting the algorithm's exploration capabilities. Furthermore, parameter tuning is crucial in optimizing metaheuristic performance. Selecting and adjusting parameters based on feedback can significantly influence the effectiveness and efficiency of the solution process.

4.2.1 Principles of metaheuristics: exploration and exploitation

Every metaheuristic optimization algorithm, despite having unique parameters, structures, and operators, fundamentally performs two main processes during the optimization process: exploration and exploitation. Exploration represents the ability of the optimizers to discover diverse solutions spread in different regions [16], and usually involves generating a diverse set of candidate solutions across the entire search space. It helps avoid getting trapped in local optima and increases the chances of finding the near-optimal solution. It is essential in the first stages of the optimization process as it has a significant impact on identifying promising regions. On the other hand, exploitation concentrates on intensifying the search within a specific sub-space (usually identified through the exploration process as likely containing the best solution) [17]. It is of higher importance as the algorithm converges to the optimal solution.

Due to their stochastic nature, a crucial element in the structure of a metaheuristic algorithm is the inclusion of random variables and the use of distribution functions. These introduce randomness in the search space exploration, allow escaping local optima, and enable local search. Some strategies for generating random numbers include creating uniformly distributed numbers, Gaussian distribution,

Monte Carlo simulations, Brownian random walks, and Lévy flights [18]. Since the selection of the distribution can significantly impact the search performance of the metaheuristics [16], a careful selection of the best-suited strategy must be performed to ensure that the algorithm is run in optimal configuration.

The exploration-exploitation balance (EEB) represents a fundamental concept of metaheuristics as it is a crucial aspect influencing performance. Too much exploration can lead to slow convergence, and excessive exploitation can result in premature convergence to suboptimal solutions [16]. Therefore, several strategies can be used to control the EEB: i) incrementally increasing the exploitation rate as the search progresses; ii) using probabilities to switch between the explorative and exploitative phases; iii) using fitness-based approaches to exploit the best-so-far solutions; iv) using different metrics (e.g., diversity) to evaluate the similarity between solutions and to direct the search; and v) using parallelization with multiple instances and different exploration-exploitation rates. Balancing these two processes is essential for the algorithm's success. Too much exploration can lead to slow convergence, and excessive exploitation can result in premature convergence to suboptimal solutions [19]. Achieving a balance between these two aspects is crucial for the effectiveness of the optimization algorithm.

Due to the stochastic nature of metaheuristic algorithms, which employ random numbers in their processes, different runs of the same algorithm may yield different solutions. This variability underscores the importance of executing a metaheuristic program multiple times to reliably approach or achieve the optimal solution. It is recommended that these algorithms be run several times to ensure that the results are not only near-optimal but also consistent, enhancing the reliability of the optimization process.

4.2.2 Classification of metaheuristic algorithms

According to Rajwar et al. [20], it is estimated that in the scientific literature, more than 540 metaheuristic algorithms have been published and developed to date, with 350 appearing in the last decade alone. Due to the abundance of metaheuristic optimization methods, numerous authors have introduced various classification schemes to categorize them. Some of the categories employed to classify the metaheuristic optimization algorithms are listed in Table 4.2. Some of these categories are employed by more than one author that proposes a classification methodology, such as the source of inspiration that is employed by Fister et al. [22] and Fausto et al. [21]. Although all the categories are focused on metaheuristic optimization algorithms, the hybridization level is exclusive to hybrid metaheuristic algorithms.

Among the existing classifications, the most used in the literature are population-based versus single-point search and the classification according to the source of inspiration. Most of the available metaheuristic algorithms are inspired by nature. The

Table 4.2: Various classifications for metaheuristic algorithms proposed in the literature.

Category	Description
Source of inspiration [21, 22]	– Nature-inspired: Methods mimicking biological or physical phenomena (e.g., particle swarm optimization and ant colony optimization) – Human-inspired: Methods based on human behavior and activities (e.g., harmony search)
Search structure [23]	– Single-point search: Uses a single solution in each iteration – Population-based search: Employs a population of solutions to explore the search space
Memory usage [23]	– Memory-based: Uses a history of previous solutions to guide the search – Memoryless: Does not store information from previous solutions
Objective function adaptation [23]	– Static: The objective function remains constant during the search – Dynamic: The objective function can change in response to specific conditions
Number of parameters [20]	– Parameter-free: No specific primary parameters – Mono-parameter: One primary parameter – Bi-parameter: Two primary parameter – Tri-parameter: Three primary parameters – Tetra-parameter: Four primary parameters – Penta-parameter: Five primary parameters Primary parameters are characteristics of each optimization algorithm. Common parameters such as population size and number of iterations are named secondary parameters.
Hybridization level [20]	Particular for hybrid algorithms – High level: Algorithms that cooperate throw well-defined interfaces while maintaining their individual identities – Low level: Deep integration where components of the algorithms are exchanged and interdependent

key features distinguishing nature-inspired algorithms from traditional algorithms include scalability, tolerance to missing data, interpretability, flexibility, optimization capability, adaptability, accessibility, and the ability to solve highly nonlinear and complex problems [20].

Nevertheless, the taxonomy and classification of metaheuristics is a complex task. A detailed discussion regarding this issue and a multi-level classification approach considering components, overall structure, and the problem being solved is proposed in [24].

4.2.3 Mathematical formulation of metaheuristic techniques

Metaheuristic techniques aim to identify solutions that approximate optimality in complex optimization scenarios, where traditional methods might not succeed. These techniques are often inspired by natural phenomena and are characterized by their ability to explore the solution space efficiently. The mathematical formulation of a metaheuristic technique typically involves the following components:

- Solution representation: The first step in formulating a metaheuristic is to define a representation for the solutions of the optimization problem. This could be a vector of decision variables, a permutation, or any other structure that contains the necessary information to describe a solution.
- Objective function: The objective function, denoted as $f(x)$, quantifies the quality of a solution x. The goal of the optimization is to minimize or maximize this function, depending on the problem.
- Initialization: The metaheuristic starts with an initial set of solutions, which can be generated randomly or using a heuristic. This initial population is denoted as $P_0 = \{x_1, x_2, \ldots, x_n\}$, where n is the number of decision variables.
- Solution update: The core of the metaheuristic involves updating the solutions based on specific rules or operators. Each optimization algorithm has its operators to perform this step, some of which are discussed later in this chapter.
- Termination criteria: The algorithm updates the candidate solutions until a termination criterion is met. This could be a maximum number of iterations, a convergence threshold or any other condition that indicates the algorithm should stop.

Different metaheuristic techniques vary in how they implement the solution update step and how they balance exploration and exploitation. The choice of metaheuristic and its parameters depends on the specific characteristics of the optimization problem.

4.2.4 Handling constraints in metaheuristic algorithms

Effectively managing constraints is crucial for generating feasible solutions that meet the requirements of an optimization problem. Different constraint-handling techniques are used in metaheuristics to address both inequality and equality constraints. This section discusses three of the most notable methods; these are the ε-constraint method, stochastic ranking, and the use of penalty functions [25].

The ε-constraint method is mainly used to handle inequality constraints in optimization problems. This method transforms the inequality constraints $g(x) \leq 0$ into a series of threshold conditions $g(x) \leq \varepsilon$, where ε is a small positive value. By converting the constraint into a more relaxed form, the algorithm can explore feasible regions of the search space that satisfy these thresholds. This approach is effective in dealing with multiple inequality constraints by breaking them down into manageable subproblems. For example, in a chemical process optimization problem, where certain operational parameters must remain below specified limits, the ε-constraint method ensures that the solutions comply with these limits by setting appropriate threshold values.

Stochastic ranking is a versatile method used for handling both inequality and equality constraints in metaheuristic algorithms. This method combines penalty functions with a probabilistic selection process to balance the objective function and the constraints. Solutions are evaluated based on a combination of these objective functions' values and the extent of their violations. With probability P_f, the ranking focuses more on the objective function than the constraints, allowing for a balanced exploration in the search process.

Stochastic ranking handles equality constraints by treating violations as a form of inequality constraint. For an equality constraint $h_j(x) = 0$, it can be managed considering $|h_j(x)| \leq \varepsilon$, where ε is a small tolerance value. This approach ensures that the solution remains close enough to the optimal solution with an acceptable deviation.

The penalty function approach is another common method for handling constraints in metaheuristic algorithms. This approach modifies the objective function to penalize infeasible solutions, thereby steering the search toward the feasible region. The modified objective function $f_{mod}(x)$ can be represented as follows:

$$f_{mod}(x) = f(x) + \sum_{i=i}^{m} p_i g_i(x) \qquad (4.1)$$

where $f(x)$ is the original objective function, $g_i(x)$ represents the ith constraint, p_i is the penalty value associates to the specific constraint, and m is the total number of constraints. The penalty factors are typically chosen based on the specific problem characteristics and may undergo dynamic adjustments throughout the search process.

Both the ε-constraint method and stochastic ranking are widely used in various fields of optimization. They help ensure that solutions are feasible and comply with necessary constraints. However, implementing these methods effectively requires careful consideration of the problem's specific characteristics and constraints.

Choosing the right constraint-handling technique is essential in practical applications. The ε-constraint method is straightforward and effective for inequality constraints but may need adaptation for equality constraints. Stochastic ranking provides a flexible approach suitable for both types of constraints but requires careful tuning of the probability parameter to achieve the desired balance between objective optimization and constraint satisfaction. The penalty function approach is versatile and can be applied to both equality and inequality constraints, but selecting appropriate penalty factors is crucial to avoid overly penalizing feasible solutions or insufficiently penalizing infeasible ones. Handling constraints is a critical aspect of metaheuristic algorithms. Properly managing these constraints ensures the feasibility and quality of the solutions, contributing to the overall effectiveness of the optimization process.

4.2.5 Issues and challenges in metaheuristic algorithms

Metaheuristic algorithms are widely used tools for addressing complex optimization problems across various fields, but they face a range of challenges that necessitate ongoing improvement and research. One major issue is scalability. As optimization problems expand in size, particularly with an increase in decision variables, the computational demands on metaheuristic algorithms intensify. These algorithms are inherently iterative and most of them are based on populations, which may increase complexity and extend computation times. The need for more iterations, which often correlates directly with the number of decision variables, increases the computational complexity and can make efficient solving of large-scale problems daunting [26].

Real-time processing presents another significant challenge. Many applications, particularly in dynamic environments such as traffic management or online transaction systems, demand real-time or near-real-time responses [27]. Metaheuristic algorithms in such scenarios need to be meticulously optimized to enhance processing speed without sacrificing accuracy. For example, in chemical engineering, effectively integrating these algorithms with real-time systems is crucial for optimizing processes like reaction kinetics, or the dynamic adjustment of process conditions based on fluctuating inputs and environmental conditions.

Parameter tuning is a considerable hurdle in deploying metaheuristic algorithms effectively. The overall performance of these algorithms heavily relies on precise parameter settings. Finding the right parameters often involves extensive empirical testing and fine-tuning, which can be resource-intensive, both in terms of time and computational power [28]. For instance, in optimizing network traffic or routing problems, the balance between exploration (searching new areas of the solution space) and exploitation (refining known good solutions) needs careful adjustment to prevent the algorithm from stagnating at local optima or wasting time in exploring unpromising regions.

Maintaining diversity within the population of solutions and avoiding premature convergence is fundamental for the success of metaheuristic algorithms. These challenges are intricately linked to the balance between exploration and exploitation. Adequate exploration ensures that the algorithm searches broadly across the solution space, which is crucial for identifying globally optimal solutions in complex landscapes. On the other hand, focused exploitation helps to intensively search around promising areas to refine solutions close to the optimal value. Achieving an optimal balance between these two aspects is key to the effectiveness of metaheuristics [29]. For example, in scheduling problems, where the task is to optimize the allocation of resources over time, maintaining diversity can help discover more efficient schedules that could be missed by a less exploratory approach.

Another critical challenge in the application of metaheuristic algorithms is the definition of an appropriate stopping criterion. A poorly defined stopping criterion can result in suboptimal performance, either by halting the algorithm too early before a good solution is found or by allowing it to run unnecessarily long, wasting computational resources. Common stopping criteria include setting a maximum number of iterations, establishing a minimum improvement threshold between iterations, and monitoring the convergence of the population in population-based algorithms. An effective stopping criterion is essential for balancing the efficiency of the algorithm with the quality of the solutions obtained [30].

Despite their widespread use in solving practical problems, the theoretical underpinnings of metaheuristic algorithms are not fully established, and more theoretical research is required. This research would help in gaining a deeper understanding of how these algorithms function and their limitations, which in turn would advance their design and enhance their robustness when applied to practical problems [29].

4.3 Overview of metaheuristic algorithms

This section briefly describes some of the most used metaheuristic algorithms employed in chemical engineering-related works. The selection of the included metaheuristic algorithms was done by conducting a search in SCOPUS using the phrase "metaheuristics AND chemical engineering." By analyzing the search results, the most prominent metaheuristic algorithms utilized in the field of chemical engineering were identified. Additionally, to ensure a comprehensive understanding, the quantities of keywords that referred to the same algorithm under different terminologies, such as "genetic algorithm" and "genetic algorithms" or "particle swarm optimization" and "particle swarm," were aggregated. This approach allowed for a clearer picture of the prevalence of each algorithm. Figure 4.1 shows the algorithms so found with the frequency they present in chemical engineering-related documents. GA refers to genetic algorithms, PSO is particle swarm optimization, SA is simulated annealing, ACO is ant colony optimization, DE is differential evolution, TS is tabu search, and FA is firefly algorithm.

The following discussion highlights these key algorithms, elaborating on their characteristics and the mechanisms employed in the iterative solution update process. It is important to note that all the algorithms discussed have multiple variants documented in the literature, reflecting their adaptability and widespread application in optimizing complex chemical engineering problems.

Also, an analysis using SCOPUS was conducted to visualize the tendency of the application of metaheuristic algorithms in chemical engineering-related areas. The search was performed using the following query: (metaheuristic* AND ("chemical engineering" OR "process engineering" OR "process optimization" OR "chemical process" OR "industrial process" OR "process systems" OR "chemical production" OR "process control")). This query aimed to capture a comprehensive range of topics within chemical engineering that employ metaheuristic techniques. The results were then ana-

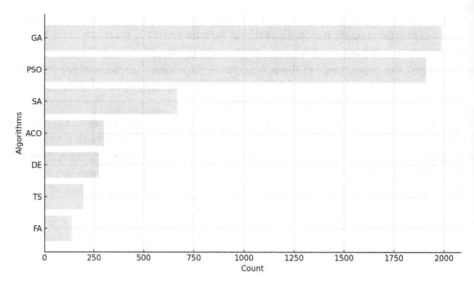

Figure 4.1: Frequency of metaheuristic algorithms in chemical engineering work's keywords.

lyzed to determine the annual number of publications over the past two decades (2003–2023). The data reveal a clear increasing trend in the number of articles, indicating a growing interest and an increment in the research activity related to the application of metaheuristics to solve problems in chemical engineering. This is graphically shown in Figure 4.2.

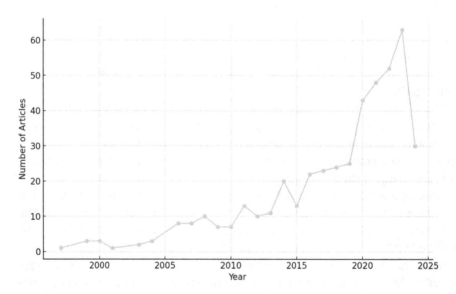

Figure 4.2: Trend of articles on metaheuristics in chemical engineering (2003–2023).

4.3.1 Genetic algorithms

GAs are a class of metaheuristic algorithms inspired by natural selection and genetics principles. Holland proposed the first version of this algorithm in 1975 [31, 32]. In GAs, a population of individuals, each representing a potential solution to a given problem, evolves over generations toward better solutions. Each individual, or chromosome, is typically encoded as a binary string, but real coding is also common in GA programs, and a fitness function $f(x)$, which measures the quality of a solution.

The process of evolution in GAs involves three main operators: selection, crossover, and mutation:

- Selection: This operator selects individuals from the current population to form a mating pool for producing the next generation. Several strategies are commonly used for selection [33]:
  - Roulette wheel selection: Each individual's selection probability is proportional to its fitness relative to the total population fitness.
  - Tournament selection: A subset of individuals is chosen randomly, and the individual with the highest fitness in this subset is selected.
  - Rank-based selection: Individuals are ranked based on their fitness, and their rank determines the selection probability.
- Crossover: Also known as recombination, this operator combines the genetic information of two-parent individuals to produce offspring. The crossover probability P_c determines the likelihood of this operation being applied to a pair of individuals. Several crossover strategies are used in GAs:
  - Single-point crossover: A random crossover point is chosen, and the genetic material is exchanged between the parents beyond that point.
  - Multipoint crossover: Multiple crossover points are chosen, and the genetic material is exchanged at these points.
  - Uniform crossover: Each gene is independently exchanged between parents with a certain probability.
- Mutation: This operator introduces random changes to the chromosomes, which helps maintain genetic diversity in the population and prevents premature convergence. Various mutation strategies include:
  - Bit-flip mutation: Each bit in the chromosome has a probability P_m of being flipped (for binary-coded GAs).
  - Swap mutation: Two genes are randomly chosen, and their values are swapped.
  - Inversion mutation: A subset of genes is chosen, and their order is reversed.

The algorithm iterates through these steps until a termination condition is met, such as reaching a maximum number of generations or achieving a desired fitness level. The final solution is the fittest individual in the last generation. Figure 4.3 shows the working mode of the standard GAs.

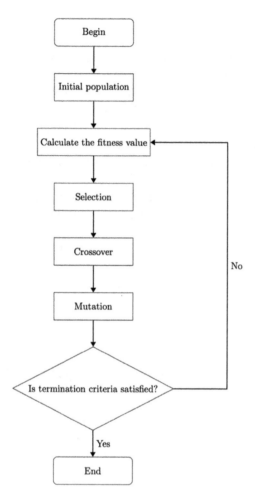

Figure 4.3: Flowchart of a standard genetic algorithm.

4.3.2 Particle swarm optimization

PSO, proposed by Kennedy and Eberhart in 1995 [34], is a technique used for global optimization. It is inspired by the social behavior of animals, such as birds flocking and fish schooling. The algorithm models the movement and interactions of individuals within a swarm, where each individual, referred to as a particle, represents a candidate solution to the optimization problem.

A distinctive feature of PSO is the method of information exchange among particles. The search for new solutions is influenced by the global best solution the entire swarm finds and the personal best solution found by each particle. PSO employs random values to perturb solutions, enabling global and local search.

The key factors influencing the generation of a new solution are encapsulated in a term known as velocity. The velocity determines how each solution is updated. As the iterations progress, the particles converge near the presumed global optimum.

The PSO algorithm is population-based, and starts by generating a set of initial solutions randomly within the decision variable bounds. The initial population, represented as a matrix X, is given by eq. (4.2).

The iterative process begins with evaluating candidate solutions in the objective function to identify the global best solution (gBest) and the personal best solution for each particle (pBest). The solutions are updated according to the eq. (4.2), where the velocity V_{t+1} is calculated stated by eq. (4.3):

$$X_{t+1} = X_t + V_{t+1} \tag{4.2}$$

$$V_{t+1} = V_t + c_1 r_1 \otimes (\text{pBest} - X_t) + c_2 r_2 \otimes (\text{gBest} - X_t) \tag{4.3}$$

with c_1 and c_2 being acceleration coefficients, r_1 and r_2 random numbers between $[0, 1]$, and $\otimes$ denoting element-wise multiplication. The velocity equation consists of momentum, cognitive, and social. The momentum term $V(t)$ contains the velocity from the previous iteration, the cognitive term $c_1 r_1 \otimes (\text{pBest} - X_t)$ reflects the influence of each particle's best solution, and the social term $c_2 r_2 \otimes (\text{gBest} - X_t)$ represents the particle's tendency to move toward the best global solution.

One issue with the original PSO algorithm is that many new solutions generated in the early iterations may fall outside the decision variable bounds. Various strategies have been employed to address this – one of the most common uses an inertia weight coefficient w that affects the first term of the velocity:

$$V_{t+1} = w V_t + c_1 r_1 \otimes (\text{pBest} - X_t) + c_2 r_2 \otimes (\text{gBest} - X_t) \tag{4.4}$$

If $w > 1$, the velocity determining the new solutions increases over iterations, resulting in divergence in the search process. The value of w should be less than one, and it is recommended to decrease from 0.9 to 0.4 over the iterations according to the following equation 4.5:

$$w = 0.9 - \frac{t}{N_{\text{iter}}} (0.9 - 0.4) \tag{4.5}$$

where N_{iter} is the total number of iterations and t is the current iteration.

The PSO algorithm proceeds with evaluating solutions, identifying pBest and gBest, calculating of w and velocities, and updating solutions until a maximum number of iterations or a termination criterion is met.

4.3.3 Simulated annealing

SA is a probabilistic optimization technique inspired by the annealing process in metallurgy. Kirkpatrick et al. introduced it in 1983 as a method to solving combinatorial optimization problems [35]. The algorithm is based on the physical process of heating and slowly cooling a material to minimize its energy state.

In SA, a solution to an optimization problem is analogous to a physical system's state, and the solution's objective function value is analogous to the system's energy. The algorithm starts with an initial solution and an initial temperature. It then iteratively explores the solution space by making small perturbations to the current solution to generate new solutions. The acceptance of a new solution is determined probabilistically, based on the change in the objective function value and the current temperature. The temperature is gradually reduced according to a cooling schedule, which decreases the probability of accepting worse solutions as the algorithm progresses.

In the SA algorithm, the process starts with an initialization step, where an initial solution s and an initial temperature T are chosen. The algorithm then enters a loop where a new solution s' is generated by making a small random change to the current solution s. The acceptance criterion is based on the Metropolis criterion, where the change in the objective function value ΔE is calculated with eq. (4.6), and the new solution s' is accepted with a probability of 1 if $\Delta E < 0$ or $\exp(-\Delta E/T)$, otherwise the temperature T is then updated according to a cooling schedule, typically by multiplying it with a cooling rate α where $0 < \alpha < 1$. The process repeats the solution perturbation, acceptance criterion, and cooling schedule steps until a termination criterion is met, such as reaching a maximum number of iterations or achieving a desired level of solution quality:

$$\Delta E = f(s') - f(s) \qquad (4.6)$$

The final solution is the best solution found during the search process. SA is known for its ability to escape local optima and explore a broader solution space, making it a popular choice for solving complex optimization problems.

4.3.4 Ant colony optimization

ACO, developed by Marco Dorigo in the early 1990s for his Ph.D. thesis, is inspired by the foraging behavior of ant colonies [36]. The algorithm has since been manipulated to propose some variants and applied to various optimization problems.

ACO is based on the idea that ants can find the shortest path between their nest and a food source by laying down pheromone trails and following them probabilistically. In the algorithm, a set of artificial ants construct solutions to an optimization problem by moving through a graph representing the problem space. The ants deposit pheromone on the graph's edges, influencing the probability of selecting a particular path or solution component.

The ACO algorithm can be described as follows. The process begins with the initialization step, where all edges are assigned an initial pheromone level. During the solution construction phase, each ant starts from a random node and selects adjacent nodes based on a probabilistic rule that considers both pheromone levels and a heuristic value. An optional local search algorithm can be applied to enhance the solutions generated by the ants. The pheromone update phase involves two steps: evaporation, where pheromone levels on all edges are reduced by a factor $(1-\rho)$, and deposition, where pheromone levels are increased on edges that are part of the best solutions found in the current iteration or the best solution overall. The algorithm repeats the solution construction and pheromone update phases until a termination criterion, such as a maximum number of iterations or a convergence threshold, is met.

The probabilistic decision rule for selecting the next node j when an ant is at node i is given as follows:

$$p_{ij} = \frac{\tau_{ij}^{\alpha} \cdot \eta_{ij}^{\beta}}{\sum_{k \in \text{allowed}} \tau_{ik}^{\alpha} \cdot \eta_{ik}^{\beta}} \tag{4.7}$$

where τ_{ij} is the pheromone level on edge from i to j, η_{ij} is the heuristic, α and β are parameters that control the relative importance of pheromone and heuristic information, and the sum in the denominator is over the set of allowed nodes that the ant can move to from node i. Figure 4.4 shows the flowchart for ACO.

4.3.5 Differential evolution

DE is a population-based optimization algorithm proposed by Storn in 1996 [37]. It is inspired by the theory of evolution, similar to GAs, but employs a less canonical approach. DE uses real numbers to represent solutions and incorporates selection, mutation and crossover operators. Unlike GAs, which use two parents for mutation, DE generates new solutions from at least three different solutions.

In the mutation step, for each solution $X_{t,i}$ at iteration t, a mutant solution vector $V_{\{t,i\}}$ is generated. In the classical version, the following equation 4.8 is used:

$$V_{t,i} = X_{t,r1} + F \cdot (X_{t,r2} - X_{t,r3}) \tag{4.8}$$

where F is a scaling factor between 0 and 1, and $r1$, $r2$, $r3$ are randomly chosen indices from the population, ensuring they differ from each other and i.

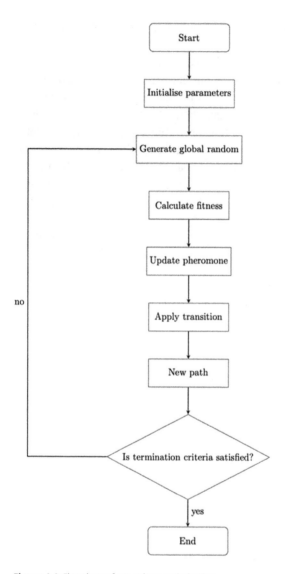

Figure 4.4: Flowchart of ant colony optimization.

During the crossover step, a trial vector $U_{t,i} = \{u_{t,i,1}, u_{t,i,2}, \ldots, u_{t,i,dim}\}$ is generated using the crossover equation 4.9:

$$u_{t,i,j} = \begin{cases} v_{t,i,j} \text{ if } \text{rand}_j < \text{Cr} \\ \text{otherwise } x_{t,i,j} \end{cases}$$

(4.9)

where Cr is the crossover probability, and $\text{rand}_j(0, 1)$ is a random number between 0 and 1 for each dimension j.

Finally, in the selection step, the trial vector $U_{t,i}$ is compared with the current solution $X_{t,i}$. The one with the better or equal objective function value is selected for the next generation. The selection is performed according to the following equation 4.10:

$$\text{if } X_{t+1,i} = U_{t,i}(U_{t,i}) \leq f(X_{t,i}), \text{ otherwise } X_{t+1,i} = X_{t,i} \tag{4.10}$$

The algorithm iterates through the mutation, crossover, and selection steps until a stopping criterion is met, such as reaching a maximum number of iterations or achieving a desired level of solution quality. The final solution is typically the best solution found during the optimization process.

4.3.6 Tabu search

TS, proposed by Glover in 1986 [38], is a heuristic method used for global optimization. It is inspired by adaptive memory strategies, allowing the algorithm to escape local optima and explore the search space effectively. The method leverages the concept of a tabu list, which prevents revisiting previously encountered solutions, thus enhancing the exploration process.

The fundamental concepts of TS include the definition of the search space, neighborhood structure, and the use of memory structures to guide the search. The search space in TS consists of all possible solutions to the optimization problem. For example, in the vehicle routing problem, the search space includes all feasible routes that meet the constraints of vehicle capacity and route duration. The neighborhood structure defines the set of solutions that can be reached from the current solution by a local move.

This algorithm involves the named tabu list, which is a short-term memory that records recent moves and prohibits reversing these moves for a certain number of iterations, known as the tabu tenure. This mechanism avoids cycling and encourages the exploration of new areas in the search space. Another important feature TS employed is the aspiration criteria, which allows overreading the tabu status of a move if specific conditions are met, such as finding a solution better than any previous solution found. This characteristic helps in the search process to escape from regions where the constraints imposed by the tabu list are very restrictive.

The TS algorithm starts with an initial solution, which can be generated randomly or through a heuristic. The process iterates through several steps until a termination criterion is met, such as a maximum number of iterations or convergence to a solution. First, the algorithm evaluates the neighborhood of the current solution. It then selects the best admissible move, which is not tabu or meets the aspiration criteria. The current solution is updated, and the move is recorded in the tabu list. The tabu list and aspiration criteria are updated as necessary. Formally, the process can be described as follows. The algorithm initializes with a chosen initial solution, sets the best solution to the initial solution, and initializes the tabu list. During each iteration, the neighborhood of the current solution is generated and all moves in the neighborhood are evaluated. The best admissible move is selected, the current solution is updated, and the move is recorded in the tabu list. If the new solution is better than the best-known solution, the best solution is updated.

Enhancements to the basic TS algorithm include intensification and diversification strategies. Intensification focuses the search on promising regions of the search space, often by exploiting elite solutions or frequently used solution components. Diversification, on the other hand, drives the search toward unexplored regions, ensuring a broad exploration of the search space. Another important strategy is strategic oscillation, which allows the search to move between feasible and infeasible regions of the search space. This is guided by adaptive penalties for constraint violations, helping to explore the boundaries of feasibility and improve solution quality.

The mathematical formulation of TS can be presented as follows. Consider the problem of minimizing the function $f(s)$ over a search space S. The TS algorithm updates the current solution s_t based on the rule shown in the following equation 4.11:

$$s_{t+1} = \arg \min_{s' \in N(s_t) \setminus T_t} f(s') \tag{4.11}$$

where $N(s_t)$ is the neighborhood of s_t and T_t is the tabu list in the iteration t. The aspiration criteria can be established as $f(s') < f(s^*)$, where s^* is the best solution found until the current iteration.

4.3.7 Firefly algorithm

The FA, developed by Xin-She Yang in 2008, is a metaheuristic optimization algorithm inspired by the flashing behavior of fireflies [39]. It is primarily used for solving continuous optimization problems, and is based on the idealized behavior of fireflies to attract each other using their bioluminescence.

The attractiveness of a firefly is determined by its brightness, which in turn is associated with the objective function of the optimization problem. The fundamental concepts of FA include the attraction mechanism, brightness function, and randomization. These elements are crucial for guiding the search process toward optimal solutions. In FA, fireflies are considered agents moving in the search space. The attractiveness of each firefly is proportional to its brightness, which decreases with distance from other fireflies. The brightness is associated with the value of the objective function, meaning that brighter fireflies represent better solutions.

The attractiveness function is defined as shown in eq. (4.12), where β_0 is the attractiveness at $r = 0$, γ is the light absorption coefficient, and r is the distance between two fireflies. As fireflies move toward the brighter ones, and their movement is influenced by both the attractiveness and randomization factor, which helps to explore the search space more effectively:

$$\beta(r) = \beta_0 e^{-\gamma r^2} \tag{4.12}$$

The flashing light of fireflies is a phenomenon observed in the summer skies in tropical and temperate regions. This light, produced by bioluminescence, serves two primary functions: attracting mating partners and potential prey. The light intensity of a firefly decreases with distance according to the inverse square law, resulting in its visibility being limited to a certain range.

The FA starts with a population of fireflies distributed randomly within the search space. Each firefly represents a candidate solution. The algorithm iterates through several steps until a termination criterion is met, such as a maximum number of iterations or convergence to a solution. Initially, the brightness of each firefly is calculated based on the objective function. The distance between every pair of fireflies is then computed. Fireflies move toward brighter ones based on their attractiveness, with their positions updated according to the following rule established by equation 4.13:

$$x_i = x_i + \beta_0 e^{-\gamma r_{ij}^2}(x_j - x_i) + \alpha(\text{rand} - 0.5) \tag{4.13}$$

The parameter γ characterizes the variation of the attractiveness, and its value is crucially important in determining the speed of convergence and the behavior of the FA algorithm. For $\gamma \to 0$, the attractiveness is constant, leading to a scenario similar to PSO. For $\gamma \to \infty$, the attractiveness becomes negligible, corresponding to a random search. The effectiveness of FA lies in adjusting γ to balance between these extremes.

4.4 Applications of metaheuristic optimization to chemical engineering

The continuous pursuit of optimal solutions to complex problems is crucial in chemical engineering. These challenges typically arise in areas such as process design and control, where traditional optimization methods may encounter difficulties due to the nonlinearity, multimodality and high dimensionality of the problem space. Metaheuristic optimization algorithms offer a robust and adaptable approach to address these difficulties, providing a valuable tool for navigating complex search spaces and avoiding local optima.

Metaheuristic algorithms have been widely applied in various chemical engineering sub-areas, including process optimization, reactor design, environmental engineering, and energy management. Their versatility and efficacy have transformed problem-solving in the field, leading to more efficient, sustainable, and innovative solutions. By exploring recent literature, this section highlights the advantages, challenges, and future directions of applying metaheuristic algorithms in chemical engineering, underscoring their potential to optimize processes, reduce costs, and minimize environmental impact.

In general, there have been many contributions by chemical engineers to meta-heuristics and their applications. As far as possible, it is important to cite papers in chemical engineering journals and/or by chemical engineers. Notable references include "Stochastic Global Optimization: Techniques and Applications in Chemical Engineering" [40] and "Optimization of Process Flowsheets through Metaheuristic Techniques" [41].

In this section, various works related to important sub-areas of chemical engineering are explored; these include parameter estimation, chemical reactor optimization, separation process, and heat exchanger network (HEN) design.

4.4.1 Parameter estimation

In chemical engineering, accurate parameter estimation is crucial for optimizing processes and ensuring operational efficiency. Accurate parameter estimation directly influences the efficiency, safety and profitability of chemical processes. The inherent complexity and nonlinearity of these systems present significant challenges in determining process parameters. In recent years, the application of metaheuristic methods has contributed to this field, offering robust and efficient solutions for parameter estimation. The adaptability and flexibility of metaheuristic approaches enable them to handle a wide range of estimation problems, making them invaluable tools for process engineers and researchers in the quest for improved process performance and understanding.

Dos Santos et al. [42] present an innovative approach to modeling monoethanolamine (MEA) using the cubic plus association equation of state. The study focuses on improving parameter estimation by integrating local search techniques with the PSO algorithm. The objective is to accurately predict the properties of MEA, which is crucial for acid gas removal processes in the petroleum and gas industries. The optimization aims to minimize the error between predicted and experimental values, subject to constraints related to the cubic plus association model parameters. This approach enhances the model's reliability and demonstrates the potential of combining metaheuristic algorithms with local search methods for parameter estimation in chemical process modeling.

Ding et al. [43] estimated the kinetic parameters for a three-component parallel reaction mechanism in biomass pyrolysis. The objective of the optimization is to minimize the error between predicted and experimental values of mass loss and mass loss rate (MLR) during pyrolysis. The study compares the accuracy and efficiency of GA and PSO under identical optimization conditions. Results indicate that PSO shows better performance in terms of convergence to the global optimum and convergence speed for the specific case studied. The optimized kinetic parameters obtained from PSO show better agreement with experimental data, especially in predicting hemicellulose, cellulose, and lignin decomposition temperature ranges. This study suggests the potential of PSO as an effective and efficient optimization tool for estimating reaction kinetic parameters in biomass pyrolysis, although more comprehensive comparisons using benchmark problems and various applications are necessary to generalize this finding. The error minimized in the study includes discrepancies between modeled and experimental cumulative mass loss and MLR. The parameter search space consists of 14 parameters related to the kinetic behavior of hemicellulose, cellulose, and lignin, including char yield, initial mass fraction, pre-exponential factor, activation energy, and reaction order. These parameters are optimized within predefined ranges, based on previous studies and practical constraints in biomass pyrolysis models.

An example where a hybrid GAs-PSO approach is used is presented by the work of Bi and Qiu [44], which introduces a method using support vector machines (SVMs) optimized by a hybrid GAs and PSO approach for predicting crude oil properties. The optimization aims to minimize the root mean square error (RMSE) between predicted and actual property values, subject to constraints on SVM parameters. The hybrid GA-PSO method shows better performance in terms of accuracy and convergence speed for the specific case studied compared to standalone GA or PSO methods. The optimized SVM model accurately predicts the true boiling point distillation curve, which is critical for oil refining. Nonetheless, the generalizability of these findings requires further testing on benchmark problems and multiple applications to account for the potential variability in performance due to different program implementations and parameter values.

Rawat et al. [45] developed a parameter estimation technique for solar photovoltaic cells using grey wolf optimization (GWO). The objective was to minimize the error between predicted and experimental values of key parameters: series resistance, shunt resistance, and diode ideality factor. The study compares GWO's performance with numerical, analytical, and hybrid methods. The results show that the proposed GWO-based method achieves lower values of RMSE, mean absolute error (MAE), and mean relative absolute error (MRAE) compared to the iterative Newton-Raphson method and the Lambert W function-based analytical method. Specifically, for the KC200GT module, the GWO approach yields an RMSE of 0.21, MRAE of 0.1, and MAE of 0.046. In comparison, the Newton-Raphson method results in an RMSE of 0.308, MRAE of 1.0651, and MAE of 0.308, while the Lambert W function-based analytical method shows an RMSE of 0.306, MRAE of 1.0651, and MAE of 0.284. This study demonstrates GWO's effectiveness in accurately estimating photovoltaic cell parameters, leading to improved modeling and performance predictions.

Roeva and Chorukova [46] performed parameter estimation for a new nonlinear mathematical model of two-stage anaerobic digestion (TSAD) of corn steep liquor using four metaheuristic algorithms: GA, FA, cuckoo search algorithm (CS), and coyote optimization algorithm (COA). The study aimed to model the process dynamics accurately in two bioreactors (BR1 and BR2). The results showed that GA achieved the lowest objective function (mean square error) value in BR2 ($J = 0.0761$), while FA provided a better trend description ($J = 0.1075$). The objective function values for CS and COA were $J = 0.0913$ and $J = 0.0940$, respectively. The parameter estimates were statistically significant, with FA and COA showing the smallest standard deviations, indicating robustness. The study concluded that FA produced the most accurate model for the TSAD process, suggesting potential for future process optimization and control applications.

Kim et al. [47] conducted parameter identification for a pseudo-two-dimensional (P2D) model of lithium-ion batteries using a novel method combining GAs and neural network cooperative optimization (GANCO). The study aimed to improve the accuracy and reliability of parameter estimation, crucial for assessing the state of health of batteries. The results indicated that GANCO outperformed traditional metaheuristic approaches like GA, PSO, and harmony search (HS), in terms of both convergence speed and accuracy. Specifically, the GANCO method reduced the RMSE of the output voltage and mean percentage error (MPE) of the parameters by significant margins, demonstrating its potential for enhancing battery management systems.

Wilberforce et al. [48] investigated the parameter estimation of a proton exchange membrane (PEM) fuel cell using five optimization algorithms: GWO, PSO, slime mould algorithm (SMA), Harris Hawk optimizer (HHO), and an artificial ecosystem-based algorithm (AEO). The study aimed to determine the optimal fuel cell parameters by minimizing the RMSE between experimental and predicted data. The results showed that the GWO algorithm provided the lowest RMSE value (0.004), indicating the highest accuracy, followed by PSO (0.005), SMA (0.006), HHO (0.007), and AEO (0.008). This study highlights the effectiveness of GWO in accurately estimating PEM fuel cell parameters compared to other methods.

The literature on the application of metaheuristic algorithms is extensive. Table 4.3 lists some works that readers may find interesting about parameter estimation.

Table 4.3: Summary of parameter estimation studies using metaheuristic algorithms.

Authors	Algorithms used	Objective function	Classification
Dos Santos et al. [49]	PSO and hybrid local search	Minimize error between predicted and experimental values	The hybrid PSO approach with local search is practical in modeling MEA with the CPA equation of state.
Ding et al. [50]	GA and PSO	Minimize error in biomass pyrolysis kinetic parameter estimation	GA, PSO Minimize error in biomass pyrolysis kinetic parameter estimation. PSO shows better performance than GA in accuracy and convergence speed for the specific case studied.

Table 4.3 (continued)

Authors	Algorithms used	Objective function	Classification
Bi and Qiu [51]	Hybrid GA-PSO	Minimize the root mean square error	The hybrid GA-PSO method shows superior performance in accuracy and convergence speed for predicting crude oil properties.
Rawat et al. [45]	Grey wolf optimization (GWO)	Minimize error between predicted and experimental values	GWO achieves lower RMSE, MAE, and MRAE compared to numerical and analytical methods for photovoltaic cell parameter estimation.
Roeva and Chorukova [46]	GA, FA, cuckoo search algorithm (CS), and coyote optimization algorithm (COA)	Minimize mean squared error	FA produced the most accurate model for TSAD process, with GA, CS, and COA also showing significant results.
Kim et al. [47]	GA and neural network cooperative optimization (GANCO)	Minimize error in parameter estimation for lithium-ion batteries	GANCO outperformed traditional metaheuristic approaches in both convergence speed and accuracy.
Wilberforce et al. [48]	GWO, PSO, slime mould algorithm (SMA), Harris Hawk optimizer (HHO), and artificial ecosystem-based algorithm (AEO)	Minimize RMSE between experimental and predicted data	GWO provided the highest accuracy in estimating PEM fuel cell parameters, followed by PSO, SMA, HHO, and AEO.
Bai and Li [52]	Hybrid cuckoo search-grey wolf optimization (CSGWO)	Minimize mean squared error	CSGWO outperforms other algorithms in identifying parameters of solid oxide fuel cells with high accuracy and convergence speed.
Huang et al. [53]	CS and PSO	Minimize error between predicted and experimental values	Combining CS and PSO enhances search capability and efficiency, especially for low-sensitivity parameter identification.

4.4.2 Reactor optimization

Optimizing chemical reactors is essential for enhancing the efficiency, safety, and profitability of industrial operations. The complexity and nonlinearity of chemical reactions often imply significant challenges in achieving optimal reactor conditions. The application of metaheuristic methods has contributed to this field, providing robust and efficient solutions for reactor optimization. Some relevant chemical reactor-related works reported in the literature are discussed below.

Further extending the application of GAs, Na et al. [54] employed a multi-objective version to optimize a microchannel reactor for Fischer-Tropsch synthesis using computational fluid dynamics. The optimization aimed to maximize C5+ productivity while minimizing temperature rise. This study highlighted the effectiveness of GAs in integrating external simulators for complex optimization tasks in reactor design. CFD simulation is computationally intensive. To address this issue in the GA-based optimization, Na et al. implemented a two-stage optimization process. Initially, the MINLP problem was solved multiple times with different crossover fractions until the integer variables stabilized. This reduced the problem to an NLP, which was further optimized in the second stage. By decomposing the problem and utilizing parallel computing, they managed to significantly reduce the number of function evaluations needed. Specifically, the GA-CFD optimizer used parallel computing with 16 cores to expedite simulations, ensuring the approach was computationally feasible even for intensive CFD models.

Zainullin et al. [39] conducted multi-criteria optimization using GAs for a catalytic reforming reactor unit. The optimization focused on maximizing the target product's octane number and yield, while minimizing the content of aromatic hydrocarbons and benzene. The results demonstrated a significant reduction in the total content of aromatic hydrocarbons and a decrease in the octane number for the specific case studied, suggesting the capability of GAs in handling multi-criteria optimization problems in reactor design. However, the performance of GAs may vary depending on the implementation and parameter values used, and further comprehensive comparisons are needed to confirm this capability across different applications.

While GAs are inherently capable of handling nonlinear search spaces, some studies have applied linearization strategies to simplify the optimization problem. Soltani et al. [55] demonstrated this approach by coupling a quasi-linear programming method with adiabatic reactor networks and GAs. This strategy involved converting discrete design variables into continuous ones, leading to near-optimal solutions that surpassed those reported in the literature. This suggests that linearization techniques combined with GAs can improve reactor design and process optimization outcomes.

Dynamic case studies have also been analyzed. For example, Bayat et al. [56] applied DE to two case studies, both steady and dynamic states, to optimize the operating conditions for an industrial fixed-bed ethylene oxide reactor. The study aims to maximize the ethylene oxide yield, while considering catalyst deactivation. The results show that yield enhancements of 1.726% and 4.22% in ethylene oxide production can be achieved by applying the first and second optimization case studies, respectively. The first optimization case study involves optimizing four parameters: the inlet molar flow rate, inlet pressure, and the temperatures of the shell and tube sides within their practical ranges. The second optimization case study follows a stepwise approach to determine optimal temperature profiles for saturated water and gas in three stages during operation. This multistage strategy allows for better adjustment to the dynamic behavior of the reactor over time, leading to higher yield improvements.

To obtain higher-quality results, some authors have proposed metaheuristic optimization problems explicitly designed to optimize chemical reactors. For example, the work of Zhang et al. [57] proposed a multi-objective dynamic DE algorithm with self-adaptive strategies named SA-MODDE. The critical components of this algorithm include parental selection, mutation strategy, parameter setting, survival selection, constraint handling, and termination criteria. The algorithm is tested on benchmark problems and chemical engineering process optimizations, including catalyst mixing policy, Lee–Ramirez bioreactor, and the alkylation process. The results demonstrate the effectiveness of SA-MODDE in handling multi-objective optimization problems in chemical engineering.

Lee et al. [58] investigated the optimization of the nonoxidative direct methane conversion process utilizing a combination of machine learning (ML) models and metaheuristic optimization algorithms. The focus was to improve the methane to olefins, aromatics, and hydrogen (MTOAH) process by maximizing hydrocarbon yields while minimizing coke formation. The study used an integrated reactor system and employed metaheuristic algorithms, including the artificial bee colony (ABC) optimization, to identify the optimal reaction conditions. The objective function used in this study aimed to simultaneously maximize hydrocarbon yield and minimize coke formation, formulated through various scoring functions. These scoring functions took into account methane conversion, hydrocarbon selectivity, and coke selectivity, penalizing undesired outcomes to steer the optimization process toward the desired reaction performance. The results demonstrated that the metaheuristic optimization, guided by the ML predictions, significantly enhanced hydrocarbon yields and reduced coke formation. For instance, the optimized reaction parameters resulted in a hydrocarbon yield of 34.6% and a reduction in coke selectivity to 11.0% under optimal conditions.

Sun et al. [62] conducted a study on predicting steady-state biogas production from waste using advanced ML and metaheuristic optimization approaches. The objective was to develop an accurate and efficient method for estimating the biogas production rate by optimizing various process variables. The study employed several metaheuristic algorithms, including the cuckoo optimization algorithm (COA), multiverse optimization (MVO), league championship algorithm (LCA), evaporation-rate water cycle algorithm (ERWCA), stochastic fractal search (SFS), and teaching–learning-based optimization (TLBO). The primary objective function was to minimize the error between predicted and actual biogas production rates, using metrics such as MAE, MSE, and coefficient of determination (R^2). These algorithms were used to optimize the input variables for an artificial neural network (ANN) model designed to predict biogas production. The results indicated that the ERWCA algorithm provided the highest prediction accuracy, achieving an R^2 of 0.9314 and 0.9302, RMSE of 0.1969 and 0.24925, and MAE of 0.1307 and 0.19591 for the training and testing datasets, respectively. This was followed by MVO, SFS, TLBO, LCA, and COA, with varying degrees of prediction accuracy. Table 4.4 shows a summary of the revised works.

Table 4.4: Summary of reactor optimization studies using metaheuristic algorithms.

Authors	Algorithms used	Objective function	Classification
Na et al. [54]	Multi-objective GA	Maximize C5+ productivity and minimize temperature rise	Effective in integrating external simulators for complex optimization tasks in reactor design.
Zainullin et al. [59]	GA	Maximize octane number and yield, minimize aromatic hydrocarbons and benzene content	Significant reduction in aromatic hydrocarbons and decrease in octane number, showcasing GAs in multi-criteria optimization.
Soltani et al. [60]	GA and quasi-linear programming	Minimize error in parameter estimation	Linearization techniques combined with GAs improve reactor design and process optimization outcomes.
Bayat et al. [56]	Differential evolution (DE)	Maximize ethylene oxide yield considering catalyst deactivation	Yield enhancements of 1.726% and 4.22% in ethylene oxide production by applying optimization case studies.
Zhang et al. [61]	SA-MODDE (self-adaptive multi-objective dynamic differential evolution)	Optimize catalyst mixing policy, Lee-Ramirez bioreactor, and alkylation process	Effective in handling multi-objective optimization problems in chemical engineering.
Lee et al. [58]	Artificial bee colony (ABC) optimization	Maximize hydrocarbon yield and minimize coke formation	Metaheuristic optimization guided by ML predictions enhanced hydrocarbon yields and reduced coke formation.
Sun et al. [62]	COA, MVO, LCA, ERWCA, SFS, and TLBO	Minimize error between predicted and actual biogas production rates	ERWCA provided the highest prediction accuracy, followed by MVO, SFS, TLBO, LCA, and COA with varying degrees of prediction accuracy.

4.4.3 Separation process optimization

The optimization of separation processes is a critical component in enhancing the efficiency and sustainability of chemical engineering operations. By fine-tuning operational parameters and equipment configurations, significant energy savings and cost reductions can be achieved. In the context of separation process optimization, this involves achieving optimal separations with minimal energy consumption. The successful application of these optimization techniques improves process efficiency and contributes to the sustainable development of the chemical processing industry by minimizing environmental impacts and resource usage. Metaheuristic strategies have been increasingly applied to optimize various distillation schemes, ranging from conventional columns to more complex and intensified systems. These strategies offer robust and efficient solutions for the complex and nonlinear problems inherent in distillation column optimization, with potential benefits in energy savings and reduced environmental impact.

In their work, Ramanathan et al. [63] explored the application of GAs and simultaneous perturbation stochastic approximation for optimizing continuous distillation columns. The study demonstrates the effectiveness of these stochastic optimization methods in determining optimal values for variables such as the number of stages, reflux ratio, and feed location in both simple and azeotropic distillation systems. These methods ultimately minimize the total annual cost (TAC) of distillation operations, while satisfying purity constraints for the products.

Similarly, Boozarjomehry et al. [64] presented a novel approach for automatically designing conventional distillation column sequences using GAs. The evaluation of different distillation column sequence alternatives is based on the TAC estimated for each sequence using short-cut methods for column design. The proposed method's efficacy is demonstrated by applying four standard benchmark problems commonly utilized in the chemical process design and optimization domain.

Vazquez-Ojeda et al. [65] optimized ethanol dehydration processes, exploring two separation sequences and employing DE coupled with rigorous Aspen Plus simulations. The study highlights the potential of using DE for optimizing complex chemical processes, and suggests that alternative separation sequences could offer more cost-effective solutions for large-scale bioethanol production.

The application of metaheuristic optimization algorithms has also extended to more efficient distillation column schemes such as heat-integrated, divided wall, and reactive distillation columns. Shahandeh et al. [66] employed a GA to optimize heat-integrated distillation columns (HIDiCs) for the separation of close boiling point mixtures, introducing a novel integer variable, the Layout number, to systematically generate more energy-efficient candidates for both internal and external HIDiCs. The study demonstrated that multivariable optimization problems could be successfully optimized by GA, resulting in significant TAC reductions compared to previously reported solutions.

The research conducted by Gutierrez-Guerra and Segovia-Hernández [67] focuses on the innovative optimization of a distillation column that incorporates phase change materials for enhanced energy storage and efficiency. The optimization employs the Boltzmann univariate marginal distribution algorithm [68], aiming to minimize the TAC of the distillation process. This approach highlights the potential of utilizing phase change materials in distillation processes to achieve significant energy savings and cost reductions, ultimately contributing to more sustainable and efficient chemical engineering practices.

In their study, García-Hernández et al. [69] explore the application of stochastic optimization techniques, specifically DE with tabu list (DETL), for optimizing the production of sustainable aviation fuel (SAF) from biobutanol derived from biomass. The research demonstrates the effectiveness of these metaheuristic optimization methods in determining optimal values for key variables such as the number of stages, reflux ratio, and feed location in distillation columns. These methods enable the minimization of TAC and environmental impact of the production process, while satisfying product purity requirements. The implementation of process intensification in the reaction and separation zones has been crucial for improving process efficiency, and significantly reducing energy consumption and operational costs.

Kruber et al. [70] explore advanced hybrid optimization methods for designing complex separation processes. The study focuses on using hybrid approaches that combine evolutionary algorithms (EAs) and multi-start grid search (MSGS), with local deterministic optimization techniques to address the nonlinear, mixed-integer programming problems encountered in chemical engineering. These methods are applied to optimize extractive distillation processes, including the selection of mass-separating agents (MSAs) and energy integration strategies. By integrating these hybrid optimization techniques, the study demonstrates the capability to effectively handle complex design variables and constraints, ultimately improving the efficiency and sustainability of the separation processes while minimizing operational costs and energy consumption. This approach highlights the potential of metaheuristic algorithms in solving challenging engineering problems, providing a robust framework for optimizing chemical processes.

This study highlights the effectiveness of process intensification and metaheuristic optimization in enhancing the production of SAF from biobutanol, providing a viable pathway for the aviation industry's shift toward more sustainable energy sources. The innovative methodology and its successful application in this research contribute to the broader goal of sustainable development within the chemical processing industry. Table 4.5 briefly summarizes some important aspects of the commented works.

Applying metaheuristic algorithms in optimizing separation processes and distillation column design has proven valuable in enhancing process efficiency, reducing costs, and promoting sustainability in chemical engineering. The versatility and functionality of these algorithms allow them to tackle a variety of optimization challenges, contributing to the development of more innovative and energy-efficient separation technologies.

4.4.4 Heat exchanger networks

In chemical engineering, the synthesis of HENs is pivotal, especially regarding energy efficiency and cost-effectiveness. The study by Behroozsarand and Soltani exemplifies this by optimizing HENs for a hydrogen plant using a coupled GA-linear programming

Table 4.5: Summary of separation process studies using metaheuristic algorithms.

Authors	Algorithms used	Objective function	Classification
Ramanathan et al. [63]	GAs and simultaneous perturbation stochastic approximation	Minimizing the total annual cost	Effective in optimizing continuous distillation columns for simple and azeotropic distillation systems
Boozarjomehry et al. [64]	GAs	Automatically design conventional distillation column sequences, minimizing total annual cost (TAC)	Efficient in designing distillation column sequences with reduced TAC
Vazquez-Ojeda et al. [71]	DE	Optimize ethanol dehydration processes for cost-effective solutions in bioethanol production	Potential for optimizing complex chemical processes and cost-effective bioethanol production
Shahandeh et al. [66]	GAs	Optimize heat-integrated distillation columns (HIDiCs) for TAC reduction	Successful in optimizing multivariable HIDiCs, achieving significant TAC reductions
Gutierrez-Guerra and Segovia-Hernández [72]	Boltzmann univariate marginal distribution algorithm	Minimize TAC of distillation processes incorporating phase change materials	Highlights the use of phase change materials for energy savings and cost reductions in distillation processes
García-Hernández et al. [69]	Differential evolution with tabu list (DETL)	Optimize production of sustainable aviation fuel (SAF) from biobutanol, minimizing TAC and environmental impact	Effective in minimizing TAC and environmental impact while satisfying product purity requirements
Kruber et al. [70]	Hybrid evolutionary algorithms (EAs), multi-start grid search (MSGS), and local deterministic optimization	Minimize CAPEX and OPEX	Capable of handling complex design variables and constraints, improving efficiency and sustainability

method in conjunction with the ASPEN HYSYS simulator [73]. This approach considers temperature and pressure effects on physical properties, phase changes, and pressure drops in heat exchangers, often neglected in traditional design methods. The study aims to enhance energy recovery by using the GA to optimize structural parameters and the LP method to handle continuous variables. Integrating ASPEN HYSYS facilitates accurate modeling of multicomponent streams and phase changes, leading to more realistic and industrially relevant HEN designs. The results underscore the significant improvements in energy efficiency and cost reduction, highlighting the synergy between metaheuristic algorithms and process simulators in designing complex HENs.

Building on this foundation, Pavao et al. [74] presented a metaheuristic approach for the multi-objective optimization of TAC and environmental impacts in medium- and large-scale HENs. This is the continuous simulated annealing method, followed by PSO. The method's ability to efficiently achieve near-Pareto fronts for four industrial-size case studies is noteworthy. The study offers solutions that optimize TAC and present configurations with low environmental impacts while maintaining competitive TAC. This emphasizes the significance of multi-criteria optimization in HEN synthesis, showcasing the method's capability to attain solutions with substantially lower environmental impacts, thereby demonstrating its potential for industrial-scale problems. Furthermore, Pavao et al. [75] extended the application of metaheuristic optimization to multi-period HEN synthesis, through a two-level method based on SA and rocket fireworks optimization (RFO). The method addresses the additional complexity and constraints associated with multi-period HEN synthesis by aiming to minimize the TAC, considering both capital and operating costs. The efficacy of this approach is validated through four case studies, with the SA-RFO algorithm showing promising results in reducing TAC and enhancing computational efficiency, thereby underscoring the value of meta-heuristic approaches in tackling the challenges of multi-period HEN synthesis.

Further advancements in HEN synthesis are demonstrated in the work of Zhang and Cui [76], who introduced a novel approach using an improved CS algorithm augmented by Lévy flights. The focus on minimizing the TAC by optimizing heat load distribution and network configuration is a testimony to the continuous pursuit of efficiency in HEN design, employing an improved CS algorithm to solve the MINLP problem for optimal HEN. Introducing a stream arrangement strategy to optimize the stream match search space exemplifies the innovative efforts to simplify the solution of large- and medium-sized HEN problems. Applying this methodology to four benchmark cases demonstrates its computational efficiency and effectiveness in identifying economically attractive energy-integrated distillation processes, highlighting the importance of efficient design methods for HEN synthesis, particularly regarding energy integration.

Similarly, the article by Aguitoni et al. [77] proposed a bi-level optimization approach for synthesizing HENs using a hybrid meta-heuristic method. The distinct utilization of SA at the upper level to optimize HEN topologies, coupled with DE at the lower level to handle continuous variables, showcases the innovative integration of different metaheuristic algorithms. The objective to minimize the TAC, considering both capital and operating costs, remains a consistent theme in pursuing cost-effective HEN design. The derived optimization model from a superstructure, accounting for stream split and non-isothermal mixing, further illustrates the complexity of

HEN synthesis. The effectiveness of this proposed method is demonstrated through four case studies, achieving better values than those reported in the literature.

Stampfli et al. [78] explored the optimization of HEN retrofit for multi-period processes in the Swiss process industry using a hybrid evolutionary algorithm. The study aimed to reduce the TAC of the HEN by optimizing both the topology and heat loads of the system. The hybrid evolutionary algorithm employed in this study combined GAs for optimizing the network topology and a DE algorithm for optimizing the heat loads. The primary objective function was to minimize the TAC, which included both capital costs and operating costs of the heat exchangers. The results demonstrated that the proposed hybrid algorithm effectively reduced the TAC by approximately 66% in an industrial case study. A summary of the commented studies is presented in Table 4.6.

Table 4.6: Summary of heat exchanger networks studies using metaheuristic algorithms.

Authors	Algorithms used	Objective function	Classification
Behroozsarand and Soltani [79]	Genetic algorithm-linear programming (GA-LP)	Enhance energy recovery by optimizing structural parameters and handling continuous variables	Significant improvements in energy efficiency and cost reduction in HEN designs
Pavao et al. [74]	Hybrid method based on SA and rocket fireworks optimization	Optimize TAC and environmental impacts in medium- and large-scale HENs	Efficient in achieving near-Pareto fronts for TAC and environmental impacts in industrial-size HENs
Pavao et al. [75]	Simulated annealing and rocket fireworks optimization (SA-RFO)	Minimize TAC considering both capital and operating costs in multi-period HEN synthesis	Promising results in reducing TAC and enhancing computational efficiency in multi-period HEN synthesis
Zhang and Cui [76]	Improved cuckoo search algorithm with Lévy flights	Minimize TAC by optimizing heat load distribution and network configuration	Effective in handling simultaneous optimization of continuous and integer variables for HEN synthesis
Aguitoni et al. [77]	Hybrid simulated annealing (SA) and differential evolution (DE)	Minimize TAC considering both capital and operating costs in HEN design	Innovative integration of SA and DE for efficient and cost-effective HEN synthesis
Stampfli et al. [78]	Hybrid evolutionary algorithm (GA and DE)	Minimize TAC by optimizing both topology and heat loads in HEN retrofit	Effectively reduced TAC by approximately 66% in an industrial case study

As these studies collectively illustrate, the application of metaheuristic algorithms in the synthesis of HENs continues to evolve, offering innovative solutions that enhance energy efficiency and cost-effectiveness. Integrating different algorithms and considering multiple objectives, including environmental impacts, highlight the multidimensional nature of optimization in this field. The advancements in computational methods allows the incorporation of real-world complexities, such as multi-period.

4.4.5 Optimization of chemical engineering systems with metaheuristic algorithms and neural networks

Integrating metaheuristic algorithms with neural networks approximation models has become a prominent approach for addressing complex optimization challenges in chemical engineering systems. Metaheuristic algorithms are recognized for their capability to explore complex search spaces. They are instrumental in identifying optimal solutions for various applications, ranging from process design and parameter estimation to operational optimization. On the other hand, neural networks approximation models excel in capturing nonlinear relationships and accurately predicting system behavior, making them invaluable in scenarios where traditional mathematical models fall short.

This synergy between metaheuristic algorithms and neural networks is advantageous in situations characterized by abundant process data and complex underlying mechanisms. For example, in chemical reactor design, neural networks can approximate reaction kinetics and heat transfer phenomena. Subsequently, metaheuristic algorithms can leverage these models to optimize reactor geometry and operating conditions, thereby enhancing efficiency and yield.

Similarly, with sufficient data, neural networks models (and other machine learning technologies) can predict the performance of any system or process, including separation processes and material development. Metaheuristic algorithms can then utilize these predictions to efficiently explore the solution space, identifying optimal settings that align with the desired objectives.

The study by Cartwright and Curteanu [1] exemplifies the potential of this integrated approach. The authors review neuro-evolutionary techniques for optimizing artificial neural networks in chemical process modeling and optimization, underscoring the benefits of combining neural networks with evolutionary algorithms. This fusion not only enhances the reliability of hybrid neuro-evolutionary methods in chemical engineering, but also showcases the practical applicability of this approach through real-world examples in the chemical industry.

Aladejare et al. [81] further illustrate the versatility of this combined methodology in their development of ANN models optimized with PSO for predicting the higher heating values of solid fuels. Their findings highlight the superior accuracy of

ANN-PSO models compared to multilinear regression models, offering a robust tool for optimizing energy content estimation in solid fuels.

In addition to these studies, the work of Zhou and Chiam [82] introduced a novel synthetic data generation strategy for data-free knowledge distillation in regression neural networks. This approach optimizes a loss function directly using predictions of student and teacher models, showcasing the potential of neural networks in enhancing the performance of regression tasks in chemical engineering.

Furthermore, the research by Al Hariri et al. [83] presents an ANN-DE optimization framework for cost optimization of a circulated permeate gap membrane distillation unit. This study demonstrates significant cost reduction for water production, highlighting the advantages of integrating neural networks with metaheuristic algorithms in optimizing membrane distillation processes.

Table 4.7 compiles studies where authors applied metaheuristic algorithms and ANN to solve chemical engineering-related problems.

Table 4.7: Summary of studies where metaheuristic algorithms and ANN are used.

Authors	Summary
Cartwright and Curteanu [80]	Reviewed neuro-evolutionary techniques for optimizing ANNs in chemical process modeling and optimization, highlighting the effectiveness of combining neural networks with evolutionary algorithms.
Khan et al. [84]	Presented an integrated framework of ANNs and metaheuristic algorithms for predicting biochar yield using biomass characteristics and pyrolysis process conditions.
Zhou and Chiam [82]	Proposed a synthetic data generation strategy for data-free knowledge distillation in regression neural networks, optimizing the loss function directly using predictions of student and teacher models.
Varol Altay et al. [85]	Proposed a hybrid ANN-grey wolf optimizer model to predict the reservoir temperature of geothermal waters in Anatolia, showing high accuracy in determining suitable drilling locations.
Al Hariri et al. [83]	Developed an ANN-DE optimization framework for cost optimization of a circulated permeate gap membrane distillation unit, demonstrating significant cost reduction for water production.
Fetimi et al. [86]	Explored emulsion liquid membrane stability optimization using a hybrid ANN-PSO model, demonstrating accurate forecasting of emulsion breaking percentages and enhancing ELM process efficiency.
Aladejare et al. [81]	Developed ANN models optimized with PSO for predicting higher heating values of solid fuels, showing higher accuracy than multiple linear regression models.

Using metaheuristic algorithms with neural networks continues to yield promising results in various chemical engineering processes. Several studies demonstrate the effectiveness of hybrid artificial neural networks and metaheuristic models in accurately predicting and optimizing complex chemical engineering systems.

4.4.6 Metaheuristic algorithms in real-world problems

The available literature on the use and application of metaheuristic algorithms to solve real-world problems in chemical engineering is limited. Nevertheless, this section discusses some relevant works that illustrate the practical applications and effectiveness of these approaches.

Larraín et al. [87] apply metaheuristic algorithms in optimizing industrial processes. This work details the use of the Strong Pareto Evolutionary Algorithm 2 (SPEA2) to enhance the efficiency and output quality of a continuous kraft pulp digester. The researchers focused on minimizing variability in the kappa number, a critical quality indicator in pulp production while maximizing the overall yield. Through a series of experiments and a detailed case study, the study demonstrated the algorithm's capability to handle complex, multi-objective optimization problems by adapting the operational parameters of the digester in real time. This work not only highlights the challenges associated with integrating such algorithms into existing industrial systems, but also showcases the potential for significant improvements in process stability and efficiency.

The work presented by Ochoa et al. [88] details the implementation of optimization-based control strategies at a pilot plant scale for the production of polyhydroxybutyrate, a biodegradable polymer. This study not only highlights the challenges associated with the variability of biological processes and the complexity of working with living microorganisms, but also showcases the effectiveness of advanced control strategies in enhancing productivity and profitability. By employing real-time optimization and soft sensors for process monitoring, the research offers valuable insights into the potential economic and efficiency benefits of these technologies in industrial settings.

Despite the potential benefits of the adoption of metaheuristic algorithms in industrial chemical engineering, its application is not widely widespread. Several factors contribute to this limited implementation. Integrating metaheuristic algorithms into existing industrial systems often requires substantial changes to control and monitoring infrastructures, which can be both costly and complex. Additionally, the inherent variability of industrial processes, particularly those involving biological systems, implies significant challenges. Metaheuristic algorithms must handle dynamic environments where process conditions and raw material qualities frequently vary, complicating their consistent application. Moreover, there is often a disparity between the theoretical optimization achieved in simulations and its practical implementation, exacerbated by the scarcity of real-time data and the computational demands of these algorithms. These issues, combined with an industry reluctance to replace traditional methods with new technologies, slow down the adoption of advanced optimization techniques in mainstream industrial applications. This situation highlights the need for more pilot studies and real-world applications that can bridge the gap between theoretical research and practical, scalable solutions.

4.5 Conclusions

Applying metaheuristic optimization algorithms in chemical engineering has proven to be a powerful and versatile approach to addressing this field's complex and multi-faceted challenges. Inspired by natural phenomena and human behavior, these algorithms offer a robust framework for navigating complex search spaces, avoiding local optima, and efficiently exploring potential solutions. Integrating metaheuristic algorithms with advanced modeling techniques, such as neural networks, further enhances their capability to tackle highly nonlinear and dynamic systems, providing accurate predictions and optimized solutions.

The studies reviewed in this chapter discuss the impact of metaheuristic optimization on various chemical engineering processes, including reactor design, separation processes, HEN synthesis, among others. By leveraging the strengths of different metaheuristic algorithms, researchers and engineers can achieve near-optimal process configurations, improved energy efficiency, and reduced environmental impact, ultimately contributing to the sustainable development of chemical engineering practices.

As the field continues to evolve, future research directions may focus on the continuous refinement and application of hybrid metaheuristic algorithms, particularly in adapting them to increasingly complex scenarios and new challenges not fully addressed by existing methods. Additionally, extending the application of metaheuristics to emerging areas such as green chemistry and the circular economy, and exploring advanced machine learning techniques to enhance the predictive capabilities of optimization models, will further broaden the impact and effectiveness of these approaches.

References

[1]   Bastien Chopard, M. T. An Introduction to Metaheuristics for Optimization. In Chopard, B., Tomassini, M. (Eds.). Springer International Publishing, 2018, pp. 1–8. https://www.ebook.de/de/product/33036871/bastien_chopard_marco_tomassini_an_introduction_to_metaheuristics_for_optimization.html.

[2]   Romero-Izquierdo, A. G., Gómez-Castro, F. I., Gutiérrez-Antonio, C., Hernández, S., Errico, M. Intensification of the alcohol-to-jet process to produce renewable aviation fuel. *Chemical*

Engineering and Processing – Process Intensification, 2021, 160, 108270. https://doi.org/10.1016/j.cep. 2020.108270.

[3] Beni, A. A., Esmaeili, A. Design and optimization of a new reactor based on biofilm-ceramic for industrial wastewater treatment. *Environmental Pollution*, 2019, 255, 113298. https://doi.org/10.1016/j. envpol.2019.113298.

[4] Diangelakis, N. A., Burnak, B., Katz, J., Pistikopoulos, E. N. Process design and control optimization: A simultaneous approach by multi-parametric programming. *AIChE Journal*, 2017, 63, 4827–4846. https://doi.org/10.1002/aic.15825.

[5] Leyva-Jiménez, F.-J., Fernández-Ochoa, Á., Cádiz-Gurrea, M. D. L. L., Lozano-Sánchez, J., Oliver-Simancas, R., Alañón, M. E., Castangia, I., Segura-Carretero, A., Arráez-Román, D. Application of response surface methodologies to optimize high-added value products developments: Cosmetic formulations as an example. *Antioxidants*, 2022, 11, 1552. https://doi.org/10.3390/antiox11081552.

[6] Campos, B. E., Dias Ruivo, T., Da Silva Scapim, M. R., Madrona, G. S., De Bergamasco, R. C. Optimization of the mucilage extraction process from chia seeds and application in ice cream as a stabilizer and emulsifier. *LWT – Food Science and Technology*, 2016, 65, 874–883. https://doi.org/10.1016/j.lwt.2015.09.021.

[7] Garcia, D. J., You, F. The water-energy-food nexus and process systems engineering: A new focus. *Computers and Chemical Engineering*, 2016, 91, 49–67. https://doi.org/10.1016/j.compchemeng.2016. 03.003.

[8] Sagonda, A. F., Folly, K. A. A comparative study between deterministic and two meta-heuristic algorithms for solar PV MPPT control under partial shading conditions. *Systems and Soft Computing*, 2022, 4, 200040. https://doi.org/10.1016/j.sasc.2022.200040.

[9] Kudela, J., Are metaheuristics worth it? A computational comparison between nature-inspired and deterministic techniques on black-box optimization problems, 2022. https://doi.org/10.48550/ARXIV. 2212.06875.

[10] Kvasov, D. E., Mukhametzhanov, M. S. Metaheuristic vs. deterministic global optimization algorithms: The univariate case. *Applied Mathematics and Computation*, 2018, 318, 245–259. https://doi.org/10.1016/j.amc.2017.05.014.

[11] Raidl, G. R., Puchinger, J., Blum, C. Metaheuristic Hybrids. In Gendreau, M., Potvin, J.-Y. (eds.). *Handbook of Metaheuristics*. Cham: Springer International Publishing, 2019, pp. 385–417. https://doi. org/10.1007/978-3-319-91086-4_12.

[12] Wolpert, D. H., Macready, W. G. No free lunch theorems for optimization. *IEEE Transactions on Evolutionary Computation*, 1997, 1, 67–82. https://doi.org/10.1109/4235.585893.

[13] Glover, F. Future paths for integer programming and links to artificial intelligence. *Computers & Operations Research*, 1986, 13, 533–549. https://doi.org/10.1016/0305-0548(86)90048-1.

[14] Sorensen, W. G., Fred, K., Encyclopedia of Operations Research and Management Science (2013).

[15] Sörensen, K. Metaheuristics – The metaphor exposed. *International Transactions in Operational Research*, 2015, 22, 3–18. https://doi.org/10.1111/itor.12001.

[16] Morales-Castañeda, B., Zaldívar, D., Cuevas, E., Fausto, F., Rodríguez, A. A better balance in metaheuristic algorithms: Does it exist? *Swarm and Evolutionary Computation*, 2020, 54, 100671. https://doi.org/10.1016/j.swevo.2020.100671.

[17] Yang, X.-S., Karamanoglu, M. *Nature-inspired Computation and Swarm Intelligence Algorithms, Theory and Applications*. Yang, X.-S.Ed. London: Academic Press, 2020, pp. 3–18. https://doi.org/10.1016/ c2019-0-00628-0.

[18] Yang, X.-S., Deb, S., Fong, S. Metaheuristic algorithms: optimal balance of intensification and diversification. *Applied Mathematics & Information Sciences*, 2014, 8, 977–983. https://doi.org/10. 12785/amis/080306.

[19] Morales-Castañeda, B., Zaldívar, D., Cuevas, E., Fausto, F., Rodríguez, A. A better balance in metaheuristic algorithms: Does it exist? *Swarm and Evolutionary Computation*, 2020, 54, 100671. https://doi.org/10.1016/j.swevo.2020.100671.

[20] Rajwar, K., Deep, K., Das, S. An exhaustive review of the metaheuristic algorithms for search and optimization: Taxonomy, applications, and open challenges. *Artificial Intelligence Review*, 2023, 56, 13187–13257. https://doi.org/10.1007/s10462-023-10470-y.

[21] Fausto, F., Reyna-Orta, A., Cuevas, E., Andrade, Á. G., Perez-Cisneros, M. From ants to whales: Metaheuristics for all tastes. *Artificial Intelligence Review*, 2020, 53, 753–810. https://doi.org/10.1007/s10462-018-09676-2.

[22] Fister, I., Yang, X.-S., Fister, I., Brest, J., Fister, D. A Brief Review of Nature-Inspired Algorithms for Optimization, 2013. https://doi.org/10.48550/ARXIV.1307.4186.

[23] Blum, C., Roli, A. Metaheuristics in combinatorial optimization. *ACM Computing Surveys*, 2003, 35, 268–308. https://doi.org/10.1145/937503.937505.

[24] Stegherr, H., Heider, M., Hähner, J. Classifying Metaheuristics: Towards a unified multi-level classification system. *Natural Computing*, 2022, 21, 155–171. https://doi.org/10.1007/s11047-020-09824-0.

[25] Lagaros, N. D., Kournoutos, M., Kallioras, N. A., Nordas, A. N. Constraint handling techniques for metaheuristics: A state-of-the-art review and new variants. *Optimization and Engineering*, 2023, 24, 2251–2298. https://doi.org/10.1007/s11081-022-09782-9.

[26] Tsai, C.-W., Chiang, M.-C., Ksentini, A., Chen, M. Metaheuristic algorithms for healthcare: open issues and challenges. *Computers and Electrical Engineering*, 2016, 53, 421–434. https://doi.org/10.1016/j.compeleceng.2016.03.005.

[27] Milan, S. T., Rajabion, L., Ranjbar, H., Navimipour, N. J. Nature inspired meta-heuristic algorithms for solving the load-balancing problem in cloud environments. *Computers & Operations Research*, 2019, 110, 159–187. https://doi.org/10.1016/j.cor.2019.05.022.

[28] Kaul, S., Kumar, Y. Nature-Inspired Metaheuristic Algorithms for Constraint Handling: Challenges, Issues, and Research Perspective. In Kulkarni, A. J., Mezura-Montes, E., Wang, Y., Gandomi, A. H., Krishnasamy, G. (Eds.). *Constraint Handling in Metaheuristics and Applications*. Singapore: Springer Singapore, 2021, pp. 55–80. https://doi.org/10.1007/978-981-33-6710-4_3.

[29] Dragoi, E. N., Dafinescu, V. Review of Metaheuristics Inspired from the Animal Kingdom. *Mathematics*, 2021, 9, 2335. https://doi.org/10.3390/math9182335.

[30] Fernández-Vargas, J. A., Bonilla-Petriciolet, A., Rangaiah, G. P., Fateen, S.-E. K. Performance analysis of stopping criteria of population-based metaheuristics for global optimization in phase equilibrium calculations and modeling. *Fluid Phase Equilibria*, 2016, 427, 104–125. https://doi.org/10.1016/j.fluid.2016.06.037.

[31] Goldberg, D. E., Goldberg, D. E. *Genetic Algorithms in Search, Optimization, and Machine Learning*. Vol. 30print. Boston: Addison-Wesley, 2012.

[32] Kramer, O. *Genetic Algorithm Essentials*. Cham: Springer International Publishing, 2017. https://doi.org/10.1007/978-3-319-52156-5.

[33] Goldberg, D. E., Deb, K. A Comparative Analysis of Selection Schemes Used in Genetic Algorithms. In *Foundations of Genetic Algorithms*. Elsevier, 1991, pp. 69–93. https://doi.org/10.1016/b978-0-08-050684-5.50008-2.

[34] Erdoğmuş, P. *Particle Swarm Optimization with Applications*. InTech, 2018.

[35] Van Laarhoven, P. J. M., Aarts, E. H. L. *Simulated Annealing: Theory and Applications*. Netherlands, Dordrecht: Springer, 1987. https://doi.org/10.1007/978-94-015-7744-1.

[36] Dorigo, M., Stützle, T. *Ant Colony Optimization*. Cambridge, Mass: MIT Press, 2004.

[37] *Differential Evolution*. Berlin/Heidelberg: Springer-Verlag, 2005. https://doi.org/10.1007/3-540-31306-0.

[38] Glover, F., Laguna, M. *Tabu Search*. Boston, MA: Springer US, 1997. https://doi.org/10.1007/978-1-4615-6089-0.

[39] Dey, N. ed. *Applications of Firefly Algorithm and Its Variants: Case Studies and New Developments*. Singapore: Springer Singapore, 2020. https://doi.org/10.1007/978-981-15-0306-1.

[40] Rangaiah, G. P. *Stochastic Global Optimization Techniques and Applications in Chemical Engineering*. Singapore: World Scientific Pub. Co, 2010.

[41] Ponce-Ortega, J. M., Hernández-Pérez, L. G. *Optimization of Process Flowsheets through Metaheuristic Techniques*. Cham: Springer International Publishing, 2019. https://doi.org/10.1007/978-3-319-91722-1.

[42] Dos Santos, L. C., Tavares, F. W., Ahón, V. R. R., Kontogeorgis, G. M. Modeling MEA with the CPA equation of state: A parameter estimation study adding local search to PSO algorithm. *Fluid Phase Equilibria*, 2015, 400, 76–86. https://doi.org/10.1016/j.fluid.2015.05.004.

[43] Ding, Y., Zhang, W., Yu, L., Lu, K. The accuracy and efficiency of GA and PSO optimization schemes on estimating reaction kinetic parameters of biomass pyrolysis. *Energy*, 2019, 176, 582–588. https://doi.org/10.1016/j.energy.2019.04.030.

[44] Bi, K., Qiu, T. An intelligent SVM modeling process for crude oil properties prediction based on a hybrid GA-PSO method. *Chinese Journal of Chemical Engineering*, 2019, 27, 1888–1894. https://doi.org/10.1016/j.cjche.2018.12.015.

[45] Rawat, N., Thakur, P., Singh, A. K., Bhatt, A., Sangwan, V., Manivannan, A. A new grey wolf optimization-based parameter estimation technique of solar photovoltaic. *Sustainable Energy Technologies and Assessments*, 2023, 57, 103240. https://doi.org/10.1016/j.seta.2023.103240.

[46] Roeva, O., Chorukova, E. Metaheuristic algorithms to optimal parameters estimation of a model of two-stage anaerobic digestion of corn steep liquor. *Applied Sciences*, 2022, 13, 199. https://doi.org/10.3390/app13010199.

[47] Kim, J., Chun, H., Baek, J., Han, S. Parameter identification of lithium-ion battery pseudo-2-dimensional models using genetic algorithm and neural network cooperative optimization. *Journal of Energy Storage*, 2022, 45, 103571. https://doi.org/10.1016/j.est.2021.103571.

[48] Wilberforce, T., Rezk, H., Olabi, A. G., Epelle, E. I., Abdelkareem, M. A. Comparative analysis on parametric estimation of a PEM fuel cell using metaheuristics algorithms. *Energy*, 2023, 262, 125530. https://doi.org/10.1016/j.energy.2022.125530.

[49] Dos Santos, L. C., Tavares, F. W., Ahón, V. R. R., Kontogeorgis, G. M. Modeling MEA with the CPA equation of state: A parameter estimation study adding local search to PSO algorithm. *Fluid Phase Equilibria*, 2015, 400, 76–86. https://doi.org/10.1016/j.fluid.2015.05.004.

[50] Ding, Y., Zhang, W., Yu, L., Lu, K. The accuracy and efficiency of GA and PSO optimization schemes on estimating reaction kinetic parameters of biomass pyrolysis. *Energy*, 2019, 176, 582–588. https://doi.org/10.1016/j.energy.2019.04.030.

[51] Bi, K., Qiu, T. An intelligent SVM modeling process for crude oil properties prediction based on a hybrid GA-PSO method. *Chinese Journal of Chemical Engineering*, 2019, 27, 1888–1894. https://doi.org/10.1016/j.cjche.2018.12.015.

[52] Bai, Q., Li, H. The application of hybrid cuckoo search-grey wolf optimization algorithm in optimal parameters identification of solid oxide fuel cell. *International Journal of Hydrogen Energy*, 2022, 47, 6200–6216. https://doi.org/10.1016/j.ijhydene.2021.11.216.

[53] Huang, S., Lin, N., Wang, Z., Zhang, Z., Wen, S., Zhao, Y., Li, Q. A novel data-driven method for online parameter identification of an electrochemical model based on cuckoo search and particle swarm optimization algorithm. *Journal of Power Sources*, 2024, 601, 234261. https://doi.org/10.1016/j.jpowsour.2024.234261.

[54] Na, J., Kshetrimayum, K. S., Lee, U., Han, C. Multi-objective optimization of microchannel reactor for Fischer-Tropsch synthesis using computational fluid dynamics and genetic algorithm. *Chemical Engineering Journal*, 2017, 313, 1521–1534. https://doi.org/10.1016/j.cej.2016.11.040.

[55] Soltani, H., Shafiei, S. Adiabatic reactor network synthesis using coupled genetic algorithm with quasi linear programming method. *Chemical Engineering Science*, 2015, 137, 601–612. https://doi.org/10.1016/j.ces.2015.06.068.

[56] Bayat, M., Hamidi, M., Dehghani, Z., Rahimpour, M. R. Dynamic optimal design of an industrial ethylene oxide (EO) reactor via differential evolution algorithm. *Journal of Natural Gas Science and Engineering*, 2013, 12, 56–64. https://doi.org/10.1016/j.jngse.2013.01.004.

[57] Zhang, X., Jin, L., Cui, C., Sun, J. A self-adaptive multi-objective dynamic differential evolution algorithm and its application in chemical engineering. *Applied Soft Computing*, 2021, 106, 107317. https://doi.org/10.1016/j.asoc.2021.107317.

[58] Lee, S. W., Gebreyohannes, T. G., Shin, J. H., Kim, H. W., Kim, Y. T. Carbon-efficient reaction optimization of nonoxidative direct methane conversion based on the integrated reactor system. *Chemical Engineering Journal*, 2024, 481, 148286. https://doi.org/10.1016/j.cej.2023.148286.

[59] Zainullin, R. Z., Zagoruiko, A. N., Koledina, K. F., Gubaidullin, I. M., Faskhutdinova, R. I. Multi-criterion optimization of a catalytic reforming reactor unit using a genetic algorithm. *Catalysis in Industry*, 2020, 12, 133–140. https://doi.org/10.1134/s2070050420020129.

[60] Soltani, H., Shafiei, S. Adiabatic reactor network synthesis using coupled genetic algorithm with quasi linear programming method. *Chemical Engineering Science*, 2015, 137, 601–612. https://doi.org/10.1016/j.ces.2015.06.068.

[61] Zhang, X., Jin, L., Cui, C., Sun, J. A self-adaptive multi-objective dynamic differential evolution algorithm and its application in chemical engineering. *Applied Soft Computing*, 2021, 106, 107317. https://doi.org/10.1016/j.asoc.2021.107317.

[62] Sun, Y., Dai, H., Moayedi, H., Nguyen Le, B., Muhammad Adnan, R. Predicting steady-state biogas production from waste using advanced machine learning-metaheuristic approaches. *Fuel*, 2024, 355, 129493. https://doi.org/10.1016/j.fuel.2023.129493.

[63] Ramanathan, S. P., Mukherjee, S., Dahule, R. K., Ghosh, S., Rahman, I., Tambe, S. S., Ravetkar, D. D., Kulkarni, B. D. Optimization of continuous distillation columns using stochastic optimization approaches. *Chemical Engineering Research and Design*, 2001, 79, 310–322. https://doi.org/10.1205/026387601750281671.

[64] Boozarjomehry, R. B., Laleh, A. P., Svrcek, W. Y. Automatic design of conventional distillation column sequence by genetic algorithm. *The Canadian Journal of Chemical Engineering*, 2009, 87, 477–492. https://doi.org/10.1002/cjce.20175.

[65] Vázquez-Ojeda, M., Segovia-Hernández, J. G., Hernández, S., Hernández-Aguirre, A., Kiss, A. A. Optimization of an Ethanol Dehydration Process Using Differential Evolution Algorithm. In *23rd European Symposium on Computer Aided Process Engineering*. Elsevier, 2013, 217–222. https://doi.org/10.1016/b978-0-444-63234-0.50037-3.

[66] Shahandeh, H., Ivakpour, J., Kasiri, N. Internal and external HIDiCs (heat-integrated distillation columns) optimization by genetic algorithm. *Energy*, 2014, 64, 875–886. https://doi.org/10.1016/j.energy.2013.10.042.

[67] Gutiérrez-Guerra, R., Segovia-Hernández, J. G. Novel approach to design and optimize heat-integrated distillation columns using Aspen Plus and an optimization algorithm. *Chemical Engineering Research and Design*, 2023, 196, 13–27. https://doi.org/10.1016/j.cherd.2023.06.015.

[68] Gutiérrez-Guerra, R., Murrieta-Dueñas, R., Cortez-González, J., Segovia-Hernández, J. G., Hernández, S., Hernández-Aguirre, A. Design and optimization of HIDiC columns using a constrained Boltzmann-based estimation of distribution algorithm-evaluating the effect of relative volatility. *Chemical Engineering and Processing: Process Intensification*, 2016, 104, 29–42. https://doi.org/10.1016/j.cep.2016.02.004.

[69] García-Hernández, A. E., Segovia-Hernández, J. G., Sánchez-Ramírez, E., Zarazúa, G. C., Araujo, I. F. H., Quiroz-Ramírez, J. J. Sustainable aviation fuel from Butanol: A Study in optimizing Economic

and Environmental impact through process intensification. *Chemical Engineering and Processing – Process Intensification*, 2024, 200, 109769. https://doi.org/10.1016/j.cep.2024.109769.

[70] Kruber, K. F., Grueters, T., Skiborowski, M. Advanced hybrid optimization methods for the design of complex separation processes. *Computers and Chemical Engineering*, 2021, 147, 107257. https://doi.org/10.1016/j.compchemeng.2021.107257.

[71] Vázquez-Ojeda, M., Segovia-Hernández, J. G., Hernández, S., Hernández-Aguirre, A., Kiss, A. A. Design and optimization of an ethanol dehydration process using stochastic methods. *Separation and Purification Technology*, 2013, 105, 90–97. https://doi.org/10.1016/j.seppur.2012.12.002.

[72] Cortez-González, J., Murrieta-Dueñas, R., Gutiérrez-Guerra, R., Segovia-Hernández, J. G., Hernández-Aguirre, A. Design and Optimization of Pressure Swing Distillation Using a Stochastic Algorithm Based in the Boltzmann Distribution. In *Computer Aided Chemical Engineering*. Elsevier, 2012, 687–691. https://doi.org/10.1016/B978-0-444-59519-5.50138-6.

[73] Behroozsarand, A., Soltani, H. Hydrogen plant heat exchanger networks synthesis using coupled Genetic Algorithm-LP method. *Journal of Natural Gas Science and Engineering*, 2014, 19, 62–73. https://doi.org/10.1016/j.jngse.2014.04.015.

[74] Pavão, L. V., Costa, C. B. B., Ravagnani, M. A. S. S., Jiménez, L. Costs and environmental impacts multi-objective heat exchanger networks synthesis using a meta-heuristic approach. *Applied Energy*, 2017, 203, 304–320. https://doi.org/10.1016/j.apenergy.2017.06.015.

[75] Pavão, L. V., Miranda, C. B., Costa, C. B. B., Ravagnani, M. A. S. S. Efficient multiperiod heat exchanger network synthesis using a meta-heuristic approach. *Energy*, 2018, 142, 356–372. https://doi.org/10.1016/j.energy.2017.09.147.

[76] Zhang, H., Cui, G. Optimal heat exchanger network synthesis based on improved cuckoo search via Lévy flights. *Chemical Engineering Research and Design*, 2018, 134, 62–79. https://doi.org/10.1016/j.cherd.2018.03.046.

[77] Aguitoni, M. C., Pavão, L. V., Antonio da Silva Sá Ravagnani, M. Heat exchanger network synthesis combining simulated annealing and differential evolution. *Energy*, 2019, 181, 654–664. https://doi.org/10.1016/j.energy.2019.05.211.

[78] Stampfli, J. A., Ong, B. H. Y., Olsen, D. G., Wellig, B., Hofmann, R. Applied heat exchanger network retrofit for multi-period processes in industry: A hybrid evolutionary algorithm. *Computers and Chemical Engineering*, 2022, 161, 107771. https://doi.org/10.1016/j.compchemeng.2022.107771.

[79] Behroozsarand, A., Soltani, H. Hydrogen plant heat exchanger networks synthesis using coupled Genetic Algorithm-LP method. *Journal of Natural Gas Science and Engineering*, 2014, 19, 62–73. https://doi.org/10.1016/j.jngse.2014.04.015.

[80] Cartwright, H., Curteanu, S. neural networks applied in chemistry. II. Neuro-Evolutionary techniques in process modeling and optimization. *Industrial & Engineering Chemistry Research*, 2013, 52, 12673–12688. https://doi.org/10.1021/ie4000954.

[81] Aladejare, A. E., Onifade, M., Lawal, A. I. Application of metaheuristic based artificial neural network and multilinear regression for the prediction of higher heating values of fuels. *International Journal of Coal Preparation and Utilization*, 2020, 42, 1830–1851. https://doi.org/10.1080/19392699.2020.1768080.

[82] Zhou, T., Chiam, K.-H. Synthetic data generation method for data-free knowledge distillation in regression neural networks. *Expert Systems with Applications*, 2023, 227, 120327. https://doi.org/10.1016/j.eswa.2023.120327.

[83] Hafiz Al Hariri, A., Khalifa, A. E., Talha, M., Awda, Y., Hasan, A., Alawad, S. M. Artificial neural network and differential evolution optimization of a circulated permeate gap membrane distillation unit. *Separation and Purification Technology*, 2024, 338, 126517. https://doi.org/10.1016/j.seppur.2024.126517.

[84] Khan, M., Ullah, Z., Mašek, O., Raza Naqvi, S., Nouman Aslam Khan, M. Artificial neural networks for the prediction of biochar yield: A comparative study of metaheuristic algorithms. *Bioresource Technology*, 2022, 355, 127215. https://doi.org/10.1016/j.biortech.2022.127215.

[85] Varol Altay, E., Gurgenc, E., Altay, O., Dikici, A. Hybrid artificial neural network based on a metaheuristic optimization algorithm for the prediction of reservoir temperature using hydrogeochemical data of different geothermal areas in Anatolia (Turkey). *Geothermics*, 2022, 104, 102476. https://doi.org/10.1016/j.geothermics.2022.102476.

[86] Fetimi, A., Dâas, A., Merouani, S., Alswieleh, A. M., Hamachi, M., Hamdaoui, O., Kebiche-Senhadji, O., Yadav, K. K., Jeon, B.-H., Benguerba, Y. Predicting emulsion breakdown in the emulsion liquid membrane process: Optimization through response surface methodology and a particle swarm artificial neural network. *Chemical Engineering and Processing – Process Intensification*, 2022, 176, 108956. https://doi.org/10.1016/j.cep.2022.108956.

[87] Larraín, S., Pradenas, L., Pulkkinen, I., Santander, F. Multiobjective optimization of a continuous kraft pulp digester using SPEA2. *Computers and Chemical Engineering*, 2020, 143, 107086. https://doi.org/10.1016/j.compchemeng.2020.107086.

[88] Ochoa, S., García, C., Alcaraz, W. Real-time optimization and control for polyhydroxybutyrate fed-batch production at pilot plant scale. *Journal of Chemical Technology and Biotechnology*, 2020, 95, 3221–3231. https://doi.org/10.1002/jctb.6500.

Mathias Neufang, Emma Pajak, Damien van de Berg, Ye Seol Lee,
and Ehecatl Antonio Del Rio Chanona*

Chapter 5
Surrogate-based optimization techniques for process systems engineering

Abstract: Optimization plays an important role in chemical engineering, impacting cost-effectiveness, resource utilization, product quality, and process sustainability metrics. This chapter broadly focuses on data-driven optimization, particularly on model-based derivative-free techniques, also known as surrogate-based optimization. The chapter introduces readers to the theory and practical considerations of various algorithms, complemented by a performance assessment across multiple dimensions, test functions, and two chemical engineering case studies: a stochastic high-dimensional reactor control study and a low-dimensional constrained stochastic reactor optimization study. This assessment sheds light on each algorithm's performance and suitability for diverse applications. Additionally, each algorithm is accompanied by background information, mathematical foundations and algorithm descriptions. Among the discussed algorithms are Bayesian optimization (BO), including state-of-the-art *trust region BO* (TuRBO), constrained optimization by linear approximation (COBYLA), the ensemble tree model optimization tool (ENTMOOT) that uses decision trees as surrogates, stable noisy optimization by branch and fit (SNOBFIT), methods that use radial basis functions (RBFs) such as dynamic coordinate search (DYCORS) and stochastic RBFs (SRBFStrategy), constrained optimization by quadratic approximations (COBYQA), as well as a few others recognized for their effectiveness in surrogate-based optimization. By combining theory with practice, this chapter equips readers with the knowledge to integrate surrogate-based optimization techniques into chemical engineering. The overarching aim is to highlight the advantages of surrogate-based optimization, introduce state-of-the-art algorithms, and provide guidance for successful implementation within process systems engineering.

Keywords: data-driven optimization, model-based optimization, derivative-free optimization, black-box optimization, chemical process engineering

Classification: 65C05, 62M20, 93E11, 62F15, 86A22

***Corresponding author: Ehecatl Antonio Del Rio Chanona**, Department of Chemical Engineering, Imperial College London, United Kingdom, e-mail: a.del-rio-chanona@imperial.ac.uk
Mathias Neufang, Department of Chemical Engineering, Imperial College London, Exhibition Rd, South Kensington, London SW7 2AZ, United Kingdom, e-mail: mathias.neufang22@imperial.ac.uk
Emma Pajak, Damien van de Berg, Department of Chemical Engineering, Imperial College London, United Kingdom
Ye Seol Lee, Department of Chemical Engineering, University College London, United Kingdom

https://doi.org/10.1515/9783111383439-005

5.1 Introduction

This chapter aims to equip process systems and chemical engineers with a comprehensive understanding of *data-driven optimization* (DDO), focusing specifically on model-based (surrogate) techniques. It explores both the theoretical foundations and the practical performance assessment of various algorithms, fostering an intuitive grasp of their behavior and efficacy. This understanding will enable readers to utilize these methods effectively. The remaining chapter takes the following structure:

This section sets the scene for the chapter, defining *derivative-free optimization* (DFO), introducing the three types of DFO methods, and exploring their application in chemical engineering.

In **Section 5.2**, readers are formally introduced to the concept of model-based DFO, where popular algorithms like *Bayesian optimization* (BO) take center stage. Background information, mathematical formulations, and algorithm descriptions are presented for a holistic overview of each method.

In **Section 5.3**, the performance assessment focuses on benchmarking model-based algorithms for unconstrained problems. The objective is to equip the reader with sufficient intuition and background information to make informed decisions when selecting algorithms.

Section 5.4 focuses on benchmarking model-based algorithms for constrained problems. The objectives mirror those in Section 5.3 but in the case of constrained black-box problems.

In **Section 5.5**, two chemical engineering optimization case studies are presented: an unconstrained, stochastic high-dimensional problem, and a stochastic low-dimensional constrained problem. This section serves to demonstrate the applicability of the surrogate-based algorithms to the process systems engineering field.

5.1.1 Background

Traditionally, optimization within chemical engineering relies on algebraic expressions or established knowledge-based models, which can be optimized by leveraging derivative information from their analytical expressions. However, the rise of digitalization, such as smart measuring devices, process analytical technology, sensor technologies, cloud platforms, and the *industrial Internet of things* (IIoT), has called for the need for optimization algorithms guided purely by the collected data, and therefore the term data-driven optimization has emerged. In complex chemical systems, it is not unusual for data collection to be feasible only through the evaluation of an expensive black-box function. This function may represent an in vitro chemical experiment with undetermined mechanisms or a costly process reconfiguration as well as an in silico simulation in the form of computational fluid dynamics or quantum mechanical calculations. Evaluating such deterministic models often relies on complex and expensive simulations that are corrupted by computational noise as unintended variations and inaccuracies happen, which makes even numerical derivatives difficult and unreliable [15]. In such instances, data-driven algorithms emerge as a solution, enabling the optimization of these systems [53].

Many engineering optimization challenges can be framed as "costly" black-box problems, which are constrained by the number of function evaluations. Engineers often construct precise models of physical systems that are either differentiable or economical to evaluate. These models can be resolved efficiently, and their solutions can be applied to the actual system. However, when gradient information or cost-effective models are unavailable, it becomes necessary to utilize efficient optimization methods that depend solely on function evaluations. The process of developing a model can be viewed as an integral part of the expensive black-box optimization procedure itself [53]. Within this context, algorithms must rise to the challenge, with optimization, *artificial intelligence* (AI), and machine learning playing pivotal roles in enabling advancements in automated control and decision-making [27]. In fact, DDO boasts a rich historical background within the realm of chemical engineering [19, 30].

Similar challenges have been the topic of research across many communities. This chapter refers to "DDO." However, literature similarly refers to this subject area as "derivative-free optimization," "gradient-free optimization," "zeroth-order optimization," "simulation-based optimization," or, more specifically within the context of process systems engineering, "black-box optimization" [19, 30]. The term "zeroth-order" alludes to algorithms that do not utilize a function's first- or second-order derivatives [57].

5.1.2 Unconstrained optimization formulation

Problem (5.1) represents the generic, unconstrained optimization formulation of interest throughout the chapter, where $f: \mathbb{R}^{n_x} \to \mathbb{R}$ describes the objective function:

$$\min_{\mathbf{x}} \quad f(\mathbf{x})$$
$$\mathbf{x} \in \mathcal{X} \subseteq \mathbb{R}^{n_x} \tag{5.1}$$

In traditional numerical optimization, given an analytical expression of f, the necessary conditions for a (local) optimal solution of Problem (5.1) would be any point or region where the gradient is zero $\nabla_x f(x) = 0$, and the sufficient condition would, in addition, have the Hessian matrix be positive semi-definite $\nabla^2_{xx} f(x) \succeq 0$. Therefore, optimization algorithms that use derivatives (e.g., Newton's method and gradient descent) seek to find regions where these conditions (particularly the necessary conditions) are met.

In DDO, things are not as "simple." Given that there is no analytical expression, algorithms must seek to explore the space in the hope of gathering adequate information about the function at hand, while at the same time using this information to optimize the function. The general assumption for practical applications in DDO is that ensuring convergence to an optimum is hard given that the function itself is unknown, and therefore termination criteria are generally set in the number of evaluations or runtime [8].

DFO, a term that will be used in this chapter interchangeably with DDO, encompasses algorithms designed to optimize functions without explicitly using derivative information. DFO methods can be broadly classified into two main categories: *model-based* derivative-free methods, also known as *surrogate-based* optimization, and *direct derivative-free* methods. Situated in-between these categories are the *finite-difference* methods, which approximate derivatives using function evaluations [57], and for second- or higher-order methods, it can be viewed as constructing a local approximation (a model) of the system. In this chapter, the focus is on surrogate-based optimization methods. A brief overview of the different methods is presented next, before focusing on model-based derivative-free (also known as surrogate-based) methods.

5.1.3 Direct methods

Direct derivative-free methods directly utilize sampled points to determine the next sampling location to approach the optimum without relying on the intermediate construction and optimization of surrogates. Many early DFO algorithms, such as the simplex (Nelder-Mead) algorithm, evolutionary (particle swarm) algorithms (which are a type of metaheuristics), as well as direct global optimization algorithms, simulated annealing, and pattern search fall into this category [53].

5.1.4 Finite-difference methods

Finite-difference methods operate on the principle that even in cases where the overall function is unknown, sampled data can be utilized to approximate derivatives. By approximating the gradient, typical gradient descent methods can be applied. These methods fall between direct derivative-free methods and surrogate-based DFO methods; they base sampling decisions directly on obtained data, similar to direct methods. However, when second-order or quasi-Newton methods are employed, they can be thought of as quadratic surrogates of the underlying true function, in some cases with iterative updates and "memory." Well-known finite difference methods include the Gradient Descent + Momentum Method, Adam, RSM Prob, and (L)BFGS. The interested reader is referred to the work on [57] for further treatment on the subject.

5.1.5 Model-based methods

In process engineering, there has been a notable surge in interest in model-based techniques, drawing upon the field's expertise in surrogate modeling, meta-modeling, and reduced-order modeling [19, 30]. Traditionally, within the process systems engineering community, model-based DFO methods are also referred to as surrogate optimization, highlighting the creation of a "surrogate" function [32, 39, 43]. As these terms are interchangeable, this chapter will refer to both throughout.

The fundamental concept of model-based DFO involves creating a model of f using data, known as a surrogate function $\hat{f}$; once this surrogate model is built it can be used to determine candidate solutions. The general framework of model-based DFO seeks to explore the function f to refine its surrogate $\hat{f}$ while at the same time finding points that optimize f. The underlying shape of f remains unknown, and the choice of model type (e.g., *radial basis functions* (RBFs), *Gaussian processes* (GPs), quadratic surrogates, neural networks, etc.) used to fit $\hat{f}$ makes some assumptions on how f behaves. Section 5.2 is devoted to the comprehensive treatment of model-based derivative-free methods.

5.1.6 Process systems engineering applications of model-based methods

Model-based DFO has been extensively used in chemical engineering, and the community has a rich history of developing state-of-the-art algorithms for various applications across a variety of fields – for instance, in quantum chemistry, using a classical state vector simulator as a surrogate model to approximate the optimization landscape in variational quantum eigensolvers, enhancing the convergence and efficiency of quantum circuit optimization [68]. In a different vein, model-based DDO has also found applications in the pharmaceutical manufacturing sector, as described in a recent paper where feasibility-driven optimization incorporates additional stages to improve both local exploitation and global exploration. This approach leads to lower costs, improved product quality, and greater process flexibility and robustness [73]. This section highlights some of the recent research and implementation of model-based DDO in chemical engineering grouped by application areas to showcase the versatility and efficacy of these approaches in addressing complex challenges across the chemical engineering domain.

Process design and flowsheeting: Process design and flowsheeting involve strategic planning and optimization of chemical processes and systems. Advanced methodologies and tools are employed to create optimized, efficient, and robust processes, emphasizing operational efficiency and environmental sustainability. For example, surrogate-based optimization strategies have been proposed for the global optimization of process flowsheets, leveraging algebraic surrogates constructed from rigorous simulations via Bayesian symbolic regression [54]; other approaches have suggested the use of GPs (kriging) to optimize nonlinear problems that include noisy implicit black box functions, such as modular process simulators, to manage noise and compensate for unavailable derivatives [11]. Other related works have focused on the use of GPs as versatile surrogate modeling tools to address feasibility constraints from non-converged simulations based on performance-risk trade-offs [64]. Another study introduces a multi-fidelity BO framework for reactor design, reducing design time for highly parameterized reactors while ensuring optimal geometry and operating conditions with experimental validation of 3D-printed reactor geometries [55]. An overview of process design via BO can be found in the following work [72].

Moving away from Bayesian approaches, other studies on specific processes demonstrate the potential of surrogate-based methods in process design, such as the optimization of a hybrid polycrystalline silicon production route, where surrogate models of key unit operations are constructed to enable entire process optimization exploring various scenarios, e.g., maximizing silicon production, minimizing operating costs, and maximizing total profit [45]. Similarly, a trust region filter framework for heat exchanger network synthesis integrates detailed shell-and-tube heat exchanger models to optimize network topology and exchanger design, including parameters such as pressure drops, shell configurations, and tube arrangements [47]. Surrogate-based optimization has also been applied to process systems for resource recovery from wastewater, integrating DFO modeling tools that incorporate classification surrogate models and address uncertainties, thereby offering holistic solutions to reduce environmental impacts in food and beverage production [64]. In addition to the above, many other surrogates have been used, from *graph neural networks* (GNNs) for granular flows [70], to symbolic regression-based surrogates for flexibility analysis [66], to quantile neural networks for two-stage stochastic optimization [63]. An overview of data-driven and hybrid models for subsequent optimization of separation processes can be found in [44].

Supply chains and planning, scheduling, and operations: Surrogate-based DDO is also a powerful tool utilized within supply chains, exemplified by a study where historical data models optimize demand response scheduling of air separation units within a nonlinear dynamic framework, ensuring dynamic feasibility and computational efficiency [36]. Another example is the extension of the DDO of bi-level mixed-integer nonlinear problems framework to tackle mixed-integer bi-level multi-follower stochastic optimization problems for integrated planning and scheduling problems under demand uncertainty [50]. In a different study on scheduling optimization in integrated chemical plants, *convex region surrogate* (CRS) models are used in mixed-integer programming frameworks, of which the effectiveness is demonstrated through its application to an industrial test case from a Praxair plant [26]. Furthermore, the automation and simulation of plant-level surrogate construction, as well as a propagation-error mitigation strategy has enabled the investigation of various levels of abstraction in surrogate modeling to enhance site-level optimization accuracy and efficiency [51]. Additionally, techniques such as optimality surrogates and DFO have been employed to address tractability challenges in large-scale, multi-level formulations, integrating hierarchical planning, scheduling, and control decisions within chemical companies [59].

Design of experiments: A recent paper illustrates high-throughput screening for *Catalytically Active Inclusion Bodies* (CatIBs), utilizing a semi-automated cloning workflow and BO to efficiently generate and screen 63 glucose dehydrogenase variants from *Bacillus subtilis*, reducing manual effort and enhancing reproducibility [69]. Similarly, BO has been utilized for the development of computation-driven materials discovery workflows, focusing on the exploration of unchartered material space [52]. Another example explores multi-objective BO in flow reactor experiments to identify the Pareto front of the optimal solution – using the *q-noisy expected hypervolume improvement* (qNEVI) acquisition function [75]. Additionally, *piecewise affine surrogate-based optimization* (PWAS) has been applied to tackle experimental planning challenges, such as optimizing Suzuki-Miyaura cross-coupling reaction conditions, benchmarked against genetic algorithms and BO variants [76].

Process dynamics and control: Surrogate DDO finds diverse applications within process control systems, for example, to create surrogates of the dynamic system by time-series modeling [41]. In a different work, artificial neural networks with rectifier units accurately represent piecewise affine functions for linear time-invariant systems within *model predictive control* (MPC) [42]. Another study enhances computational efficiency, noise resilience, and control action smoothness in optimal dynamic product transitions using a data-driven Bayesian approach [65]. Additionally, an innovative application of online feedback optimization with GP regression mitigates plant-model mismatch in compressor station operations, resulting in reduced power consumption despite incomplete plant knowledge [60].

Methodological developments: Finally, looking at methods-based applications of surrogate-based optimization, a recent publication introduces a new DDO algorithm using *support vector machines* (SVMs) to tackle numerical infeasibilities within *differential algebraic equations* (DAEs), showcasing its effectiveness across diverse case studies, including complex scenarios in reaction engineering, such as the thermal cracking of natural gas liquids [40]. Another approach uses decision-focused surrogate modeling, which aims to address computationally challenging nonlinear optimization problems in real-time settings, validated through nonlinear process case studies such as reactors and heat exchangers [67]. Surrogate models also find applications in global optimization, exemplified by a novel algorithm tailored for solving linearly constrained mixed-variable problems, employing a piecewise affine surrogate of the objective function alongside an exploration function utilizing *mixed-integer linear program* (MILP) solvers to search the feasible domain [62].

Two works explore optimization methodologies. One employs surrogate-based branch-and-bound algorithms for simulation-based optimization, ensuring consistent convergence to optimal solutions despite variability in initialization, sampling, and surrogate model selection [61]. Another develops a method for BO of mixed-integer nonlinear programming problems [71]. A final example is a data-driven coordination framework for enterprise-wide optimization, which maintains organizational autonomy while outperforming conventional distributed optimization methods across various case studies [58].

As it is clear from this section, the process systems engineering community has a rich history of utilizing and advancing model-based (surrogate-based) DFO techniques.

This chapter has two main objectives. The first objective is to serve as an introduction to newcomers to the field, to introduce the cornerstone concepts behind model-based derivative-free methods, commonly termed surrogate-based optimization, as well as to present the main advantages, shortcomings, and general framework. A second objective of this book chapter is to offer a framework for objectively comparing different model-based derivative-free methods, and particularly focus on process systems engineering applications. While an exhaustive and all-encompassing benchmark is beyond the scope of this book chapter, benchmarking results have been provided for a selection of functions, both in the constrained and unconstrained case, as well as process systems engineering examples. A particular effort has been made to share all the associated code from this benchmarking, outlining its use, in the appended GitHub repository. The authors hope this will inspire practitioners to use the code base to compare different algorithms to meet their specific goals.

5.2 Model-based derivative-free optimization

As introduced in Section 5.1.5, model-based methods utilize surrogate models of the objective function to inform their updates, thereby guiding the optimization process by only leveraging sampled data without relying on explicit derivative information [37]. However, sampling in many cases does not yield the true function value $f(\mathbf{x})$, because of measurement noise. The feedback of the sampled system is represented by y, which consists of the objective function value $f(\mathbf{x})$ and may be blurred by measurement noise ε as follows:

$$y = f(\mathbf{x}) + \varepsilon \tag{5.2}$$

It is generally assumed that ε is a random variable, and in many cases, this assumption is extended to be normally distributed, i.e., $\varepsilon \sim \mathcal{N}(0, \sigma_\varepsilon)$. The sampled data set $\mathcal{D}$ containing n_d sampled positions and corresponding system feedback $(\mathbf{x}, y)$ are depicted in this chapter as $\mathcal{D} = \{(\mathbf{x}^{(1)}, y^{(1)}), (\mathbf{x}^{(2)}, y^{(2)}), \dots, (\mathbf{x}^{(n_d)}, y^{(n_d)})\}$. This leaves $X \in \mathbb{R}^{n_x \times n_d}$ as the matrix containing the data from the inputs (decision variables), and $\mathbf{y} \in \mathbb{R}^{n_d}$ as the vector containing the data from the output (sampled objective function values possibly corrupted by noise):

$$\mathbf{x} = \begin{bmatrix} x_1 \\ x_2 \\ \vdots \\ x_{n_x} \end{bmatrix}, \quad X = \begin{bmatrix} \left(\mathbf{x}^{(1)}\right)^T \\ \left(\mathbf{x}^{(2)}\right)^T \\ \vdots \\ \left(\mathbf{x}^{(n_d)}\right)^T \end{bmatrix} = \begin{bmatrix} x_1^{(1)} & \cdots & x_{n_x}^{(1)} \\ x_1^{(2)} & \cdots & x_{n_x}^{(2)} \\ \vdots & \ddots & \vdots \\ x_1^{(n_d)} & \cdots & x_{n_x}^{(n_d)} \end{bmatrix}, \quad \mathbf{y} = \begin{bmatrix} y^{(1)} \\ y^{(2)} \\ \vdots \\ y^{(n_d)} \end{bmatrix}$$

For simplicity, the output dimension is set as $n_y = 1$. However, this is revisited later on in the case of constrained problems.

In most model-based optimization algorithms, two sequential optimizations are conducted. The first one is responsible for the surrogate (model) building step, which generally relies on building a surrogate (model) function that minimizes a likelihood function between the data and the surrogate function (commonly a least squares). This creates an approximation of f denoted as $\hat{f}$. The second optimization optimizes $\hat{f}$ to find the next best candidate. Solely optimizing the surrogate function $\hat{f}$ might lead to overexploitation, and not enough exploration, and therefore algorithms include some exploration components into their routine. Different algorithms propose different ways to add this exploration element; for example, BO, which uses GPs as surrogates, leverages the prediction of the uncertainty to sample points that are promising but also unknown. Other algorithms, directly sample points that are meant to be used in the surrogate building step to obtain more information from the function f. The subsequent sections describe the different algorithms and how they handle this exploration-exploration dilemma.

In summary, at each iteration, a surrogate function is created through some optimization procedure, and then this surrogate function is optimized to find the best candidate point(s) for the next iteration. It should also be noted that DFO does not preclude the use of conventional derivative-based solvers for the optimization of surrogate models [53]. Notably, for model-based DFO to prove effective, the building of the surrogate and its subsequent optimization should be computationally less expensive than sampling the true function [38].

The choice of the most appropriate surrogate model is contingent upon the specific characteristics of the system at hand, and a variety of surrogates can achieve state-of-the-art performance; some notable examples:

– Algorithms that construct quadratic surrogates, such as *constrained optimization by quadratic approximation* (COBYQA) [83] and *convex quadratic trust-region optimizer* (CUATRO) [53]
– Approaches utilizing GPs, as seen in BO [25]
– Techniques relying on radial basis functions such as DYCORS and SRBFStrategy [9, 23]
– The use of *decision trees* (DTs) as surrogates such as in *ensemble tree model optimization tool* (ENTMOOT) [49]
– Strategies that leverage basis functions to construct surrogates [13, 28, 30]

In the next section, the general workflow for surrogate-based optimization is presented.

5.2.1 Workflow for model-based DFO

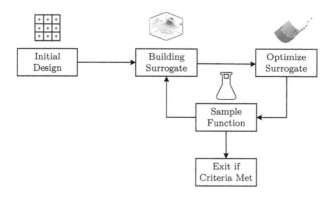

Figure 5.1: Fundamental workflow of model-based DFO.

The fundamental workflow of model-based DFO is illustrated in Figure 5.1, with further details on each stage below:

- **Initial design:** This stage entails the initial set of experiments or samples aimed at gathering essential information about the objective function. Techniques commonly employed include *design of experiments* (DoE), *Latin hypercube sampling* (LHS), factorial designs, or space-filling designs in general, which explore the input space to capture its behavior and characteristics.
- **Build surrogate:** This stage uses the available data to build a surrogate model of the objective function. Surrogates can take various forms, including machine learning techniques such as GPs, DTs, and neural networks. It is noteworthy that surrogates do not have to be machine learning models; for example, quadratic surrogates, which approximate the objective function using quadratic polynomials can be used. For instance, algorithms like COBYQA utilize quadratic surrogates to iteratively optimize the objective function.
- **Optimize surrogate:** This step uses the surrogate to find the next best point(s) to sample. One straightforward option is to optimize the surrogate (a model of the objective function) and sample there. However, more often than not the surrogate is not simply optimized and sampled, but some exploration component is added. This involves a trade-off between exploration (sampling in unexplored regions) and exploitation (sampling in regions that are likely to contain the optimal solution).

- **Sample function:** This phase involves querying the "true" system and evaluating the objective function to obtain new data. These samples are later used to update the surrogate, and at the same time represent optimal (or near-optimal) solutions at termination.
- **Termination:** Termination occurs when the optimization problem-specific criterion is met, such as reaching a predefined number of iterations, achieving a satisfactory solution, or exhausting computational resources.

5.2.2 Local versus global surrogates

Surrogate optimization approaches can be broadly classified into local and global approaches. Global approaches proceed by constructing a surrogate model based on all their samples and constructing a single (flexible) surrogate for the whole decision space. The optimization of this surrogate and the next sample is also allowed to be anywhere within the problem constraints, without it having to be close to the current (or any previous) sampled point. Several practical implementations rely on neural networks [17, 56], GPs [12, 46], RBFs [7, 34], or a combination of basis functions [28, 30] to create the surrogate model.

By contrast, local approaches seek to maintain an accurate approximation of the original optimization problem within a trust region, whose position and size are adapted iteratively. This procedure entails updating or reconstructing the surrogate model as the trust region moves around, and it benefits from a well-developed convergence theory providing sufficient conditions for local optimality in unconstrained and bound-constrained problems [14]. However, it is important to highlight that to guarantee convergence these methods rely on a substantial number of function evaluations, which might be too expensive for costly optimization problems, and therefore are rarely used in expensive black-box optimization problems.

Handling constraints for DFO problems is still an active field of research [20, 31, 35, 79], and in some instances trust regions have been found beneficial, particularly in safe exploration [29, 48].

In the following sections, the main ideas behind different surrogate-based optimization algorithms are presented. These are intended to give a high-level understanding to the audience, and the interested reader is encouraged to go into additional material and references for a complete treatment of the different methodologies.

5.2.3 Bayesian optimization (BO)

BO is a surrogate-based optimization strategy that relies on a probabilistic model as the surrogate. This probabilistic surrogate is used to predict the outcome of the objective function at unobserved points and then uses an acquisition function to determine the next point to evaluate. This approach balances exploration (trying points where little is known about the function) and exploitation (trying points where the model predicts a high value).

The process begins with initial observations of the objective function, which are used to build a probabilistic model, often a GP. This model is then used to predict the outcome of the function at any untried point. The acquisition function, such as *expected improvement* (EI), *probability of improvement* (PI), or upper confidence bound, uses these predictions to decide which point to evaluate next. This point is then evaluated on the actual objective function, and the result is used to update the probabilistic model. This cycle continues until a satisfactory solution is found or the computational budget is exhausted. As the reader can appreciate, BO follows the standard model-based DFO framework that is seen in Figure 5.1.

5.2.3.1 BO: surrogate model

BO relies on a probabilistic model, usually a GP. A GP is a stochastic process that, instead of defining a specific function, specifies a distribution over functions, where any finite set of function values has a joint Gaussian distribution. This makes GPs particularly useful for tasks like regression, interpolation, uncertainty quantification, and BO.

A GP is fully characterized by its mean function $m(\mathbf{x})$ and covariance function (or kernel) $k(\mathbf{x}, \mathbf{x}')$. For a given set of input points $\mathbf{x}$, the corresponding GP is defined as:
-   Mean function: $m(\mathbf{x})$ describes the expected value of the process at any point $\mathbf{x}$. Typically, it is assumed to be zero, assuming that the dataset is standardized. This is the value that the GP will predict as an expected value far away from any data point.
-   Covariance function: $k(\mathbf{x}, \mathbf{x}')$ specifies how the process covaries between any two points $\mathbf{x}$ and $\mathbf{x}'$. It captures the similarity between inputs and determines how much influence nearby points have on each other. Common choices include the squared exponential kernel, Matérn kernel, and rational quadratic kernel.

Given that a GP is a probabilistic model of the objective function $f \sim \mathcal{GP}(m(\cdot), k(\cdot, \cdot))$, once a set of data points, $\mathcal{D} = \{X, \mathbf{y}\} = \{(\mathbf{x}^{(i)}, y^{(i)})\}_{i=1}^{n_d}$, has been sampled, including input points X and their corresponding objective function values $\mathbf{y}$, the joint distribution between the sampled points and the objective function value to be predicted at a new point $\mathbf{x}^{(\text{new})}$ can be defined as follows:

$$p(f, \mathbf{y}) = \mathcal{GP}\left(\begin{bmatrix} f \\ \mathbf{y} \end{bmatrix}; \begin{bmatrix} m(\mathbf{x}^{(\text{new})}) \\ \mathbf{m}(X) \end{bmatrix}, \begin{bmatrix} K(\mathbf{x}^{(\text{new})}, \mathbf{x}^{(\text{new})}) & k^{\mathsf{T}}(X, \mathbf{x}^{(\text{new})}) \\ k(X, \mathbf{x}^{(\text{new})}) & K(X, X) + \sigma_n^2 I \end{bmatrix}\right) \tag{5.3}$$

where $\mathbf{m}(X) = [m(\mathbf{x}^{(1)}), \dots, m(\mathbf{x}^{(n_d)})]^{\mathsf{T}}$ is the mean function evaluated at each data point, $\sigma_n^2 I$ is the observation noise covariance, and K is the covariance of the samples. It is therefore possible to find the probability of f given the data by marginalization:

$$p(f|\mathbf{y}) = \mathcal{GP}(f; \mu_{\mathcal{D}}(\cdot), \sigma_{\mathcal{D}}^2(\cdot, \cdot)) \tag{5.4}$$

where $\mu_{\mathcal{D}}$ is the expected value and $\sigma_{\mathcal{D}}^2$ is the variance of the prediction of f at a new point $\mathbf{x}^{(\text{new})}$:

$$\mu_{\mathcal{D}} = m(\mathbf{x}^{(\text{new})}) + k^{\mathrm{T}}(X, \mathbf{x}^{(\text{new})})[K(X,X) + \sigma_n^2 I]^{-1}(\mathbf{y} - \mathbf{m}(X)) \tag{5.5}$$

$$\sigma_{\mathcal{D}}^2 = K(\mathbf{x}^{(\text{new})}, \mathbf{x}^{(\text{new})}) - k^{\mathrm{T}}(X, \mathbf{x}^{(\text{new})})[K(X,X) + \sigma_n^2 I]^{-1} k(X, \mathbf{x}^{(\text{new})}) \tag{5.6}$$

5.2.3.2 BO: surrogate optimization

Given a GP that models the objective function, BO uses this model to pick the next point. One straightforward way to conduct this would be to minimize the expected value of the predicted function, i.e., $\min_X \mu_{\mathcal{D}}(\mathbf{x})$. However, given that the objective function is modeled by a probabilistic model (e.g., a GP) it is possible to use the uncertainty predicted by the model to search for the next point to sample.

In BO the *acquisition function* $\mathcal{A}$ is used to determine the next point to evaluate in the search space. It quantifies the utility or potential of evaluating a point $\mathbf{x}$ based on the current probabilistic model. Popular acquisition functions include PI, EI, and *lower confidence bound* (LCB). The choice of acquisition function depends on the trade-off between exploration (sampling in regions of uncertainty) and exploitation (sampling where the surrogate model predicts high values).

One easy-to-understand acquisition function is the LCB, where at each iteration, after the GP is built the surrogate optimization step solves the following problem:

$$\min_{\mathbf{x}} \mathcal{A}^{lcb}(\mathbf{x}) = \min_{\mathbf{x}} \mu_{\mathcal{D}}(\mathbf{x}) - \gamma \sigma_{\mathcal{D}}(\mathbf{x}) \tag{5.7}$$

where γ is a hyperparameter of the algorithm that determines how much exploration versus exploitation is preferred. From an empirical perspective, it can be observed that this acquisition function minimizes the expected value (i.e., $\mu\mathcal{D}(\mathbf{x})$) while at the same time encouraging the exploration of regions with uncertainty that could yield an even better solution (i.e., the term $\gamma \sigma_{\mathcal{D}}(\mathbf{x})$).

In this way by iteratively updating the surrogate model (e.g., a GP) and selecting points to evaluate based on minimizing the acquisition function, BO explores the search space and converges to optimal or near-optimal solutions while minimizing the number of evaluations of the true objective function.

5.2.3.3 BO algorithms

This chapter uses various BO algorithms in its benchmarking procedure. Two of these are standard variants of BO, the first, GPyOpt [25], is an off-the-shelf implementation based on [21, 80, 81] and the second is an in-house implementation. Additionally, the state-of-the-art *trust region BO* (TuRBO) algorithm was included in the high-dimensional case study [78]. One key challenge of the standard version of BO with GPs is its efficacy in high-dimensional settings. This shortcoming arises because standard BO uses a single, uniform model to represent the entire search space. Such a model fails to capture the diverse characteristics of different regions and overly focuses on exploring new areas across the entire search space, rather than efficiently targeting the most promising regions. The TuRBO algorithm addresses these issues by adopting a local probabilistic approach for global optimization in large-scale, high-dimensional problems. Instead of relying on a single global model, TuRBO fits multiple local models within dynamically adjusted trust regions, focusing on promising areas of the search space. This method leverages an implicit bandit approach to allocate samples efficiently across these local models.

5.2.4 Ensemble Tree MOdel optimization tool (ENTMOOT)

The ENTMOOT is a framework designed to perform BO using DT-based surrogate models [49]. In each iteration, ENTMOOT approximates the black-box function using a *gradient-boosted DT* (GBDT) model from LightGBM [82]. GBDTs are captured in a basic fashion in the subsequent section, followed by a description of their embedding into the BO framework.

5.2.4.1 ENTMOOT: surrogate model

DTs are supervised learning algorithms that can be interpreted as a piecewise constant approximation of the underlying function [4], in this case, the objective function. As the name suggests DTs have a tree-like structure composed of nodes, leaves, and branches. The topmost node in a DT is the root node representing the entire dataset. The root node is split into decision nodes, representing the split of the data based on certain conditions. Terminal nodes are the leaf nodes, representing the final output and no further splitting occurs at these nodes. All nodes are connected through branches, which represent the outcome of a decision rule applied to a former node.

When adapted to a regression task as required for this study, a DT can be depicted as an optimization problem that embeds the search for optimal splitting parameters that minimize the mean-squared error loss function $\mathcal{L}^{MSE}$ at each node. Thereby the overall optimization problem for a DT yields

$$\min_T \sum_{n \in \text{nodes}} \left(\frac{N_L(n)}{N_n} \mathcal{L}_L^{MSE}(n) + \frac{N_R(n)}{N_n} \mathcal{L}_R^{MSE}(n) \right) \tag{5.8}$$

with T representing the structure of the tree composed of n nodes. Each node n governs the split of N_n samples with $N_L(n)$ and $N_R(n)$ being the numbers of samples in the left and right subsets for node n, respectively. $\mathcal{L}_L^{MSE}(n)$ and $\mathcal{L}_R^{MSE}(n)$ are the mean-squared errors for the left and right subsets for node n with $\mathcal{L}^{MSE}$ as

$$\mathcal{L}_S^{MSE} = \frac{1}{|S|} \sum_{i \in S} (y_i - ys)^2 \tag{5.9}$$

Here, $|S|$ is the number of samples in subset S, y_i is the actual target value of the ith sample, and ys is the mean target value for subset S.

Gradient boosting in the context of GBDTs refers to the construction of a series of trees with a new tree added to the ensemble after each iteration. Each tree attempts to improve upon the performance of the ensemble thus far, minimizing the error between objective function observation y_i and prediction $f_{\text{entmoot},i}$.

The residuals r_i are defined as the negative gradients of the loss function (5.9):

$$r_i = -\frac{\delta \mathcal{L}^{MSE}(y_i, f_{\text{entmoot},i})}{\delta f_{\text{entmoot},i}} \tag{5.10}$$

The new tree is trained to predict these residuals.

For each observation i the residuals $r_i = y_i - f_{\text{entmoot},i}$ are calculated and a new DT $h_m(x)$ is trained to predict the residuals from the previous step. The new tree's predictions are added to the ensemble:

$$f_{\text{entmoot},i} = f_{\text{entmoot},i}^{\text{previous}} + vh_m(x_i) \tag{5.11}$$

with v being the learning rate controlling the contribution of each tree. The final prediction is the sum of all contributions from trees $m \in M$:

$$f_{\text{entmoot}} = f_{\text{entmoot},0} + \sum_{m=1}^{M} vh_m(x) \tag{5.12}$$

5.2.4.2 ENTMOOT: surrogate optimization

Following the BO paradigm, ENTMOOT leverages an acquisition that balances exploration and exploitation to determine the next sampled point. To do this, the surrogate model's prediction is combined with an uncertainty measure, reflecting varying degrees of trust in the data points. Since DTs do not come with uncertainty quantification as GPs do, [49] uses the uncertainty measure $a(x)$ to quantify the model uncertainty using the distance to the closest point $\mathbf{x}_d$ in data set $\mathcal{D}$ to quantify the confidence of model predictions:

$$a(\mathbf{x}) = \min_{d \in \mathcal{D}} \|\mathbf{x} - \mathbf{x}_d\|_p \tag{5.13}$$

where p determines the metric used to determine the distance. Thebelt et al. [49] used implementations of Euclidean or Manhattan distance metrics and discuss advantages and shortcomings in their paper.

This combination forms the acquisition function, which is then optimized to identify the best candidate for the optimal point in the current iteration. The following acquisition function represents the LCB acquisition function as seen in eq. (5.7) adjusted to the elements of GBDTs. It is a simplified version as presented in [49] that captures the main elements:

$$\min_{\mathbf{x}} \mathcal{A}^{\text{lcb,entmoot}}(\mathbf{x}) = \min_{\mathbf{x}} f_{\text{entmoot}}(\mathbf{x}) - \gamma\, a(\mathbf{x}) \tag{5.14}$$

f_{entmoot} refers to the tree model prediction in eq. (5.12), capturing how eq. (5.14) exploits the underlying surrogate model to find promising areas in the search space. Variable $a(\mathbf{x})$ as introduced in eq. (5.13) handles exploration and quantifies the degree of uncertainty expected from prediction f_{entmoot}. $\gamma \in \mathbb{R}$ balances exploitation and exploration and is a hyperparameter depending on the application as described previously in Section 5.2.3.2. Full details of this algorithm can be found in [49].

5.2.5 Constrained optimization by linear approximation (COBYLA)

First introduced in 1994 by M.J.D. Powell, a pioneer of computational mathematics [33], the *constrained optimization by linear approximations* (COBYLA) algorithm iteratively refines solutions for nonlinearly constrained optimization problems by leveraging linear approximations of the objective (and any constraint) functions [6]. COBYLA employs a trust region bound to limit changes to the variables. The algorithm dynamically adjusts the trust region radius based on predicted improvements to both the objective function and feasibility conditions. Additionally, it utilizes a merit function to compare the effectiveness of different variable vectors in improving the shape of the simplex, while ensuring adherence to constraints.

5.2.5.1 COBYLA: surrogate model

This algorithm solves the following problem at every iteration to determine the next best point to sample:

$$\min_{\mathbf{x}\in\mathbb{R}^{n_x}} \quad f_{\text{cobyla}}(\mathbf{x})$$
$$\text{s.t.} \quad c_{\text{cobyla},i}(\mathbf{x}) \geq 0, \quad i=1,2,\ldots,m \tag{5.15}$$

As mentioned, COBYLA utilizes a linear approximation to construct a surrogate of the objective function. Hence, eq. (5.15), presents the linear programming problem, which is iteratively solved, yielding a vector of variables, $\mathbf{x}$, where $f_{\text{cobyla}}(\mathbf{x})$ represents the surrogate and $f(\mathbf{x})$ is the true objective function. Similarly, $c_{\text{cobyla},i}(\mathbf{x})$ is a set of unique linear functions that serve as surrogates to the true constraints, $c_{\text{cobyla},i}(\mathbf{x})$.

At every iteration, COBYLA models the objective and the constraint functions with linear interpolants, which consists of $n_x + 1$ points that are updated along the iterations.

5.2.5.2 COBYLA: surrogate optimization

After constructing the initial surrogate, $f_{\text{cobyla}}(\mathbf{x})$, and a given center point $\mathbf{x}_k$, the algorithm iteratively enhances the linear approximation by adjusting the trust region radius, ρ, and evaluating potential solutions to determine the next variable vector, $\mathbf{x}_*$. Initially, it is verified whether $\mathbf{x}_k$ is the optimal vertex and ensures the simplex is acceptable. Subsequently, the trust region condition on $\mathbf{x}_*$ is given as follows:

$$\|\mathbf{x}_* - \mathbf{x}_k\|_2 < \frac{1}{2}\rho \tag{5.16}$$

If feasible, the surrogate is now minimized by x_*, subject to the trust region condition and linear constraints $c_{\text{cobyla},i}(x)$. In cases where multiple possible $\mathbf{x}_*$ exist, the vector that yields the lowest value of $\|x_* - x_k\|_2$ is selected.

To deal with constraints the COBYLA algorithm uses a merit function. This merit function, denoted as $\Phi(\mathbf{x})$, combines both the objective function $f(\mathbf{x})$ and the constraint functions $c_i(\mathbf{x})$ into a single scalar value:

$$\Phi_{\text{cobyla}}(\mathbf{x}) = f(\mathbf{x}) + \left[\max_{i=1,\ldots,m} c_i(\mathbf{x})\right]_+ \tag{5.17}$$

The merit function serves a dual purpose: the first term, $f(\mathbf{x})$, ensures that solutions are evaluated based on their ability to optimize the objective function. The second term, $[\max c_i(\mathbf{x}): i=1, 2, \ldots, m]_+$, captures the magnitude of constraint violations. By considering the maximum violation among all constraints, the merit function guides the optimization towards solutions that not only optimize the objective function but also satisfy the constraints as closely as possible.

During the optimization process, COBYLA aims to minimize the merit function $\Phi_{\text{cobyla}}(\mathbf{x})$ by iteratively adjusting the variable vector $\mathbf{x}$ within the trust region bounds and evaluating potential solutions. By minimizing the merit function, COBYLA effectively balances the trade-off between optimizing the objective function and satisfying the constraints, ultimately guiding the search toward feasible and optimal solutions.

5.2.6 Constrained optimization by quadratic approximation (COBYQA)

COBYQA builds on Powell's COBYLA, presenting a similar model-based DFO method, but instead utilizing quadratic approximations. COBYQA constructs quadratic models of objective and constraint functions using derivative-free symmetric Broyden updates, enabling efficient optimization without explicit derivatives. It dynamically adjusts its trust-region radius and incorporates a geometry-improving procedure to enhance numerical stability. Importantly, COBYQA strictly adheres to bound constraints, ensuring robustness in various engineering and industrial applications. Unlike alternative methods like *sequential-quadratic-programming DFO* (SQPDFO), COBYQA directly handles inequality constraints without introducing additional slack variables, maintaining efficiency and consistency throughout optimization iterations.

5.2.6.1 COBYQA – surrogate model

In COBYQA, the quadratic surrogate is constructed by approximating the objective $f(\mathbf{x})$ and constraints $c_i(\mathbf{x})$ functions within a trust region defined by a radius Δ_k around the current iteration $\mathbf{x}_k$ – where k represents the current iteration. The quadratic model $f_{\text{cobyqa}}(\mathbf{x})$ and $c_{\text{cobyqa},i}(\mathbf{x})$, defined in eqs. (5.18) and (5.19), respectively, are formed using function evaluations at selected points within this region, capturing the curvature of the functions and providing a refined estimate of their behavior [83]:

$$f_{\text{cobyqa}}(\mathbf{x}) = f(\mathbf{x}_k) + \mathbf{g}_k^{\mathsf{T}}(\mathbf{x} - \mathbf{x}_k) + \frac{1}{2}(\mathbf{x} - \mathbf{x}_k)^{\mathsf{T}} B_k(\mathbf{x} - \mathbf{x}_k) \tag{5.18}$$

$$c_{\text{cobyqa},i}(\mathbf{x}) = c_i(\mathbf{x}_k) + \nabla c_i(\mathbf{x}_k)^{\mathsf{T}}(\mathbf{x} - \mathbf{x}_k) + \frac{1}{2}(\mathbf{x} - \mathbf{x}_k)^{\mathsf{T}}\nabla^2 c_i(\mathbf{x}_k)(\mathbf{x} - \mathbf{x}_k), \quad i=1, 2, \ldots, m \tag{5.19}$$

where $\mathbf{g}_k \in \mathbb{R}^{n_x}$ is the surrogate of the gradient of the objective function and $B_k \in \mathbb{R}^{n_x \times n_x}$ is a positive definite symmetric matrix representing the surrogate Hessian of the objective function. Similarly, $\nabla c_i(\mathbf{x}_k)$ is the surrogate of the gradient of the ith constraint function, and $\nabla^2 c_i(\mathbf{x}_k)$ is the surrogate of its Hessian matrix.

5.2.6.2 COBYQA: surrogate optimization

COBYQA optimizes the quadratic surrogate to identify the next iterate $\mathbf{x}_{k+1}$ by minimizing a merit function $\Phi_{cobyqa,k}(\mathbf{x})$ within the trust region. The merit function, defined by eq. (5.20) combines the objective and constraint violations, weighted by penalty parameters, guiding the optimization process. The trust region, denoted by $\mathcal{B}(\mathbf{x}_k, \Delta_k)$, is given as follows:

$$\Phi_{cobyqa,k}(\mathbf{x}) = f_{cobyqa,k}(\mathbf{x}) + \sum_{i=1}^{m} \rho_i [c_{i,cobyqa,k}(\mathbf{x}) - \Delta_k]_+ \tag{5.20}$$

$$\mathcal{B}(\mathbf{x}_k, \Delta_k) = \{\mathbf{x}: \| \mathbf{x} - \mathbf{x}_k \| \leq \Delta_k\} \tag{5.21}$$

Here, ρ_i represents penalty parameters associated with the constraint functions.

5.2.7 Local search with quadratic models (LSQM)

Local search with quadratic models (LSQM) is a naive surrogate-based optimization method studied in this chapter. It constructs a quadratic surrogate model based on sampled data and optimizes it within a trust region. LSQM is advantageous as both the model construction and optimization are convex problems. However, the algorithm is inherently exploitative; hence it lacks any exploration element.

5.2.7.1 LSQM: surrogate model

LSQM builds surrogates of the form:

$$f_{lsqm}(\mathbf{x}; Q, c, b) = \mathbf{x}^T Q \mathbf{x} + c^T \mathbf{x} + b \tag{5.22}$$

where $Q \in \mathbb{R}^{n_x \times n_x}$ is symmetric, $c \in \mathbb{R}^{n_x}$, and $b \in \mathbb{R}$. Given the dataset $\mathcal{D} = \{(\mathbf{x}^{(i)}, y^{(i)})\}_{i=1}^{n_d}$, where values obtained from the objective function are denoted as $y^{(i)} \leftarrow f(\mathbf{x}^{(i)})$. The following least squares problem is formulated to estimate Q, c, and b:

$$\min_{Q,c,b} \sum_{i=1}^{n_d} \left(f_{lsqm}(Q, c, b; \mathbf{x}^{(i)}) - y^{(i)} \right)^2 \tag{5.23}$$

This is a convex optimization problem that can be easily solved. Furthermore, if semi-positive-definiteness is imposed on Q (i.e., $Q \succeq 0$) then the ensuing surrogate optimization problem is also convex.

5.2.7.2 LSQM: surrogate optimization

Once the quadratic surrogate model is constructed, the following optimization problem is formulated to find the next point to sample and evaluate:

$$\min_{\mathbf{x}} \; f_{\text{lsqm}}(\mathbf{x}; Q, c, b) = \min_{\mathbf{x}} \; \mathbf{x}^T Q \mathbf{x} + c^T \mathbf{x} + b \tag{5.24}$$

N.B. The bias term b can be omitted from the optimization as it will not influence the optimal point. In summary, LSQM presents a simple surrogate-based optimization algorithm that is used as a baseline in this chapter.

5.2.8 Convex quadratic trust-region optimizer (CUATRO)

CUATRO is a quadratic model-based trust region method first introduced in [53]. It works similarly to COBYQA and LSQM with some key distinctions. CUATRO is implemented in the Python framework for convex optimization CVXPY [22], which allows explicit handling of surrogate constraints. Additionally, CUATRO is developed with the peculiarities of chemical engineering applications in mind: it includes heuristic routines for sample-efficient and safe exploration, as well as dimensionality reduction techniques to exploit latent structure in the solution space to scale to thousands of variables [74].

5.2.8.1 CUATRO: surrogate model

Similarly to LSQM, CUATRO builds objective and constraint surrogates of the following form:

$$f_{\text{cuatro}}(\mathbf{x}; Q, \mathbf{p}, r) = \mathbf{x}^T Q \mathbf{x} + \mathbf{p}^T \mathbf{x} + r \tag{5.25}$$

$$c_{\text{cuatro},i}(\mathbf{x}; Q_i, \mathbf{p}_i, r_i) = \mathbf{x}^T Q_i \mathbf{x} + \mathbf{p}_i^T \mathbf{x} + r_i \tag{5.26}$$

The objective $f(\cdot)$ is again trained via regular least squares regression. This time, however, convexity is enforced by constraining Q to be positive semi-definite:

$$\min_{Q \succeq 0, \, \mathbf{p}, \, r} \; \sum_{i=1}^{n_d} \left(f_{\text{cuatro}}(\mathbf{z}^{(i)}; Q, \mathbf{p}, r) - y^{(i)} \right)^2 \tag{5.27}$$

Instead of training separate constraint surrogates using least squares regression on each set of constraint evaluations, the default implementation of CUATRO performs convex quadratic discrimination: it finds the ellipsoid that minimizes the total distance to the discrimination boundary of all falsely classified samples. The interested reader is referred to [58] for the convex formulation.

5.2.8.2 CUATRO: surrogate optimization

The surrogate optimization step follows that of COBYQA, with the exception that box bounds ($[\mathbf{x}_L, \mathbf{x}_U]$) and the constraint surrogate $c(\,\cdot\,)$ are included as explicit constraints resulting in a convex semidefinite program:

$$
\begin{aligned}
\min_{\mathbf{x} \in \mathcal{B}(\mathbf{x}_k, \delta_k)} \quad & f_{\text{cuatro}}(\mathbf{x}; \cdot) \\
\text{s.t.} \quad & c_{\text{cuatro}}(\mathbf{x}; \cdot) \le 0 \\
& \mathcal{B}(\mathbf{x}_k, \delta_k) = \{\mathbf{x} \in [\mathbf{x}_L, \mathbf{x}_U] \,|\, \|\mathbf{x} - \mathbf{x}_k\|_F^2 \le \delta_k^2\}
\end{aligned}
\tag{5.28}
$$

5.2.8.3 CUATRO: improving high-dimensional performance

Van de Berg et al. [74] introduced CUATRO-pls, an extension to CUATRO that performs dimensionality reduction to find a lower-dimensional subspace over which to perform surrogate updates. At each iteration, *partial least squares* (PLS) regression identifies the linear projection matrix $M \in \mathbb{R}(n_z \times n_x)$ and reconstruction matrix $R = (M^{\mathsf{T}} M) M^{\mathsf{T}}$ that best predict the outputs $\mathbf{y}$ of all samples within the original trust region $\mathcal{B}(\,\cdot\,)$ after projection to their linear embedding Z, such that $Z = MX$ and $X \approx RZ$. The surrogates are then trained in the reduced dimensional dataset Z:

$$
\min_{Q \succeq 0, \, \mathbf{p}, \, r} \quad \sum_{i=1}^{n_d} \left(f_{\text{cuatro}}(\mathbf{z}^{(i)}; Q, \mathbf{p}, r) - y^{(i)} \right)^2
\tag{5.29}
$$

Surrogate fitting and minimization are also performed in the subspace. To conserve convexity, the equivalent trust region constraint in the original space $(R\mathbf{z} - \mathbf{x}_k)^{\mathsf{T}} (R\mathbf{z} - \mathbf{x}_k) \le r^2$ is replaced with a heuristic trust region in embedding space centered around $\mathbf{z}_k = M\mathbf{x}_k$, where an effective radius is defined as the distance from $\mathbf{z}_k$ to the furthest projected sample as $\delta^2 = \max_{\mathbf{z} \in Z} (\mathbf{z} - \mathbf{z}_k)^{\mathsf{T}} (\mathbf{z} - \mathbf{z}_k)$:

$$
\begin{aligned}
\min_{\mathbf{z} \in \mathcal{B}(\mathbf{z}_k, \delta_k)} \quad & f_{\text{cuatro}}(\mathbf{z}; \cdot) \\
\text{s.t.} \quad & c_{\text{cuatro}}(\mathbf{z}; \cdot) \le 0
\end{aligned}
\tag{5.30}
$$

The minimization candidate is evaluated and the trust region is updated after reconstruction to the original space. This methodology introduces a crucial choice in the form of the embedding dimensionality n_{pls}. Van de Berg et al. [74] show that a default underestimation of $n_{pls} = 2$ works quite well in the absence of information about the intrinsic embedding. This makes intuitive sense, as the method reduces to a search in the most informative linear combination of dimensions, which would have merit as an idea on its own.

5.2.9 Stable noisy optimization by branch and fit (SNOBFIT)

Stable noisy optimization by branch and fit (SNOBFIT) by Huyer and Neumaier [13] is a surrogate-based optimization algorithm that combines global and local search. To do so SNOBFIT branches the search space to create smaller sub-spaces. Subsequently, surrogate models are fitted within the sub-spaces to obtain information about promising areas of the objective function. Candidate points generated in these sub-spaces are sorted into classes indicating whether a local or global aspect of SNOBFIT has determined the respective subspace. There are three classes for local candidates and two classes for global candidates. Technically, "local" aspects refer to an exploitative strategy leveraging so-called *safeguarded nearest neighbors* (SNNs) to promising function evaluations. Accordingly, "global" aspects refer to an explorative strategy applied to gain information about unexplored sub-spaces. The subsequent part of this section starts by explaining how surrogate models are built to then delve into the usage of local and global information to navigate the optimization of the objective function.

5.2.9.1 SNOBFIT: surrogate model

For fast local convergence, SNOBFIT handles local search from the location $\mathbf{x}^{best}$ of the best-so-far objective function value with a full quadratic model. To create such a model, the number of points in a local area N must exceed the dimensions of the problem n by Δn, $N \geq n + \delta n$. Δn are called the previously introduced SNNs and can go up to n. The SNNs are technically a set of previous function evaluation points that are consulted as support points for the intended local surrogate model fit. The procedure to determine the SNNs for a point uses coordinate-wise comparison of previous function evaluation positions guided by a threshold to ensure diversity and to promote exploration. After each iteration of SNOBFIT, the SNNs for the locations of new evaluations are determined and the SNNs for previous evaluated locations are updated. For an in-depth description of the SNN determination the interested reader is referred to section 3 in [13].

Subsequently, a local model around each point $\mathbf{x}$ is fitted with the aid of the SNNs:

$$q(x) := f_{best} + \mathbf{g}^{\mathrm{T}}(\mathbf{x} - \mathbf{x}^{best}) + \frac{1}{2}(\mathbf{x} - \mathbf{x}^{best})^{\mathrm{T}}G(\mathbf{x} - \mathbf{x}^{best}) \tag{5.31}$$

which corresponds to a Taylor approximation around the best observed objective function value f_{best}. Gradient $\mathbf{g}$ and Hessian G are determined based on the SNNs as described previously. For a more in-depth description of this approach, the reader is referred to [13]. Depending on the user's preferences regarding the number of local optimization candidates, the previously described approach can be repeated, with adaptations described in [13].

5.2.9.2 SNOBFIT: surrogate optimization

Different from other algorithms previously discussed, SNOBFIT generates a batch of new candidate locations rather than only one. These candidate locations are generated by optimizing the surrogate models around the best-so-far evaluation from the previous section. The optimization of the quadratic surrogate model around this point is described in Section 5.2.7.2. Depending on the user's preference, the number of such exploitative generated candidates can be increased to include more candidates from SNN-supported local surrogates.

The global aspect of SNOBFIT fills the remaining spots in the batch of candidate points for the succeeding iteration. Such candidates are generated from so far unexplored regions of the search space. For a box $[x_l, x_u]$ with corresponding point previously observed point x the candidate point z is generated by

$$z := \begin{cases} \frac{1}{2}(x_l + x) & \text{if } x - x_l > x_u - x \\ \frac{1}{2}(x + x_u) & \text{otherwise,} \end{cases} \qquad (5.32)$$

If after applying eq. (5.32) there are still spots in the batch to be filled by global candidates, random sampling within the bounds $[x_l, x_u]$ is used. Typically this happens during the initialization of the algorithm or the early iterations of the algorithm when not enough points are yet observed to generate the local quadratic models.

5.2.10 Dynamic coordinate search using response surface models (DYCORS)

This chapter considers three algorithms that are based on RBF surrogates by Regis and Shoemaker [18]: DYCORS, SOPStrategy, and SRBFStrategy. RBFs are versatile mathematical tools used in surrogate modeling, characterized by their radial symmetry around a center point and employed to approximate complex functions. These functions are flexible, capable of capturing nonlinearities and adapt well to irregularly sampled data points, making them suitable for various applications in optimization and machine learning [77]. While this section focuses extensively on detailing DYCORS, the interested reader is referred to the literature for comprehensive insights into SOPStrategy [23] and SRBFStrategy [9, 16].

The DYCORS framework represents a sophisticated approach for surrogate-based optimization, tailored for *high-dimensional, expensive, and black-box functions* (HEB). DYCORS integrates elements from the *dynamically dimensioned search* (DDS) algorithm [10] into a surrogate-based optimization context, specifically leveraging RBF surrogates. DYCORS is particularly effective for high-dimensional optimization problems due to its dynamic search strategy, which balances exploration and exploitation. By progressively focusing the search and using a surrogate model to guide the selection of trial points, DYCORS can efficiently navigate the complex landscape of black-box functions.

5.2.10.1 DYCORS: surrogate model

In the context of DYCORS and other RBF algorithms discussed, the interpolation model employs RBFs to approximate the objective function based on known data points. Given n_d distinct points $\mathbf{x}^{(1)}, \ldots, \mathbf{x}^{(n_d)} \in \mathbb{R}^{n_x}$ with corresponding function values $f(\mathbf{x}^{(1)}), \ldots, f(\mathbf{x}^{(n_d)})$, the RBF interpolant is formulated as

$$s_n(\mathbf{x}) = \sum_{i=1}^{n_d} \lambda_i \phi(\| \mathbf{x} - \mathbf{x}^{(i)} \|) + p(\mathbf{x}), \quad \mathbf{x} \in \mathbb{R}^{n_x}, \tag{5.33}$$

where ϕ is an RBF kernel, such as the cubic form $\phi(r) = r^3$. Other kernels like the thin plate spline and Gaussian can also be used. The coefficients $\lambda_i \in \mathbb{R}$ and $p(\mathbf{x})$ represent a linear polynomial in n_x variables.

The matrix $\Phi \in \mathbb{R}^{n_d \times n_d}$ is defined by $\Phi_{ij} = \phi(\| \mathbf{x}^{(i)} - \mathbf{x}^{(j)} \|)$, and $P \in \mathbb{R}^{n_d \times (n_x+1)}$ is constructed such that its ith row is $[1, (\mathbf{x}^{(i)})^T]$. The cubic RBF model that interpolates the points $(\mathbf{x}^{(1)}, f(\mathbf{x}^{(1)})), \ldots, (\mathbf{x}^{(n_d)}, f(\mathbf{x}^{(n_d)}))$ is obtained by solving the system:

$$\begin{bmatrix} \Phi & P \\ P^T & 0 \end{bmatrix} \begin{bmatrix} \lambda \\ c \end{bmatrix} = \begin{bmatrix} F \\ 0 \end{bmatrix}, \tag{5.34}$$

where $F = (f(\mathbf{x}^{(1)}), \ldots, f(\mathbf{x}^{(n_d)}))^T$, $\lambda = (\lambda_1, \ldots, \lambda_{n_d})^T \in \mathbb{R}^{n_d}$, and $c = (c_1, \ldots, c_{n_x+1})^T \in \mathbb{R}^{n_x+1}$ are the coefficients for the RBF and linear polynomial $p(\mathbf{x})$, respectively. This coefficient matrix is invertible if and only if $\text{rank}(P) = n_x+1$.

5.2.10.2 DYCORS: surrogate optimization

The optimization of the surrogate model within the DYCORS framework involves the dynamic coordinate search (DCS), which integrates several key steps to efficiently navigate HEB objective functions:

Perturbation probability: A subset of the coordinates of the current best solution x_{best} is perturbed to generate trial points. The probability p_{select} of perturbing a coordinate is given by a strictly decreasing function $\phi(n_x)$. This function ensures that the number of perturbed coordinates decreases as the algorithm progresses, leading to a more localized search.

Generating trial points: For each trial point, the selected coordinates are perturbed by adding normally distributed random variables with mean zero and a standard deviation σ_n, known as the step size. The set of coordinates to be perturbed, $I_{perturb}$, is chosen randomly in each iteration, ensuring a diverse exploration of the search directions.

Selection of the next iterate: From the set of generated trial points, the next iterate is selected using criteria that balance exploration and exploitation. In the DYCORS-LMSRBF algorithm, for instance, the selection is based on a weighted score combining the estimated function value from the RBF surrogate and the distance from previously evaluated points. This balance encourages both the discovery of new regions and the refinement of promising areas.

Evaluation and update: The objective function f is evaluated at the selected trial point. This new data point is then incorporated into the surrogate model, updating it to reflect the most recent information. This iterative updating process ensures that the surrogate model continuously improves in accuracy.

Adaptive step size adjustment: The step size σ_n is adjusted based on the success of the iterations. Parameters such as the number of consecutive successful iterations $C_{success}$ and the number of consecutive failed iterations C_{fail} are monitored. Depending on these parameters and optional thresholds, the step size is either increased or decreased, facilitating an adaptive search strategy.

Convergence criteria: The optimization process continues until a predefined convergence criterion is met. Common criteria include a maximum number of function evaluations, a maximum number of iterations, or a tolerance threshold for changes in the objective function value. The best solution found during the search is then reported as the optimal solution.

Specific implementations in DYCORS: The DYCORS algorithm has two specific variants: DYCORS *local metric stochastic* RBF (DYCORS-LMSRBF) and DYCORS-DDSRBF. In the performance assessment carried out in this chapter, the DYCORS-LMSRBF variant is used. This variant has been shown to effectively balance exploration and exploitation by selecting the iterate based on a combination of the RBF surrogate value and the distance from previously evaluated points. The nuances of the variants are briefly detailed below:

- **DYCORS-LMSRBF:** This variant is a modification of the LMSRBF algorithm. It uses the dynamic coordinate search strategy and selects the iterate based on a weighted score of the RBF surrogate value and the distance criterion. The perturbation probability function $\phi(n_x)$ is chosen to ensure an initial high probability of perturbing coordinates, which decreases logarithmically as the number of iterations increases.
- **DYCORS-DDSRBF:** This variant incorporates the RBF surrogate into the DDS algorithm. It maintains the DCS strategy but uses the surrogate model to enhance the efficiency and accuracy of the search.

5.3 Surrogate optimization methods performance assessment

In addition to the theoretical presentation, a performance assessment was conducted to evaluate the different model-based algorithms highlighted in this chapter. This assessment is supplemented with the associated code, such that the interested reader can further explore this performance assessment and the algorithms studied.

5.3.1 Performance assessment procedure

For this comparative assessment, the algorithms to be assessed along with the test functions have been defined. The set containing the different algorithms is called , $\mathbb{A}$ and the set containing the different test functions as . $\mathbb{F}$ This chapter conducts a performance assessment for unconstrained black-box problems as well as for constrained black-box problems. In the unconstrained case the functions and algorithms are as follows:

$$a \in \mathbb{A}, \mathbb{A} = \{\text{LSQM, SNOBFIT, SRBF, DYCORS, SOP, COBYLA,}$$

$$\text{COBYQA, CUATRO, BO, ENTMOOT}\}$$

$$f \in \mathbb{F}, \mathbb{F} = \{\text{Ackley, Levy, Rosenbrock, Quadratic}\}.$$

Therefore every algorithm $a \in \mathbb{A}$ is assessed on every objective function $f \in \mathbb{F}$ and its performance is compared relative to the other optimization algorithms on the same function. The procedure allocates each algorithm a budget of n_e function evaluations and the optimization of f is conducted five times per algorithm to account for algorithm and function evaluation stochastic factors. For a given algorithm, the trajectories of function values y_k with $k = 1, 2, \ldots, n_e$ are stored in five vectors each of length n_e, where only the best-so-far values within a trajectory are stored. The evaluation budget is given proportional to the dimension of the function; where n_x denotes dimensions, the function evaluation budgets are allocated as follows, 20 function evaluations for $n_x = 2$, 50 function evaluations for $n_x = 5$, and 100 function evaluations for $n_x = 10$.

In addition to displaying all the performance figures for each algorithm, a quantitative metric is provided for each algorithm's performance for each function. It is important to note that this performance assessment is not based on the final objective value that the algorithms arrive at. Instead, the assessment is based on the algorithms' respective trajectories; trajectories offer a more robust and less arbitrary measure for comparison. Equation (5.35) defines the normalized scoring metric.

To account for random initialization resulting in sub-optimal objective function values, as well as the need for using initial sampling to build the first surrogate model before starting the optimization (see Figure 5.1) the best point found so far does not start counting from the very first evaluation, but rather after the fist n_c evaluations. For $n_x = 2$ this is $n_c = 5$, for $n_x = 5$ this is $n_c = 10$ and for $n_x = 10$ this is $n_c = 15$. By this, the effective length of the trajectory considered for the benchmarking reduces to n, yielding $n = 15$, $n = 40$, and $n = 85$ for $d = 2$, $d = 5$, and $d = 10$, respectively.

For example, when LSQM is benchmarked on the two-dimensional Levy function LSQM will do a total of 20 function evaluations. Initially, $n_x + 1 \rightarrow 3$ evaluations are done to construct the surrogate (implementation details like these can be found in the GitHub repository). After LSQM builds the initial surrogate it will perform two more function evaluations for a total of five, which corresponds to $n_c = 5$, before our benchmarking procedure starts counting. From iterations 6 to 20, the best point found so far will be compared to the other algorithm's performance (noting that averages over a total of 5 runs are what are compared).

The specific scoring metrics used to deliver a quantitative score are the trajectories of length n, and they are compared using the quotient $r_{k,a}^{(n_x)}$, which represents the relative performance of algorithm a on function f for iteration k with n_x input dimensions:

$$r_{k,a}^{(n_x)} = \frac{y_k - y_{k,a}^{mean}}{y_k - y_k^*}, \qquad 0 \le r_{k,a}^{(n_x)} \le 1 \qquad (5.35)$$

Note, for simplicity, the superscript (n_x) is only shown on the left-hand side, whereas all variables presented account for the respective dimension. Here, y_{ka}^{mean} denotes the mean value of the objective function f along the trajectory achieved by algorithm a given n_x input dimensions. This mean is computed at a specific trajectory position k and is derived from the averaging of five optimization runs.

y_k represents the maximum (worst) function value for iteration k out of all algorithms investigated. In contrast, y_k^* is the lowest (best) function value achieved and therefore denotes the best performance by any algorithm on iteration k. To summarize, the closer y_{ka}^{mean} is to y_k^* – and hence the closer $r_{ka}^{(d)}$ is to 1 – the better a performed in iteration k compared to other algorithms.

Finally, $p_a^{(n_x)}$ as in eq. (5.36) represents the overall performance of algorithm a on objective function f with a given budget n for d input dimensions. Again, the closer $p_a^{(n_x)}$ is to 1, the better a performed relative to the algorithms in $\mathbb{A}$:

$$p_a^{(n_x)} = \frac{\sum_{k=1}^{n} r_{ka}^{(n_x)}}{n}, \qquad 0 \le p_a^{(n_x)} \le 1 \qquad (5.36)$$

Geometrically, this score represents how low (good) an algorithm's trajectory is, normalized by the best and worst performance among the algorithms.

5.3.2 Unconstrained performance assessments

This section presents the performance assessment of the model-based algorithms presented in Section 5.2 for unconstrained optimization problems. The section is structured as follows before a conclusion is drawn for the unconstrained case:

- **Mathematical objective functions:** Objective functions for the respective optimization problems are presented.
- **2D trajectory plots:** For qualitative evaluation, two-dimensional plots are presented that help gain intuition on the algorithm's behavior.
- **1D convergence plots:** For quantitative analysis, one-dimensional plots to observe the algorithm's convergence.
- **Performance tables:** Tables summarizing the performance in benchmarking numbers following the procedure described in Section 5.3.1.

5.3.2.1 Mathematical objective functions

The objective functions used in the unconstrained benchmarking are the Ackley function as proposed by David H. Ackley in 1987 [5], the Levy function by Paul Levy in 1954 [1], the Rosenbrock function by Howard H. Rosenbrock in 1960 [2], and a quadratic mildly ill-conditioned function. The former two are multimodal functions meaning they have multiple local minima, whereas the latter two are unimodal functions, denoting their single global minimum. Mathematical expressions can be found in the following equations:

- Ackley:

$$f(\mathbf{x}) = -a\exp\left(-b\sqrt{\frac{1}{n_x}\sum_{i=1}^{n_x}x_i^2}\right) - \exp\left(\frac{1}{n_x}\sum_{i=1}^{n_x}\cos(cx_i)\right) + a + \exp(1) \tag{5.37}$$

- Levy:

$$f(\mathbf{x}) = \sin^2(\pi w_1) + \sum_{i=1}^{n_x-1}\left[(w_i-1)^2(1+10\sin^2(\pi w_i+1))\right] + (w_{n_x}-1)^2(1+\sin^2(2\pi w_{n_x}))$$

$$\tag{5.38}$$

– Rosenbrock:

$$f(\mathbf{x}) = \sum_{i=1}^{n_x-1} \left[100(x_{i+1} - x_i^2)^2 + (1-x_i)^2 \right] \tag{5.39}$$

– Quadratic:

$$f(x) = \sum_{i=1}^{n_x} \left[(i \cdot x_i)^2 + \left(\frac{ai}{n_x} \right) x_i x_{n_x} \right] \tag{5.40}$$

For all functions $\mathbf{x} = (x_1, x_2, \ldots, x_{n_x})$ is the n_x-dimensional input vector. For the Ackley function a, b, and c are constants set to $a = 20$, $b = 0.2$, and $c = 2\pi$, respectively. For the Levy function $w_i = 1 + \frac{x_i-1}{4}$ for $i = 1, 2, \ldots, n_x$. In the quadratic function $a = 1.9$, $x = (x_1, x_2, \ldots, x_{n_x})$. The respective case studies are summarized in Table 5.1. N.B., that all case studies in this section are deterministic, a study of noise has been conducted in [53]. Furthermore, both chemical engineering case studies involved an element of stochasticity.

Table 5.1: Case studies for the unconstrained benchmarking. The input dimensions are $n_x = 2, 5, 7$ on domain $\mathbf{x} \in [-5, 5]^{n_x}$ with global minimum $f(\mathbf{x}^*) = 0$.

Case study	Topology	Global minimum
Ackley function	Non-convex	$\mathbf{x}^* = (0, \ldots, 0)$
Levy function	Non-convex	$\mathbf{x}^* = (1, \ldots, 1)$
Rosenbrock function	Non-convex	$\mathbf{x}^* = (1, \ldots, 1)$
Quadratic function	Convex, ill-conditioned	$\mathbf{x}^* = (0, \ldots, 0)$

5.3.2.2 Results: convergence plots

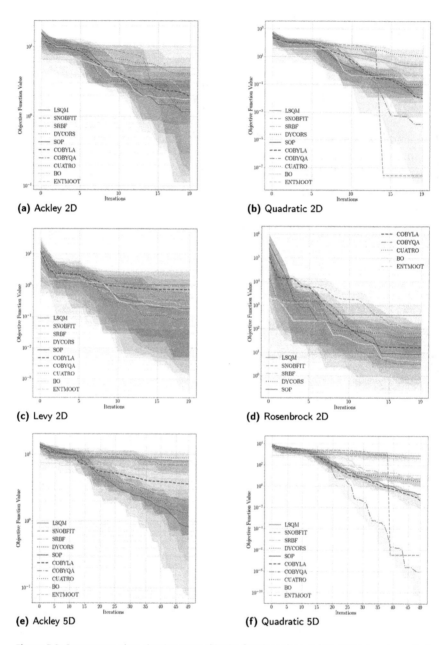

Figure 5.2: Convergence plots, showing mean objective function values and 10–90% intervals enveloping trajectories over 10 repetitions from 10 different starting points. Subcaptions indicate the test function and input dimensionality.

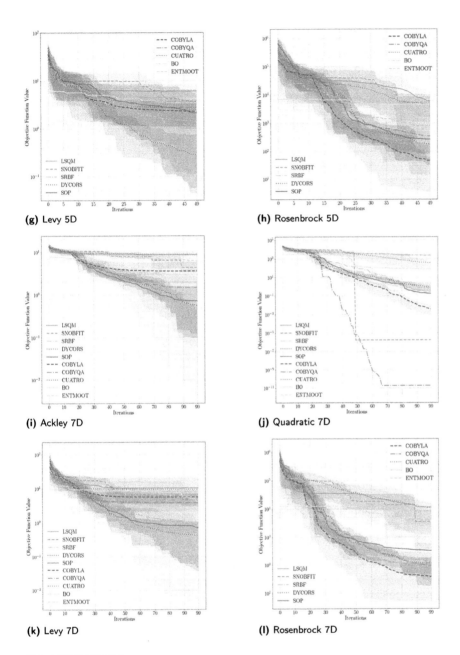

(g) Levy 5D

(h) Rosenbrock 5D

(i) Ackley 7D

(j) Quadratic 7D

(k) Levy 7D

(l) Rosenbrock 7D

Figure 5.2 (continued)

5.3.2.3 Results: trajectory plots

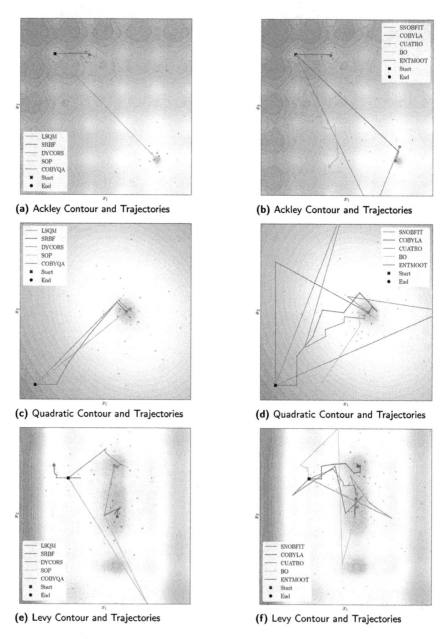

(a) Ackley Contour and Trajectories

(b) Ackley Contour and Trajectories

(c) Quadratic Contour and Trajectories

(d) Quadratic Contour and Trajectories

(e) Levy Contour and Trajectories

(f) Levy Contour and Trajectories

Figure 5.3: Two-dimensional contour plots with exemplary optimization trajectories for the unconstrained case for each test function. The lines connect best-so-far evaluations and the scattered points show remaining evaluations.

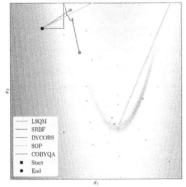

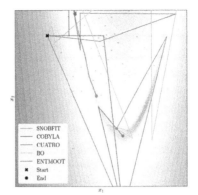

(g) Rosenbrock Contour and Trajectories

(h) Rosenbrock Contour and Trajectories

Figure 5.3 (continued)

5.3.2.4 Results: tables

Table 5.2: Quantitative algorithm scores in descending order for unconstrained optimization benchmarking.

	Ackley	Levy	Multimodal	Rosenbrock	Quadratic	Unimodal	All
DYCORS							
D2	0.93	1.00	0.96	0.89	1.00	0.95	0.95
D5	0.94	1.00	0.97	0.87	0.95	0.91	0.94
D7	0.95	1.00	0.98	0.93	0.95	0.94	0.96
All	0.94	1.00	0.97	0.90	0.97	0.93	0.95
SRBF							
D2	0.96	0.96	0.96	0.93	0.98	0.95	0.96
D5	0.94	0.86	0.90	0.97	0.91	0.94	0.92
D7	0.92	0.92	0.92	0.96	0.93	0.94	0.93
All	0.94	0.91	0.93	0.95	0.94	0.94	0.93
SOP							
D2	0.87	0.93	0.90	0.97	0.99	0.98	0.94
D5	1.00	0.58	0.79	0.89	0.97	0.93	0.86
D7	1.00	1.00	1.00	0.96	0.92	0.94	0.97
All	0.96	0.84	0.90	0.94	0.96	0.95	0.92

Table 5.2 (continued)

	Ackley	Levy	Multimodal	Rosenbrock	Quadratic	Unimodal	All
BO							
D2	1.00	0.91	0.95	1.00	1.00	1.00	0.98
D5	0.87	0.53	0.70	0.85	0.94	0.90	0.80
D7	0.76	0.38	0.57	0.79	0.85	0.82	0.69
All	0.88	0.61	0.74	0.88	0.93	0.91	0.82
COBYLA							
D2	0.79	0.43	0.61	0.86	0.92	0.89	0.75
D5	0.74	0.74	0.74	1.00	1.00	1.00	0.87
D7	0.74	0.56	0.65	1.00	1.00	1.00	0.83
All	0.76	0.58	0.67	0.95	0.97	0.96	0.81
COBYQA							
D2	0.83	0.00	0.41	0.88	0.81	0.84	0.63
D5	0.88	0.57	0.72	0.85	0.92	0.89	0.81
D7	0.89	0.72	0.80	0.94	0.86	0.90	0.85
All	0.86	0.43	0.65	0.89	0.86	0.88	0.76
ENTMOOT							
D2	0.55	0.88	0.71	0.74	0.73	0.74	0.73
D5	0.46	0.80	0.63	0.96	0.44	0.70	0.66
D7	0.18	0.27	0.22	0.63	0.00	0.31	0.27
All	0.40	0.65	0.52	0.78	0.39	0.58	0.55
CUATRO							
D2	0.08	0.44	0.26	0.87	0.00	0.43	0.35
D5	0.00	0.72	0.36	0.45	0.12	0.28	0.32
D7	0.00	0.16	0.08	0.00	0.37	0.18	0.13
All	0.03	0.44	0.23	0.44	0.16	0.30	0.27
LSQM							
D2	0.28	0.26	0.27	0.41	0.71	0.56	0.41
D5	0.09	0.16	0.12	0.05	0.21	0.13	0.12
D7	0.02	0.00	0.01	0.22	0.04	0.13	0.07
All	0.13	0.14	0.13	0.23	0.32	0.27	0.20
SNOBFIT							
D2	0.00	0.45	0.22	0.00	0.12	0.06	0.14
D5	0.02	0.00	0.01	0.00	0.00	0.00	0.00
D7	0.18	0.17	0.18	0.17	0.41	0.29	0.23
All	0.06	0.21	0.14	0.06	0.18	0.12	0.13

5.3.2.5 Results and discussion: mathematical unconstrained functions

This section presents the conclusions and discussion of results from the unconstrained synthetic benchmarking. The performance assessment considers varying dimensionality and includes both unimodal and multimodal test functions. The interpretations drawn here are based on the quantitative benchmarking ranking and trajectory plots. It is important to note that all conclusions and results are relative to the specific set of algorithms included in this performance assessment. Overall, DYCORS emerged as the best-performing algorithm in the unconstrained synthetic benchmarking. SRBF secured the second-best position, demonstrating consistent and robust performance. On the other hand, SNOBFIT was the poorest-performing algorithm, with LSQM ranking as the second poorest (Figure 5.2 and Table 5.2). It is important to highlight that very competitive algorithms were benchmarked, so poor performance does not mean an algorithm is bad. Additionally, four test functions across three dimensions is not an exhaustive combination, therefore readers are encouraged to draw only directional conclusions, and even make their own assessment if they wish to use these algorithms for a specific application. Furthermore, Figure 5.3 shows erratic behavior from some of the algorithms. This highlights the difference between gradient-based algorithms, when derivatives are available, and derivative-free methods, which have to explore and "learn" the function as well as optimize it. Algorithm-specific conclusions have been grouped and presented below.

BO: The in-house implementation of BO was the best-performing algorithm on the 2D unimodal test and the Ackley function in 2D. However, a decline in performance was observed with increasing dimensionality of the test functions.

COBYQA: COBYQA performed the worst on the 2D Levy function. It showed better performance on unimodal test functions than multimodal ones, with performance increasing with dimensionality across all test functions except the ill-conditioned quadratic function.

LSQM: LSQM was the second worst-performing algorithm in the unconstrained synthetic category. It performed slightly better on unimodal test functions, with the best performance observed on the Rosenbrock function.

COBYLA: COBYLA demonstrated strong overall performance, excelling in unimodal test functions over multimodal ones. It achieved the best results on the 2D and 7D unimodal tests, particularly on the Rosenbrock and ill-conditioned quadratic functions.

CUATRO: CUATRO performed the worst in Ackley D5 and D7, ill-conditioned quadratic 2D, and Rosenbrock 7D, but exhibited good performance in the Rosenbrock 2D test.

DYCORS: DYCORS emerged as the top-performing algorithm in the unconstrained synthetic benchmarking and was the best among the three RBF-based Shoemaker et al. algorithms. It excelled in the Levy test function for D2, D5, and D7, as well as in the ill-conditioned quadratic D2 test. Increasing dimensionality did not significantly affect its performance within the tested range.

ENTMOOT: ENTMOOT had similar performance in both unimodal and multimodal tests, excelling in the Rosenbrock D5 function but performing the worst in the ill-conditioned quadratic D7. Its performance decreased in all test functions except for the Rosenbrock function.

SNOBFIT: SNOBFIT was the worst-performing algorithm in the unconstrained synthetic category, with the poorest results in ill-conditioned quadratic D2, Rosenbrock D2, and D5, Levy D5, and being the worst in unimodal D5.

SOP: SOP showed strong performance across both unimodal and multimodal test functions, with performance not significantly impacted by increasing dimensionality. It was the best in Levy D7, ill-conditioned quadratic D5, and D7, and had the best overall performance in multimodal D7.

SRBF: SRBF was a solid all-rounder and the second-best algorithm in the overall unconstrained synthetic benchmarking. Although it did not rank first in any single category, it scored over 0.9 in all but the Levy test function in D5. It had a very small standard deviation in performance, and increasing dimensionality did not significantly impact its results.

5.4 Surrogate-based constrained optimization

Constrained optimization is pivotal in process optimization, as constraints play a critical role in ensuring safety, operational, and environmental constraints. From a safety standpoint, constraints ensure the process operates within safe operating limits, preventing potential hazards such as chemical leaks, fires, or explosions. Operationally, they maintain process efficiency and prevent equipment breakdown by establishing clear boundaries for variables such as flow rates, reaction times, and equipment utilization. By doing so, constraints mitigate equipment damage, minimize downtime, and ensure consistent product quality. Finally, environmental constraints are critical and equally vital, reducing the impact of chemical processes on the environment. Highlighting their significance, in practical terms, every process invariably operates at the limit of at least one constraint – this is driven by the demands of a competitive global economy, which pushes processes to achieve peak performance. Hence, it is imperative to develop algorithms that are capable of optimizing such problems whilst also satisfying constraints.

Given the importance of constraints, this section will outline data-driven model-based optimization algorithms that can deal with constrained problems. It will also present the theory behind the algorithms as well as a comparative study for the reader to develop an intuition and understanding of the different algorithms.

5.4.1 Problem formulation for constrained surrogate-based optimization

Recall the general optimization formulation in eq. (5.1) with modifications for a constrained system:

$$
\begin{aligned}
\min_{\mathbf{x}} \quad & f(\mathbf{x}) \\
\text{s.t.} \quad & \mathbf{g}(\mathbf{x}) \leq \mathbf{0} \\
& \mathbf{x} \in \mathbb{R}^{n_x} \\
& f: \mathbb{R}^{n_x} \to \mathbb{R} \\
& \mathbf{g}: \mathbb{R}^{n_x} \to \mathbb{R}^{n_g}
\end{aligned}
\tag{5.41}
$$

The challenge arises from the lack of prior knowledge regarding the locations where the constraint equations $\mathbf{g}(\cdot)$ are satisfied. The handling of constraints varies depending on the safety criticality of the system. For instance, when sampling a simulation, the primary objective is to ensure that the resulting solution complies with the constraints. However, when sampling a physical reactor, the goal might be to minimize both the number and severity of constraint violations during sampling. In scenarios such as simulations without safety considerations, an intuitive way to deal with constraints might be via the penalty method where violations are penalized using the Lp-norm:

$$
\min_{\mathbf{x}} f(\mathbf{x}) + \rho \sum_{j=1}^{n_g} (\max(0, g_j(\mathbf{x})))^p
\tag{5.42}
$$

where ρ constitutes the penalty parameter. However, the choice of penalty parameter ρ is not trivial: if too small, the returned solution might accept small constraint violations in favor of increases in $f(\cdot)$; if too big, the problem might become ill-conditioned, and difficult to solve with model-based methods.

Typically, it is assumed that both $f(\mathbf{x})$ and $\mathbf{g}(\mathbf{x})$ can only be sampled. Regarding $\mathbf{g}(\mathbf{x})$, there are two scenarios: in the first, constraints may be violated during optimization as long as the optimal candidate meets the constraints. In the second scenario, constraints cannot be violated during the optimization sequence or by the optimal candidate. The first is referred to as constrained optimization, and the second is safe optimization. Safe optimization is particularly relevant in real-world chemical engineering problems where, for example, samples could represent experiments conducted in a reactor. For instance, if the constraint $T \leq 500$ (K) represents the safe operating region that avoids runaway reactions, safe optimization is most suitable for ensuring process safety and stability.

In what follows, the focus shifts to the more interesting "safe" scenario of minimizing both the number and severity of constraint violations. This necessitates the adoption of a distinct category of explicit constraint-handling methods.

5.4.2 Explicit constraint handling methods

All the constrained DFO solvers studied in this performance assessment handle constraints *explicitly*. These methods not only construct surrogates of the objective function f but also for the constraint evaluations: $g_j(\mathbf{x}) \approx \tilde{g}_j(\mathbf{x})$. These surrogate models are then incorporated into the minimization step as constraints, ensuring that the next sample is expected to satisfy the constraints before the evaluation. To address uncertainties that arise from inexact surrogates and noisy constraint evaluations, some considered methods also employ trust regions $T(\cdot)$ to restrict the update step to the vicinity of the current (feasible) candidate, where the algorithms are more confident in the accuracy of its surrogates:

$$
\begin{aligned}
\min_{\mathbf{x} \in \mathbb{R}^{n_x}} \quad & f(\mathbf{x}) \\
\text{s.t.} \quad & \mathbf{g}(\mathbf{x}) \le 0 \\
& \mathbf{x} \in T(\mathbf{x}_c)
\end{aligned}
\tag{5.43}
$$

where $\mathbf{x}_c$ is a "safe" center point, and $\mathbf{g}$ could either map a different surrogate for each constraint or map a single surrogate on the maximum constraint violation. The methods benchmarked mostly differ in the type of surrogate $\mathbf{g}$ employed.

In the following sections, the algorithms for constrained model-based DFO are outlined – specifically, how the existing algorithms are modified (if in any way) to account for constraints.

5.4.3 Constrained surrogate optimization methods

CUATRO, COBYLA, and COBYQA account for constraints by construction; see Sections 5.2.8, 5.2.5, and 5.2.6, respectively, for more details. However, not all model-based derivative-free surrogates account for constraint by construction, for example, BO. In the following section, we will describe how constraints can be incorporated into the BO framework. This approach can also serve as a template for incorporating constraints in other surrogate-based optimization methodologies.

5.4.3.1 Constraint handling in Bayesian optimization

In *constrained BO* (CBO), constraints can be handled through various methods. A common approach involves constructing surrogates of the constraints, typically in the form of GPs. Since the constraints are unknown but can be sampled like the objective function, surrogates can be built following a similar procedure to that of the objective function.

There are two usual ways to utilize these surrogate constraints. The first approach involves penalizing constraint violations in the objective function. This can be done by either multiplying the objective function by one minus the probability of violation or adding the constraint violation as a penalty. The second common approach involves using the surrogate constraints to formulate a constrained optimization problem and solving it using constrained optimization algorithms, such as interior point methods or sequential quadratic programming. Specifically, given sampled inputs $X = \left[\mathbf{x}^{(1)}, \ldots, \mathbf{x}^{(n_d)}\right]^{\mathsf{T}}$, BO constructs a GP model $\mathcal{GP}_f$ to predict the outputs $\mathbf{y}_f = \left[f(\mathbf{x}^{(1)}), \ldots, f(\mathbf{x}^{(n_d)})\right]$. Similarly, a GP $\mathcal{GP}_{g_i}$ can be constructed to model each constraint, provided it can be measured, resulting in $\mathbf{y}_{g_i} = \left[g_i(\mathbf{x}^{(1)}), \ldots, g_i(\mathbf{x}^{(n_d)})\right]$. Consequently, the following optimization problem can be formulated:

$$
\begin{aligned}
\min_{\mathbf{x}\in\mathbb{R}^{n_x}} \quad & \mathcal{A}_{\mathcal{GP}_f}(\mathbf{x}) \\
\text{s.t.} \quad & \mathbf{g}_{\mathcal{GP}}(\mathbf{x}) \le \mathbf{0}
\end{aligned}
\tag{5.44}
$$

where $\mathcal{A}_{\mathcal{GP}_f}$ denotes the acquisition function using the GP model of the objective function (e.g., $\mu_f(\mathbf{x}) - \gamma\sigma_f^2(\mathbf{x})$), and $\mathbf{g}_{\mathcal{GP}}$ denotes the use of the GPs that model the constraints to ensure feasibility. One straightforward way to incorporate them is to use the mean of the GP for each constraint:

$$
\begin{aligned}
\min_{\mathbf{x}\in\mathbb{R}^{n_x}} \quad & \mathcal{A}_{\mathcal{GP}_f}(\mathbf{x}) \\
\text{s.t.} \quad & \mu_{\mathcal{GP}_i}(\mathbf{x}) \le 0 \quad \text{for } i=1,\ldots n_g
\end{aligned}
\tag{5.45}
$$

While this can be effective, using only the mean of a GP defeats the purpose of this surrogate and it could be replaced by less computationally expensive surrogates such as RBFs. One way in which the variance term can be incorporated is as a back off to add an extra layer of safety, for example:

$$
\begin{aligned}
\min_{\mathbf{x}\in\mathbb{R}^{n_x}} \quad & \mathcal{A}_{\mathcal{GP}_f}(\mathbf{x}) \\
\text{s.t.} \quad & \mu_{\mathcal{GP}_i}(\mathbf{x}) + \sigma_{\mathcal{GP}_i}(\mathbf{x}) \le 0 \quad \text{for } i=1,\ldots n_g
\end{aligned}
\tag{5.46}
$$

Finally, if safe exploration is important a trust region can be incorporated [46]. In this book chapter, the implementation follows the version in eq. (5.45) to maintain consistency with the rest of the methods described.

5.4.4 Performance assessment of constrained surrogate-based optimization algorithms

For the constrained performance assessment of black-box problems the functions and algorithms are as follows:

$$
a \in \mathbb{A}, \ \mathbb{A} = \{\text{CBO, COBYLA, COBYQA, CUATRO}\}
$$

$f \in \mathbb{F}$, $\mathbb{F} = \{$Constrained Rosenbrock, Constrained Quadratic, Constrained Matyas$\}$.

Therefore every algorithm $a \in \mathbb{A}$ is assessed on every constrained problem $f \in \mathbb{F}$ and its performance is compared relative to the other optimization algorithms on the same constrained black-box optimization problem.

5.4.4.1 Mathematical objective functions

The objective functions used in this subsection are again the ill-conditioned quadratic function and the Rosenbrock function presented in Section 5.3.2.1, as well as the Matyas function proposed by Matyas in 1965 [3]. The input dimension chosen for this section is $n_x = 2$. Subsequently, the objective functions and the constraints can be presented as follows:

- Rosenbrock:

$$\min \quad f(x_1, x_2) = (1 - x_1)^2 + 100(x_2 - x_1^2)^2$$
$$\text{s.t.} \quad g(x_1, x_2) = x_1 + 1.27 - 2.83x_2 + 0.69x_2^2 \tag{5.47}$$

- Quadratic:

$$\min \quad f(x_1, x_2) = x_1^2 + 0.95x_1x_2 + 5.9x_2^2$$
$$\text{s.t.} \quad g(x_1, x_2) = 1.5x_1 + 0.6 - x_2 \tag{5.48}$$

- Matyas:

$$\min \quad f(x_1, x_2) = 0.26(x_1^2 + x_2^2) - 0.48x_1x_2$$
$$\text{s.t.} \quad g(x_1, x_2) = 6.31x_1 + 3.60 - x_2 \tag{5.49}$$

It can be observed that when the algorithms commence evaluating alongside the constraint, there are very small constraint violations, e.g., 0.0008. Therefore, the threshold of 0.001 was chosen to determine a constraint violation as such. This means that when the constraint is violated when $g(\mathbf{x}) > 0$ it is counted as a violation when $g(\mathbf{x}) > 0.001$.

5.4.4.2 Results – convergence plots

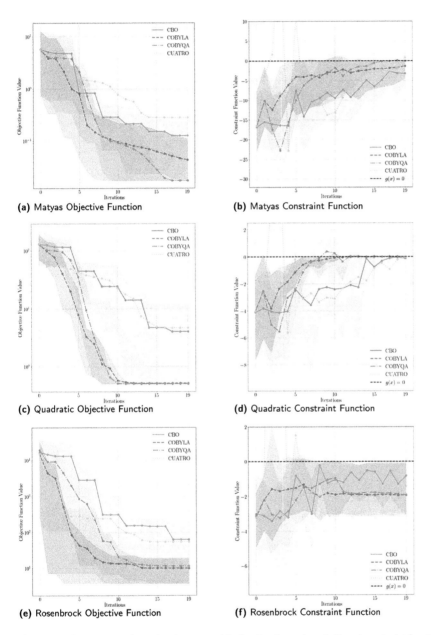

(a) Matyas Objective Function

(b) Matyas Constraint Function

(c) Quadratic Objective Function

(d) Quadratic Constraint Function

(e) Rosenbrock Objective Function

(f) Rosenbrock Constraint Function

Figure 5.4: Convergence plots, showing mean objective function values (left) and constraint function values (right) and 90–10% intervals enveloping trajectories over 10 repetitions from 10 different starting points with a budget (trajectory length) of 20 evaluations. The dashed line (right) indicates the boundary of constraint violation for $g(x) > 0$. The dimensionality of the problem is two input dimensions and one output dimension for objective and constraint function.

5.4.4.3 Results: trajectory plots

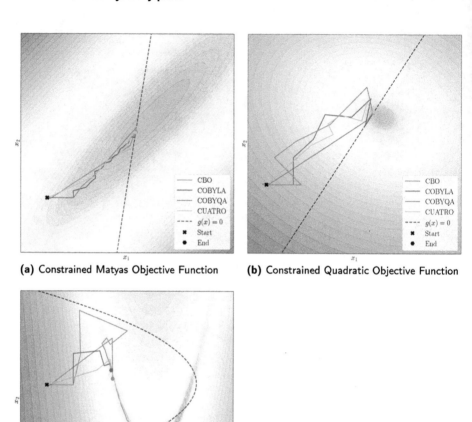

(a) Constrained Matyas Objective Function   **(b)** Constrained Quadratic Objective Function

(c) Constrained Rosenbrock Objective Function

Figure 5.5: Two-dimensional trajectory plots with constraint overlay for each constrained test function. The lines show connection of mean best-so-far evaluations of 10 repetitions per algorithm from a shared starting point and a budget of 60 evaluations.

5.4.4.4 Results: tables

Table 5.3: Benchmarking table for the constrained optimization.

	Rosenbrock	Quadratic	Matyas	Average
COBYQA	0.98 | 94.75% | 0.29	1.00 | 88.75% | 0.41	1.00 | 87.50% | 1.26	0.99 | 90.33% | 0.65
COBYLA	1.00 | 94.39% | 0.45	1.00 | 95.12% | 0.10	0.92 | 92.20% | 1.20	0.97 | 93.90% | 0.58
CBO	0.00 | 87.64% | 0.25	0.04 | 97.09% | 0.14	0.60 | 92.91% | 0.46	0.21 | 92.55% | 0.28
CUATRO	0.49 | 87.66% | 8.50	0.00 | 89.92% | 3.05	0.00 | 74.24% | 8.85	0.16 | 83.94% | 6.80

N.B.: The cells are to be read as p_a | constraint satisfaction | average constraint violation.

5.4.4.5 Results and discussion: mathematical constrained functions

For the constrained synthetic benchmarking, in addition to the trajectory plots and normalized algorithm scores, the constraint violations are considered through constraint violation trajectory plots. As in the unconstrained case, the algorithms' performance is observed for different test functions (Figures 5.4–5.5, and Table 5.3). It is important to note that all conclusions and results are relative to the specific set of algorithms included in this performance assessment. Algorithm-specific conclusions have been grouped and presented below.

CBO: The in-house implementation of CBO overall performed poorly, with the worst performance on the Rosenbrock function. The trajectory plots for the Matyas and ill-constrained quadratic test functions show the largest standard deviation among all algorithms. Additionally, the objective function value remains roughly constant for the first five iterations across all test functions before improving, likely due to the algorithm using these iterations to construct its GP surrogate.

COBYLA: COBYLA performs best on the ill-constrained quadratic and Rosenbrock functions, demonstrating very strong performance on the Matyas function as well.

COBYQA: COBYQA performs best on the Matyas function and shares the top performance with COBYLA on the ill-constrained quadratic function, with identical trajectories in the quantitative ranking region. It also performs very well on the Rosenbrock function. The ill-constrained quadratic constraint violation plot shows a small constraint violation by this algorithm.

CUATRO: CUATRO exhibits the poorest performance on the ill-constrained quadratic and Matyas functions, with reasonable performance on the Rosenbrock function. Among all constraint violation plots, CUATRO demonstrates the most severe constraint violations. Furthermore, its first five iterations have a standard deviation of zero for all test functions, likely due to an in-algorithm-imposed random sampling during these initial iterations.

5.5 Chemical engineering case studies

In addition to the benchmarking procedure based on mathematical functions, the performance of the algorithms was tested on two chemical engineering case studies. These case studies serve to evaluate the effectiveness of these algorithms in optimizing complex engineering systems. As the reader will discover in the results section, the performance of algorithms can vary significantly between mathematical case studies and real-world engineering problems. This discrepancy can be attributed to a variety of factors, including the fact that real-world problems generally present symmetries, and highly correlated variables which synthetic case studies generally do not. Examining the strengths and limitations of each algorithm in different contexts will provide insights into their suitability for various types of optimization problems and guide the selection of the most appropriate method for a given application.

Regarding the algorithms used for the unconstrained controller tuning case study, there are a few changes compared to those used for the unconstrained synthetic case study presented previously in Section 5.3.2. ENTMOOT and LSQM have proven to be too computationally expensive in higher dimensions, which make them unsuitable for the controller tuning case study with 32 dimensions. Regarding the BO framework, the in-house implementation that was used in the synthetic case study has been swapped to make way for two state-of-the-art implementations by GPyOpt [25] and TuRBO [78]. Furthermore, CUATRO-pls was added to showcase the dimensionality reduction capabilities of this add-on to the base version of CUATRO.

5.5.1 PID controller tuning problem

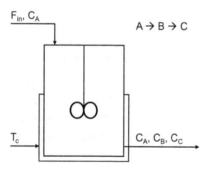

Figure 5.6: Diagram of a continuous stirred tank reactor (CSTR) with PID control.

This 32-dimensional case study presents a classical chemical engineering problem involving the dynamic control of a *continuous stirred tank reactor* (CSTR) equipped with a cooling jacket (Figure 5.6). In this system, the inlet flow rate and the cooling jacket temperature are the manipulated variables, and the reactor temperature is the control variable. Operating under a constant volume steady-state, the CSTR facilitates reactions where reactant A is converted into product B, which subsequently transforms into product C.

The objective is to maintain the control variable at its set point. This requires balancing the inherent nonlinearity and complexity of the system. To achieve this, a *proportional-integral-derivative* (PID) controller is employed to adjust the manipulated variables based on the deviation of the control variable from its set point. The performance of the control system is assessed by the objective function, defined as the system error. The error is a sum of the deviation of the control variable from its set point with a penalty for any changes in control action.

The optimization problem is high-dimensional, with 32 controller gain variables. The purpose of this engineering case study is to assess the potential of the benchmarked algorithms in optimizing high-dimensional, complex chemical processes.

5.5.1.1 System definition

The rate of change of concentration of species A (C_A) in the reactor is given by

$$\frac{dC_A}{dt} = \frac{F_{in}}{V}(C_{A_f} - C_A) - r_A \tag{5.50}$$

where the reaction rate r_A for the reaction $A \rightarrow B$ is defined as

$$r_A = k_{0,AB} \exp\left(-\frac{E_{AB}}{RT}\right) C_A \tag{5.51}$$

The rate of change of temperature (T) in the reactor is given by

$$\frac{dT}{dt} = \frac{F_{in}}{V}(T_f - T) + \frac{\Delta H_{AB}}{\rho C_p} r_A + \frac{\Delta H_{BC}}{\rho C_p} r_B + \frac{UA}{V\rho C_p}(T_c - T) \tag{5.52}$$

where the reaction rate r_B for the reaction $B \rightarrow C$ is defined as

$$r_B = k_{0,BC} \exp\left(-\frac{E_{BC}}{RT}\right) C_B \tag{5.53}$$

- T_c: Temperature of cooling jacket (K)
- F_{in}: Inlet flow rate (m³/s)
- C_i (for $i \in \{A, B, C\}$): Concentration of species i (mol/m³)
- T: Temperature in CSTR (K)
- T_f: Feed temperature (K)
- C_{A_f}: Feed concentration of A (mol/m³)
- V: Reactor volume (m³)
- ρ: Density of mixture (kg/m³)

- C_p: Heat capacity of mixture (J/kg · K)
- UA: Overall heat transfer coefficient (W/K)
- ΔH_j (for $j \in \{AB, BC\}$): Heat of j reaction (J/mol)
- E_j (for $j \in \{AB, BC\}$): Activation energy for j reaction (J/mol)
- $k_{0,j}$ (for $j \in \{AB, BC\}$): Pre-exponential factor for j reaction (1/s)

5.5.1.2 PID controller benchmarking results

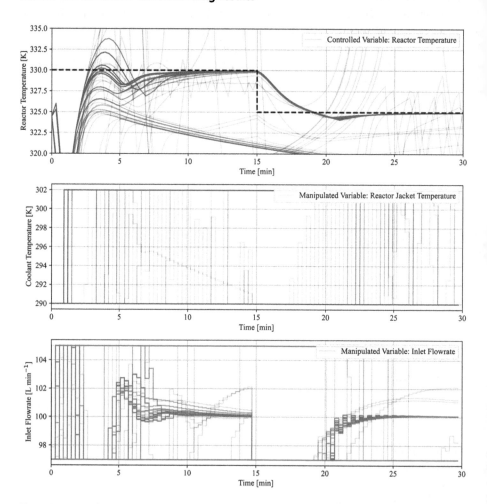

Figure 5.7: Example training trajectories of PID controller.

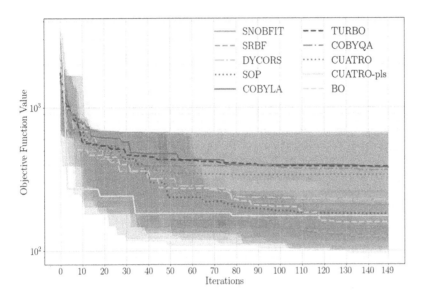

Figure 5.8: Trajectories for the CSTR-PID case study. The budget (trajectory length) is 150 evaluations. Thick lines represent the mean performance on each iteration over 10 repetitions and shaded regions indicate the 90–10%- interval enveloping trajectories for all repetitions.

Table 5.4: Benchmarking ranking for CSTR-PID case study.

Algorithm	Overall performance (all)	Rank
CUATRO-pls	1.00	1
SRBF	0.83	2
BO	0.75	3
SOP	0.73	4
DYCORS	0.64	5
SNOBFIT	0.61	6
CUATRO	0.32	7
COBYQA	0.17	8
TURBO	0.08	9
COBYLA	0.00	10

In this study, the performance of several optimization algorithms on a PID controller tuning case was evaluated. The algorithms tested included CUATRO-pls, SRBF, CO-BYLA, and TuRBO. The results indicate that CUATRO-pls was the best-performing algorithm, followed by SRBF. COBYLA was the worst-performing algorithm, with TuRBO being the second worst (Figure 5.7 and Table 5.4).

To demonstrate the process of training the PID controller by continuously updating the controller gain variables, an exemplar set of training trajectory graphs has been included – these trajectories are derived from a CUATRO-pls optimization run.

These graphs show the optimization process over time, where each algorithm was allowed 150 function evaluations. At each function evaluation, the algorithm updated the set of controller gains, resulting in different control trajectories (Figure 5.8).

There is one graph for each manipulated variable and one for the control variable. By overlaying these trajectories, a visual representation of the PID controller becoming more effective at maintaining the control variable at its set point is provided. The earlier, more poorly-tuned trajectories are shown in a fainter color, while the darker colors depict the later trajectories as the optimization progresses. It can be observed that the darker trajectories more closely follow the set point, indicating improved performance of the PID controller.

5.5.2 Williams-Otto benchmark problem

As a case study, the classical Williams-Otto benchmark problem is examined. This involves a CSTR, which is supplied with two streams of pure components, A and B, having respective mass flow rates F_A and F_B. The reactor operates at a steady state with a temperature T_r. Within the reactor, the chemical reactions between these reagents yield two primary products, P and E, through a series of reactions that also produce an intermediate compound C, and a by-product G. The reactions can be summarized as follows:

$$A + B \rightarrow C$$
$$B + C \rightarrow P + E$$
$$C + P \rightarrow G$$

For brevity, the complete set of mass balance equations and kinetic rate equations for this reaction system, as detailed by [24], are not reproduced here.

5.5.2.1 Williams-Otto benchmarking results

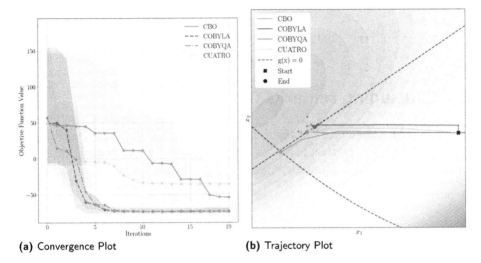

(a) Convergence Plot   **(b)** Trajectory Plot

Figure 5.9: Visualization for performance on Williams-Otto benchmarking problem: (a) Convergence plot, showing mean objective function values (thick lines) and 90–10% - intervals over 10 repetitions from 10 different starting points with a budget (trajectory length) of 20 evaluations. (b) Williams-Otto contour with dashed constraint-line and a single trajectory for each algorithm from an exemplary shared starting point. The lines show best-so-far evaluation positions and the scattered points show remaining evaluation positions.

Table 5.5: Benchmarking table for the constrained optimization of Williams-Otto.

	p_a	Feasible samples	Mean violation
COBYQA	1.00	81.75%	0.0062
COBYLA	1.00	84.28%	0.0044
CBO	0.00	90.86%	0.0085
CUATRO	0.14	80.10%	0.0125

Based on the performance scores, the best-performing algorithms are COBYQA and CO-BYLA with a perfect score of 1.0, followed by CUATRO with a score of 0.14. On the other end of the spectrum, the in-house implementation of CBO is the worst-performing algorithm with a score of 0.0 (Figure 5.9 and Table 5.5).

5.6 Code

The source code for this project including benchmarking routines and algorithms can be found at:

https://github.com/OptiMaL-PSE-Lab/DDO-4-ChemEng

5.7 Concluding remarks

This chapter has presented the application and evaluation of various model-based DFO algorithms for chemical processes, emphasizing the utility of data-driven approaches in enhancing optimization performance. The key insights from this study can be summarized as follows.

5.7.1 Performance comparison of algorithms

The performance evaluation of several optimization algorithms, including CUATRO-pls, SRBF, COBYLA, and TuRBO, highlighted significant differences in their effectiveness. CUATRO-pls emerged as the best-performing algorithm, particularly in the context of tuning a PID controller for a CSTR. SRBF followed closely, while COBYLA and TuRBO lagged behind; additionally, SNOBFIT performed much better in this case study. This ranking indicates that while traditional algorithms like COBYLA still hold value, newer and more sophisticated approaches such as CUATRO-pls and SRBF can offer substantial improvements in optimization tasks. Interestingly, some algorithms, such as CUATRO and SNOBFIT performed significantly differently in test functions compared to chemical engineering case studies, highlighting that assessing the performance of algorithms in engineering applications must be done in addition to a screening over their performance in traditional mathematical test functions.

5.7.2 Benchmarking on mathematical functions

The study employed several well-known mathematical objective functions, including the Ackley, Levy, Rosenbrock, and a quadratic ill-conditioned function, to benchmark the algorithms. These functions, selected for their diverse characteristics – ranging from multimodal to unimodal – provided a testbed for evaluating algorithm performance. The results indicated that the algorithms' effectiveness varied with the complexity and nature of the objective functions. For example, CUATRO-pls performed exceptionally well on unimodal functions like Rosenbrock but showed mixed results on multimodal functions like Ackley and Levy, and similar conclusions can be drawn for other algorithms.

5.7.3 General observations and future work

The benchmarking results highlighted the importance of algorithm selection based on the specific characteristics of the optimization problem at hand. Surrogate models, particularly those based on BO and RBFs, demonstrated robust performance across various scenarios. However, the study also highlighted areas for future exploration, such as the integration of noise in optimization processes and the application of these algorithms to more complex, real-world chemical engineering problems.

Finally, this study reaffirms the potential of surrogate-based optimization algorithms in enhancing the efficiency and accuracy of optimization in chemical processes for specific applications. The comparative analysis provides a reference for selecting appropriate algorithms, while the insights gained pave the way for future advancements and applications in the field. The continuous evolution of these algorithms promises further improvements, making DDO an indispensable tool in chemical engineering.

References

[1]   Lévy, P. Le mouvement brownien. *Mémorial Des Sciences Mathématiques*, 126, 1954.
[2]   Rosenbrock, H. H. An automatic method for finding the greatest or least value of a function. *The Computer Journal*, 2024, 3, 175–184. ISSN: 0010-4620. 1460-2067. Mar. 1960.
[3]   Matyas, A. Random Optimization. *Automation and Remote Control*, 1965, 26, 246–252.
[4]   Breiman, L., Friedman, J. H., Olshen, R. A., Stone, C. J. *Classification And Regression Trees*. 1984.
[5]   Ackley, D. H. *A Connectionist Machine for Genetic Hillclimbing*. In Mitchell, T. M. (Ed.). Springer US, Boston, MA, 2024, ISBN: 978–1-4612-9192-3 978–1-4613-1997-9. 1987.
[6]   Powell, M. J. D. *In Advances in Optimization and Numerical Analysis*. In Gomez, S., Hennart, J.-P. (Eds.). Springer, Netherlands, Dordrecht, 1994, pp. 51–67. ISBN: 978–90-4814358-0 978–94-015-8330-5. 2024.
[7]   Gutmann, H.-M. A radial basis function method for global optimization. *Journal of Global Optimization*, 2024, 19, 201–227. ISSN: 09255001. 2001.
[8]   Neumaier, A. Complete search in continuous global optimization and constraint satisfaction. *Acta Numerica*, 2024, 13, 271–369. ISSN: 0962-4929. 1474-0508. May. 2004.
[9]   Regis, R. G., Shoemaker, C. A. A stochastic radial basis function method for the global optimization of expensive functions. *INFORMS Journal on Computing*, 2024, 19, 497–509. ISSN: 1091-9856. 1526-5528. Nov. 2007.
[10]   Tolson, B. A., Shoemaker, C. A. dynamically dimensioned search algorithm for computationally efficient watershed model calibration. *Water Resources Research*, 2024, 43, 2005WR004723. ISSN: 0043-1397. 1944-7973. Jan. 2007.
[11]   Caballero, J. A., Grossmann, I. E. An algorithm for the use of surrogate models in modular flowsheet optimization. *AIChE Journal*, 2024, 54, 2633–2650. ISSN: 0001-1541. 1547-5905. Oct. 2008.
[12]   Caballero, J. A., Grossmann, I. E. An algorithm for the use of surrogate models in modular flowsheet optimization. *AIChE Journal*, 2024, 54, 2633–2650. ISSN: 0001-1541. 1547-5905. Oct. 2008.
[13]   Huyer, W., Neumaier, A. SNOBFIT – stable noisy optimization by branch and fit. *ACM Trans. Math. Software*, 2024, 35, 1–25. ISSN: 0098-3500. 1557-7295. July. 2008.

[14] Conn, A. R., Scheinberg, K., Vicente, L. N. Global convergence of general derivative-free trust-region algorithms to first and second-order critical points. *SIAM Journal on Optimization*, 2024, 20, 387–415. ISSN: 1052-6234. 1095-7189. Jan. 2009.

[15] Moré, J. J., Wild, S. M. Benchmarking derivative-free optimization algorithms. *SIAM Journal on Optimization*, 2024, 20, 172–191. ISSN: 1052-6234. 1095-7189. Jan. 2009.

[16] Regis, R. G., Shoemaker, C. A. Parallel stochastic global optimization using radial basis functions. *INFORMS Journal on Computing*, 2024, 21, 411–426. ISSN: 1091-9856. 1526-5528. Aug. 2009.

[17] Henao, C. A., Maravelias, C. T. Surrogate-based superstructure optimization framework. *AIChE Journal*, 2024, 57, 1216–1232. ISSN: 0001-1541. 1547-5905. May. 2011.

[18] Regis, R. G., Shoemaker, C. A. Combining radial basis function surrogates and dynamic coordinate search in high-dimensional expensive black-box optimization. *Engineering Optimization*, 2024, 45, 529–555. ISSN: 0305-215X. 1029-0273. May. 2013.

[19] Biegler, L. T., Lang, Y.-D., Lin, W. Multi-scale optimization for process systems engineering. *Computers & Chemical Engineering*, 2023, 60, 17–30. ISSN: 00981354. Jan. 2014.

[20] Boukouvala, F., Ierapetritou, M. G. Derivative-free optimization for expensive constrained problems using a novel expected improvement objective function. 2014, 60.

[21] González, J., Longworth, J., James, D. C., Lawrence, N. D. *Bayesian Optimization for Synthetic Gene Design*, 2024. arXiv: 1505.01627 [stat]. May 2015.

[22] Diamond, S., Boyd, S. *CVXPY: A Python-Embedded Modeling Language for Convex. Optimization*, 2024, arXiv: 1603.00943 [math]. June 2016.

[23] Krityakierne, T., Akhtar, T., Shoemaker, C. A. SOP: Parallel surrogate global optimization with Pareto center selection for computationally expensive single objective problems. *Journal of Global Optimization*, 66, 2024, 417–437. ISSN: 0925-5001. 1573-2916. Nov. 2016.

[24] Mendoza, D. F., Graciano, J. E. A., Dos Santos Liporace, F., Le Roux, G. A. C. Assessing the reliability of different real-time optimization methodologies. *The Canadian Journal of Chemical Engineering*, 2024, 94, 485–497. ISSN: 0008-4034. 1939-019X. Mar. 2016.

[25] The GPyOpt authors. *GPyOpt: A Bayesian Optimization Framework in Python*, 2016.

[26] Zhang, Q., Grossmann, I. E., Sundaramoorthy, A., Pinto, J. M. Data-driven construction of convex region surrogate models. *Optimization and Engineering*, 2024, 17, 289–332. ISSN: 1389-4420. 1573-2924. June 2016.

[27] Bogle, I. D. L. A perspective on smart process manufacturing research challenges for process systems engineers. *Engineering*, 2023, 3, 161–165. ISSN: 20958099. Apr. 2017.

[28] Boukouvala, F., Floudas, C. A. ARGONAUT: algorithms for global optimization of coNstrAined grey-box compUTational problems. *Optimization Letters*, 2024, 11, 895–913. ISSN: 1862-4472. 1862-4480. June 2017.

[29] Sorek, N., Gildin, E., Boukouvala, F., Beykal, B., Floudas, C. A. Dimensionality reduction for production optimization using polynomial approximations. *Computational Geosciences*, 2023, 21, 247–266. ISSN: 1420-0597. 1573-1499. Apr. 2017.

[30] Wilson, Z. T., Sahinidis, N. V. The ALAMO approach to machine learning. *Computers & Chemical Engineering*. 2023, 106, 785–795. ISSN: 00981354. Nov. 2017.

[31] Audet, C., Conn, A. R., Le Digabel, S., Peyrega, M. A progressive barrier derivative-free trust-region algorithm for constrained optimization. *Computational Optimization and Applications*, 2024, 71, 307–329. ISSN: 0926-6003. 1573-2894. Nov. 2018.

[32] Bhosekar, A., Ierapetritou, M. Advances in surrogate based modeling, feasibility analysis, and optimization: A review. *Computers & Chemical Engineering*, 2023, 108, 250–267. ISSN: 00981354. Jan. 2018.

[33] Buhmann, M. D., Fletcher, R., Iserles, A., Toint, P. Michael J. D. Powell. 29 July 1936 – 19 April 2015. *Biographical Memoirs of Fellows of the Royal Society*, 2024, 64, 341–366. ISSN: 0080-4606. 1748–8494. June 2018.

[34]   Costa, A., Nannicini, G. RBFOpt: An open-source library for black-box optimization with costly function evaluations. *Mathematical Programming Computation*, 2024, 10, 597–629. ISSN: 1867-2949. 1867–2957. Dec. 2018.

[35]   Eason, J. P., Biegler, L. T. Advanced trust region optimization strategies for glass box **/** black box models. *AIChE Journal*, 2023, 64, 3934–3943. ISSN: 0001-1541. 1547-5905. Nov. 2018.

[36]   Tsay, C., Baldea, M., Shi, J., Kumar, A., Flores-Cerrillo, J. *Computer Aided Chemical Engineering*. Elsevier, 2024, pp. 1273–1278. 2018. ISBN: 978-0-444-64241-7.

[37]   Larson, J., Menickelly, M., Wild, S. M. Derivative-free optimization methods. *Acta Numerica*, 2023, 28, 287–404. ISSN: 0962-4929. 1474-0508. May 2019.

[38]   Larson, J., Menickelly, M., Wild, S. M. Derivative-free optimization methods. *Acta Numerica*, 2024, 28, 287–404. ISSN: 0962-4929. 1474-0508. arXiv: 1904 . 11585 [math] May 2019.

[39]   Schweidtmann, A. M., Huster, W. R., Lüthje, J. T., Mitsos, A. Deterministic global process optimization: accurate (Single-Species) properties via artificial neural networks. *Computers & Chemical Engineering*, 2023, 121, 67–74. ISSN: 0098-1354. Feb. 2019.

[40]   Beykal, B., Onel, M., Onel, O., Pistikopoulos, E. N. A data-driven optimization algorithm for differential algebraic equations with numerical infeasibilities. *AIChE Journal*, 2024, 66, e16657. ISSN: 0001-1541. 1547–5905. Oct. 2020.

[41]   Bradford, E., Imsland, L., Zhang, D., Del Rio Chanona, E. A. Stochastic data-driven model predictive control using Gaussian processes. *Computers & Chemical Engineering*. 2024, 139, 106844. ISSN: 00981354. Aug. 2020.

[42]   Karg, B., Lucia, S. Efficient representation and approximation of model predictive control laws via deep learning. *IEEE Transactions on Cybernetics*, 2024, 50, 3866–3878. ISSN: 2168-2267. 2168-2275. Sept. 2020.

[43]   Kim, S. H., Boukouvala, F. Machine learning-based surrogate modeling for datadriven optimization: a comparison of subset selection for regression techniques. *Optimization Letters*, 2023, 14, 989–1010. ISSN: 1862-4472. 1862-4480. June 2020.

[44]   McBride, K., Sanchez Medina, E. I., Sundmacher, K. Hybrid semi-parametric modeling in separation processes: a review. *Chemie Ingenieur Technik*, 2024, 92, 842–855. ISSN: 0009-286X. 1522-2640. July 2020.

[45]   Ramírez-Márquez, C., Martín-Hernández, E., Martín, M., Segovia-Hernández, J. G. Surrogate based optimization of a process of polycrystalline silicon production. *Computers & Chemical Engineering*. 2024, 140, 106870. ISSN: 00981354. Sept. 2020.

[46]   Chanona, E. A. D. R., Petsagkourakis, P., Bradford, E., Graciano, J. E. A., Chachuat, B. Real-time optimization meets Bayesian optimization and derivativefree optimization: a tale of modifier adaptation. *Computers & Chemical Engineering*, 2024, 147, 107249. ISSN: 00981354. Apr. 2021.

[47]   Kazi, S. R., Short, M., Biegler, L. T. A trust region framework for heat exchanger network synthesis with detailed individual heat exchanger designs. *Computers & Chemical Engineering*. 2024, 153, 107447. ISSN: 00981354. Oct. 2021.

[48]   Petsagkourakis, P., Chachuat, B., Antonio Del Rio-Chanona, E. *Safe Real-Time Optimization Using Multi-Fidelity Gaussian Processes in 2021 60th IEEE Conference on Decision and Control (CDC)*. IEEE, Austin, TX, USA, 2024, pp. 6734–6741. Dec. 2021. ISBN: 978-1-66543-659-5.

[49]   Thebelt, A. *et al.* ENTMOOT: A framework for optimization over ensemble tree models. *Computers & Chemical Engineering*. 2024, 151, 107343. ISSN: 00981354. arXiv: 2003.04774 [cs, math, stat]. Aug. 2021.

[50]   Beykal, B., Avraamidou, S., Pistikopoulos, E. N. Data-driven optimization of mixed-integer bi-level multi-follower integrated planning and scheduling problems under demand uncertainty. *Computers & Chemical Engineering*, 2024, 156, 107551. ISSN: 00981354. Jan. 2022.

[51]   Ma, K. *et al.* Data-driven strategies for optimization of integrated chemical plants. *Computers & Chemical Engineering*, 2024, 166, 107961, ISSN: 00981354. Oct. 2022.

[52] Mroz, A. M., Posligua, V., Tarzia, A., Wolpert, E. H., Jelfs, K. E. Into the unknown: how computation can help explore uncharted material space. *Journal of the American Chemical Society*, 2024, 144, 18730–18743. ISSN: 0002-7863. 1520-5126. Oct. 2022.

[53] van de Berg, D. *et al.* Data-driven optimization for process systems engineering applications. *Chemical Engineering Science*, 2023, 248, 117135. ISSN: 00092509. Feb. 2022.

[54] Forster, T., Vázquez, D., Guillén-Gosálbez, G. Algebraic surrogate-based process optimization using Bayesian symbolic learning. *AIChE Journal*, 2024, 69, e18110. ISSN: 0001-1541. 1547–5905. Aug. 2023.

[55] Savage, T., Basha, N., McDonough, J., Matar, O. K., Del Rio Chanona, E. A. Multi-fidelity data-driven design and analysis of reactor and tube simulations. *Computers & Chemical Engineering*, 2024, 179, 108410. ISSN: 00981354. Nov. 2023.

[56] Savage, T., Basha, N., McDonough, J., Matar, O. K., Del Rio Chanona, E. A. Multi-fidelity data-driven design and analysis of reactor and tube simulations. *Computers & Chemical Engineering*, 2024, 179, 108410. ISSN: 00981354. Nov. 2023.

[57] Shi, H.-J. M., Qiming Xuan, M., Oztoprak, F., Nocedal, J. On the numerical performance of finite-difference-based methods for derivative-free optimization. *Optimization Methods & Software*, 2023, 38, 289–311. ISSN: 1055-6788. 1029-4937. Mar. 2023.

[58] Van De Berg, D., Petsagkourakis, P., Shah, N., Del Rio-Chanona, E. A. Data-driven coordination of subproblems in enterprise-wide optimization under organizational considerations. *AIChE Journal*, 2024, 69, e17977. ISSN: 0001-1541. 1547-5905. Apr. 2023.

[59] van de Berg, D., Shah, N., Del Rio-Chanona, E. A. *Hierarchical Planning-SchedulingControl – Optimality Surrogates and Derivative-Free Optimization*, 2024. arXiv: 2310.07870 [cs, math]. Oct. 2023.

[60] Zagorowska, M. *et al.* Online feedback optimization of compressor stations with model adaptation using Gaussian process regression. *Journal of Process Control*, 2024, 121, 119–133. ISSN: 09591524. Jan. 2023.

[61] Zhai, J., Boukouvala, F. Surrogate-based branch-and-bound algorithms for simulation-based black-box optimization. *Optimization and Engineering*, 2024, 24, 1463–1491. ISSN: 1389-4420. 1573-2924. Sept. 2023.

[62] Zhu, M., Bemporad, A. *Global and Preference-based Optimization with Mixed Variables Using Piecewise Affine Surrogates*, 2024. arXiv: 2302 . 04686 [cs, math]. June 2023.

[63] Alcántara, A., Ruiz, C., Tsay, C. *A Quantile Neural Network Framework for Two-stage Stochastic Optimization*, 2024. arXiv: 2403.11707 [math]. Mar. 2024.

[64] Durkin, A., Otte, L., Guo, M. Surrogate-based optimisation of process systems to recover resources from wastewater. *Computers & Chemical Engineering*, 2024, 182, 108584. ISSN: 00981354. Mar. 2024.

[65] Flores-Tlacuahuac, A., Fuentes-Cortés, L. F. A data-driven bayesian approach for optimal dynamic product transitions. *AIChE Journal*, 2024, 70, e18428. ISSN: 0001-1541. 1547-5905. June 2024.

[66] Forster, T., Vázquez, D., Moreno-Palancas, I. F., Guillén-Gosálbez, G. Algebraic surrogate-based flexibility analysis of process units with complicating process constraints. *Computers & Chemical Engineering*, 2024, 184, 108630. ISSN: 00981354. May 2024.

[67] Gupta, R., Zhang, Q. Data-driven decision-focused surrogate modeling. *AIChE Journal*, 2024, 70, e18338. ISSN: 0001-1541. 1547–5905. Apr. 2024.

[68] Gustafson, E. J. *et al. Surrogate Optimization of Variational Quantum Circuits*, 2024. arXiv: 2404.02951 [cond-mat, physics:physics, physics:quant-ph]. Apr. 2024.

[69] Helleckes, L. M. *et al.* High-throughput screening of catalytically active inclusion bodies using laboratory automation and Bayesian optimization. *Microbial Cell Factories*, 2024, 23, 67. ISSN: 1475-2859. Feb. 2024.

[70] Jiang, Y., Byrne, E., Glassey, J., Chen, X. Integrating graph neural network-based surrogate modeling with inverse design for granular flows. *Industrial & Engineering Chemistry Research*, 2024, 63, 9225–9235. ISSN: 0888-5885. 1520-5045. May 2024.

[71]   Morlet-Espinosa, J., Flores-Tlacuahuac, A. A Bayesian optimization approach for data-driven mixed-integer nonlinear programming problems. *AIChE Journal*, 2024, e18448. ISSN: 0001-1541. 1547-5905. June 2024.

[72]   Paulson, J. A., Tsay, C. *Bayesian Optimization as a Flexible and Efficient Design Framework for Sustainable Process Systems,* 2024. arXiv: 2401.16373 [cs, math]. Jan. 2024.

[73]   Tian, H., Ierapetritou, M. G. A surrogate-based framework for feasibility-driven optimization of expensive simulations. *AIChE Journal*, 2024, 70, e18364. ISSN: 0001-1541. 1547-5905 May 2024.

[74]   van de Berg, D., Shah, N., Del Rio-Chanona, A. *Computer Aided Chemical Engineering*. In Manenti, F., Reklaitis, G. V. (Eds.). Elsevier, 2024, pp. 3193–3198. Jan. 2024.

[75]   Zhang, J., Sugisawa, N., Felton, K. C., Fuse, S., Lapkin, A. A. Multi-objective bayesian optimisation using q -noisy expected hypervolume improvement (q NEHVI) for the Schotten–Baumann reaction. *Reaction Chemistry and Engineering*, 2024, 9, 706–712. ISSN: 2058-9883. 2024.

[76]   Zhu, M. *et al. Discrete and Mixed-Variable Experimental Design with Surrogate-Based Approach* 2024. Apr. 2024.

[77]   A Radial Basis Function Method for Global Optimization.

[78]   Eriksson, D., Pearce, M., Gardner, J., Turner, R. D., Poloczek, M. Scalable Global Optimization via Local Bayesian Optimization.

[79]   Gardner, J. R., Kusner, M. J., Jake, G. Bayesian Optimization with Inequality Constraints.

[80]   Gonzalez, J., Dai, Z., Hennig, P., Lawrence, N. Batch Bayesian Optimization via Local Penalization.

[81]   Gonzalez, J., Osborne, M., Lawrence, N. D. GLASSES: Relieving The Myopia Of Bayesian Optimisation.

[82]   Ke, G. *et al*. LightGBM: A Highly Efficient Gradient Boosting Decision Tree.

[83]   Ragonneau, T. M. Model-Based Derivative-Free Optimization Methods and Software.

Daniel Rangel-Martínez and Luis A. Ricardez-Sandoval*

Chapter 6
Data-driven techniques for optimal and sustainable process integration of chemical and manufacturing systems

Abstract: One of the major challenges in the field of Process Systems Engineering (PSE) is the development of an integral approach to perform the optimization in chemical and manufacturing plants. This approach is complex as it consists of the simultaneous optimization of multiple hierarchical tasks that occur at different temporal and spatial scales in chemical plants. Challenges include the disparity in timescales among the different tasks, different operational, logical and tactical constraints, different objectives or goals, and the curse of dimensionality. To handle these situations, the continuously growing field of artificial intelligence (AI) has developed tools that can be used to reduce or overcome the complexity that these challenges pose to the integration. The application of AI techniques is largely supported by the growing volume of data that can be used for building data-driven models. In this work, a review of the literature in PSE that makes use of data-driven techniques to perform the optimal process integration in chemical and manufacturing plants is presented. The tasks involved in the optimal process integration discussed in this work include integration of process design and control, planning and scheduling, scheduling and control and planning, scheduling and control. Key insights combined with an outlook are presented and aim to provide the reader with a perspective of the current state of the art in this field.

Keywords: data-driven models, optimal process integration, enterprisewide optimization, machine learning

Acknowledgment: The financial support provided by Consejo Nacional de Humanidades, Ciencias y Tecnologias (CONAHCYT), is acknowledged.

*Corresponding author: Luis A. Ricardez-Sandoval**, Department of Chemical Engineering, University of Waterloo, 200 University Ave W, Waterloo, ON, N2L 3G1 Canada, e-mail: laricard@uwaterloo.ca
Daniel Rangel-Martínez, Department of Chemical Engineering, University of Waterloo, 200 University Ave W, Waterloo, ON, N2L 3G1 Canada

https://doi.org/10.1515/9783111383439-006

6.1 Introduction

Enterprisewide optimization (EWO) is an emerging field in process systems engineering (PSE) that aims to integrate the multiple decision-making processes in chemical engineering systems that work separately but depend on each other [1]. The desire for this integration has been among the interests of academics and industrial practitioners aimed at remaining competitive in complex and demanding markets [2]. A growing set of tools such as the Internet of things (IoT), 5G network, and sensors in the chemical industry is providing the necessary resources for new techniques that can ease the integration of processes of supply, manufacturing, and distribution. The integration of tasks involving different levels of the process (i.e., design, planning, scheduling, and control) is the most common form of approaching EWO. Each one of these tasks (or levels) happens at different temporal and spatial scales in chemical and manufacturing plants. The most typical integrated operational aspects include planning and scheduling (iPS), design and scheduling (iDS), scheduling and control (iSC), and planning, scheduling, and control (iPSC). The integration of the different decision-making processes consists of the interconnection between optimization and mathematical programs at the different levels through the influence that each decision variable exerts over other levels. This influence is communicated through the state and output variables and is translated as constraints and performance measures to other decision levels.

The literature has shown that the optimization of integrated systems provides solutions of higher quality than the standard sequential solution of independent decision-level problems [3]. Moreover, infeasible scenarios and constraint violations can be avoided as the communication between the independent tasks are likely to result in a feasible and economically attractive operation. Nevertheless, a computational cost is often needed to handle the complexity that the integration of multiple optimization programs demand. This has been the main focus point addressed in the literature, i.e., the reduction in the computational effort without losing solution quality. In this way, new techniques that take advantage of the growing available resources from processes to generate models from data are under development; this is referred to as data-driven models. With the use of data-driven methods the field of EWO promises a reduction in costs, an increase in economic performance and in customer satisfaction, thus resulting in a more sustainable operation.

The recent increase in access and production of data has propelled the development of multivariate statistical methods that can extract useful information from historical data. This knowledge is key to define heuristics from phenomena, formulate predictive and classification models, and even generate new data. Among the most attractive data-driven techniques, machine learning (ML) has received great attention from the industry; in particular, attention has focused on key areas of deep learning, i.e., supervised, unsupervised, and reinforcement learning [4]. Artificial neural networks (ANNs) have found a niche in engineering applications due to their capacity to model complex nonlinear processes. Although some disadvantages exist in their implementation, their presence and use is expected to increase in the future [5]. Other methods such as support vector machines (SVMs), decision trees (DTs), and random forests (RFs) have also made an important impact on the processing of industrial data [6]. Additional modeling techniques supported with data include Monte Carlo simulations, piecewise affine models [7], and other data-driven surrogate models.

These tools have taken advantage of the existing historical and new (fresh) plant data produced online along with process simulations. The generated models have improved optimization methods in many ways, e.g., the reduction in complexity of the optimization problems and thus the reduction in computational costs, as well as the identification of attractive solutions compared to those generated from state-of-the-art tools and methods [8]. Figure 6.1 shows the usual general path to build and integrate a data-driven method into an optimization program. Different challenges arise in the implementation of these steps, which depend on the particular characteristics of each case study. In Figure 6.1, five general steps are used to define the integration of a data-driven method into the optimization process. The first three steps correspond to the training of the model, which could be an ML or a deep learning method. The first and second steps consist of the collection and preprocessing of the data for training the model, respectively. The main challenges from these steps consist of having access to the data and filtering it to remove information that will bias the model, e.g., outliers or out-of-date data points. In the third step, the training requires a hyperparameter exploration to ensure that the final model can generalize the data set, without under- or overfitting. The set of hyperparameters from this tuning includes learning rates, the architecture of the model, and training epochs, among others [9]. The connection between the data-driven methods is problem-dependent and should be defined for each application, although in general terms, the model can be embedded in the optimization problem as a constraint or as an initial value. After these

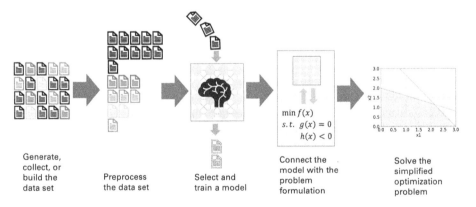

Generate, collect, or build the data set

Preprocess the data set

Select and train a model

Connect the model with the problem formulation

$$\min f(x)$$
$$s.t.\ g(x) = 0$$
$$h(x) < 0$$

Solve the simplified optimization problem

Figure 6.1: General path for building and incorporating a data-driven method into an optimization problem.

steps, the optimization should be simplified and ready to be approached with classic mathematical programming formulations. Although the introduction of data-driven models into EWO is still in a considerable early stage, and far from the widespread level of implementation that exist to address single-task problems, e.g., process design or process control, the current (recent) results reported in the literature are promising and provide high expectations for the commercial adoption of these methods in the near future [10].

The tools that are currently used in EWO are computationally intensive due to the complex interconnections that exist between the multiple hierarchical tasks, e.g., taking into account the effects that a decision exerts over other tasks. Moreover, dynamic operability and feasibility under multiple uncertainty realizations taking place at different scales are also key factors that must be considered when integrating these tasks. These aspects that typically emerge in EWO increase the complexity of the resulting problems that often require a significant computational effort. Data-driven methods are an attractive option to reduce the complexity of integrating these multiple hierarchical tasks due to their capacity to predict and classify in short turnaround times, a large number of actions or events. These features are mainly implemented in the prediction of values from decision variables and the feasibility analysis of the regions in the decision space, respectively, and help to reduce the computational burden related to these tasks. Nevertheless, developing data-driven models that provide this information to the integrated optimization formulations is a major challenge. This is mainly due to the data collection (or generation) that is required to train the resulting surrogate models. However, embedding these models within complex optimization formulations has become an attractive option for the following reasons: a) Once trained, a data-driven model provides fast responses that can reduce the computational effort, b) data-driven methods allow the generalization of complex systems and provides a range of options to model data (e.g., ANNs, DTs, and SVMs), and c) the training of these models allows to account for more realistic scenarios, e.g., account for parametric uncertainties or model-plant mismatch.

This chapter aims to provide a general perspective of the state of the art on the implementation of data-driven models on different classes of process integration in chemical engineering and manufacturing systems. An initial review on process integration with standard methods is presented to highlight the current challenges in the field and to describe common general approaches, along with their respective limitations and results. This is followed by a review on the current literature on data-driven models applied to different process integration in chemical and manufacturing systems, which aims to provide insights on the contributions of these surrogate models into the current optimal integration processes. This section is organized according to the type of integration, i.e., iDS, iPS, iSC, and iPSC. A brief description of the methodologies, the performance of the implementations and the type of systems considered to demonstrate their potential is provided. Conclusions from this work and an outlook of this emergent field are presented at the end. To the authors' knowledge, this work aims to fill a gap in the literature since there are no contributions that exclusively review the impact of data-driven models in optimal process integration for chemical and manufacturing systems.

6.2 State of the art in process integration

The integration of multiple decision levels aims to improve the operability of the process and ensures that relevant objectives in the process are working together towards a common goal [11]. In this section, the current literature on works that review the state of the art of process systems integration is presented to give an initial viewpoint of the benefits, limitations and challenges in this area. The studies discussed in this section correspond to previous perspectives of different integrations reported in the literature, specifically iPSC, iDSC, iSC, and iPS. Although extensive literature on these topics is available, e.g., [12–16], only a few works are presented to show the aspects that give rise to the interest on applying data-driven models into optimal process integration methods. These aspects include challenges such as the increase in complexity mainly due to the rise in dimensionality on the optimization problem, disparity in time and spatial scales, and consideration of uncertainty and external perturbations.

6.2.1 Reviews on process integration

In this section, the most recent reviews on process integration are introduced and discussed. This section aims to explore the limitations and challenges of the current approaches available to perform optimal process integration. As mentioned above, the number of reviews presented in this section is limited because the aim is to outline the challenges in process integrations. First, works integrating two tasks are presented followed by studies that perform the integration of three tasks.

Maravelias and Sung [17] present a study on the integration of planning and scheduling; they highlighted the importance of taking simultaneous decisions and to consider the interdependence of both problems to achieve global optimality. The authors discussed the advantages of integrating planning and scheduling, the different approaches that exist in the literature, and the most common challenges, which include the increase in complexity of the problem and the different time grids that need to be handled by the problem. In that work, the authors indicated that this integration aims to obtain a feasible and *near-optimal* production solution arguing that finding the optimal solution for the problem implies a large computational burden. The authors presented different alternatives that exist in the literature for reduction in the problem's complexity, e.g., relaxed and aggregated scheduling formulations, off-line surrogate models, and hybrid modeling. The work presents a general classification for the solution of integrated production planning and scheduling, i.e., hierarchical, iterative, and full-space solutions. Each approach is extended with descriptions of different works from the literature that employed such approach. Challenges and new opportunities such as parallel computing and hybrid methods are discussed to show the future possible directions of development.

Dias and Ierapetritou discussed the integration of scheduling and control and the influence of process uncertainties in this integration [18]. The authors classified uncertainties in different classes as in [19, 20]; these classes are broadly used for describing approaches in the literature for process control and scheduling. The authors showed that, for process control, there is still no general approach that can be implemented online, guarantee robust constraint satisfaction and compute solutions with a low level of conservatism. The methods for approaching scheduling under uncertainty include reactive scheduling, robust scheduling, stochastic scheduling, and fuzzy programming scheduling. A motivating example that does not consider uncertainty is provided to demonstrate the implications of integrating scheduling and control problems. The authors pointed out that the problem would become intractable without the use of decomposition techniques.

Regarding the integration of scheduling and control, the authors presented the most common approach, i.e., the use of full-scale dynamic models of the process as constraints in the scheduling problem. With this method, the control problem (formulated as a Mixed Integer Dynamic Optimization (MIDO) problem) is reduced to a Mixed Integer Nonlinear Program (MINLP) through the implementation of numerical discretization techniques [21]. Challenges for this integration are associated with the computational complexity of handling large realistic problems. Even more complex is to solve the problem in an online mode, which in most cases becomes impractical [22]. Handling disturbances at the scheduling level using an integrated framework is difficult; and studies addressing this challenge are scarce in the literature [23, 24]. Approaches that handle uncertainty already exist but find some problems when applied to real scenarios due to the complexity and the difficulty of the implementation, i.e., formulating the integrated problem and solving the problem in acceptable turnaround times.

In their review of integration methods for design, scheduling, and control, Pistikopoulos et al. commented on recent contributions on integration of scheduling and control, and simultaneous design and control [25]. Notable approaches such as superstructures, MIDO, and hybrid methods that combine model predictive control (MPC) with online optimization techniques were highlighted. The review also discussed the relevance of the effects of the decisions made from one task on another and the need for a method that considers multiple problems as a single task. Decisions are classified as long-, mid-, and short-term corresponding to design, scheduling, and control, respectively.

The authors focused on discussing the application of mid-term optimization. They argued that the usual approach in real plants consists in solving the operational scheduling assuming the fixation of the design of the plant and the ideal performance of the regulatory and supervisory control. Thus, the long- and short-term decisions may remain unchanged during operation. Nevertheless, for the integration of scheduling and control, assessing the effect of disturbances is key to test the performance of the control system; also, maintaining the set points provided by the scheduling process is one of the major challenges. The dependencies of the set points on the schedule and the interaction that exist between these two levels (i.e., design and control) have motivated studies in this direction. Moreover, the authors presented a framework for simultaneous design, scheduling, and control aiming to address the current lack of specialized software designed for this task.

Andrés-Martinez and Ricardez-Sandoval [10] provide a review of the different methodologies that have been developed for the integration of planning and scheduling, scheduling and control, and planning, scheduling, and control. For the integration of planning and scheduling, the authors highlighted that the time grid representation often affects the complexity of the problem in the number of integer variables and the general convexity of the model. Decomposition and relaxation techniques are preferred for the integration of planning and scheduling because they reduce the complexity of the problem at the expense of solving the problem using most likely an iterative process. Nevertheless, for larger problems, a higher computational effort is required to achieve a solution. Challenges that largely involve the computational complexity reduction of the problem through good initialization values and the use of linear space representation or multiparametric programming are also discussed.

The authors pointed out that uncertainty has not been widely considered for iSC but options include two-stage stochastic programming, chance-constrained, and robust optimization; however, they also indicated that the adoption of such methods may result in complex and more expensive optimization problems. For iPSC the authors noted that most contributions are an extension of the tools proposed for the two previous integration strategies. The large computational costs associated with these problems makes them currently unsuitable for online calculations. Furthermore, the authors briefly discussed the potential of ML integrated with current optimization techniques to reduce the computational effort needed to solve problems integrating planning, scheduling, and control decisions simultaneously. Among some attractive features are that ML methods are useful to handle high-dimensional uncertainties and the wide availability of open source and free software.

Table 6.1 summarizes the challenges for the integration of two and three tasks discussed in the reviews presented in this section. Moreover, this table also presents the current techniques used to deal with the challenges in the area. It appears inevitable that, even for the integration of two tasks involving complex systems, a relaxation, simplification, or decomposition method must be considered to make problems tractable for large-scale chemical and manufacturing systems. Moreover, for the integration of three tasks, the available strategies become an extension of those proposed for the integration of two tasks with the additional complexity from the third task. Methods that reduce the complexity of the problem by making assumptions like the ideal performance of the control system or the absence of disturbances are reduced in complexity but may drift away from the real scenarios. The increasing complexity that this integration gives rise to is due to different factors that include the increase in dimensionality, the difference in timescales from one task to another, and the effects of uncertainty at different levels. Industrial scenarios and online implementations will demand a more intensive computational effort, making the problem not viable for real-world applications. To alleviate the computational effort, an alternative to the mathematical modeling of the process is the development of models using data-driven methods. These methods could take advantage of the available historical data to build reliable process models that can serve to gain insights into the operability of the process and decrease the computational bottlenecks while performing the optimal process integration.

Table 6.1: Challenges and current methods in process integration systems.

Scheme	Challenges of existing approaches in the literature	Methods used in the literature
Two tasks	– Different time grids in the problem – Large computational times required to find solutions – No guarantee of robust constraint satisfaction – No guarantee of low conservatism in the solution – Disturbances are difficult to handle – Formulating the integrated problem is complicated – Uncertainty has not been addressed properly	– Relaxation and aggregation techniques – Stochastic and fuzzy programming as optimization methods – Decomposition methods – Reduction of the problem through numerical and discretization techniques – Provide good initialization values to the optimization method
Three tasks	– Each task exerts influence on the others, which increases the complexity of the problem – Different time grids in the problem for short-, mid-, and long-term decisions – Disturbances in one task will have an effect on other tasks – There is a lack of specialized software for approaching these problems – Online implementations are unsuitable at the moment	– Approach the problem by fixing decisions and making assumptions to reduce the complexity of the problem – Propose frameworks for representation, modeling and solution of integrated problems – Strategies from two-process integration are used for this scheme

6.2.2 Data-driven techniques in PSE

Data-driven modeling encompasses techniques that can be used to gain insights and analyze the existing connections between state variables in a system [26]. An advantage of these techniques is that the process behavior is obtained from the data, which may include the physical dynamics of the system. Naturally, their source of information is data generated from simulations or historical data from the process. The field of data-driven modeling includes other disciplines such as ML, artificial intelligence (AI), data mining, and computational intelligence; these areas are similar but differ in their techniques for data treatment for deriving the models. The particular case of the application of AI, and its subfield ML, in decision-making methods in process engineering has seen an increasing influence in recent years due to the increase in data sets coming from past and present applications [27]. Specifically, the developments of sensors, IoT, and the availability of historical data have an important role in the widespread of AI. The surrogate models generated from these techniques avoid the use of expensive calculations, thus making them attractive to perform an efficient interaction between different decision levels in the optimal integration process. For instance, a model derived from one task can be embedded as a constraint for a second model such that the complexity of the optimization method is reduced in the search space to only feasible solutions for both tasks [28]. As discussed in Section 6.3.1, this approach has been used in many applications and arguably could also be extended to integrate more than two-level tasks.

Thon et al. [29] provide a general classification of AI in process engineering, which consists of predictive modeling, process optimization, fault detection, and process control. That study highlighted the use of ANNs as a feature of high interest due to their capacity to generalize complex nonlinear functions. Nevertheless, other techniques from ML are used for developing models as they might be easier to train or because the features of the method are adequate for certain problems. For instance, DTs and RFs provide clear insight into how decisions are made compared to ANNs; SVMs need fewer volumes of data to create a classification model compared to ANNs. Moreover, the literature shows a tendency to use hybrid methods that bring together the features of both analytical methods with data-driven techniques for designing current structures from process engineering. For instance, ML methods are used to extract information from images or large data sets using ANNs for their use in the construction of models. On the other hand, in applications where not too much data is available for training ML models, there is the option to use available heuristic models to generate data from ideal scenarios, which is then incorporated with historical (actual) data to improve the fidelity and confidence of those models.

Overall, the use of data-driven methods looks promising for process integration since they could reduce the need for expensive computational resources and are effective to capture the dynamics of complex systems and provide relatively highly accurate predictions in short turnaround times. Nevertheless, there are limitations related to the availability and quality of data, as well as in the hyperparameter tuning during the training phase.

6.3 Data-driven methods for optimal process integration

In this section, the literature review of the integration of data-driven methods into different levels of the decision-making process is presented: namely, the integration of 1) design and control, 2) planning and scheduling, 3) scheduling and control, and 4) planning, scheduling, and control. In this work, a data-driven modeling method is defined as a technique where a model is generated with a data set to predict values or classify parameters in the process. The aim of the resulting model is to reduce computational expenses in the optimization process in the integration of two or more decision processes.

Figure 6.2 shows the chronological distribution of the studies published in the literature, which have made use of data-driven techniques for the integration of two- or three task-levels. Note the recent visible interest in the use of these methods for process integration. Most of the studies reported consist of integrating the scheduling process with either the task from a higher decision level (i.e., planning) or lower decision

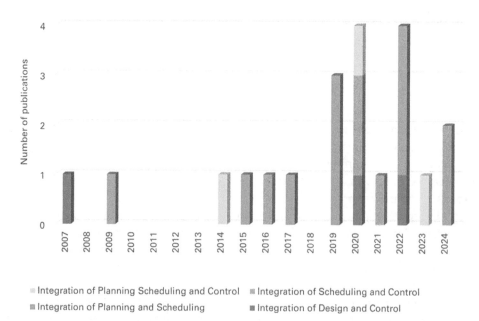

Figure 6.2: Distribution of publications on process integration with data driven from the literature found for this work.

level (i.e., process control). Thus, hierarchical integrations of contiguous levels in the process are preferred, e.g., planning with scheduling and scheduling with control. It was observed that the integration of more than two tasks remains a challenge; very few articles attempting this task were found, presumably due to the complexity of the problem formulation associated with the integrated scheme [30–32].

The four task levels discussed above are classified in two categories: integration of two decision levels and integration of three decision levels. The contributions from each category are discussed next.

6.3.1 Integration of two decision levels

In this section, works that attempt the integration of two decision levels aided by data-driven methods are presented. The works are classified considering the type of integration attempted, i.e., design and control, planning and scheduling, and scheduling and control. Moreover, it was observed that standard solution methods (i.e., with no data-driven methods) have a similar preference for two-task integrations over three tasks. Since the process for integrating more than two decision levels is, in general terms, an extension of the methods used for the integration of two levels, the latter methods are presented first as a preface to more complex problems.

6.3.1.1 Integration of design and control

The integrated optimization problem of design and control conveys the challenges associated with the formulations from both levels, i.e., an MINLP for the design problem and a MIDO for the control problem [33]. Different methods in the literature have attempted to find a systematic solution for these types of problems, e.g., [33–36], where the complexity and nonlinearity of the problem are alleviated through the reduction of the search space (e.g., trust region methods). Due to the difference in nature of both problems, the integration is complex and multidimensional [37, 38]. Moreover, the timescale is different as the control problem is solved over a pre-specified time horizon whereas the design decisions are time-independent variables that remain constant during operation.

For the solution of the control task alone, the use of techniques like reinforcement learning (RL) has returned promising results [39, 40]. This learning method consists of training an intelligent agent that reacts to disturbances in the environment where it is trained. The work of Dutta and Upreti [41] examines the recent infiltration of actor-critic (AC) methods (an RL methodology) in the process control industry. The authors recognize a continuous increase in the use of AC methods for set point tracking and disturbance rejection in real-time implementations. In abnormal events, AC methods might be useful to bring a process back to a desired operating state without prior knowledge. It is expected that the use of RL methods to approach control problems will extend to process integration because of their capacity to make decisions considering multiple sources of information. The following works present different ways on how this can be accomplished.

Sachio et al. [42] presented an integration of process design and control that uses RL to substitute the controller in a bi-level optimization problem where the control task (inner level) is embedded in the design problem (outer level). Once the solution of the nonlinear optimal control problem is computed, the design problem integrates this solution as a constraint. The problem takes advantage of the possibility to output a mixed integer linear function from the agent, which can be easily integrated into the design task and solved using mixed-integer programming (MIP) methods. The method was tested in two case studies; the first involves a tank design (without reaction) whereas the second case considers the optimal design of a CSTR. For the first case study, the RL agent was able to handle disturbances and outperformed a standard proportional-derivative (PD) controller and a multiparametric MPC. Moreover, the authors initialized the agent with a pre-training using data from a standard proportional-integral (PI) controller, which reduced the training of the model. In the second case study, the controller was able to maintain control over the concentration of a component while satisfying operational constraints. Although the performance was deemed acceptable for both cases, the authors mentioned that there is no guarantee of finding a global solution since RL is an approximate dynamic programming technique.

Mendiola-Rodriguez and Ricardez-Sandoval [43], proposed an integration strategy for process design and control to a system that is subject to uncertainty through an RL model. The methodology, which consists of an agent trained with an AC method, is applied to an anaerobic digestion system (ADS) involved in the processing of Tequila Vinasses. The agent provides control and design actions at each step of a discretized horizon. Open-loop control was implemented in those systems. Although both design and control decisions are made at every time interval, the process is set to only take the first design decision at the beginning of the process and keep it fixed for the rest of the transient operation. In the case of the control actions, these are collected and executed at every time interval in the horizon. The method is compared with three ADS scenarios where the response and final solution of the RL agent are compared with a sequential approach of the design and control problem. Computational tests showed that the RL agent showed a more stable and smoother performance with satisfactory results. The authors highlighted that the final policy from the RL method should be verified as its learning could have been driven to situations that are not desired for the process, e.g., unsafe practices or infeasible scenarios, to achieve higher rewards.

Egea et al. [44] made a comparison between three surrogate-based optimization solvers and six different optimization methods for an integrated design and control multimodal optimization problem. The surrogate-based global optimization solvers considered in this work were: 1) radial basis function (RBF) interpolation algorithm, 2) the efficient global optimization (EGO) algorithm, and 3) the stable noisy optimization by branch and fit (SNOBFIT) method. On the other hand, the benchmark methods included the direct search local method, sequential quadratic programming, the generalization pattern search algorithm, and the heuristic population-based stochastic approach, among others. For the sake of brevity, the other methods are not defined because they are hybrid methods that combine several strategies; the reader is referred to their contribution for more detail [44].

A wastewater treatment plant simulation is used as a testing facility to compare the results from both approaches. The surrogate-based global optimization solvers use data points collected from different evaluations of the objective functions to make informed decisions on the search of static variables of the process design, operating conditions, and the controllers' parameters. On the other hand, the benchmark solvers only rely on standard evaluations of the original model, for instance, the simplex method, the heuristic population-based stochastic approach, and evolution strategies. This condition reduces the space search and derives local optimal solutions. All the solvers were run 10 times with different initialization values to obtain a mean value for each solver. Results showed that RBF and SNOBFIT found the global optimal solution for the control problem in the wastewater treatment plant whereas the other methods were only capable of finding a local optimum. Moreover, a reduction in time was accomplished by the methods with surrogate models, which was expected due to the reduction in the complexity of the optimization problem.

RL is a well-known technique that is gaining popularity in process control, and its applicability can be extended to other decision-making processes, including process flowsheet design [45–47]. The other techniques presented have demonstrated their capacity to compete with standard methods and have returned promising results. While the performance reported in the studies is acceptable, the solutions tend to be suboptimal. Also, the studies that reported the use of ANNs indicated that the learning has to be validated as the final result can diverge to different outcomes, i.e., perform in a different way than expected. Table 6.2 shows a summary of the specific challenges that were addressed in the works discussed above for integration of design and control.

6.3.1.2 Integration of planning and scheduling

The integration of planning and scheduling is a topic of interest in the field of supply chain management because it guarantees the timely execution of tasks in the manufacture of the required production. The difference between planning and scheduling lies in the length of the time horizon; while planning deals with the allocation of resources for the next weeks or months, the scheduling task is reduced to the organization of production in the following hours or days. In this sense, the first obstacle in

Table 6.2: Challenges and highlights from data-driven methods in design and control integration.

Challenges in the design and control integration	Highlights in the implementation of data-driven models
− Complexity of the control problem, usually a nonlinear dynamic problem − Disturbances in the control task are difficult to handle − Constraints of the design and control problem need to be met simultaneously − Integrated problem takes large computational times to provide solutions	Benefits: − ANNs can model the behavior of the controller under different scenarios − The learning method of RL can be used to avoid constraint violation and fulfillment of the objectives − Data-driven models need shorter computational times
	Limitations: − Optimality not guaranteed − Deep learning models are black-boxes that need to be meticulously assessed before implementation

the integration of planning and scheduling is the difference in timescales, which means that the problem requires solutions at different timescales within the horizon. Moreover, planning usually focuses on economic profit maximization, while scheduling tends to search for the feasibility of the process to execute the manufacturing tasks without issue or on reducing the makespan of the process [48].

Due to the major differences between planning and scheduling, the usual approach is to separate the problems and solve them in a sequential fashion. However, this often results in an inefficient allocation of resources and thus, a suboptimal solution to the problem or in the worst case, an infeasible production planning strategy [49]. The usual solution proposed to this situation is the integration of both tasks in one optimization problem; nevertheless, the solution method remains as a major challenge because of the large size of the problem that can become intractable or impractical to implement. The problem becomes even more complex if several variables are subject to some type of uncertainty. Data-driven methods remain an option to make the integration practical by reducing the complexity of the problem and providing feasible and optimal (or close-to-optimal) solutions for task production. The following methodologies show how the integration of models generated with data alleviates the complexity of the problem and enables access to near-optimal solutions.

The use of multiple AI methods to generate less expensive models was observed recursively in the literature showing the growing interest in this field for approaching complex optimization problems. In the work of Dias and Ierapetritou [28] different classification methods borrowed from ML are used to identify feasible regions of a scheduling problem to complement the solution of the planning level of the integrated problem. The evaluated ML methods were the DTs, SVMs, and ANNs. The generated model with each method was translated into algebraic equations and incorporated into the planning optimization problem as a classifier of production targets. The space of production targets for the scheduling horizon is uncertain whereas the classifier will partition the regions as feasible and infeasible. Each classifier was trained using historical data and tested using the feasibility metrics proposed in [50] to test the performance of each classifier when searching for the feasible regions in the uncertain space. The data is assumed to be already preprocessed in terms of data processing and feature selection for the ML implementation. The feasible space found by the classifiers is then used in the search for the optimal solution to the planning problem by making a considerable reduction of the search space. As a benchmark, the authors used two non-data-driven methodologies to compare the performance of the three classifiers.

The generated models (i.e., three data-driven and the two benchmark methods) were tested in three problems; each one more complex than the other in terms of model-dimension, i.e., 2, 3, and 7 dimensions. The experiments showed that ANNs provided the best results in terms of identifying the feasibility region in the domain of each problem with an overall 2.92%, 2.1%, and 7.33% of error. Although the error is relatively small, the authors mentioned that it is still a considerable error that might be attended in future improvements of this method. The solution obtained from the five generated methods showed that the proposed approach with data-driven methods became better than the benchmark methods. This work also indicated that the increase in the complexity of the problems (dimensionality) did not affect the computational time of response of the ML methods because they are fixed functions that toned not perform complex calculations.

Badejo et al. proposed a methodology that integrates the scheduling problem into the tactical and operational planning, motivated by the large volumes of data that are available, and that could help measure the feasibility of schedules [51]. A monolithic model is used to address the tactical and operational planning problems in which feasibility constraints from the scheduling problem are incorporated. The feasibility constraints are interpreted from design variables through a surrogate model; these variables include production facility data, production recipe, and the scheduling horizon. Three ML methods are assessed to generate the surrogate models, which are two variants of SVM, and a linear model called linear expanded model. Feasibility accuracy quantification, proposed by Wang and Lerapetritou [50], was used to measure the quality of the surrogate models. Two case studies are used to highlight the advantages of the framework; the first is a multi-facility problem used to show the versatility of the method presented, whereas the second case is a high-dimensional problem used to show the robustness of the solution quality and the time complexity. For both cases, the solutions were acceptable although not superior to the solution provided by a full-space model, i.e., a global optimum was not achieved. The main advantages of the integration of surrogate models include the acceptable quality of the solution and the reduction in the computational times of the optimization process.

Casazza and Caselli [52] proposed the integration of a data-driven method (specifically from ML) that reduces the space search of the optimization problem by providing feasible schedules into the planning problem. The methodology consists of a DT that is embedded in the optimization program, which is then solved by general-purpose optimization tools. The DT is a classification model that uses historical data to be trained and defines if a schedule instance is feasible or infeasible. Then the model is encoded as an equation that substitutes the scheduling constraints in the overall formulation. The features that have a relevant influence on the DT were the number of jobs per time slot, the number of jobs per machine, the processing times of each job, and time slots of overlapping in machines for a certain job. Since the trained DT works as a white box model, it could be encoded into a mathematical programming formulation, which can replace the scheduling constraints in an integrated planning and scheduling formulation.

The methodology is tested and compared against the results from the integrated planning and scheduling formulation, i.e., with no constraint substitution. Note that the data used to train the DT consisted of multiple single time-slot scheduling instances and needed to be labeled as feasible or infeasible. It was observed that the integrated problem with no data-driven method could find about 64% of the feasible instances during testing while the data-driven integration could find 100% of the feasible scenarios. In terms of the accuracy in the objective value, a difference with the optimal value of less than 1% was observed. Results from this study showed that the model was highly effective in approximating the feasibility of a given schedule, albeit not finding the optimal value in the final solution.

Evolutionary algorithms, and in particular genetic algorithms (GA), have been extensively used for decision-making processes in chemical engineering due to their ability to handle multiple objectives simultaneously [53]. Dao et al. [54] presented a modified GA, which reduces the variability of the solution on the integration of planning and scheduling problems, which the authors argue to be variable due to its dynamic nature. The modified GA can gain insights about the problem dynamics from the experience collected during the evolution to create more robust solutions. The method consists of the translation of valuable information from one run of generation to another, each one called civilization, thus substituting the random initialization of a civilization, which is intrinsic in the GA method for the evolution of the model. The information that is passed from one civilization to the other corresponds to the best chromosomes generated in the previous civilization. Without the randomness provided by the new chromosomes, the method suffers from premature convergence into local optima but this is avoided by a principle proposed by the authors, which consists of dropping out chromosomes that have been passed from one civilization to another. The initialization from one civilization to another allows the modified GA to enhance learning by using previous experience. Moreover, the translated knowledge can be preprocessed to provide better performance to the next civilization.

The methodology is used in a case study from [55] where a manufacturing company with three production lines and 50 products is solved for the planning and scheduling for the next month. A comparison was made with three different commercial solvers: a) pattern search solver (PSS), b) simulated annealing (SA), and c) GA. The proposed framework showed a considerable reduction of computational times compared to the three respective solvers, i.e., PSS: 35.4%, SA: 87.7%, and GA: 43%. The solution quality from the solvers was also outperformed by the proposed method; the solutions from the modified GA were PSS: 27.3%, SA: 62.3%, and GA: 80.2% higher than the benchmark methods. The authors also highlighted the importance of appropriate tuning of the information translated from one civilization to another, arguing that this is key to avoid local optimal solutions.

Agent-based methods consist of the development of intelligent agents that are autonomous and can make decisions through interaction with other agents [56]. These methods are useful in the field of PSE for their capacity to do process analysis, process monitoring, process performance, and process prediction [57]. In the work from Chu et al. [58], a framework for integrating the planning and scheduling tasks for a system under production uncertainty is proposed. The method consists of a bi-level formulation with a mixed-integer linear programming (MILP) solver (upper level) for the planning task that contains an agent-based reactive scheduling method [59] (lower level). The agent generates cutting-plane constraints that are integrated into the upper-level problem for generating corrected production quantities at the planning level. The framework is implemented under an iterative algorithm, which solves the planning program while the agent produces the schedule of the production under uncertainty. If, at any given iteration, the solution provided from the upper level is not feasible due to different constraint violations, then a cutting-plane constraint is generated by the lower level and integrated into the planning model. Subsequent iterations are then performed until all constraints from both levels are satisfied. The agent-based method implements different agents in the scheduling process to define the plane constraints with a Monte Carlo simulation method. The framework is tested on two case studies, and the authors highlighted that less time was needed to solve the problem since the planning problem is solved only one time at every iteration.

Surrogate-based methods are an alternative to expensive simulations in the optimization of complex systems, which most of the time are intractable or require lengthy simulations to provide a function evaluation [60]. These methods are easy to implement and they work by mapping, or regressing the input-output relationships from a more complex model [61]. The work of Ma et al. [62] presents a data-driven surrogate-based optimization method for an industrial site that addresses short-term market changes and long-term maintenance plans in the planning and scheduling operation. The authors defined two surrogate models, namely a plant-level and a unit-level surrogate model. At the plant level, a model is created for each process in the plant, whereas at the unit level, a surrogate model is defined for a group of processes that are highly related. The surrogate models aim to find the relations between the inputs of the plant (corresponding to the recycle ratio and the upstream flow proper-ties) and the outputs, which are the downstream flow properties. Both models learn the input-output relations from different levels of abstraction of the industrial site. The models were embedded independently as constraints in the optimization problem and compared with each other; the results showed that the unit-level surrogate model took longer to be solved than the plant-level model (10 min and 0.5 min, respectively) due to a larger number of variables.

Overall, both models showed accurate predictions but the unit-level model showed an inferior solution because of the inaccuracy of specific intermediate varia-bles, which influenced the final solution. On the other hand, the plant-level model, although less accurate, could find higher objective function values because the feasi-ble solution domain was not cut off due to inaccuracies in the surrogate models. The authors highlighted that the use of surrogate models is convenient for modeling at high levels of abstraction, i.e., to create a model of the entire plant instead of creating multiple surrogate models (one per unit operation or multiple related unit opera-tions). Although the latter provides a more detailed modeling of the plant, it is prone to error propagation through the other surrogate models. Additionally, the authors also indicated that reducing the accuracy of the surrogate model reduces the risk of cutting off feasible solutions that are near the boundaries of the feasible region.

Sung and Maravelias also used a surrogate model to determine the feasible region of production to complete the production planning while schedules are solved using a rolling horizon method to achieve the production targets [63]. The methodology con-sists of creating a surrogate model based on the identification of the feasible regions of the scheduling task in the space of the production demands. The surrogate model is determined by limiting a convex hull of the scheduling feasible region with the pro-duction amount. The delimitation is made gradually, separating the feasible and in-feasible regions, until a set of polyhedrons (infeasible) and polytopes (feasible) are de-fined. This method is an extension of a previous work [64], where the model could provide a convex approximation of the feasible region; the proposed methodology aimed to extend this capability to capture non-convex regions as well. Moreover, it is possible to combine the model with a rolling horizon technique to enhance the identi-fication of feasible solutions. A limitation of the method is that it might misclassify a feasible section as infeasible if very specific conditions apply; the opposite could also happen, i.e., an infeasible solution misplaced within the feasible region. The method was tested and compared with other methods including rolling horizon with linear relaxation, rolling horizon with convex approximation, and full space method. The results showed a good performance of the proposed models since the latter exhibited a gap of less than 1% compared to the optimal objective value.

In the work by Ehm [65], the author presented a methodology for defining feasi-ble AND/OR graphs on disassembly planning operations and machine scheduling for multiple heterogeneous products. This method aims to reduce the makespan in the process of disassembling multiple products simultaneously using the generated AND/

OR graphs. The data set used to train the graphs to generate the disassembly planning and machine scheduling models is generated through an artificial disassembly structure generation method. This method creates transition matrices generated with the existing relations from the disassembly process data, and then they are used to model and solve the planning and scheduling problems.

The previous method was tested in 18 different process configurations with 20 random instances each. From the 360 instances, a total of 283 could be solved to optimality; of the remaining cases, some could not be solved within the time limit given to the method and some others left more than 50% of the instances unsolved. Results showed a dependency on the complexity of the problem, i.e., a larger number of jobs resulted in more complex problems to be solved. An industrial case study is presented, which involves the disassembly of two types of forklift trucks where the solution showed an improvement of 18% in the reduction of the makespan in comparison to conventional process planning and machine scheduling approaches. For the integration of their methodology in this scenario, slight changes were needed to adapt the flexible workstations. Although there was an important degree of simplification, reaching an optimal solution became more challenging as the problem size increased. For this reason, the authors proposed the use of methods to decompose large-scale problems.

In their work, Beykal et al. presented a simultaneous modeling and optimization framework that integrates medium-term planning and short-term scheduling in problems under demand uncertainty [66]. Their methodology used mixed-integer bi-level multi-follower programming and data-driven optimization. The authors aimed to guarantee feasibility in the bi-level optimization of the problem that integrates planning and scheduling. The upper level of the problem refers to the planning problem while the lower level represents the scheduling problem. The authors use and extend a framework called DOMINO, which is a data-driven modeling optimization strategy that approximates bi-level problems to single-level problems [67]. The algorithm collects data from the upper level and uses it to generate deterministic global solutions from the lower lever. This data is then used to generate optimal or near-optimal solutions from the bi-level problem. An outstanding feature of the framework is that it provides only feasible solutions, which is considered one of the main challenges in the integration of planning and scheduling. The methodology was tested on three state-task networks (STN) of different sizes using multiple data-driven optimization solvers with DOMINO to compare performance. The results showed the capability of identifying feasible solutions and the capacity to handle complex systems with thousands of constraints and hundreds of continuous and binary variables. Nevertheless, the model was not compared with other optimizers outside DOMINO to analyze the quality of the solution.

Figure 6.3 shows the usual circumstances that justify the use of data-driven methods for the optimal integration of two tasks based on the insights gained from current studies available in the literature. The special capacity of data-driven methods to generalize the joint complexity of multiple tasks at the same time is probably the main reason why data-driven methods are an attractive alternative. This complexity can be

reflected in several situations like the existence of multiple timescales, multiple sources of uncertainty, constraints, and objectives. Individually, these features are already
a challenge to model and solve for classic optimization techniques. As a group, these
features require complex techniques to model the nonlinear formulations often required to describe a chemical engineering process. Also, when the computational capacity is limited, data-driven techniques can be used to simplify the optimization
problem as it was discussed in this section. Naturally, the existence of reliable and
well-processed data is a requirement for the development of the models.

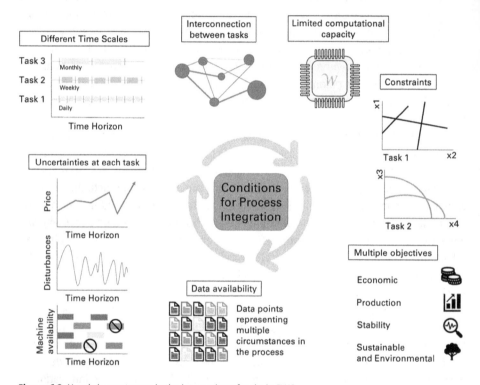

Figure 6.3: Usual circumstances in the integration of tasks in EWO.

A data-driven model, which aims to model these circumstances accurately, demands a
data set that contains the outcomes of multiple scenarios or events that can take place
in the process, e.g., uncertainties and disturbances. When data is limited a generative
model or a simulation (e.g., plant mechanistic model) is useful to compensate the lack
of data and to validate the resulting models. Clearly, the multiplication of elements
from each condition presented in Figure 6.3 makes the integration more challenging
than a single-task problem. Moreover, the variety in the nature of the elements at
each condition does not allow integrating them and approaching them as a larger en-

tity. These circumstances are also recurrent in the following sections corresponding to the integration of scheduling and control and the three-task process integration.

The implementation of data-driven methods in the integration of planning and scheduling tasks has proven to bring important contributions to the optimization process with some conditions that are worth highlighting. Although standard decomposition methods (hierarchical and iterative methods [17]) can reduce the complexity of the problem, the reduction comes with the cost of reducing the chances of finding the globally optimal solution (against the full-space approach) [54]. Data-driven methods bring another option that maintains the integrity of the problem but can reduce its complexity in more than one way. In the literature, these options included the use of ML methods and surrogate models developed in each work according to their needs and the resources available. Hybrid methodologies that integrate data-driven methods to reduce the complexity of computationally expensive stages of the optimization problem are more common than pure data-driven methods that approach the whole integration and optimization. Moreover, it was observed that the main use of data-driven models was the identification of feasible and infeasible spaces in one task by either classifying data points as feasible and infeasible or by mapping the search space and identifying infeasible and feasible regions.

The main disadvantages of these methods are directly related to the data used to generate the models. Unlabeled data might take a long time to be processed, and sometimes this cannot be automatized but done by a human resource. If no historical data is available, then a generative method has to be developed using the physical laws that govern the process of generating the data. A disadvantage of this approach is that these data sets might not incorporate extraordinary events produced by unexpected disturbances or uncertainties. Nevertheless, when enriched data is available with a good variety of scenarios that include uncertainties in the process, accurate models can be generated. Accounting for the uncertainties means an important reduction in the complexity of the problem. Furthermore, it is important to note that the computational savings that the integration of data-driven models presented usually come at the cost of a gap with the optimal value, which might be large or small depending on the quality of the model. Considering these aspects, the trade-off between the inaccuracies of the data-driven methods against the reduction of the computational burden is worth evaluating in terms of practicability. Table 6.3 presents a summary of the challenges along with the features from the data-driven methods that are useful to approach planning and scheduling problems.

6.3.1.3 Integration of scheduling and control

The need for approaches that simultaneously consider the key decision variables from scheduling and control is justified by the increasing sources of information of endogenous and exogenous events taking place in the process during operation. Ex-

Table 6.3: Challenges and highlights from data-driven methods in planning and scheduling integration.

Challenges in the planning and scheduling integration	Highlights in the implementation of data-driven models
– Planning and scheduling evolve at different timescales – Search space of both tasks is considerably large – Often approached as a nested problem with a highly constrained domain – Avoid schedule infeasibility is often a key topic – Endogenous and exogenous sources of uncertainty demands reactive scheduling processes	Benefits: – ML classifier methods can be used to define a region (or a schedule) as feasible or infeasible – Surrogate models can substitute and reduce the constraints set – Classifiers could be used to interpret aspects from the process, e.g., DT or RF – Surrogate models can find complex correlations
	Limitations: – Preprocessing of data is often needed – Optimality not guaranteed – Performance depends on an adequate tuning parameter strategy – Models prone to errors if they are not well-calibrated

amples of these include the growing sensory data that can be collected from the process, the demand and production of a larger variant of products or the changes in the prices in the market defined by external factors. Moreover, changes in these parameters are registered more frequently during the horizon, which leads to the need for reactive methods that can avoid the infeasibility of a schedule or the missing of product quality and safety requirements. Both scheduling and control are dependent on each other but their purpose in the process is different; while the scheduling process defines the batches, starting times, and state reference, the control system aims to keep the process on target and close to its desired set points [68]. Also, they often evolve at different timescales just as the planning and scheduling problems.

The usual approach for the integration of the scheduling and control tasks is the decomposition of the problem to reduce its complexity [69]. Then, the scheduling problem can be solved as an MILP problem whereas the control problem is often posed as a MIDO problem. The disadvantage of this approach, whether is solved sequentially or iteratively is that for the latter there is a low probability of finding the optimal value because the variables from one problem are fixed while the other is solved, and for the former the computational burden is high [70]. Due to these reasons, there is a constant search for new methods for solving the integrated problem that can reduce the complexity, particularly at the control level where dynamic decisions are key. Data-driven methods arise as an alternative for mitigating the complexity of the problem, for example, by modeling the dynamics that are involved in the control problem.

Kück et al. aimed to deal with the dynamic effects that have a negative impact on production scheduling and process control through the use of a data-driven method [71]. The method consists of a simulation model that is used to produce a schedule, which is continuously updated with real-time system state data to adjust the schedule or to generate a new one. The authors proposed a data-driven Simulation-Based Optimization (SBO) method to keep continuous track of the intensive dynamics of the production systems. This method allows a permanent evaluation of the effect of parameter changes in real time. The work presents an SBO method that allows continuous changes of the simulation model showing the current state of the manufacturing system in real time. To preserve a suitable computational effort, the work suggests the integration of a heuristic method that could reduce the scheduling effort in case there are abrupt changes in the system – this consists of the use of sections of a previously known schedule. The control task is performed considering the continuous adaptation of the simulation model using the same real-time data that is utilized to update the scheduling task. Essentially, a continuous flow of data feeds both systems and continuously updates the optimization processes using previous and new data efficiently to reduce computational costs.

The model was tested on a semiconductor manufacturing facility where the method achieved improved results compared to optimization systems that do not allow the continuous updating of data from the system. The authors highlighted that their model was challenged by the inherent dynamic nature of the manufacturing system, which included machine failures. Nevertheless, the model was capable of making better decisions compared to the benchmark used. They stated that using real-time data provides an advantage to enhance the decision-making process.

A method for the integration of scheduling and control in systems under uncertainty is presented by Scholz [1]. It consists of the use of two surrogate models to build a representation of the system's dynamics. The first model consists of a classification model built with an ANN that is used for identifying the feasible space for operation. The second surrogate model consists of another ANN used to build an approximation of state and input variables for the system. Both models are trained with data generated either from an MPC or a tube-based MPC. The use of these two sources of data for training the surrogate model was for comparison purposes, i.e., compare the performance of standard MPC with another strategy that considers disturbances affecting the system (i.e., tube-based MPC) in a system that is under uncertainty. Both the surrogate and the classification model are used to approximate the feasible region of the operation of the control problem and the functions of the state values, state vectors and input vectors of the system. The models captured relevant information from both tasks and are integrated into the scheduling problem, which is solved to local optimality.

The methodology is implemented in two case studies: a continuous-stirred tank reactor (CSTR) and a spray dryer system for milk powder. Results showed a fast solution (less than 100 s) of the integrated framework for both strategies (standard and tube-based) in both case studies. Moreover, the predictions from the ANNs for the scheduling and control problems were considered satisfactory as there was a mismatch from the actual cost values of less than 0.7%. The latter may be due to inaccuracies of the surrogate models and disturbances from the process. Uncertainty was handled better with the surrogate model trained with the tube-based MPC while the other model failed to meet some requirements from each case study.

Simkoff and Baldea [72] proposed a data-driven model used to represent nonlinear process dynamics in systems with long timescales and subject to uncertain parameters. The approach aims to attend the needs of a chlor-alkali process which is constantly subject to deregulated electricity markets, thus the price changes very often. To reduce the computational effort of the highly nonlinear first principles dynamic model, the authors made use of the Hammerstein-Wiener model proposed by Kelley et al. [73]. For this case in particular, piecewise linear functions were used as inputs and output variables for the models that were used to compute the process state of the system. The objective function aims to minimize the economic expenses by building schedules that consider variables of interest from the control task, namely the temperature and power consumption. The models for both variables were built using data produced from a simulation of the system and were used to predict the dynamic response.

The methodology was tested on ideal scenarios without uncertainty to present a nominal solution to the problem and compare the advantages of solving the problem with the reduced-order model against the full-order model. Although the solution quality is less than 1% lower in the reduced model, the time required to solve it was 35 times less than with the full-order model. For the problem subject to uncertainty in product demand and price of electricity, the authors considered a two-stage stochastic programming approach with the reduced-order model embedded. Four different price distributions and two demand distributions were defined for the uncertain parameters. A comparison with two other formulations, which are solved once the uncertainties are realized, was made to evaluate the quality of the solution. When comparing the solution from the two formulations with the proposed method, the latter displayed a better performance in reducing the economic expenses, i.e., between 1% and 9% improvement. Moreover, the computational time was four times lower than the time required for solving the deterministic problem using a full-order model.

In a follow-up work, Simkoff and Baldea [74] proposed another data-driven model based on the Hammerstein-Wiener structure to represent relevant dynamics for the scheduling process in the integration of scheduling and control. The integrated framework was modified to reduce the number of binary variables in the problem and thus reduce the computational expenses. The method was implemented to design the scheduling and control of a multiproduct batch polymerization reactor that operates in a wheel-type mode. The objective function aimed to maximize profits and reduce storage costs. The scheduling decisions are made over an extended time horizon and are used to determine the set points of a nonlinear process control system. Different models were trained using the data set available in [73], using the molecular weight distribution set point as the input variable and the flow of the coolant and the flow of the initiator of the reaction as the output (manipulated) variables. The authors highlighted some general limitations of these trained models: a) the deficit in an accurate method to capture penalties for constraint violations due to the use of soft (and not hard) constraints; b) the assumption that the process was approximated at steady state, which produces conservative approximations, and c) the quality of the model is subject to the characteristics of the training data set.

The model was tested using two case studies, both involving the same multiproduct polymerization reactor. In the second case, the process was subject to demand and processing capacity uncertainty. In the first case, it was demonstrated that the main attribute of the integration of the model was the considerable reduction of the computational time (two orders of magnitude), whereas in the second case, the authors highlighted that the solutions were superior compared to the Hammerstein-Wiener model structure alone. Note that in the Hammerstein-Wiener approach the problem did not converge even after 21,600 s of simulation. On the other hand, the method proposed in this study found a solution after 37.28 s.

Santander and Baldea [75] proposed a scheduling methodology that takes into consideration the effect of disturbances and control dynamics on batch process scheduling. The methodology is divided into two parts: the first makes use of historical data to define the relationship between the effect of the multiple factors that affect the processing time like disturbances, uncertainties, and operating conditions. The authors defined time relationships and probability distributions to capture the effect of these variables in the scheduling process. Moreover, the data describes the effects that exogenous and endogenous uncertainty produce on the schedule. From these data, the relations between the disturbances and the effects on the time are identified. The second part integrates the relations obtained from the first part into a two-stage stochastic scheduling problem. The first-stage decisions are solved considering endogenous uncertainties, process disturbances, and operating constraints whereas the second-stage considers the demand realization (exogenous uncertainty). The resulting framework that integrates the surrogate models reduces the complexity of the integration from a MIDO problem into an MILP.

The framework was implemented in two case studies. In the first case, the control and scheduling decisions from the proposed framework led to a more conservative solution compared to the nominal result as the prediction took into consideration different possible disturbances. Nevertheless, the economic objective value was larger than that of the nominal solution because the schedule was completed and delivered on the expected time. For the second case study, the plant from [76] is adopted and solved with the proposed framework. Results were similar in the sense that the solutions were conservative but infeasibilities were avoided and a larger profit was reached because the production was completed according to the schedule.

Zhuge and Ierapetritou [68] proposed a piecewise affine (PWA) identification technique that can handle multidimensional nonlinear functions, used to model the dynamics of a system. The PWA model is then integrated as linear constraints into the scheduling level to reduce the complexity of the problem, leaving an MILP-integrated problem. To search for the optimal control actions that can handle disturbances on the system, a fast MPC was proposed to be used with the PWA; this framework is compared with an implementation of a multiparametric MPC (mp-MPC) proposed in a previous study [77]. The framework is tested on the integration of scheduling and control of two case studies: a single input single output (SISO) CSTR and a multiple input multiple output (MIMO) CSTR. The solution of the latter case study was compared with the original integrated problem [78] resulting in reduced computational time and the objective values (economic profit) were nearly the same. The former case study (MIMO CSTR) was used to compare the features of mp-MPC with fast MPC. Results showed that mp-MPC exponentially increased the complexity of the problem with the number of dimensions on the state and manipulated variables, whereas fast MPC was shown to reduce the dimensionality of the problem.

Caspari et al. [79] make a comparison between top-down and bottom-up paradigms in which the former uses a data-driven model to predict the dynamics of the process and optimize an economic objective function. The data-driven model consists of a scale-bridging model (SBM), which can be trained with data from different sources, e.g., historical data, dedicated experiments, or data generated by simulation of a dynamic model. Typically, the information exchange between the upper and lower-level problem, which is needed for complementing the optimization process requires a large computational effort that makes the approach intractable. For this reason, the SBM is integrated to predict the dynamic behavior of the process and minimize the economic objective function. For the bottom-up integration, the authors used an economic NMPC (eNMPC) to solve the nonlinear dynamic optimization problem.

Both methodologies were tested on an air separation unit (ASU) with four different scenarios that consider possible moderate and extreme fluctuation in prices as well as unplanned maintenance events. A comparison between the two models showed the advantages and disadvantages of both methods although they enable overall economic improvements. The data-driven-based method was capable of maintaining constant dynamic stability; also, the problem could be solved over a longer time horizon. Moreover, the method was implemented in an online fashion. In the case of the eNMPC-based model, the process was more accurate because it made precise calculations compared to the approximations made by the data-driven model. Hence, the eNMPC was able to find better economic improvements. Nevertheless, in terms of computational effort, the eNMPC approach demanded a considerable amount of time, which made it impractical for online implementations.

From the review of these works, it can be observed that data-driven methods are useful for generalizing the complex dynamics of the control task in the integration of scheduling and control. This reduces the complexity of the optimization problem and allows faster solutions than standard methods. Note that the integration approach is different and dependent on the resources available to train the learning methods.

Contrary to the previous section (planning and scheduling) these frameworks were hybrid methods, where data-driven models are integrated in some fashion into current optimization solvers, i.e., no optimization process consists of a pure ML or surrogate method. Among the advantages of these methodologies is the capability to respond in a short time allowing an online implementation of the framework, a situation that is most of the time impractical for standard methods. Also, an important feature is the capacity to approach exogenous and endogenous uncertainties and disturbances. Naturally, the training set to develop the models should include data from the realization of uncertainties and external events affecting the process.

It was observed that models built this way could ensure the feasibility of the problem although the global optimal value may be attainable. This represents one of the main disadvantages of data-driven models and is repetitive from the previous integration, namely the quality of the approximation provided by the model is dependent on the quality of the data used to train the model. Then, for processes with multiple abnormal events (i.e., disturbances or uncertainties), a low model performance is expected and thus, a direct impact on the quality of the solution. Although there are some disadvantages related to the characteristics of the data used, data-driven methods have the potential to reduce the complexity of the integrated optimization problems, thus making them attractive for large-scale applications. Table 6.4 presents a summary of the different challenges collected from the works presented in this section along with the features that the data-driven methods contribute to approach such problems.

Table 6.4: Challenges and highlights from data-driven methods in scheduling and control integration.

Challenges in the scheduling and control integration	Highlights in the implementation of data-driven models
– Tasks have different goals and operate at different time-scales – Exogenous and endogenous uncertainty impacts the decision-making process, e.g., market prices or disturbances – Real time optimization is computationally expensive for this integration – Information exchange between the two tasks is expensive – Modeling the nonlinear dynamics of the controller results in non-convex problems	Benefits: – Data-driven models can generalize the controller's online decisions – Reduction in computational times allows corrections in the process, even in real time – Model classification can be used to specify feasible regions – Data from simulations (e.g., MPC) can be used to train models
	Limitations: – Optimality not guaranteed – Models can be inaccurate or overly conservative – Data for training often limits the model's predictability capacity

6.3.2 Integration of three levels

In this section, multiple methods developed for the integration and solution of three decision-making levels are presented, i.e., planning, scheduling, and control. The literature shows very few studies aimed to address these problems in comparison to the integration of two tasks decision levels. As mentioned above, the process for integrating more than two decision levels is, in general terms, an extension of the methods used for the integration of two levels.

The three task-level integration is one of the main objectives of EWO but it comes with multiple challenges that limit their widespread in the field. These challenges include the complexity in formulating one problem that merges the three tasks and the computational burden and time needed to solve the optimization problem. The differences in the characteristics of each task include the timescale, the type of state variables and the dimensions of each search domain; thus, finding a feasible solution is a complex challenge in itself. Although the addition of surrogate methods alleviates the complexity of the problem, this is still a complex problem to formulate and to solve using state-of-the-art computer software. This in some way justifies the low number of works on three task-level integration. However, the approaches proposed thus far provide key insights and are key to better understanding the advantages and limitations of data-driven models for this type of optimal process integration.

Berg et al. [30] proposed the assessment of two techniques for the integration of planning, scheduling and control in a multisite, multiproduct case study. The techniques considered are optimality surrogates and derivative-free optimization (DFO). For the optimality surrogate technique, the authors generated a model with an ANN of the lower decision level problem using supervised learning techniques. The model could map different combinations of variables in the control process (lower level) with the best combination of variables corresponding to the planning and scheduling (upper level). The model is then embedded as a constraint into the upper level of the formulation, i.e., the planning and scheduling formulation. The ANN provides the values for the control variables given a decision set from the planning and scheduling. On the other hand, the DFO aims to fully determine the upper-level variables and finding the values to minimize the objective function; then the lower level is solved as a single-level problem.

The techniques are tested individually and simultaneously on a hierarchical integrated planning-scheduling-control problem. Several integration schemes were assessed in this work to observe the quality evaluation at each stage, i.e., scheduling only, then scheduling and control, and finally, planning, scheduling, and control. The integrated problem minimizes economic objective at the planning level, augmented by the optimal scheduling and control costs, the minimization of the makespan, and the determination of the specific variables that minimize energy costs and processing times. The different instances where the method was tested showed the capacity of these methodologies to enhance the quality of the solution when compared to conventional sequential solution approaches. Nevertheless, the authors highlighted that the complexity of the implementation, and the solution itself are largely dependent on the size of each layer in the problem formulation. Moreover, sampling, training, hyperparameter tuning, and validation became highly involved in the quality of the resulting model.

Dias and Lerapetritou [31] proposed the integration of planning, scheduling, and control using two building blocks: the first integrates the scheduling and control aspects whereas the second integrates planning and scheduling. Data-driven feasibility analysis and surrogate models are used as constraints in the problem formulation to reduce the complexity of the problem. The former consists of a grey-box type formulation where some of the equations (including constraints and objective functions) of the problem formulation are not available and thus, substituted with models built with plant data. The surrogates consisted of ANNs or regression models used to predict the behavior of the system. The method for integrating the three tasks consists of a sequence of steps: in the first step, the scheduling formulation for the scheduling-control task and the planning formulation for the planning-scheduling-control task are performed. Then, the unknown constraints (including feasibility constraints) as well as the information needed from the surrogate models are identified. The surrogates are built through simulations of the control problem where the inputs are state and input variables and the outputs are the predicted decision variables. Furthermore, the surrogate models for both the first and second integration are included in the optimization problem, which is then solved using local solvers.

The framework is tested using an ASU of three plants that produce nitrogen gas, oxygen gas and liquid nitrogen where the fluctuation in electricity prices is considered to optimize the production of the gases in economic terms. Results showed an overall decrease in the economic expenses in the long term, achieving a 3.38% reduction in comparison to a baseline case with fixed production values. The authors mentioned that, although their model could be complex, it meets the level of complexity given in industrial-sized problems that exhibit high dimensionality. They mentioned that their work consists of the combination of a data-driven method with classical optimization techniques, which should be gradually adopted for industrial applications. Finally, it was highlighted that the surrogate models and the optimization process were made under the assumption that there was no uncertainty in any parameter of the system or in the data used to train the models.

Chu and You [32] proposed two methods to reduce the complexity of the integrated problems through the use of surrogate models. The first method uses a surrogate model to predict the processing costs depending on the processing times; this model substitutes the set of nonlinear constraints from the dynamic part of the integrated problem. A bi-level optimization model is developed, which considers the planning and scheduling optimization while adding the surrogate model that represents the dynamic aspects of the process. To reduce the complexity of the first approach, the second method consists of another surrogate model, which is developed for the scheduling task, by decomposing the problem and adding a representation of the optimal value function into the integrated problem. This framework makes use of data from the scheduling scenario and predicts the production costs based on the production quantities; the surrogate model is then used to solve the planning problem. For both cases, a feasibility exploration method was proposed such that the surrogate models only provide feasible solutions for the dynamics and scheduling tasks. Challenges in building the surrogate models include the large computational times needed to generate the data that is required due to the complexity of the problems.

The framework was tested on two case studies and showed notable advantages in the reduction of computational time and the quality of the solutions compared to the benchmark methods. In particular, the solution showed an advantage reducing the economic costs of the process by more than 15% when compared to the values obtained from a full-space (monolithic) optimization approach.

Although the addition of data-driven methods into the integration of the planning, scheduling, and control tasks reduced the computational expenses associated with the search of an optimal solution, the main issue, which is inherent to the integration of the three tasks remains, i.e., the complexity of the problem. The more likely advantage of the data-driven methods on these integrations is that the problems can be tractable and practical from an industrial point of view, compared to standard approaches that would require large or even prohibitive computational calculations. Also, this integration allows obtaining the promised advantages from the integration of the three tasks, which is a better solution than those provided by the sequential methodologies.

It is clear that new challenges arise in this field of data-based modeling, which is considered in an early stage of development. Challenges include the accurate modeling of the process, the sampling of data, and hyperparameter tuning. Moreover, uncertainty is an important factor in the incorporation of these methods into real industrial scenarios as it further increases the complexity of the problem. Since these frameworks are extensions of the approaches discussed in previous sections, observing similar issues to those that appeared in the integration of two tasks was expected. Consequently, the inclusion of an additional task directly impacts the complexity of the integrated formulation. Nevertheless, the trade-off between solution quality and the complexity of the frameworks seems to be fair and deserves more attention from the research community. Table 6.5 presents a summary of the different challenges collected from the works presented in this section along with the features that the data-driven methods contribute to tackle such problems.

Table 6.5: Challenges and highlights from data-driven methods in planning, scheduling, and control integration.

Challenges in the planning, scheduling and control integration	Highlights in the implementation of data-driven models
– Tasks rely in different timescales and spatial scales – Search space is often extremely large – Uncertainty in one task impacts the remaining tasks – Number of constraints is often large	Benefits: – More than one data-driven model may be used in the formulation – Reduction of computational burden – Quality of the solution could bring major benefits to the operation, e.g., increase in profits and reduce CO_2 emissions
	Limitations: – Data for training needs to consider a larger search space – No data-driven model built to comprise three tasks has been reported – Problem formulation is still perceived as complex

6.4 Effectiveness of data-driven methods in addressing process integration challenges

Figure 6.4 shows a general schematic representation of the multiple tasks that are often considered in optimal process integration, along with their respective features. Each level of integration shows key features considered for optimization for that particular task. The main difference between them, as mentioned before, is the temporal scales at which each of these tasks operates. For this representation, the timescales reduce from top to bottom for the planning, scheduling, and control levels; note that the timescales used in the figure are only a general representation and are not specific for these tasks, i.e., other scales can be used, for instance, weeks (planning), hours (scheduling), and seconds (control). The process design task is shown on the left side as it remains fixed during operation of the other tasks but this is not necessary for all cases. Each of these tasks is updated at different timescales, being the control task updated more frequently. Along with this difference in timescales, each task includes individual constraints, objective functions, and decision variables that will affect the other decision-level tasks during operation. That is, the optimal conditions of the entire system are subject to the state of the tasks, which might not necessarily be at an optimal state. Consequently, performing optimal process integration of the entire system becomes a highly complex problem.

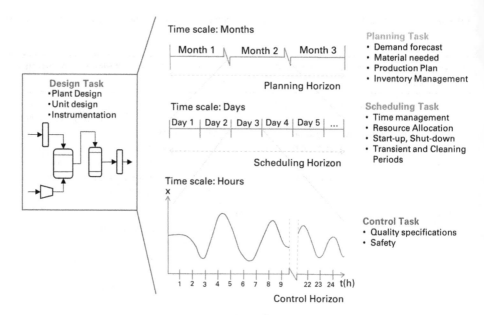

Figure 6.4: Schematic representation of integral process integration.

The application of data-driven methods in optimal process integration has shown to be effective at different levels but at the same time dependent on different factors. The level of reduction of complexity of a problem is related to the prediction performance of the model, which at the same time is dependent on how well the data reflects the behavior of the systems. Problems related to incomplete or small data sets, extensive preprocessing of the data before use, and the need to produce the data using alternative generation methods are among the most common obstacles observed for the generation of the models in this field. Nevertheless, these issues do not impede the integration but are challenges that can be overcome in different ways; for instance, by employing human resources to label data, building transformation pipelines to preprocess complex data sets or developing generative models to create new data. In all the cases, the integration exhibited advantages over the standard methods related to the quality of the solutions, in some of them the solution was closer or at the optimal value but other cases reported the solution of intractable problems. The common use of data-driven methods for reducing the dimensionality in control problems or to ensure feasibility in planning and scheduling supports this argument. Another advantage is the fast response of the surrogate methods that leads to shorter computational times when solving the optimization problems, thus allowing online implementations of the integrated methods.

Conversely, one of the most recurrent issues reported in the literature was the difficulty in creating models that could provide an accurate representation of the processes, which points to a lack of a standardization method. Similar to other implementations in AI, each model is unique and dependent on the available data and the characteristics of the problem statement [80]. Thus, the development of standardized methodologies of data-driven models' integration is limited by the features that vary from problem to problem. Moreover, a key feature of these methods is that scaling the problem does not affect the time response of the data-driven models because these are not related to the complexity of the problem, i.e., once the model is trained, it returns predictions in short periods independently of the problem's size. The literature shows a preference for hybrid methods that incorporate current optimization techniques with the data-driven models obtained specifically from each problem. Although the addition of data-driven methods into the integration of tasks in PSE is effective and often provides satisfactory results, it still faces important challenges related to data generation that do not allow a faster and ready integration into the optimization field.

Figure 6.5 depicts the level of impact that data-driven methods have on the challenges presented in Table 6.1 in accordance with the discussions presented above. Notable achievements that were recurrent on multiple contributions are listed as attainable challenges. It can be stated that the ability of data-driven methods to generalize complex nonlinear systems results in most of the features listed in this category; additionally the fast response provided by these models make them suitable for online implementation. Also, there were multiple challenges that would depend on the particular characteristics of each problem, for instance, the quality of the data or the generalization capacity of the model. These are listed as conditioned in the table and are related to the restrictions of the problem and the performance of the resulting data-driven model. These challenges are related to the nature of the data-driven methods as approximators, which, contrary to analytical modeling methods, aim to generalize a data set and provide the statistically best possible answer. Furthermore, the list of challenges shows those aspects where the addition of the data-driven models could not make a significant impact or may not improve the state of the art as is the

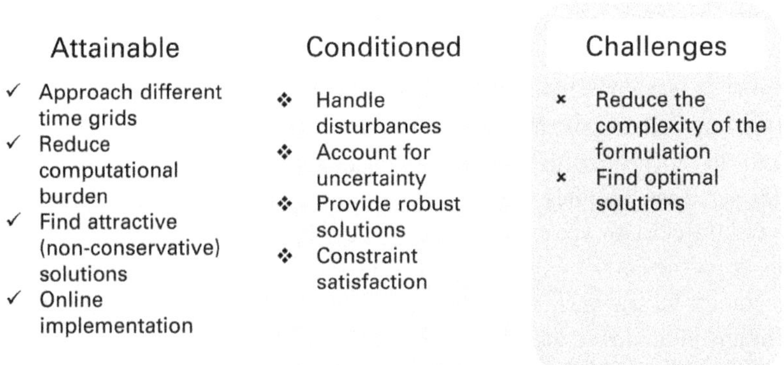

Attainable	Conditioned	Challenges
✓ Approach different time grids	❖ Handle disturbances	✗ Reduce the complexity of the formulation
✓ Reduce computational burden	❖ Account for uncertainty	✗ Find optimal solutions
✓ Find attractive (non-conservative) solutions	❖ Provide robust solutions	
✓ Online implementation	❖ Constraint satisfaction	

Figure 6.5: The level of impact that data-driven models have on the challenges presented in Table 6.1.

case of finding an optimal solution. On the other hand, the complexity of the final integration is a challenge that is intrinsic to the formulation of the problem where data-driven methods cannot exert a relevant influence.

6.5 Outlook

The number of reported studies in process integration using data-driven models is scarce at the moment. While Figure 6.2 shows an increase from 2019 until 2024, a continuous growth cannot be predicted at the moment due to the little amount of data. Instead, the data suggest that this field is at a very early stage. This is not the case for data-driven methods in process optimization at the independent tasks, i.e., design, planning, scheduling, and control; the number of publications in this area is continuously growing at a fast pace [80–83] due to advances in fields like ML. It is expected that around 2030–2035, ML will have a wider impact and penetration in industrial applications given its current state [27]. Similar to ML methodologies, the growth of the field of process integration with data-driven methods is probably dependent on other fields that may provide more data of higher quality. Examples of these fields include sensing technologies, IoT, generative methods, and 5G networks, all related to the generation, transmission, and processing of data sets [84, 85]. Considering this, the impact of the individual growth of the mentioned fields should extend to process integration gradually.

The challenges that arise in process integration with data-driven models are the same from the standard integration methods, i.e., the complexity of the problem formulation. Note that this is not related to the implementation of the data-driven models but to the problems themselves. Data-driven methods are used to address another major challenge in process integration, which is the curse of dimensionality that these problems often exhibit. Thus, the implementation of these methods aims to reduce the impact of dimensionality and allow current techniques to work without major difficulties. The data-driven models can be classified depending on their purpose in the optimization process; we propose two broad classes, which were observed in the literature: a) feasibility-based data-driven models, which were used to reduce the search space by defining the feasible regions of it and were seen mostly (but not exclusively) in the integration of planning and scheduling tasks; b) behavioral/dynamic data-driven based models, used to capture the dynamics of the process and provide optimal (or close to optimal) decisions. These two classes were useful to reduce the impacts of the curse of dimensionality and leverage the capacities of the optimization methods, for instance, by allowing them to account for uncertainty or by allowing an online implementation.

In the future, it is expected that a more diversified range of implementations of data-driven techniques will emerge in the field of process integration in chemical engineering and other areas of relevance such as health, agriculture, renewable energies and smart grid [86–90]. ML techniques, for instance, keep expanding their capacities to new industrial implementations. Supervised and unsupervised learning methods find a niche in processing and acquiring insight from historical data, while the intelligent agents from RL are being used to substitute control systems (e.g., [91, 92]). These methods, as those shown in this study, face the fact of not guaranteeing optimal solutions on decision-making processes but consistently reduce the computational times needed to provide a solution. More complex methods used in other branches of AI such as natural language processing or image processing could be adapted and tested in industrial decision-making processes. Nevertheless, a balance between performance, computational burden, and solution quality should serve as a guideline for future developments.

References

[1] Scholz Dias, L. M. "Enterprise-Wide Optimization: Integrating Planning, Scheduling and Control Problems Using Feasibility Analysis and Surrogate Models," Ph.D., Rutgers The State University of New Jersey, School of Graduate Studies, United States – New Jersey, 2019. Accessed: Jan. 31, 2024. [Online]. Available: https://www.proquest.com/docview/2342583958/abstract/C965257F5C84B5FPQ/1

[2] Gani, R. et al. Challenges and Opportunities for Process Systems Engineering in a Changed World. In Yamashita, Y., Kano, M. (Eds.). *Computer Aided Chemical Engineering* Vol. 49. in 14 International Symposium on Process Systems Engineering, Vol. 49. Elsevier, 2022. pp. 7–20. doi: 10.1016/B978-0-323-85159-6.50002-6.

[3] Bernal, D. E., Ovalle, D., Liñán, D. A., Ricardez-Sandoval, L. A., Gómez, J. M., Grossmann, I. E. Process Superstructure Optimization through Discrete Steepest Descent Optimization: A GDP Analysis and Applications in Process Intensification. In Yamashita, Y., Kano, M. (Eds.). *Computer Aided Chemical Engineering* Vol. 49. 14 International Symposium on Process Systems Engineering, Vol. 49 Elsevier, 2022. pp. 1279–1284. doi: 10.1016/B978-0-323-85159-6.50213-X.

[4] Schweidtmann, A. M. et al. Machine learning in chemical engineering: A perspective. *Chemie Ingenieur Technik*, 2021, 93(12), 2029–2039. doi: 10.1002/cite.202100083.

[5] Himmelblau, D. M. Applications of artificial neural networks in chemical engineering. *Korean Journal of Chemical Engineering*, Jul. 2000, 17(4), 373–392. doi: 10.1007/BF02706848.

[6] Dobbelaere, M. R., Plehiers, P. P., Van de Vijver, R., Stevens, C. V., Van Geem, K. M. Machine learning in chemical engineering: strengths, weaknesses, opportunities, and threats. *Engineering*, Sep. 2021, 7(9), 1201–1211. doi: 10.1016/j.eng.2021.03.019.

[7] Alur, R., Singhania, N. Precise Piecewise Affine Models from Input-output Data. In *Proceedings of the 14th International Conference on Embedded Software*. New Delhi India: ACM, Oct. 2014, pp. 1–10. doi: 10.1145/2656045.2656064.

[8] Silver, D. et al. Mastering the game of Go without human knowledge, *Nature*, Oct. 2017, 550(7676), 354–359. doi: 10.1038/nature24270.

[9] Géron, A. *Hands-on Machine Learning with Scikit-Learn and TensorFlow: Concepts, Tools, and Techniques to Build Intelligent Systems*. First edition, Beijing ; Boston: O'Reilly Media, 2017.

[10] Andrés-Martínez, O., Ricardez-Sandoval, L. A. Integration of planning, scheduling, and control: A review and new perspectives. *The Canadian Journal of Chemical Engineering*, 2022, 100(9), 2057–2070. doi: 10.1002/cjce.24501.

[11] Harjunkoski, I., Nyström, R., Horch, A. Integration of scheduling and control – Theory or practice? *Computers and Chemical Engineering*, Dec. 2009, 33(12), 1909–1918. doi: 10.1016/j.compchemeng.2009.06.016.

[12] Sharifzadeh, M. Integration of process design and control: A review. *Chemical Engineering Research and Design*, Dec. 2013, 91(12), 2515–2549. doi: 10.1016/j.cherd.2013.05.007.

[13] Phanden, R. K., Jain, A., Verma, R. Integration of process planning and scheduling: A state-of-the-art review. *International Journal of Computer Integrated Manufacturing*, Jun. 2011, 24(6), 517–534. doi: 10.1080/0951192X.2011.562543.

[14] Li, X., Gao, L., Zhang, C., Shao, X. A review on integrated process planning and scheduling. *International Journal of Manufacturing Research*, Jan. 2010, 5(2), 161–180. doi: 10.1504/IJMR.2010.03163.

[15] Baldea, M., Harjunkoski, I. Integrated production scheduling and process control: A systematic review. *Computers and Chemical Engineering*, Dec. 2014, 71, 377–390. doi: 10.1016/j.compchemeng.2014.09.002.

[16] Wang, L., Shen, W., Hao, Q. An overview of distributed process planning and its integration with scheduling. *International Journal of Computer Applications in Technology*, Jan. 2006, 26(1–2), 3–14. doi: 10.1504/IJCAT.2006.010076.

[17] Maravelias, C. T., Sung, C. Integration of production planning and scheduling: Overview, challenges and opportunities. *Computers and Chemical Engineering*, Dec. 2009, 33(12), 1919–1930. doi: 10.1016/j.compchemeng.2009.06.007.

[18] Dias, L. S., Ierapetritou, M. G. Integration of scheduling and control under uncertainties: Review and challenges. *Chemical Engineering Research and Design*, Dec. 2016, 116, 98–113. doi: 10.1016/j.cherd.2016.10.047.

[19] Pistikopoulos, E. N. Uncertainty in process design and operations. *Computers and Chemical Engineering*, Jun. 1995, 19, 553–563. doi: 10.1016/0098-1354(95)87094-6.

[20] Li, Z., Ierapetritou, M. Process scheduling under uncertainty: Review and challenges. *Computers and Chemical Engineering*, 32(4–5), Art. no. 4–5, Apr. 2008. doi: 10.1016/j.compchemeng.2007.03.001.

[21] Andrés-Martínez, O., Palma-Flores, O., Ricardez-Sandoval, L. A. Optimal control and the Pontryagin's principle in chemical engineering: History, theory, and challenges. *AIChE Journal*, 2022, 68(8), e17777. doi: 10.1002/aic.17777.

[22] Harmonosky, C. M. Implementation issues using simulation for real-time scheduling, control, and monitoring. *Institute of Electrical and Electronics Engineers (IEEE)*, 1990.

[23] Andrés-Martínez, O., Ricardez-Sandoval, L. A. A switched system formulation for optimal integration of scheduling and control in multi-product continuous processes. *Journal of Process Control*, Oct. 2021, 106, 94–109. doi: 10.1016/j.jprocont.2021.08.017.

[24] Andrés-Martínez, O., Ricardez-Sandoval, L. A. A nested online scheduling and nonlinear model predictive control framework for multi-product continuous systems. *AIChE Journal*, 2022, 68(5), e17665. doi: 10.1002/aic.17665.

[25] Pistikopoulos, E. N., Diangelakis, N. A. Towards the integration of process design, control and scheduling: Are we getting closer? *Computers and Chemical Engineering*, Aug. 2016, 91, 85–92. doi: 10.1016/j.compchemeng.2015.11.002.

[26] Solomatine, D., See, L. M., Abrahart, R. J. Data-Driven Modelling: Concepts, Approaches and Experiences. In Abrahart, R. J., See, L. M., Solomatine, D. P. (Eds.) *Practical Hydroinformatics: Computational Intelligence and Technological Developments in Water Applications*. Water Science and Technology Library. Berlin, Heidelberg: Springer,, 2008, pp. 17–30. doi: 10.1007/978-3-540-79881-1_2.

[27] Venkatasubramanian, V. The promise of artificial intelligence in chemical engineering: Is it here, finally? *AIChE Journal*, 65(2), Art. no. 2, Feb. 2019. doi: 10.1002/aic.16489.

[28] Dias, L. S., Ierapetritou, M. G. Data-driven feasibility analysis for the integration of planning and scheduling problems. *Optimization and Engineering*, Dec. 2019, 20(4), 1029–1066. doi: 10.1007/s11081-019-09459-w.

[29] Thon, C., Finke, B., Kwade, A., Schilde, C. Artificial intelligence in process engineering. *Advanced Intelligent Systems*, 2021, 3(6), 2000261. doi: 10.1002/aisy.202000261.

[30] van de Berg, D., Shah, N., Del Rio-Chanona, E. A. Hierarchical planning-scheduling-control – Optimality surrogates and derivative-free optimization. *arXiv*, Oct. 11 2023. doi: 10.48550/arXiv.2310.07870.

[31] Dias, L. S., Ierapetritou, M. G. Integration of planning, scheduling and control problems using data-driven feasibility analysis and surrogate models. *Computers and Chemical Engineering*, Mar. 2020, 134, 106714. doi: 10.1016/j.compchemeng.2019.106714.

[32] Chu, Y., You, F. Integrated planning, scheduling, and dynamic optimization for batch processes: MINLP model formulation and efficient solution methods via surrogate modeling. *Industrial & Engineering Chemistry Research*, Aug. 2014, 53(34), 13391–13411. doi: 10.1021/ie501986d.

[33] Rafiei-Shishavan, M., Ricardez-Sandoval, L. A. A Stochastic Approach for Integration of Design and Control under Uncertainty: A Back-off Approach Using Power Series Expansions. In Espuña, A., Graells, M., Puigjaner, L. (Eds.). *Computer Aided Chemical Engineering* Vol. 40. 27 European Symposium on Computer Aided Process Engineering, Vol. 40 Elsevier, 2017. pp. 1861–1866. doi: 10.1016/B978-0-444-63965-3.50312-3.

[34] Rafiei, M., Ricardez-Sandoval, L. A. Integration of design and control for industrial-scale applications under uncertainty: A trust region approach. *Computers and Chemical Engineering*, Oct. 2020, 141, 107006. doi: 10.1016/j.compchemeng.2020.107006.

[35] Rafiei, M., Ricardez-Sandoval, L. A. A trust-region framework for integration of design and control. *AIChE Journal*, 2020, 66(5), e16922. doi: 10.1002/aic.16922.

[36] Tian, Y., Pappas, I., Burnak, B., Katz, J., Pistikopoulos, E. N. Simultaneous design & control of a reactive distillation system – A parametric optimization & control approach. *Chemical Engineering Science*, Feb. 2021, 230, 116232. doi: 10.1016/j.ces.2020.116232.

[37] Iftakher, A. *et al.* RD-toolbox: A computer aided toolbox for integrated design and control of reactive distillation processes. *Computers and Chemical Engineering*, Aug. 2022, 164, 107869. doi: 10.1016/j.compchemeng.2022.107869.

[38] Bahakim, S. S., Mehta, S., Ahmed, H., Gaspar, E., Ricardez-Sandoval, L. Integration of design and control using efficient PSE approximations. *IFAC-PapersOnLine*, Jan. 2015, 48(8), 894–899. doi: 10.1016/j.ifacol.2015.09.083.

[39] Kiumarsi, B., Vamvoudakis, K. G., Modares, H., Lewis, F. L. Optimal and Autonomous Control Using Reinforcement Learning: A Survey. *IEEE Transactions on Neural Networks & Learning Systems*, Jun. 2018, 29(6), 2042–2062. doi: 10.1109/TNNLS.2017.2773458.

[40] Buşoniu, L., De Bruin, T., Tolić, D., Kober, J., Palunko, I. Reinforcement learning for control: Performance, stability, and deep approximators. *Annual Reviews in Control*, Jan. 2018, 46, 8–28. doi: 10.1016/j.arcontrol.2018.09.005.

[41] Dutta, D., Upreti, S. R. "A survey and comparative evaluation of actor-critic methods in process control," *The Canadian Journal of Chemical Engineering*, vol. 100, no. 9, pp. 2028–2056, 2022, doi: 10.1002/cjce.24508.

[42] Sachio, S., Mowbray, M., Papathanasiou, M. M., Del Rio-chanona, E. A., Petsagkourakis, P. Integrating process design and control using reinforcement learning. *Chemical Engineering Research and Design*, Jul. 2022, 183, 160–169. doi: 10.1016/j.cherd.2021.10.032.

[43] Mendiola-Rodriguez, T. A., Ricardez-Sandoval, L. A. Integration of design and control for renewable energy systems with an application to anaerobic digestion: A deep deterministic policy gradient framework. *Energy*, 2023, 274. doi: 10.1016/j.energy.2023.127212.

[44] Egea, J. A., Vries, D., Alonso, A. A., Banga, J. R. Global optimization for integrated design and control of computationally expensive process models. *Industrial & Engineering Chemistry Research*, Dec. 2007, 46(26), 9148–9157. doi: 10.1021/ie0705094.

[45] Mann, V., Sales-Cruz, M., Gani, R., Venkatasubramanian, V. *eSFILES*: Intelligent process flowsheet synthesis using process knowledge, symbolic AI, and machine learning. *Computers and Chemical Engineering*, Feb. 2024, 181, 108505. doi: 10.1016/j.compchemeng.2023.108505.

[46] Göttl, Q., Tönges, Y., Grimm, D. G., Burger, J. Automated flowsheet synthesis using hierarchical reinforcement learning: Proof of concept. *Chemie Ingenieur Technik*, 2021, 93(12), 2010–2018. doi: 10.1002/cite.202100086.

[47] Göttl, Q., Grimm, D. G., Burger, J. Automated synthesis of steady-state continuous processes using reinforcement learning. *Front Chem Sci Eng*, Feb. 2022, 16(2), 288–302. doi: 10.1007/s11705-021-2055-9.

[48] Grossmann, I. E., "Discrete Optimization Methods and their Role in the Integration of Planning and Scheduling".

[49] Menon, K. G., Fukasawa, R., Ricardez-Sandoval, L. A. Integration of planning and scheduling for large-scale multijob multitasking batch plants. *Industrial and Engineering Chemistry Research*, Jan. 2024, 63(2), 1039–1054. doi: 10.1021/acs.iecr.3c02408.

[50] Wang, Z., Ierapetritou, M. A novel feasibility analysis method for black-box processes using a radial basis function adaptive sampling approach. *AIChE Journal*, 2017, 63(2), 532–550. doi: 10.1002/aic.15362.

[51] Badejo, O., Ierapetritou, M. Integrating tactical planning, operational planning and scheduling using data-driven feasibility analysis. *Computers and Chemical Engineering*, May. 2022, 161, 107759. doi: 10.1016/j.compchemeng.2022.107759.

[52] Casazza, M., Ceselli, A. Heuristic Data-Driven Feasibility on Integrated Planning and Scheduling. In Paolucci, M., Sciomachen, A., Uberti, P. (Eds.). *Advances in Optimization and Decision Science for Society, Services and Enterprises: ODS, Genoa, Italy, September 4-7, 2019*. AIRO Springer Series. Cham: Springer International Publishing, 2019, pp. 115–125. doi: 10.1007/978-3-030-34960-8_11.

[53] Gupta, S. K., Ramteke, M. Applications of Genetic Algorithms in Chemical Engineering II: Case Studies. In Valadi, J., Siarry, P. (Eds.). *Applications of Metaheuristics in Process Engineering*. Cham: Springer International Publishing, 2014. pp. 61–87. doi: 10.1007/978-3-319-06508-3_3.

[54] Dao, S. D., Abhary, K., Marian, R. An improved genetic algorithm for multidimensional optimization of precedence-constrained production planning and scheduling. *Journal of Industrial Engineering International*, Jun. 2017, 13(2), 143–159. doi: 10.1007/s40092-016-0181-7.

[55] Dao, S. D., Marian, R. Modeling and Optimisation of Precedence-Constrained Production Sequencing and Scheduling for Multiple Production Lines Using Genetic Algorithms. *CTA*, 2(6), Jun. 2011. doi: 10.17265/1934-7332/2011.06.009.

[56] Macal, C. M., North, M. J. Agent-based modeling and simulation. *Proceedings of the 2009 Winter Simulation Conference (WSC)*, Dec. 2009, 86–98. doi: 10.1109/WSC.2009.5429318.

[57] Gao, Y., Shang, Z., Kokossis, A. Agent-based intelligent system development for decision support in chemical process industry. *Expert Systems with Applications*, Oct. 2009, 36(8), 11099–11107. doi: 10.1016/j.eswa.2009.02.078.

[58] Chu, Y., You, F., Wassick, J. M., Agarwal, A. Integrated planning and scheduling under production uncertainties: Bi-level model formulation and hybrid solution method. *Computers and Chemical Engineering*, Jan. 2015, 72, 255–272. doi: 10.1016/j.compchemeng.2014.02.023.

[59] Relvas, S., Matos, H. A., Barbosa-Póvoa, A. P. F. D., Fialho, J. Reactive Scheduling Framework for a Multiproduct Pipeline with Inventory Management. *Industrial and Engineering Chemistry Research*, Aug. 2007, 46(17), 5659–5672. doi: 10.1021/ie070214q.

[60] Koziel, S., Ciaurri, D. E., Leifsson, L. Surrogate-Based Methods. In Koziel, S., Yang, X.-S. (Eds.). *Computational Optimization, Methods and Algorithms*. Studies in Computational Intelligence. Berlin, Heidelberg: Springer, 2011, pp. 33–59. doi: 10.1007/978-3-642-20859-1_3.

[61] McBride, K., Sundmacher, K. Overview of Surrogate Modeling in Chemical Process Engineering. *Chemie Ingenieur Technik*, 2019, 91(3), 228–239. doi: 10.1002/cite.201800091.

[62] Ma, K. *et al.* Data-driven strategies for optimization of integrated chemical plants. *Computers and Chemical Engineering*, Oct. 2022, 166, 107961. doi: 10.1016/j.compchemeng.2022.107961.

[63] Sung, C., Maravelias, C. T. A projection-based method for production planning of multiproduct facilities. *AIChE Journal*, 2009, 55(10), 2614–2630. doi: 10.1002/aic.11845.

[64] Sung, C., Maravelias, C. T. An attainable region approach for production planning of multiproduct processes. *AIChE Journal*, 2007, 53(5), 1298–1315. doi: 10.1002/aic.11167.

[65] Ehm, F. A data-driven modeling approach for integrated disassembly planning and scheduling. *The Journal of Remanufacturing*, Jul. 2019, 9(2), 89–107. doi: 10.1007/s13243-018-0058-6.

[66] Beykal, B., Avraamidou, S., Pistikopoulos, E. N. Data-driven optimization of mixed-integer bi-level multi-follower integrated planning and scheduling problems under demand uncertainty. *Computers and Chemical Engineering*, Jan. 2022, 156, 107551. doi: 10.1016/j.compchemeng.2021.107551.

[67] Beykal, B., Avraamidou, S., Pistikopoulos, I. P. E., Onel, M., Pistikopoulos, E. N. DOMINO: Data-driven Optimization of bi-level Mixed-Integer Nonlinear Problems. *Journal of Engineering and Optimization*, Sep. 2020, 78(1), 1–36. doi: 10.1007/s10898-020-00890-3.

[68] Zhuge, J., Ierapetritou, M. G. An integrated framework for scheduling and control using fast model predictive control. *AIChE Journal*, 2015, 61(10), 3304–3319. doi: 10.1002/aic.14914.

[69] Valdez-Navarro, Y. I., Ricardez-Sandoval, L. A. A Novel Back-off Algorithm for Integration of Scheduling and Control of Batch Processes under Uncertainty. *Industrial and Engineering Chemistry Research*, Dec. 2019, 58(48), 22064–22083. doi: 10.1021/acs.iecr.9b04963.

[70] Rodríguez Vera, H. U., Ricardez-Sandoval, L. A. Integration of Scheduling and Control for Chemical Batch Plants under Stochastic Uncertainty: A Back-Off Approach. *Industrial and Engineering Chemistry Research*, Mar. 2022, 61(12), 4363–4378. doi: 10.1021/acs.iecr.1c04386.

[71] Kück, M., Ehm, J., Hildebrandt, T., Freitag, M., Frazzon, E. M. Potential of data-driven simulation-based optimization for adaptive scheduling and control of dynamic manufacturing systems. *2016 Winter Simulation Conference (WSC)*, Dec. 2016, 2820–2831. doi: 10.1109/WSC.2016.7822318.

[72] Simkoff, J. M., Baldea, M. Stochastic Scheduling and Control Using Data-Driven Nonlinear Dynamic Models: Application to Demand Response Operation of a Chlor-Alkali Plant. *Industrial and Engineering Chemistry Research*, May. 2020, 59(21), 10031–10042. doi: 10.1021/acs.iecr.9b06866.

[73] Kelley, M. T., Pattison, R. C., Baldick, R., Baldea, M. An efficient MILP framework for integrating nonlinear process dynamics and control in optimal production scheduling calculations. *Computers and Chemical Engineering*, Feb. 2018, 110, 35–52. doi: 10.1016/j.compchemeng.2017.11.021.

[74] Simkoff, J. M., Baldea, M. Parameterizations of data-driven nonlinear dynamic process models for fast scheduling calculations. *Computers and Chemical Engineering*, Oct. 2019, 129, 106498. doi: 10.1016/j.compchemeng.2019.06.023.

[75] Santander, O., Baldea, M. Control-aware batch process scheduling. *Computers and Chemical Engineering*, Sep. 2021, 152, 107360. doi: 10.1016/j.compchemeng.2021.107360.

[76] Ierapetritou, M. G., Floudas, C. A. Effective Continuous-Time Formulation for Short-Term Scheduling. 1. Multipurpose Batch Processes. *Industrial and Engineering Chemistry Research*, Nov. 1998, 37(11), 4341–4359. doi: 10.1021/ie970927g.

[77] Zhuge, J., Ierapetritou, M. G. Integration of scheduling and control for batch processes using multi-parametric model predictive control. *AIChE Journal*, 2014, 60(9), 3169–3183. doi: 10.1002/aic.14509.

[78]   Flores-Tlacuahuac, A., Grossmann, I. E. Simultaneous Cyclic Scheduling and Control of a Multiproduct CSTR. *Industrial and Engineering Chemistry Research*, Sep. 2006, 45(20), 6698–6712. doi: 10.1021/ie051293d.

[79]   Caspari, A., Tsay, C., Mhamdi, A., Baldea, M., Mitsos, A. The integration of scheduling and control: Top-down vs. bottom-up. *Journal of Process Control*, Jul. 2020, 91, 50–62. doi: 10.1016/j. jprocont.2020.05.008.

[80]   Rangel-Martinez, D., Nigam, K. D. P., Ricardez-Sandoval, L. A. Machine learning on sustainable energy: A review and outlook on renewable energy systems, catalysis, smart grid and energy storage. *Chemical Engineering Research and Design*, Oct. 2021, 174, 414–441. doi: 10.1016/j. cherd.2021.08.013.

[81]   Weichert, D., Link, P., Stoll, A., Rüping, S., Ihlenfeldt, S., Wrobel, S. A review of machine learning for the optimization of production processes. *The International Journal of Advanced Manufacturing Technology*, 104(5–8), Art. no. 5–8, Oct. 2019. doi: 10.1007/s00170-019-03988-5.

[82]   Chen, Y., Yuan, Z., Chen, B. Process optimization with consideration of uncertainties – An overview. *Chinese Journal of Chemical Engineering*, Aug. 2018, 26(8), 1700–1706. doi: 10.1016/j.cjche.2017.09.010.

[83]   Ning, C., You, F. Optimization under uncertainty in the era of big data and deep learning: When machine learning meets mathematical programming. *Computers and Chemical Engineering*, Jun. 2019, 125, 434–448. doi: 10.1016/j.compchemeng.2019.03.034.

[84]   Ejaz, W. *et al.* Internet of Things (IoT) in 5G Wireless Communications. *IEEE Access*, 2016, 4, 10310–10314. doi: 10.1109/ACCESS.2016.2646120.

[85]   Moraru, A., Pesko, M., Porcius, M., Fortuna, C., Mladenic, D. Using Machine Learning on Sensor Data. *CIT. Journal of Computing and Information Technology*, 18(4), Art. no. 4, Feb. 2011. doi: 10.2498/cit.1001913.

[86]   Gondal, F. K., Shahzad, S. K., Jaffar, M. A., Iqbal, M. W. "A Process Oriented Integration Model for Smart Health Services," *IASC*, vol. 35, no. 2, pp. 1369–1386, 2022, doi: 10.32604/iasc.2023.028407.

[87]   Kong, K. G. H. *et al.* Towards data-driven process integration for renewable energy planning. *Current Opinion in Chemical Engineering*, Mar. 2021, 31, 100665. doi: 10.1016/j.coche.2020.100665.

[88]   Tan, S., De, D., Song, W.-Z., Yang, J., Das, S. K. Survey of Security Advances in Smart Grid: A Data Driven Approach. *IEEE Communications Surveys and Tutorials*, 2017, 19(1), 397–422. doi: 10.1109/COMST.2016.2616442.

[89]   Patrón, G. D., Ricardez-Sandoval, L. Economically optimal operation of recirculating aquaculture systems under uncertainty. *Computers & Electronics in Agriculture*, May. 2024, 220, 108856. doi: 10.1016/j.compag.2024.108856.

[90]   Kamali, S., Ward, V. C. A., Ricardez-Sandoval, L. Closed-loop operation of a simulated recirculating aquaculture system with an integrated application of nonlinear model predictive control and moving horizon estimation. *Computers & Electronics in Agriculture*, Jun. 2023, 209, 107820. doi: 10.1016/j.compag.2023.107820.

[91]   Petsagkourakis, P., Sandoval, I. O., Bradford, E., Zhang, D., Del Rio-Chanona, E. A. Reinforcement learning for batch bioprocess optimization. *Computers and Chemical Engineering*, Feb. 2020, 133, 106649. doi: 10.1016/j.compchemeng.2019.106649.

[92]   Pan, E., Petsagkourakis, P., Mowbray, M., Zhang, D., Del Rio-Chanona, E. A. Constrained model-free reinforcement learning for process optimization. *Computers and Chemical Engineering*, Nov. 2021, 154, 107462. doi: 10.1016/j.compchemeng.2021.107462.

Antonio Flores-Tlacuahuac*

Chapter 7
Applications of Bayesian optimization in chemical engineering

Abstract: This study examines the utilization of Bayesian optimization in the field of chemical engineering, with a specific focus on the optimization of process design and dynamic product transition tasks. Through the introduction of a data-driven Bayesian approach, this research addresses the challenges inherent in dealing with complex system dynamics and uncertain measurements within the processing industry. The Bayesian optimization algorithm provides a computationally efficient solution that does not rely on intricate mathematical models, rendering it well-suited for real-world applications characterized by the prevalence of noisy data. Through the analysis of three case studies, the proposed approach demonstrates its efficacy in identifying optimal transition trajectories and fulfilling product composition requirements while ensuring the implementation of smooth control actions. The results obtained underscore the value of Bayesian optimization as a valuable tool for optimizing various aspects of chemical processes, including catalyst design, drug formulations, bioprocess parameters, materials design, energy system efficiency, and the reduction of environmental impact. In summary, this work highlights the potential of Bayesian optimization to enhance decision-making processes and drive innovation in the realm of chemical engineering applications.

Keywords: Bayesian optimization, nonlinear systems, noisy measurements, dynamic optimization, data-driven systems

7.1 Introduction

Optimization methods are employed with the objective of identifying the optimal solution to a given problem within specified constraints. These methods can be broadly classified into two categories based on solution approaches: deterministic and heuristic methods. Deterministic optimization methods, as discussed by Nocedal, Biegler, and Grossmann [1–3], have been extensively utilized in addressing engineering problems. They are grounded in a robust theoretical framework, drawing upon centuries of advancements in pure and applied mathematics. The widespread application of deterministic methods gained momentum with the emergence of digital computing facil-

*Corresponding author: Antonio Flores-Tlacuahuac, Instituto Tecnológico y de Estudios Superiores de Monterrey, Institute of Advanced Materials for Sustainable Manufacturing, Av. Eugenio Garza Sada 2501, Monterrey, Nuevo León 64849, México, e-mail: antonio.flores.t@tec.mx

https://doi.org/10.1515/9783111383439-007

ities in the mid-twentieth century. Presently, we possess a comprehensive understanding of the scope and limitations of deterministic optimization tools. They are particularly well-suited for addressing continuous nonlinear systems of moderate size, where the term size denotes the number of decision variables and associated constraints. However, challenges arise when dealing with discrete decision variables, as deterministic optimization methods may not be as advantageous in such scenarios. The introduction of a significant number of discrete variables can give rise to highly combinatorial problems, complicating the search for even a local optimal solution. Nevertheless, there is a burgeoning interest in leveraging quantum computing facilities to tackle the resolution of large combinatorial optimization problems, as highlighted by Bernal et al. [4].

On the other hand, heuristic optimization is an approach to problem-solving and optimization that involves the use of heuristics, which are practical and intuitive strategies or rules of thumb, rather than strictly following formal mathematical methods [5]. Unlike deterministic optimization methods that aim to find an optimal solution with a high level of certainty, heuristic optimization methods are more flexible and often trade optimality for efficiency and feasibility in finding good solutions. Heuristic optimization algorithms are designed to handle complex problems, particularly those with large search spaces, high dimensionality, or combinatorial aspects where exhaustive search or deterministic methods may be impractical. Common examples of heuristic optimization algorithms include genetic algorithms [6], simulated annealing [7], ant colony optimization [8], and particle swarm optimization [9]. These methods are widely used in various fields such as engineering, operations research, computer science, and artificial intelligence, where finding an exact optimal solution may be computationally infeasible or time-consuming.

Bayesian optimization is generally considered a data-driven probabilistic or surrogate model-based optimization approach [10, 11], which falls under the broader category of heuristic optimization. Unlike deterministic optimization methods that aim to find an optimal solution with certainty, Bayesian optimization incorporates probabilistic models to guide the search for the optimum. In Bayesian optimization, a probabilistic surrogate model is built from plant measurements to approximate the objective function, which represents the unknown true function being optimized. This surrogate model is iteratively updated based on the measured data points, and a balance is maintained between exploring new regions of the search space and exploiting the current best-known solutions. Bayesian optimization often incorporates an acquisition function that guides the selection of new points to evaluate in the search space. The probabilistic nature of Bayesian optimization allows it to efficiently handle complex and expensive objective functions, as it adapts its search based on the uncertainty in the model.

Bayesian optimization is a powerful optimization technique that is particularly useful for optimizing complex and expensive black-box functions [12–16]. There has been a growing interest in adopting a data-driven approach for the development of reliable surrogate models, subsequently employing formal Bayesian statistics for system optimization [10, 17–19]. Consequently, data science [20], materials design [21, 22], hyper-parameter tuning in modern machine learning strategies [23], and the optimal design of experiments [24] have been prominent fields where Bayesian optimization (BO) has been applied. Specifically, applications of BO in the PET chemical recycling process [25], computational fluid dynamics [26], tuning of process controllers [27], process control [28, 29], heating, ventilation, and air conditioning (HVAC) plants [30], chemical reactor design using computational fluid dynamics [31], extending BO tools to deploy parallel computer architectures [32], the design of complex distillation columns [33], and for dynamic optimization product transitions [34], among other applications, have been recently published.

The objective of this work is to offer a comprehensive overview of potential applications of Bayesian optimization in data-driven chemical engineering applications, emphasizing both its advantages and disadvantages. Three practical chemical engineering case studies are addressed to show the advantages of the data-driven optimization of industrial systems. Consequently, a full literature review on this subject is not provided. Nevertheless, diligent efforts have been made to underscore the majority of significant research contributions. Any work of this nature inevitably reflects the bias of the author's research field. Therefore, we have emphasized applications in Process System Engineering.

7.2 Bayesian optimization methodology

Bayesian optimization is a probabilistic model-based approach for optimization of expensive, black-box functions. It is particularly well-suited for scenarios where evaluating the objective function is time-consuming, expensive, or involves real-world experiments. The main idea behind Bayesian optimization is to sequentially build a surrogate probabilistic model of the objective function and use this model to make informed decisions about where to sample the function next.

7.2.1 Main steps in Bayesian optimization

In the following, we provide an overview of the key components and steps involved in Bayesian optimization, as shown in Figure 7.1:
– **Objective function:** We start by defining a target objective function that depends on the decision variables. The convergence behavior of the Bayesian optimization approach heavily depends on the shape of the objective function.
– **Initial design:** The optimization process starts with an initial design, typically a small set of randomly or quasi-randomly chosen points that represent the values

of the decision variables and the objective function, to initialize the surrogate model.

- **Surrogate model:** A probabilistic surrogate model is used to approximate the behavior of the objective function in terms of the decision variables. Gaussian processes (GPs) are commonly employed for this purpose. Being a probabilistic model, the GP model provides not only predictions of the objective function values but also estimates of uncertainty associated with those predictions.
- **Acquisition function:** The acquisition function is a utility function that guides the selection of the next point to evaluate. It balances exploration (sampling in regions with high uncertainty) and exploitation (sampling in regions with potentially high objective values). Common acquisition functions include probability of improvement (PI), expected improvement (EI), and upper confidence bound (UCB).
- **Objective function evaluation:** The selected point from the acquisition function is evaluated in the true objective function, providing a new data point. The surrogate model is then updated to incorporate this new information.
- **Convergence criteria:** The optimization process continues until a predefined convergence criterion is met. This criterion could be a maximum number of iterations, reaching a certain level of precision, or other user-defined conditions.
- **Optimal solution:** The final output is the point or set of points that are identified as the optimal solution(s) of the objective function within the specified search space.

Bayesian optimization has been successfully applied in various fields, including machine learning hyperparameter tuning, robotics, materials science, and engineering optimization problems. Its ability to efficiently explore and exploit the search space makes it a valuable tool for problems, where objective function evaluations are resource-intensive.

7.2.2 Limitations and challenges of BO

Currently, there exist certain limitations that constrain the widespread application of Bayesian optimization in the era of data-driven systems. Many of these shortcomings are subjects of ongoing research efforts.

- **High-dimensional spaces:** Bayesian optimization becomes less efficient as the dimensionality of the search space increases. The curse of dimensionality can lead to a sparse and unreliable surrogate model.
- **Scalability:** Scaling Bayesian optimization to handle a large number of evaluations or parallel evaluations can be challenging. The computational cost of updating the surrogate model and optimizing the acquisition function may become prohibitive.

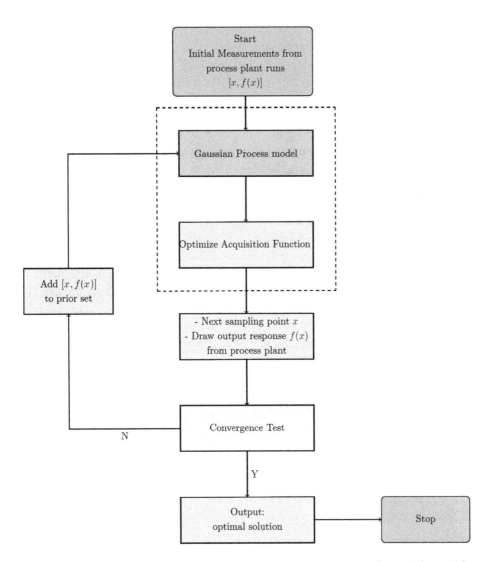

Figure 7.1: The diagram depicts the data-driven Bayesian optimization approach. It begins with an initial set of process measurements, from which a probabilistic surrogate model is constructed. Subsequently, the optimization of an acquisition function leads to updates in the values of the decision variables. The purpose of the acquisition function is to direct the search toward optimal conditions, progressively improving the likelihood of identifying the best optimal conditions with each iteration. Following numerous iterations, an optimal solution may be found, or the iterations may reach their limit. In instances where the optimal solution remains elusive, the information gathered from the current iteration (i.e., objective function and decision variables) is collected and utilized to develop a new surrogate model approximation. This aids in deducing new values for the decision variables. The dashed box encompasses the two critical steps of the Bayesian optimization strategy.

- **Sensitivity to initial conditions:** The performance of Bayesian optimization can be sensitive to the choice of hyperparameters and the initial design of points. Poor initial conditions may lead to suboptimal results.
- **Limited exploration:** Bayesian optimization can be conservative and may exploit regions that seem promising based on past evaluations. However, it may not explore enough to discover potentially better regions, especially if the surrogate model is not accurate.
- **Non-convexity:** Like many optimization algorithms, Bayesian optimization struggles with non-convex and multimodal objective functions. It may get stuck in local optima and fail to find the global optimum.
- **Noise handling:** While Bayesian optimization is designed to handle noisy objective functions, it may struggle in situations where the noise level is high or poorly understood. Modeling and accounting for noise can be challenging.
- **Modeling assumptions:** The performance of Bayesian optimization heavily depends on the adequacy of the surrogate model. If the underlying assumptions about the function smoothness or shape do not hold, the model might provide inaccurate predictions.
- **Black-box function limitation:** Bayesian optimization assumes that the objective function is a black box, meaning it does not exploit any structural information about the function. This limitation can be restrictive when such information is available.
- **Limited theoretical understanding:** Despite its empirical success, there is still limited theoretical understanding of Bayesian optimization in certain aspects, such as its convergence properties or behavior in specific scenarios.

Having outlined the present constraints of Bayesian optimization, it is noteworthy to mention that significant progress has been made in addressing these limitations. For instance, advancements have been achieved in handling high-dimensional problems [35, 36], integrating derivative information [37], and establishing convergence behavior [38].

7.3 Potential applications in chemical engineering

In the following, we provide an overview of some of the specific areas or problems within chemical engineering where Bayesian optimization has been applied. As previously stated, this overview is far from complete since we concentrated in some areas.

1. Process optimization: Bayesian optimization is used to optimize chemical processes, adjusting parameters to maximize the desired output (yield, purity, etc.) while minimizing costs or resource consumption. An application regarding the optimal steady-state design for the separation of ethanol and water in complex distillation columns is shown in [33], while optimal dynamic product transitions has been assed in [34]. Other works related to process optimisation tasks have also been reported and can be found elsewhere [14, 25, 26, 39–45].

2. Catalyst design: Designing optimal catalysts for chemical reactions is a complex task. Bayesian optimization helps in efficiently exploring the vast design space to identify catalyst compositions that enhance reaction rates and selectivity [46–49].

3. Drug formulation: In pharmaceutical engineering, Bayesian optimization is applied to optimize drug formulations. It helps in finding the right combination of ingredients and their concentrations to enhance drug efficacy and minimize side effects [50–52].

4. Bioprocess optimization: Bayesian optimization is employed in optimizing bioprocess parameters for the production of biofuels, enzymes, and pharmaceuticals. This includes optimizing conditions such as temperature, pH, and nutrient concentrations in bioreactors [53–58].

5. Materials design: In chemical engineering related to materials science, Bayesian optimization aids in the design of new materials with specific properties. This could involve optimizing compositions for polymers, alloys, or other materials with the desired characteristics [59–65].

6. Energy system optimization: Bayesian optimization is applied to optimize the performance of energy systems in chemical plants. This includes optimizing the use of energy resources, heat exchanger networks, and overall energy efficiency [66–68].

7. Environmental impact reduction: Chemical engineers use Bayesian optimization to minimize the environmental impact of chemical processes. This involves optimizing processes to reduce waste production, energy consumption, and emissions [69–71].

8. Parameter tuning in computational models: Bayesian optimization is used to tune parameters in computational models used in chemical engineering simulations. This helps in improving the accuracy of models and reducing the need for time-consuming trial-and-error approaches [72–74].

9. Control system optimization: Bayesian optimization is applied in optimizing control systems for chemical processes, ensuring stable and efficient operation under varying conditions [75–77].

10. Sensor placement and data collection: Bayesian optimization aids in the optimal placement of sensors in chemical processes, improving the efficiency of data collection and monitoring [78–80].

7.4 Background and theory

In this section, we will delve into the methodology behind the establishment of GP models and the formulation of data-driven optimization strategies using the Bayesian optimization approach. We will proceed under the assumption that process data or measurements are available, acknowledging the potential presence of noise. Furthermore, it is presumed that users have meticulously chosen appropriate decision variables and have defined the shape or mathematical structure of the objective function to the best of their abilities. It is imperative to highlight that in the subsequent discussion on constructing GP models, there is an implicit assumption that only a single output variable is derived from the GP approach. While this may not necessarily be viewed as a limitation, particularly when employing only one objective function (i.e., the output of the GP model), to the best of our knowledge, there is currently no definitive answer regarding the efficient modelling of multiple output GPs, as it remains an active research topic [81, 82].

7.4.1 Gaussian process models

A GP model will be utilized to conduct black-box modeling of the system response, encompassing process uncertainty and accounting for noisy measurements. It should be stressed that the following description of a GP is limited and that it only covers a subset of possible GPs. Full descriptions of GPs can be found elsewhere [10, 18].

A GP $f(\cdot)$ is characterized in terms of a mean μ, process variance σ^2 and a Kernel function $K(x, x^*)$, such that a finite collection of $f = [f(x_1), f(x_2), \ldots, f(x_n)]$ follows a multivariable Gaussian distribution, i.e.,

$$f \sim \mathcal{N}(\mathbf{1}\mu, \sigma \mathbf{K}) \tag{7.1}$$

where $\mathbf{1}$ is a vector with n ones, and $\mathbf{K}$ is the correlation matrix with its element $\mathbf{K}_{i,j} = K(x_i, x_j)$.

7.4.1.1 Gaussian kernel function

A one-dimensional Gaussian exponential kernel $K(x_i, x_j)$ is expressed as follows:

$$K(x_i, x_j) = e^{-\theta(x_i - x_j)^2} \tag{7.2}$$

where θ is a kernel parameter that controls the correlation strength between points i and j of vector x. Similarly, an m-dimensional Gaussian kernel is expressed as follows:

$$K(x_i, x_j) = \exp\left[-\sum_{k=1}^{m} \theta_k \left(x_i^k - x_j^k\right)^2\right] \tag{7.3}$$

which is simply a series of multiplication of the one-dimensional Gaussian kernel for each feature. Here, we have the kernel parameters $\theta = [\theta_1, \theta_2, \ldots, \theta_m]$.

7.4.1.2 Training a GP

To draw the values of μ, σ^2, and θ, we use an optimization formulation known as the maximum likelihood estimation. In this setting, the likelihood L of observing the labels $(y_1, y_2, \ldots, y_n)$ of the training instances $(x_1, x_2, \ldots, x_n)$ is given as follows:

$$L(y|\mu, \sigma^2, \theta) = \frac{1}{\sqrt{(2\pi\sigma^2)^n |K|}} \exp\left[-\frac{1}{2\sigma^2} (y - 1\mu)^T K^{-1} (y - 1\mu)\right] \tag{7.4}$$

where $y = [y_1, y_2, \ldots, y_n]$ and K is the correlation matrix of the training instances. However, in practice, the logarithm of the Likelihood L is maximized in order to avoid round-off errors:

$$\ln(L) = -\frac{n}{2}\ln(2\pi) - \frac{n}{2}\ln(\sigma^2) - \frac{1}{2}\ln(|K|) - \frac{1}{2\sigma^2}(y - 1\mu)^T K^{-1}(y - 1\mu) \tag{7.5}$$

the optimal values of μ and σ^2 can be easily obtained by setting the first derivative with respect to these decision variables, using the above equation. Hence,

$$\mu = \left(1^T K^{-1} 1\right)^{-1} 1^T K^{-1} y \tag{7.6}$$

$$\sigma^2 = \frac{1}{n}(y - 1\mu)^T K^{-1}(y - 1\mu) \tag{7.7}$$

The optimal value of θ can be drawn from the solution of the following unconstrained optimization problem:

$$\theta = \text{argmax}_\theta\left[-\frac{n}{2}\ln\sigma^2 - \frac{1}{2}\ln(|K|)\right] \tag{7.8}$$

7.4.1.3 Forecasting using a GP model

Upon training a GP model, the intention is to employ it for output forecasting. This process can be executed as follows: to forecast the value of f^* at the specific point x^*, we initially establish the joint distribution of f^* and the observed values of the training instances x^*:

$$\begin{pmatrix} y \\ f^* \end{pmatrix} \theta \sim \mathcal{N}\left(\mu, \sigma^2 \begin{pmatrix} K & k^* \\ k^{*T} & 1 \end{pmatrix}\right) \tag{7.9}$$

where k^* is a correlation vector between the testing and training values, where the ith element is cast as $k_i^* = K(x^*, x_i)$. Finally, the distribution of f^* conditioned on y is derived from their joint distribution, where the mean and variance values are given as follows:

$$\mu^* = \mu + k^{*T} K^{-1}(y - 1\mu) \tag{7.10}$$

$$\Sigma^* = \sigma^2 \left(1 - k^{*T} K^{-1} k^*\right) \tag{7.11}$$

Hence, the distribution of f^* conditioned on y, written as $f^* | y \sim \mathcal{N}(\mu^*, \Sigma^*)$ fully characterizes the GP forecasting at x^*.

7.4.2 Acquisition function

Various acquisition functions leverage Bayesian inference in different ways to achieve a balance between exploration and exploitation. For example, EI and UCB explicitly utilize uncertainty estimates from Bayesian inference to determine subsequent sampling locations. For a more thorough understanding of the derivation and optimization of different acquisition functions, please refer to [11]. Considering that our study heavily relies on the EI function as its primary objective, we offer a detailed explanation of its mathematical definition. Furthermore, we elucidate how the EI function is derived and optimized to select the next promising measurement point. EI is widely recognized as an acquisition function in Bayesian optimization, evaluating the potential improvement in the objective function by examining specific points within the search space.

7.4.2.1 Mathematical definition

$$EI(x) = \mathbb{E}[\max(f(x^*) - f(x), 0)] \tag{7.12}$$

where $f(x^*)$ is the best objective function value observed so far among all evaluated points, $f(x)$ is the predicted objective function value at point x, obtained from the surrogate model (typically a GP), and $\mathbb{E}$ denotes the expected value operator.

7.4.2.2 EI computational procedure

The process to compute EI values involves several sequential steps:
1. Given a point x within the search space, deploy the GP model to calculate the mean (μ) and standard deviation (σ^2) values.
2. Compute the standardized improvement (z) value using the formula:

$$z = \frac{\mu - f^*}{\sigma^2 + \in}$$
(7.13)

where $\in$ represents a small tolerance value introduced to prevent division by zero.
3. Compute the cumulative distribution (Φ) and probability density (ϕ) functions of the standard normal distribution.
4. Determine the EI value according to the formula:

$$EI(x) = (\mu - f^*)\, \Phi\left(\frac{\mu - f^*}{\sigma^2}\right) + \sigma^2 \phi\left(\frac{\mu - f^*}{\sigma^2}\right)$$
(7.14)

It is essential to highlight that a higher posterior value of the first term on the equation's right-hand side emphasizes the Exploitation phase, while a larger posterior value of the second term encourages the Exploration phase. For a comprehensive derivation of the aforementioned EI defining equation, refer to [11].

7.4.2.3 Best next sampling value

Once EI values are computed for various points within the search space, selecting the optimal next sampling value involves maximizing the EI. The steps to identify the best next sampling value proceed as follows:
1. Employ the previously described procedure for computing EI values to generate EI values for a set of candidate points within the search space, utilizing the surrogate model predictions and the equations defining the EI function.
2. Select the point corresponding to the maximum EI value as the next value to be evaluated within the objective function. This step can be carried out by a simple search procedure or using an unconstrained optimization algorithm.

7.5 Applications of BO in process system engineering

In this section, we present three case studies in chemical engineering to demonstrate the application of Bayesian optimization for the data-driven optimization of these systems. In this context, data-driven implies that a first principles mathematical model will not be employed to find optimal operating regions. Instead, utilizing simulation or experimental environments, information pertaining to the decision variables and the response of the system under consideration will be utilized to incrementally construct a probabilistic model. This model serves as a guide toward optimal regions. As we do not have access to a genuine industrial setting to gather experimental or process data, we have utilized widely deployed simulation tools to capture plant data information. However, it is important to note that this does not imply the use of a first-principles model for optimization calculations. The purpose of the simulation tools is to replicate plant behavior. Details (and software) about how to implement the above procedure can be found in a recent publication of our research group [33].

7.5.1 Binary distillation column

The industrial separation of gas and liquid mixtures stands as one of the most prevalent unit operations in the processing industry. Distillation, despite its typically substantial energy requirements, remains the primary separation technology extensively employed for this purpose. Consequently, distillation operations present a noteworthy and challenging prospect for energy reduction through the exploration of optimal operating conditions. By continuously monitoring data, the achievement of this objective can be pursued, and the optimal operation of distillation processes can be realized and implemented through online feedback control systems.

To demonstrate the advantages of Bayesian optimization for the data-driven black-box optimization of a typical unit operation, we will consider the separation of the benzene-toluene binary system. The feed stream consists of 50 kmol/h of an equimolar mixture at 100 °C and 5 bar. As shown in Figure 7.2, the column features 16 trays and the feed stream is located on tray 8.

The column is operated at 1 bar and the drop pressure is 0.5 bar/tray. The aim of the distillation configuration is to draw pure components, as shown in Figure 7.2. Therefore, the output-controlled variables are chosen as the high purity of the benzene (X_B) and toluene (X_T) at the distillate and bottoms streams, respectively, while the manipulated variables are the distillate flowrate (D) and the reflux ratio (R). To this aim, we have assumed that the industrial behavior of this system can be well represented by deploying the ASPEN process simulation facility.

In this case study, we will assume that our intention is to determine the optimal values for the distillate and reflux ratio, with the objective of minimizing the following objective function:

$$\text{Min}(\Omega) = \left(X_B^t - X_B\right)^2 + \left(X_T^t - X_T\right)^2 \tag{7.15}$$

where the superscript t stands for target values (specified as $X_B^t = X_T^t = 0.99$). The aforementioned objective function serves as a convex metric designed to signify our intention to operate the column in close proximity to the target product compositions. To

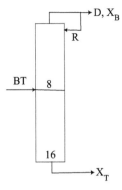

Figure 7.2: Conventional distillation column flowsheet for separation of the benzene (*B*)-toluene (*T*) binary mixture into pure components. The output-controlled variables are the mol fraction of distillate benzene (X_B) and toluene bottoms (X_T), while the input-manipulated variables are the distillate (*D*) and reflux ratio (*R*).

handle the solution of this task, we have imposed the following box constraints: $D \in$ [5, 25] and $R \in$ [3, 8].

Starting from initial arbitrary values of D and R (shown as values for the first iteration), results in Table 7.1 depict the convergence behavior of the Bayesian optimization approach. As shown there, in a few iterations, the optimization approach swiftly finds the D and R optimal operation values, as noted by the value of the objective function. We wish to emphasize that, despite not being presented, comparable convergence behavior was observed when utilizing alternative feasible starting points.

Table 7.1: Binary distillation column convergence behavior of the Bayesian optimization approach.

Iter	D	R	X_B	X_T	Ω
1	10	3	0.9887	0.6221	0.135352
2	23.1219	5.8895	0.9977	0.9281	0.00389
3	21.6	6.3006	0.9981	0.8788	0.012431
4	25	8	0.9935	0.9935	2.45×10^{-5}

The initial guessed values of D and R are the employed values for the first iteration.

7.5.2 Petlyuk column

The Petlyuk separation column is a type of distillation column that incorporates an internal heat exchange process to improve energy efficiency [83]. As shown in Figure 7.3, the key feature of a Petlyuk column is the introduction of intermediate heat exchangers and side strippers within the column. These side strippers help in redistributing heat along the column height, allowing for better temperature control and reducing the temperature difference between the top and bottom of the column. This internal heat exchange improves the separation efficiency and reduces the energy requirements for the distillation process.

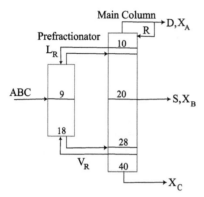

Figure 7.3: Typical flowsheet of a Petlyuk distillation column for the separation of the ABC ternary mixture into pure components. The separation system consists of coupled pre-fractionator and main columns. The aim of the flowsheeet is to draw pure components by using an intermediate side-stream (S), taken from the main column, in addition to the distillate and bottom streams. The distillate flowrate (D), reflux ratio (R), liquid flowrate (LR) (fed from the main to the pre-fractionator column) and vapor flowrate (VR) (fed from the main to the pre-fractionator column) are the set of manipulated variables, while the distillate (X_A), bottoms (X_C), and side-stream (X_B) product purity are the controlled variables.

The Petlyuk column is particularly useful in separating close-boiling components, where traditional distillation columns may face challenges. It is often employed in the petrochemical and chemical industries for the separation of multicomponent mixtures into essentially pure components. The design and operation of a Petlyuk column can be complex, and it requires careful consideration of thermodynamics, heat transfer, and fluid dynamics. While Petlyuk distillation can offer energy savings, it may also require advanced control strategies to optimize its performance [84–87]. It is worth noting that the application of Petlyuk distillation columns may vary, depending on the specific needs of a given separation process, and they are not as widely used as traditional distillation columns.

In this case study, using the complex Petlyuk distillation configuration, we will address the problem of drawing the optimal values of the manipulated variables, such that essentially pure components are separated in each one of the output product streams. We will take advantage of the fact that the Petlyuk distillation configuration provides two additional degrees of freedom to search for optimal operating conditions. Therefore, the manipulated variables are: distillate flow rate (D), reflux ratio (R), liquid flowrate (L_R) (fed from the main to the pre-fractionator column), and vapor flowrate (V_R) (fed from the main to the pre-fractionator column). The product purity of each one of the three components is set as follows: $X_i \geq 0.99$; i = A, B, C. Therefore, a high-purity separation operation is specified. The stream fed to the pre-fractionator column is a 45.4 kmol/h equimolar saturated mixture liquid of n-heptane (A), n-hexane (B), and n-heptane (C) at 30 psia [88, 89]. Both columns are operated at 2 atm, neglecting trays drop pressure. To mimic plant behavior, ASPEN models of the Petlyuk distillation configuration was deployed.

In a manner akin to the case study of the binary distillation column, we employed the subsequent objective function that incorporates the deviation from the target product compositions:

$$Min(\Omega) = \left(X_A^t - X_A\right)^2 + \left(X_B^t - X_B\right)^2 + \left(X_C^t - X_C\right)^2 \qquad (7.16)$$

Additionally, the following box constraints on the manipulated variables were enforced: D [10, 15], R [4, 10], L_R [5, 25], and V_R [5, 30]; target product purities were set as follows: $X_A^t = X_B^t = X_C^t = 0.99$.

The performance of the data-driven Bayesian optimization approach is shown in Table 7.2. Initially, a feasible starting point is chosen. We carried out a trial-and-error design process to find some promising initial values of the decision variables D, R, L_R, and V_R. Those initial values correspond to the values reported in Table 7.2 for the first iteration ($D = 10$, $L_R = 8$, $R = 4$, $V_R = 10$). As seen from the results of the first iteration, those input values give rise to product compositions far from the target values, except for n-hexane. This fact is also reflected in the value of the objective function. Then, we used the Bayesian optimization approach for fine tuning the input variables and to operate under optimal operation conditions. It took only nine iterations of the Bayesian optimization approach to find optimal operating conditions. As indicated in Table 7.2, the Bayesian optimization approach exhibits a predominant focus on exploitation during the initial six iterations. This is attributed to the rapid decrease in the objective function value. Nevertheless, in the seventh iteration, a shift toward exploration of a new optimal region occurs, without a corresponding improvement in the objective function. Consequently, the approach reverts to the best promising optimality region and remains there until achieving a favorable value for the objective function.

Table 7.2: Petlyuk column convergence behavior of the Bayesian optimization approach.

Iter	D	L_R	R	V_R	X_A	X_B	X_C	Ω
1	10	8	4	10	0.999735	0.630622	0.722999	0.200536
2	10.0385	8.2372	8.2562	24.1184	0.9998	0.6397	0.7293	0.190752
3	12.2127	13.2694	7.2868	23.605	0.9998	0.8105	0.8403	0.054703
4	12.2262	13.3923	7.3879	23.203	0.9998	0.8121	0.8414	0.053797
5	13.2761	15.6385	6.7391	23.8837	0.9999	0.8833	0.8954	0.0204
6	14.0878	18.8037	9.8242	25.412	0.9999	0.9386	0.9421	0.005027
7	10.2051	22.4264	6.9878	25.9514	0.9999	0.6779	0.7559	0.152259
8	15	16.5195	10	24.8173	0.9838	0.9819	0.9984	1.7365×10^{-4}
9	15	17.1973	10	28.6687	0.9936	0.9916	0.9979	7.9329×10^{-5}

The initial guessed values of D, R, L_R, and V_R are the employed values for the first iteration.

7.5.3 Methyl-methacrylate polymerization reactor dynamic-grade transition

In the context of polymerization reactors, a grade refers to a specific type or variant of polymer product with distinct properties and characteristics [90–92]. Polymers can have different grades based on various factors, including molecular weight, molecular weight distribution, composition, and other relevant properties. These differences in grades make polymers suitable for various applications with specific performance requirements. The choice of polymer grade can impact factors such as strength, flexibility, transparency, and thermal resistance. Manufacturers often produce a range of grades to meet diverse market demands and application needs. Dynamic-grade transitions in polymerization reactors involve the ability to switch between these different grades during the production process, providing flexibility to adapt to changing market requirements.

In this part, we will show how Bayesian optimization can be deployed for carrying out optimal dynamic optimization between two different grade products in a polymerization reactor. To this aim, we will address the non-isothermal bulk polymerization of methyl-methacrylate (MMA) in a continuous-stirred tank reactor. As shown in Figure 7.4, continuous streams of MMA and initiator are mixed and reacted to produce a polymer featuring a nominal grade. A cooling water stream is also provided for temperature regulation purposes.

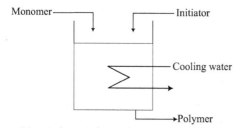

Figure 7.4: Flowsheet of the methyl-methacrylate continuous-stirred tank in methyl-methacrylate polymerization reactor. The dynamic optimization decision variables are the volumetric flowrates of the monomer, initiator, and cooling water [L/s], while the target variable is the molecular weight distribution of the output polymer product during dynamic-grade transition operations.

The non-isothermal dynamic mathematical model of the free radicals MMA polymerization is given as follows [93]:

$$\frac{dC_m}{dt} = \frac{Q\left(C_m^f - C_m\right)}{V} - \left(K_p + K_{fm}\right)C_m P_0 \tag{7.17}$$

$$\frac{dC_i}{dt} = \frac{\left(Q_i C_i^f - Q C_i\right)}{V} - K_i C_i \tag{7.18}$$

$$\frac{dT}{dt} = \frac{Q\left(T^f - T\right)}{V} + \frac{\Delta H_r K_p C_m P_0}{\rho C_p} - \frac{UA\left(T - T_j\right)}{\rho C_p V} \tag{7.19}$$

$$\frac{dT_j}{dt} = \frac{Q_w\left(T_j^f - T_j\right)}{V_j} + \frac{UA\left(T - T_j\right)}{\rho_w C_{pw} V_j} \tag{7.20}$$

$$\frac{dD_0}{dt} = \frac{-Q D_0}{V} + (0.5 K_{tc} + k_{td}) P_0^2 + K_{fm} C_m P_0 \tag{7.21}$$

$$\frac{dD_1}{dt} = \frac{-Q D_1}{V} + M_m (K_p + K_{fm}) C_m P_0 \tag{7.22}$$

where reaction kinetics are given as follows:

$$K_i = A_i e^{-\frac{E_i}{RT}}, i = p, i, tc, td, fm \tag{7.23}$$

$$P_0 = \sqrt{\frac{2f^* K_i C_i}{K_{td} + K_{tc}}} \tag{7.24}$$

in the above model, description C stands for concentration [kmol/m^3], T is temperature [K], Q is volumetric flowrate [L/s], V is volume [m^3], D is dead polymer, U is heat-transfer coefficient [kJ/(h-K-m^2)], A is the heat exchanger area [m^2], ΔH_r is the reaction heat [kJ/kmol], ρ is density [kg/m^3], C_p is the heat capacity [kJ/(kg K)], and f^* is the initiator efficiency. The subscripts m, i, w, j, 0, and 1 stand for monomer, initiator, cooling water, jacket, zero, and one, respectively. The superscript f stands for feed stream. Additional information regarding nominal operating conditions, thermodynamics, and kinetic information is provided in Table 7.3.

Table 7.3: Methyl-methacrylate continuous stirred tank reactor data, including nominal operating conditions, thermodynamics, and kinetics information.

Parameter	Value	Units
C_m^f	6.4678	kmol/m^3
C_i^f	8	kmol/m^3
T^f	366	K
T_f^w	293.2	K
U	720	kJ/(h-K-m^2)
A	2	m^2
V	0.1	m^3
V_j	0.02	m^3

Table 7.3 (continued)

Parameter	Value	Units
ρ	866	kg/m^3
ρ_w	1,000	kg/m^3
C_p	2	kJ/(kg K)
C_{pw}	4.2	kJ/(kg K)
M_m	100.12	kg/kmol
f^*	0.58	
R	8.314	kJ/(kmol K)
ΔH_r	57,800	kJ/kmol
E_p	1.8283e04	kJ/kmol
E_i	1.2877e05	kJ/kmol
E_{fm}	7.4478e04	kJ/kmol
E_{tc}	2.9442e03	kJ/kmol
E_{td}	2.9442e03	kJ/kmol
A_p	1.77e09	m^3/(kmol h)
A_i	3.792e18	h^{-1}
A_{fm}	1.0067e15	m^3/(kmol h)
A_{tc}	3.8223e10	m^3/(kmol h)
A_{td}	3.1457e11	m^3/(kmol h)

As shown in Figure 7.4, in this case study, we will draw optimal grade transition trajectories between two grades using the monomer, initiator, and water flowrates as manipulated or decision variables. The output target variable is the polymer molecular weight distribution. An initial MMA polymer-grade product featuring 60,000 units as polymer molecular weight value is manufactured under the following operations conditions: monomer flowrate $(Q) = 1.714780$, initiator flowrate $(Q_i) = 0.000318$, and cooling water flowrate $(Q_w) = 0.219593$, all in [L/s]. In addition, the following bounds on the manipulated variables were set: $Q \in [0, 3]$, $Q_i \in [1 \times 10^{-5}, 1]$, and $Q_w \in [0, 1.5]$. The aim of the dynamic-grade transition operation is to take the reactor from this initial product grade and to optimally take it to the process operating point featuring a grade polymer value of 40,000 units. Moreover, a sampling rate of 1 min was deployed. The numerical integration of the polymerization reactor dynamic model was carried out using the LSODA numerical integrator, embedded within the SciPy package.

After some trials, we came to the conclusion that the following form of the target objective function (Ω) is suitable for performing the optimal-grade dynamic transition calculations:

$$\text{Min}(\Omega) = \lambda \left(\frac{M_w^t - M_w}{M_w^t} \right)^2 \qquad (7.25)$$

where M_w stands for polymer molecular weight distribution, λ is a weighting function, and the superscript t stands for target value. The above equation is just a simple convex metric for measuring the approximation error between the target and calculated values of the molecular weight distribution. The quadratic term helps to improve convergence behavior. We found that when setting $\lambda = 10$, optimal dynamic transition can be easily realized, and leads to improvement in the minimization of the above objective function.

In Figure 7.5, the dynamic behavior of the objective function (Ω) is displayed. As can observed, after an initial deviation from small values of the objective function, the Bayesian optimization approach swiftly finds the right values of the manipulated variables, leading to acceptable dynamic polymer-grade transition performance. Figure 7.6 depicts the dynamic behavior of the output-controlled variable and the dynamic behavior of the three input manipulated variables. It should be stressed that the results depicted in Figure 7.6 strongly depend on drawing swift process measurements of the polymer molecular weight distribution. Nowadays, fast online measurement of molecular weight distribution takes around 15 min, making infeasible the results shown in Figure 7.6. However, to cope with this issue, we assume that online polymer viscosity measurements can be deployed to infer the target output variable using a correlation between viscosity and molecular weight distribution. Usually, online polymer viscosity measurements are drawn in around one second. Finally, we would like to comment that although different initial values of the decision variables can be used in-principle, it makes sense to deploy as initial conditions the nominal value of the manipulated variables of the initial steady state.

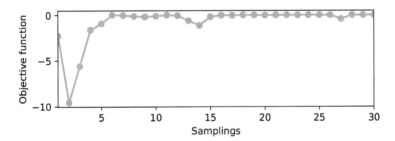

Figure 7.5: Objective function behavior as time function. After an initial deviation, the Bayesian optimization approach swiftly finds the optimal operating region and stays close to this region for the remaining iterations.

7.6 General remarks

–   While the Bayesian optimization framework is capable of addressing process uncertainties, it is important to note that this study has not delved into the analysis of the impact of uncertainty on the optimal results. Comprehensive details on how this analysis can be conducted can be found elsewhere [33, 34].

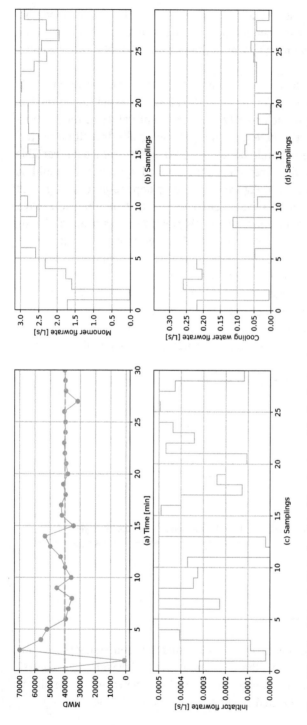

Figure 7.6: Optimal dynamical behavior of the (a) molecular weight distribution, (b) monomer, (c) initiator, and (d) cooling water flowrates.

- In this study, we have limited ourselves to predicting only a single output from the GP. It is essential to note that this restriction does not pose an inherent disadvantage to GP models. To address the prediction of multiple outputs, one can explore the use of multitask Gaussian models. However, it is worth mentioning that dealing with multitask Gaussian models is more complex. Currently, the efficient resolution of multitask Gaussian models stands as an active research topic, as highlighted in the literature [81, 82].
- As evident from the three case studies discussed in this work, our approach to initiating the search for optimal conditions involved using only a single measurement. The objective was to demonstrate the robustness of the Bayesian approach, emphasizing its ability to identify optimal operating conditions, even with just a single initial measurement. While in practical scenarios, multiple measurement points are typically available and can be employed to construct a more accurate surrogate model, there are situations, such as in the evaluation of expensive objective functions (i.e., quantum mechanical calculations [94–96]), where only a limited set of measurements may be available.
- This study does not delve into the examination of certain issues, including: (a) inequality constraints [97], (b) discrete-continuous systems [98, 99], (c) multiobjective optimization [40, 100, 101], and (d) parallel computer implementation of the optimization approach [32, 102–104]. While there have been reported advances in addressing some of these concerns, a universally reliable approach to handle general versions of these research problems is currently lacking. Additionally, an intriguing avenue for exploration involves extending the Bayesian optimization approach to quantum computing environments.

7.7 Conclusions

In summary, the utilization of Bayesian optimization in chemical engineering, specifically within process system engineering, presents a promising approach for the data-driven optimization of intricate systems. The findings presented in this study demonstrate the effectiveness of Bayesian optimization in achieving optimal operating conditions and dynamic product transitions without relying on complex mathematical models. Through the application of Bayesian optimization, engineers can efficiently navigate the design space, fine-tune decision variables, and enhance process performance while accounting for uncertainties and noisy measurements. The capability of Bayesian optimization to balance exploration and exploitation strategies facilitates swift convergence to optimal solutions, as illustrated in the discussed case studies. Overall, this work underscores the importance of integrating Bayesian optimization techniques into chemical engineering practices to streamline decision-making processes, enhance system efficiency, and foster innovation within the field of process system engineering.

References

[1] Nocedal, J., Wright, S. *Numerical Optimization*. Springer, 2006, pp. 664.

[2] Biegler, L. Nonlinear programming: concepts, algorithms. *Siam*, 2010, 416.

[3] Grossmann, I. E. *Advanced Optimization for Process Systems Engineering*. Cambridge University Press, 2021.

[4] Bernal, D. E., Ajagekar, A., Harwood, S. M., Stober, S. T., Trenev, D., You, F. Perspectives of quantum computing for chemical engineering. *AIChE Journal*, 2022, 68, e17651. https://doi.org/10.1002/aic.17651.

[5] Goldberg, D. E. *Genetic Algorithms in Search, Optimization, and Machine Learning*. Addison-Wesley, 1989.

[6] Gutiérrez-Antonio, C., Briones-Ramírez, A., Jiménez-Gutiérrez, A. Optimization of Petlyuk sequences using a multi objective genetic algorithm with constraints. *Computers and Chemical Engineering*, 2011, 35, 236–244. https://doi.org/10.1016/j.compchemeng.2010.10.007.

[7] Wang, Y., Bu, G., Wang, Y., Zhao, T., Zhang, Z., Zhu, Z. Application of a simulated annealing algorithm to design and optimize a pressure-swing distillation process. *Computers and Chemical Engineering*, 2016, 95, 97–107. https://doi.org/10.1016/j.compchemeng.2016.09.014.

[8] Gebreslassie, B. H., Diwekar, U. M. Efficient ant colony optimization for computer aided molecular design: Case study solvent selection problem. *Computers and Chemical Engineering*, 2015, 78, 1–9. https://doi.org/10.1016/j.compchemeng.2015.04.004.

[9] Montain, M. E., Blanco, A. M., Bandoni, J. A. Optimal drug infusion profiles using a particle swarm optimization algorithm. *Computers and Chemical Engineering*, 2015, 82, 13–24. https://doi.org/10.1016/j.compchemeng.2015.05.026.

[10] Garnett, R. *Bayesian Optimization*. Cambridge University Press, 2023.

[11] Liu, P. *Bayesian Optimization: Theory and Practice Using Python*. Apress, 2023.

[12] Negrellos-Ortiz, I., Flores-Tlacuahuac, A., Gutiérrez-Limón, M. Product dynamic transitions using a derivative-free optimization trust-region approach. *Industrial & Engineering Chemistry Research*, 2016, 55, 8586–8601. https://doi.org/10.1021/acs.iecr.6b00268.

[13] Liu, Y., Tang, L., Liu, C., Su, L., Wu, J. Black box operation optimization of basic oxygen furnace steelmaking process with derivative free optimization algorithm. *Computers and Chemical Engineering*, 2021, 150, 107311. https://doi.org/10.1016/j.compchemeng.2021.107311.

[14] Ma, K., Sahinidis, N. V., Bindlish, R., Bury, S. J., Haghpanah, R., Rajagopalan, S. Data-driven strategies for extractive distillation unit optimization. *Computers and Chemical Engineering*, 2022, 167, 107970. https://doi.org/10.1016/j.compchemeng.2022.107970.

[15] Van de Berg, D., Savage, T., Petsagkourakis, P., Zhang, D., Shah, N., Del Rio-chanona, E. A. Data-driven optimization for process systems engineering applications. *Chemical Engineering Science*, 2022, 248, 117135. https://doi.org/10.1016/j.ces.2021.117135.

[16] Paulson, J. A., Tsay, C. Bayesian optimization as a flexible and efficient design framework for sustainable process systems. *arXiv*, preprint, 2024.

[17] Mockus, J., Eddy, W., Mockus, A., Mockus, L., Reklaitis, G. *Bayesian Heuristic Approach to Discrete and Global Optimization: Algorithms, Visualization, Software, and Applications*. Springer, 1997.

[18] Carl Edward Rasmussen, C. K. I. W. *Gaussian Processes for Machine Learning*. The MIT Press, 2006.

[19] Hennign, P., Osborne, M. A., Kersting, H. *Probabilistic Numerics*. Cambridge University Press, 2022.

[20] Francesco Archetti, A. C. *Bayesian Optimization and Data Science*. Springer, 2019.

[21] Packwood, D. *Bayesian Optimization for Material Science*. Springer, 2017.

[22] Wang, K., Dowling, A. W. Bayesian optimization for chemical products and functional materials. *Current Opinion in Chemical Engineering*, 2022, 36(100728). https://doi.org/10.1016/j.coche.2021.100728.

[23] Nguyen, Q. *Bayesian Optimization in Action*. Manning Publications, 2023.

[24] Nguyen, Q. *Experimentation for Engineers*. Manning Publications, 2023.

[25] Urm, J. J., Choi, J. H., Kim, C., Lee, J. M. Techno-economic analysis and process optimization of a PET chemical recycling process based on Bayesian optimization. *Computers and Chemical Engineering*, 2023, 179, 108451. https://doi.org/10.1016/j.compchemeng.2023.108451.

[26] Begall, M. J., Schweidtmann, A. M., Mhamdi, A., Mitsos, A. Geometry optimization of a continuous millireactor via CFD and Bayesian optimization. *Computers and Chemical Engineering*, 2023, 171, 108140. https://doi.org/10.1016/j.compchemeng.2023.108140.

[27] Coutinho, J. P., Santos, L. O., Reis, M. S. Bayesian Optimization for automatic tuning of digital multi-loop PID controllers. *Computers and Chemical Engineering*, 2023, 173, 108211. https://doi.org/10.1016/j.compchemeng.2023.108211.

[28] Paulson, J. A., Makrygiorgos, G., Mesbah, A. Adversarially robust Bayesian optimization for efficient auto-tuning of generic control structures under uncertainty. *AIChE Journal*, 2022, 68, e17591. https://doi.org/10.1002/aic.17591.

[29] Sorourifar, F., Makrygirgos, G., Mesbah, A., Paulson, J. A. A data-driven automatic tuning method for MPC under uncertainty using constrained Bayesian optimization. *IFAC-PapersOnLine*, 2021, 54, 243–250. https://doi.org/10.1016/j.ifacol.2021.08.249, 16th IFAC Symposium on Advanced Control of Chemical Processes ADCHEM 2021.

[30] Lu, Q., González, L. D., Kumar, R., Zavala, V. M. Bayesian optimization with reference models: A case study in MPC for HVAC central plants. *Computers and Chemical Engineering*, 2021, 154, 107491. https://doi.org/10.1016/j.compchemeng.2021.107491.

[31] Park, S., Na, J., Kim, M., Lee, J. M. Multi-objective Bayesian optimization of chemical reactor design using computational fluid dynamics. *Computers and Chemical Engineering*, 2018, 119, 25–37. https://doi.org/10.1016/j.compchemeng.2018.08.005.

[32] González, L. D., Zavala, V. M. New paradigms for exploiting parallel experiments in Bayesian optimization. *Computers and Chemical Engineering*, 2023, 170(108110). https://doi.org/10.1016/j.compchemeng.2022.108110.

[33] Pérez-Ones, O., Flores-Tlacuahuac, A. A stochastic data-driven Bayesian optimization approach for intensified ethanol–water separation systems. *Chemical Engineering and Processing - Process Intensification*, 2024, 197, 109708. https://doi.org/10.1016/j.cep.2024.109708.

[34] Flores-Tlacuahuac, A., Fuentes-Cortés, L. F. A Data-Driven Bayesian Approach for Optimal Dynamic Product Transitions. *American Institute of Chemical Engineers Journal 2024*, Accepted for publication.

[35] Eriksson, D., Pearce, M., Gardner, J. R., Turner, R., Poloczek, M. *Scalable Global Optimization via Local Bayesian Optimization*. 2019. https://doi.org/10.48550/arXiv.1910.01739.

[36] Kandasamy, K., Schneider, J., Poczos, B. High Dimensional Bayesian Optimisation and Bandits via Additive Models. 2015; https://doi.org/10.48550/arXiv.1503.01673.

[37] Wu, J., Poloczek, M., Wilson, A. G., Frazier, P. I. Bayesian Optimization with Gradients. 2017; https://doi.org/10.48550/arXiv.1703.04389.

[38] Bull, A. D. Convergence rates of efficient global optimization algorithms. 2011; https://doi.org/10.48550/arXiv.1101.3501.

[39] Spinti, J. P., Smith, P. J., Smith, S. T., Díaz-Ibarra, O. H. Atikokan Digital Twin, Part B: Bayesian decision theory for process optimization in a biomass energy system. *Applied Energy*, 2023, 334, 120625. https://doi.org/10.1016/j.apenergy.2022.120625.

[40] Wang, X., Jiang, B. Multi-objective optimization for fast charging design of lithium-ion batteries using constrained Bayesian optimization. *Journal of Power Sources*, 2023, 584, 233602. https://doi.org/10.1016/j.jpowsour.2023.233602.

[41] Luo, Y., Wang, Z., Srinivasan, P., Vlachos, D. G., Ierapetritou, M. Process Design and Bayesian Optimization of 5-Hydroxymethylfurfural Hydrodeoxygenation. In Kokossis, A. C., Georgiadis, M. C., Pistikopoulos, E. (Eds.). *33rd European Symposium on Computer Aided Process Engineering*. Computer

Aided Chemical Engineering; Elsevier, 2023. Vol. 52, pp. 2185–2191. https://doi.org/10.1016/B978-0-443-15274-0.50348-6.

[42] Eugene, E. A., Jones, K. D., Gao, X., Wang, J., Dowling, A. W. Learning and optimization under epistemic uncertainty with Bayesian hybrid models. *Computers and Chemical Engineering*, 2023, 179, 108430. https://doi.org/10.1016/j.compchemeng.2023.108430.

[43] Fezai, R., Malluhi, B., Basha, N., Ibrahim, G., Choudhury, H. A., Challiwala, M. S., Nounou, H., Elbashir, N., Nounou, M. Bayesian optimization of multiscale kernel principal component analysis and its application to model Gas-to-liquid (GTL) process data. *Energy*, 2023, 284, 129221. https://doi.org/10.1016/j.energy.2023.129221.

[44] Basha, N., Ibrahim, G., Choudhury, H. A., Challiwala, M. S., Fezai, R., Malluhi, B., Nounou, H., Elbashir, N., Nounou, M. Bayesian-optimized Neural Networks and their application to model gas-to-liquid plants. *Gas Science and Engineering*, 2023, 113, 204964. https://doi.org/10.1016/j.jgsce.2023.204964.

[45] Wang, A., Ye, H., Yang, Y., Dong, H. Bayesian optimization of HDPE copolymerization process based on polymer product-process integration. *Polymer*, 2024, 292, 126554. https://doi.org/10.1016/j.polymer.2023.126554.

[46] Ramirez, A., Lam, E., Gutierrez, D. P., Hou, Y., Tribukait, H., Roch, L. M., Copéret, C., Laveille, P. Accelerated exploration of heterogeneous CO2 hydrogenation catalysts by Bayesian-optimized high-throughput and automated experimentation. *Chem Catalysis*, 2024, 4, 100888. https://doi.org/10.1016/j.checat.2023.100888.

[47] Peng, J., Damewood, J. K., Karaguesian, J., Gómez-Bombarelli, R., Shao-Horn, Y. Navigating multimetallic catalyst space with Bayesian optimization. *Joule*, 2021, 5, 3069–3071. https://doi.org/10.1016/j.joule.2021.11.011.

[48] Lim, S., Lee, H., Bae, S., Shin, J. S., Kim, D. H., Lee, J. M. Bayesian Optimization for Automobile Catalyst Development. In Yamashita, Y., Kano, M. (Eds.). *14th International Symposium on Process Systems Engineering*. Computer Aided Chemical Engineering Elsevier, 2022, Vol. 49, pp. 1213–1218. https://doi.org/10.1016/B978-0-323-85159-6.50202-5.

[49] Cherif, A., Atwair, M., Atsbha, T. A., Zarei, M., Duncan, I. J., Nebbali, R., Sen, F., Lee, C. J. Enabling low-carbon membrane steam methane reforming: Comparative analysis and multi-objective NSGA-II-integrated Bayesian optimization. *Energy Convers Management*, 2023, 297, 117718. https://doi.org/10.1016/j.enconman.2023.117718.

[50] Miftahurrohmah, B., Iriawan, N., Wulandari, C., Dharmawan, Y. S. Individual Control Optimization of Drug Dosage Using Individual Bayesian Pharmacokinetics Model Approach. In *Procedia Computer Science*. Vol. 161, pp. 593–600. https://doi.org/10.1016/j.procs.2019.11.161The Fifth Information Systems International Conference, 23–24 July 2019 Surabaya, Indonesia 2019

[51] Fernandez, M., Caballero, J. Chapter 4 – Genetic Algorithm Optimization of Bayesian-Regularized Artificial Neural Networks in Drug Design. In Puri, M., Pathak, Y., Sutariya, V. K., Tipparaju, S., Moreno, W. (Eds.). *Artificial Neural Network for Drug Design, Delivery and Disposition*. . Boston: Academic Press, 2016, pp. 83–102. https://doi.org/10.1016/B978-0-12-801559-9.00004-1.

[52] Hickman, R. J., Aldeghi, M., Häse, F., Aspuru-Guzik, A. Bayesian optimization with known experimental and design constraints for chemistry applications††Electronic supplementary information (ESI) available. *Digital Discovery*, 2022, 1, 732–744. https://doi.org/10.1039/d2dd00028h.

[53] Yoshida, K., Watanabe, K., Chiou, T.-Y., Konishi, M. High throughput optimization of medium composition for Escherichia coli protein expression using deep learning and Bayesian optimization. *Journal of Bioscience & Bioengineering*, 2023, 135, 127–133. https://doi.org/10.1016/j.jbiosc.2022.12.004.

[54] Cheng, Y., Bi, X., Xu, Y., Liu, Y., Li, J., Du, G., Lv, X., Liu, L. Machine learning for metabolic pathway optimization: A review. *Computational and Structural Biotechnology Journal*, 2023, 21, 2381–2393. https://doi.org/10.1016/j.csbj.2023.03.045.

[55]   Liu, C., Ji, C., Han, C., Gu, C., Dai, J., Sun, W., Wang, J. Application of Bayesian Optimization in HME
       Batch Concentration Process. In Kokossis, A. C., Georgiadis, M. C., Pistikopoulos, E. (Eds.). *33rd
       European Symposium on Computer Aided Process Engineering*. Computer Aided Chemical Engineering
       Elsevier, 2023, Vol. 52, pp. 1551–1557. https://doi.org/10.1016/B978-0-443-15274-0.50247-X.
[56]   Kaya, E. Y., Ali, I., Ceylan, Z., Ceylan, S. Prediction of higher heating value of hydrochars using
       Bayesian optimization tuned Gaussian process regression based on biomass characteristics and
       process conditions. *Biomass and Bioenergy*, 2024, 180, 106993. https://doi.org/10.1016/j.biombioe.
       2023.106993.
[57]   Signori-Iamin, G., Santos, A. F., Mazega, A., Corazza, M. L., Aguado, R. J., Delgado-Aguilar,
       M. Bayesian-optimized random forest prediction of key properties of micro-/nanofibrillated
       cellulose from different woody and non-woody feedstocks. *Industrial Crops and Products*, 2023, 206,
       117719. https://doi.org/10.1016/j.indcrop.2023.117719.
[58]   Alruqi, M., Sharma, P., Agbulut, Ü. Investigations on biomass gasification derived producer gas and
       algal biodiesel to power a dual-fuel engines: Application of neural networks optimized with
       Bayesian approach and K-cross fold. *Energy*, 2023, 282, 128336. https://doi.org/10.1016/j.energy.
       2023.128336.
[59]   Honarmandi, P., Attari, V., Arroyave, R. Accelerated materials design using batch Bayesian
       optimization: A case study for solving the inverse problem from materials microstructure to process
       specification. *Computational Materials Science*, 2022, 210, 111417. https://doi.org/10.1016/j.commatsci.
       2022.111417.
[60]   Khatamsaz, D., Vela, B., Singh, P., Johnson, D. D., Allaire, D., Arróyave, R. Multiobjective materials
       bayesian optimization with active learning of design constraints: Design of ductile refractory multi-
       principal-element alloys. *Acta Materialia*, 2022, 236, 118133. https://doi.org/10.1016/j.actamat.2022.
       118133.
[61]   Hanaoka, K. Comparison of conceptually different multi-objective Bayesian optimization methods
       for material design problems. *Materials Today Communications*, 2022, 31, 103440. https://doi.org/10.
       1016/j.mtcomm.2022.103440.
[62]   Hanaoka, K. Bayesian optimization for goal-oriented multi-objective inverse material design.
       iScience, 2021, 24, 102781. https://doi.org/10.1016/j.isci.2021.102781.
[63]   Thelen, A., Zohair, M., Ramamurthy, J., Harkaway, A., Jiao, W., Ojha, M., Ishtiaque, M. U., Kingston,
       T. A., Pint, C. L., Hu, C. Sequential Bayesian optimization for accelerating the design of sodium metal
       battery nucleation layers. *Journal of Power Sources*, 2023, 581, 233508. https://doi.org/10.1016/j.jpows
       our.2023.233508.
[64]   Ishii, A., Kikuchi, S., Yamanaka, A., Yamamoto, A. Application of Bayesian optimization to the
       synthesis process of BaFe2(As,P)2 polycrystalline bulk superconducting materials. *Journal of Alloys
       and Compounds*, 2023, 966, 171613. https://doi.org/10.1016/j.jallcom.2023.171613.
[65]   Folch, J. P., Lee, R. M., Shafei, B., Walz, D., Tsay, C., Van der Wilk, M., Misener, R. Combining multi-
       fidelity modelling and asynchronous batch Bayesian Optimization. *Computers and Chemical
       Engineering*, 2023, 172, 108194. https://doi.org/10.1016/j.compchemeng.2023.108194.
[66]   Yu, C., Li, Y., Liu, Y., Ge, L., Wang, H., Luo, Y., Pan, L. Dispatch of highly renewable energy power
       system considering its utilization via a data-driven Bayesian assisted optimization algorithm.
       Knowledge-Based Systems, 2023, 281, 111059. https://doi.org/10.1016/j.knosys.2023.111059.
[67]   Dong, C., Huang, G., Cai, Y., Cheng, G., Tan, Q. Bayesian interval robust optimization for sustainable
       energy system planning in Qiqihar City, China. *Energy Economics*, 2016, 60, 357–376. https://doi.org/
       10.1016/j.eneco.2016.10.012.
[68]   Chen, D., Wu, W., Chang, K., Li, Y., Pei, P., Xu, X. Performance degradation prediction method of
       PEM fuel cells using bidirectional long short-term memory neural network based on Bayesian
       optimization. *Energy*, 2023, 285, 129469. https://doi.org/10.1016/j.energy.2023.129469.

[69]   Eladly, A. M., Abed, A. M., Aly, M. H., Salama, W. M. Enhancing circular economy via detecting and recycling 2D nested sheet waste using Bayesian optimization technique based-smart digital twin. *Results in Engineering*, 2023, 20, 101544. https://doi.org/10.1016/j.rineng.2023.101544.

[70]   Liu, M., Wen, Z., Zhou, R., Su, H. Bayesian optimization and ensemble learning algorithm combined method for deformation prediction of concrete dam. *Structures*, 2023, 54, 981–993. https://doi.org/10.1016/j.istruc.2023.05.136.

[71]   Chien, C.-F., Hong Van Nguyen, T., Li, Y.-C., Chen, Y.-J. Bayesian decision analysis for optimizing in-line metrology and defect inspection strategy for sustainable semiconductor manufacturing and an empirical study. *Computers & Industrial Engineering*, 2023, 182, 109421. https://doi.org/10.1016/j.cie.2023.109421.

[72]   Joy, T. T., Rana, S., Gupta, S., Venkatesh, S. Fast hyperparameter tuning using Bayesian optimization with directional derivatives. *Knowledge-Based Systems*, 2020, 205, 106247. https://doi.org/10.1016/j.knosys.2020.106247.

[73]   Wojciuk, M., Swiderska-Chadaj, Z., Siwek, K., Gertych, A. Improving classification accuracy of fine-tuned CNN models: Impact of hyperparameter optimization. *Heliyon*, 2024, e26586. https://doi.org/10.1016/j.heliyon.2024.e26586.

[74]   Sadoune, H., Rihani, R., Marra, F. S. DNN model development of biogas production from an anaerobic wastewater treatment plant using Bayesian hyperparameter optimization. *Chemical Engineering Journal*, 2023, 471, 144671. https://doi.org/10.1016/j.cej.2023.144671.

[75]   Baheri, A., Bin-Karim, S., Bafandeh, A., Vermillion, C. Real-time control using Bayesian optimization: A case study in airborne wind energy systems. *Control Engineering Practice*, 2017, 69, 131–140. https://doi.org/10.1016/j.conengprac.2017.09.007.

[76]   Cai, D., Liu, W., Ji, L., Shi, D. Bayesian optimization assisted meal bolus decision based on Gaussian processes learning and risk-sensitive control. *Control Engineering Practice*, 2021, 114, 104881. https://doi.org/10.1016/j.conengprac.2021.104881.

[77]   Yamakage, S., Kaneko, H. Design of adaptive soft sensor based on Bayesian optimization. *Case Studies in Chemical and Environmental Engineering*, 2022, 6, 100237. https://doi.org/10.1016/j.cscee.2022.100237.

[78]   Dao, F., Zeng, Y., Qian, J. Fault diagnosis of hydro-turbine via the incorporation of bayesian algorithm optimized CNN-LSTM neural network. *Energy*, 2024, 290, 130326. https://doi.org/10.1016/j.energy.2024.130326.

[79]   Bansal, S., Cheung, S. H. On the Bayesian sensor placement for two-stage structural model updating and its validation. *Mechanical Systems & Signal Processing*, 2022, 169, 108578. https://doi.org/10.1016/j.ymssp.2021.108578.

[80]   Lin, X., Chowdhury, A., Wang, X., Terejanu, G. Approximate computational approaches for Bayesian sensor placement in high dimensions. *Information Fusion*, 2019, 46, 193–205. https://doi.org/10.1016/j.inffus.2018.06.006.

[81]   Makrygiorgos, G., Bonzanini, A. D., Miller, V., Mesbah, A. Performance-oriented model learning for control via multi-objective Bayesian optimization. *Computers and Chemical Engineering*, 2022, 162, 107770. https://doi.org/10.1016/j.compchemeng.2022.107770.

[82]   Basha, N., Kravaris, C., Nounou, H., Nounou, M. Bayesian-optimized Gaussian process-based fault classification in industrial processes. *Computers and Chemical Engineering*, 2023, 170, 108126. https://doi.org/10.1016/j.compchemeng.2022.108126.

[83]   Alstad, V., Halvorsen, I. J., Skogestad, S. Optimal Operation of a Petlyuk Distillation Column: Energy Savings by Over-fractionating. In Barbosa- Póvoa, A., Matos, H. (Eds.). *European Symposium on Computer-Aided Process Engineering-14*. Computer Aided Chemical Engineering Elsevier, 2004, Vol. 18, pp. 547–552. https://doi.org/10.1016/S1570-7946(04)80157-3.

[84]   Segovia-Hernández, J. G., Hernández, S., Femat, R., Jiménez, A. Dynamic Control of a Petlyuk Column via Proportional-integral Action with Dynamic Estimation of Uncertainties. In Kraslawski, A.,

Turunen, I. (Eds.). *European Symposium on Computer Aided Process Engineering-13*. Computer Aided Chemical Engineering. Elsevier, 2003, Vol. 14, pp. 515–520. https://doi.org/10.1016/S1570-7946(03) 80167-0.

[85] Dwivedi, D., Halvorsen, I. J., Skogestad, S. Control structure selection for three-product Petlyuk (dividing wall) column. *Chem Eng Process Process Intensifi*, 2013, 64, 57–67. https://doi.org/10.1016/j. cep.2012.11.006.

[86] Carranza-Abaíd, A., González-García, R. A Petlyuk distillation column dynamic analysis: Hysteresis and bifurcations. *Chemical Engineering and Processing - Process Intensification*, 2020, 149, 107843. https://doi.org/10.1016/j.cep.2020.107843.

[87] Zumoffen, D., Molina, G., Nieto, L., Basualdo, M. Systematic Control Approach for the Petlyuk Distillation Column. *IFAC Proceedings Volumes*, 2011, 44, 8552–8557. https://doi.org/10.3182/ 20110828-6-IT-1002.02220,. 18th IFAC World Congress.

[88] Hernandez, S., Jimenez, A. Design of energy-efficient Petlyuk systems. *Computers and Chemical Engineering*, 1999, 23, 1005–1010.

[89] Ramírez-Corona, N., Jiménez-Gutiérrez, A., Castro-Aguero, A., Rico-Ramírez, V. Optimum design of Petlyuk and divided-wall distillation systems using a shortcut model. *Chemical Engineering Research and Design*, 2010, 88, 1405–1418.

[90] Flores-Tlacuahuac, A., Saldívar-Guerra, E., Ramírez-Manzanares, G. Grade transition dynamic simulation of HIPS polymerization reactors. *Computers and Chemical Engineering*, 2005, 30, 357–375. https://doi.org/10.1016/j.compchemeng.2005.10.002.

[91] Wang, Y., Biegler, L. T., Ostace, G. S., Majewski, R. A. Optimal polymer grade transitions for fluidized bed reactors. *Journal of Process Control*, 2020, 88, 86–100. https://doi.org/10.1016/j.jprocont.2020. 02.001.

[92] Shi, J., Biegler, L. T., Hamdan, I., Wassick, J. Optimization of grade transitions in polyethylene solution polymerization process under uncertainty. *Computers and Chemical Engineering*, 2016, 95, 260–279. https://doi.org/10.1016/j.compchemeng.2016.08.002.

[93] Flores-Tlacuahuac, A., Biegler, L. T. Integrated control and process design during optimal polymer grade transition operations. *Computers and Chemical Engineering*, 2008, 32, 2823–2837.

[94] Kou, H., Zhang, Y., Pueh Lee, H. Dynamic optimization based on quantum computation-A comprehensive review. *Computers & Structures*, 2024, 292, 107255. https://doi.org/10.1016/j.comp struc.2023.107255.

[95] Hong, -Y.-Y., Rioflorido, C. L. P. P., Zhang, W. Hybrid deep learning and quantum-inspired neural network for day-ahead spatiotemporal wind speed forecasting. *Expert Systems with Applications*, 2024, 241, 122645. https://doi.org/10.1016/j.eswa.2023.122645.

[96] Borujeni, S. E., Nannapaneni, S., Nguyen, N. H., Behrman, E. C., Steck, J. E. Quantum circuit representation of Bayesian networks. *Expert Systems With Applications*, 2021, 176, 114768. https://doi. org/10.1016/j.eswa.2021.114768.

[97] Paulson, J. A., Lu, C. COBALT: COnstrained Bayesian optimizAtion of computationaLly expensive greybox models exploiting derivaTive information. *Computers and Chemical Engineering*, 2022, 160, 107700. https://doi.org/10.1016/j.compchemeng.2022.107700.

[98] Morlet-Espinosa, J., Flores-Tlacuahuac, A. A Bayesian optimization approach for data-driven mixed-integer non-linear programming problems. *American Institute of Chemical Engineers Journal 2024*, In revision.

[99] Garrido-Merchán, E. C., Hernández-Lobato, D. Dealing with categorical and integer-valued variables in Bayesian Optimization with Gaussian processes. *Neurocomputing*, 2020, 380, 20–35. https://doi. org/10.1016/j.neucom.2019.11.004.

[100] Hao, Z., Caspari, A., Schweidtmann, A. M., Vaupel, Y., Lapkin, A. A., Mhamdi, A. Efficient hybrid multiobjective optimization of pressure swing adsorption. *Chemical Engineering Journal*, 2021, 423, 130248. https://doi.org/10.1016/j.cej.2021.130248.

[101] Motoyama, Y., Tamura, R., Yoshimi, K., Terayama, K., Ueno, T., Tsuda, K. Bayesian optimization package: PHYSBO. *Computer Physics Communications*, 2022, 278, 108405. https://doi.org/10.1016/j. cpc.2022.108405.

[102] Tran, A., Sun, J., Furlan, J. M., Pagalthivarthi, K. V., Visintainer, R. J., Wang, Y. pBO-2GP-3B: A batch parallel known/unknown constrained Bayesian optimization with feasibility classification and its applications in computational fluid dynamics. *Computer Methods in Applied Mechanics and Engineering*, 2019, 347, 827–852. https://doi.org/10.1016/j.cma.2018.12.033.

[103] Kitahara, M., Dang, C., Beer, M. Bayesian updating with two-step parallel Bayesian optimization and quadrature. *Computer Methods in Applied Mechanics and Engineering*, 2023, 403, 115735. https://doi. org/10.1016/j.cma.2022.115735.

[104] Chandra, R., Tiwari, A. Distributed Bayesian optimisation framework for deep neuroevolution. *Neurocomputing*, 2022, 470, 51–65. https://doi.org/10.1016/j.neucom.2021.10.045.

Seyed Reza Nabavi, Zhiyuan Wang, and Gade Pandu Rangaiah*

Chapter 8
Sensitivity assessment of multi-criteria decision-making methods in chemical engineering optimization applications

Abstract: This chapter assesses the sensitivity of multi-criteria decision-making (MCDM) methods to modifications within the decision or objective matrix (DOM) in the context of chemical engineering optimization applications. Employing eight common or recent MCDM methods and three weighting methods, this study evaluates the impact of three specific DOM alterations: linear transformation of an objective (LTO), reciprocal objective reformulation (ROR), and the removal of alternatives (RAs). Comprehensive analysis of results obtained for six applications reveals that the weights generated by the entropy method are more sensitive to the examined modifications compared to the criteria importance through intercriteria correlation (CRITIC) and standard deviation (StDev) methods. Compared to LTO and RA, ROR is found to have the largest effect on the ranking of alternatives. Moreover, certain MCDM methods, namely, gray relational analysis (GRA) without any weights, multi-attributive border approximation area comparison (MABAC), combinative distance-based assessment (CODAS), and simple additive weighting (SAW) with entropy or CRITIC weights, and CODAS, SAW, and technique for order of preference by similarity to ideal solution (TOPSIS) with StDev weight are more robust to DOM modifications. This investigation not only corroborates the findings from our recent study, but also offers insights into the stability and reliability of MCDM methods in the context of chemical engineering applications. However, to generalize these findings, further studies are required on other MCDM methods and chemical engineering applications.

Keywords: multi-criteria decision-making, multi-objective optimization, chemical engineering

*Corresponding author: Gade Pandu Rangaiah**, School of Chemical Engineering, Vellore Institute of Technology, Vellore 632014, Tamil Nadu, India; Department of Chemical and Biomolecular Engineering, National University of Singapore, Singapore 117585, Singapore, e-mail: chegpr@nus.edu.sg
Seyed Reza Nabavi, Department of Applied Chemistry, Faculty of Chemistry, University of Mazandaran, Babolsar, Iran
Zhiyuan Wang, Department of Computer Science, DigiPen Institute of Technology Singapore, Singapore 139660, Singapore

https://doi.org/10.1515/9783111383439-008

8.1 Introduction

Multi-criteria decision-making (MCDM) is a powerful tool to assist decision-makers navigate complex scenarios, where they must select one of the alternatives available, considering multiple criteria simultaneously. These criteria often conflict with one another, adding to the decision-making challenge. In recent years, MCDM has garnered considerable interest across a diverse range of fields, such as logistics [1, 2], healthcare [3–5], finance [6, 7], sustainable agriculture [8, 9], product design [10, 11], artificial intelligence [12, 13], and chemical engineering [14–17].

As depicted in Figure 8.1, MCDM is oftentimes viewed as the step following multi-objective optimization (MOO) [18]. In applications involving multiple conflicting objectives, MOO produces a set of Pareto-optimal solutions (also referred to as non-dominated solutions). However, implementing all these solutions is impractical in real-world applications, thus necessitating the selection of only one solution using MCDM. In the context of MCDM and throughout this chapter, the term "criteria" is used interchangeably with "objectives," and "alternatives" with "solutions."

Previous works by Pamučar and Ćirović [19], Mufazzal and Muzakkir [20], Wang et al. [21], Shih [22], and Dehshiri and Firoozabadi [23] showcased that MCDM methods struggle with maintaining consistency with respect to linearly transformed changes of measurement unit (which commonly occurs in chemical engineering), equivalent objective formulation (e.g., in a thermal cracking process, maximization of ethylene selectivity can also be expressed as the minimization of 1/selectivity or 1 – selectivity) and/or addition/removal of alternatives (that could happen in case MOO did not find all Pareto-optimal solutions in the first run, and subsequently, additional solutions were added in later runs). The sensitivity due to addition or removal of some alternatives is also known as rank-reversal phenomenon in MCDM literature.

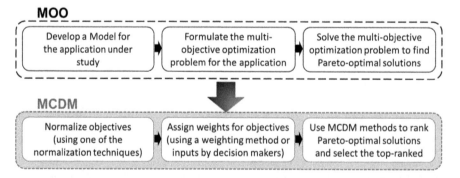

Figure 8.1: Main steps of MOO and MCDM.

This chapter is a sequel to our recent study [24], wherein we examined the sensitivity of MCDM methods to various perturbations within the decision or objective matrix (DOM). Applications covered in that study are from engineering in general. Here, we focus on the sensitivity assessment of MCDM methods for chemical engineering optimization applications. Findings of the present study will help us to corroborate and consolidate those in Nabavi et al. [24]. To this end, eight common or recent MCDM methods are selected to evaluate across six chemical engineering applications. The selected MCDM methods, in alphabetical sequence, are the combinative distance-based assessment (CODAS), complex proportional assessment (COPRAS), gray/grey relational analysis (GRA), multi-attributive border approximation area comparison (MABAC), multiobjective optimization on the basis of ratio analysis (MOORA), preference ranking on the basis of ideal-average distance (PROBID), simple additive weighting (SAW) and technique for order of preference by similarity to ideal solution (TOPSIS). To determine the weight of each objective within the DOM, three weighting techniques are employed in conjunction with the chosen MCDM methods. These are criteria importance through intercriteria correlation (CRITIC), entropy, and standard deviation (StDev). The sensitivity analysis incorporates three specific types of modifications to the DOM, that is, linear transformation of an objective (LTO), reciprocal objective reformulation (ROR), and removal of alternatives (RA).

The structure of the remaining sections of this chapter is as follows: Section 8.2 provides an overview of the selected MCDM methods, the weighting techniques, the modifications applied to DOM, and the metrics for sensitivity analysis. Section 8.3 details the chemical engineering optimization applications considered in this study. Section 8.4 presents the results and their discussion; it includes limitations of the present study and possible future studies. Finally, Section 8.5 concludes the chapter, summarizing the key findings and suggesting avenues for future research.

8.2 Methodologies

8.2.1 MCDM methods

The chosen eight MCDM methods are outlined in this section. These are chosen for the following reasons: SAW for simplicity of its principle and popularity, proposed two to four decades ago and have been popular (namely, GRA, TOPSIS, COPRAS, and MOORA) or recent methods (namely, CODAS, MABAC, and PROBID). Further, these methods have been used in recent studies [25, 26]. These are outlined below in alphabetical sequence.

CODAS method, developed by Ghorabaee et al. [27], finds the performance score of each alternative by calculating both the Euclidean and Taxicab distances from the negative ideal solution. The performance score of an alternative increases with its distance from the negative ideal solution, with Euclidean distance serving as the principal metric. In instances where the performance scores of two alternatives are remarkably similar, the Taxicab distance is utilized as the secondary metric to distinguish between them. For a detailed explanation of the CODAS method, including its procedural steps and mathematical equations, refer Ghorabaee et al. [27].

COPRAS method, conceived by Zavadskas and Kaklauskas [28], employs a systematic approach to rank alternatives based on their relative importance. In this, the first step is normalizing the original DOM using sum normalization. It is followed by multiplying each objective value by its corresponding weight to produce a weighted normalized DOM. Subsequently, the method calculates the sums of these weighted normalized values for objectives under maximization and minimization categories separately. The two sums are then utilized to ascertain the relative importance of each alternative according to specific equations of the method. See Zavadskas and Kaklauskas [28] for details on the COPRAS method, including its steps and mathematical equations.

GRA method, grounded in the gray/grey system theory [29], has several variants documented in the MCDM literature. The specific variant we reference, as outlined by Martinez-Morales et al. [30], is notable for its independence from user inputs (e.g., weight for each objective). This particular GRA has three steps: the DOM normalization using max-min technique, the definition of an ideal alternative, and the calculation of the gray relational coefficient (GRC), which effectively measures the similarity between each alternative and the ideal alternative. The alternative having the highest GRC is recommended. For details on this variant of GRA, including its steps and mathematical equations, see Martinez-Morales et al. [30].

MABAC method, introduced by Pamučar and Ćirović [19], centers on calculating the distance of each alternative from the border approximation area, which is derived from the product of the weighted normalized values of each objective. Like the other MCDM methods, it begins with the construction of a weighted normalized DOM from the original DOM. This is followed by the determination of the border approximation area for each objective. Finally, the distances between each alternative and the border approximation areas are computed. The alternative furthest from these areas is identified as the leading optimal solution. For an in-depth understanding of the MABAC method, including its steps and mathematical formulations, refer to the work of Pamučar and Ćirović [19].

MOORA method, proposed by Brauers and Zavadskas [31], is extensively applied across various fields, including chemical engineering, to address MCDM problems. It begins with constructing a weighted normalized DOM from the original DOM. Then, the performance score for each alternative is calculated by deducting the aggregate of objectives to be minimized (i.e., cost objectives) from the sum of objectives to be maximized (i.e., benefit objectives). The alternative having the highest performance score is the recommended solution. For a comprehensive understanding of the MOORA method, including its procedure and mathematical equations, refer Brauers and Zavadskas [31].

PROBID method, proposed by Wang et al. [21], distinguishes itself by assessing alternatives against a spectrum of ideal solutions, alongside consideration of the mean solution. This spectrum includes the most positive ideal solution followed by subsequent tiers (second-most positive ideal solution, third-most positive ideal solution, etc.), till the most negative ideal solution. The PROBID method involves calculating the distances between each alternative and these ideal solutions, and then integrating these distances with the distance from the mean solution to generate the overall performance score. For understanding this method, including its steps and mathematical equations, see Wang et al. [21].

SAW method, pioneered by Fishburn [32] and MacCrimmon [33], stands as one of the most straightforward MCDM methods. It involves converting the original DOM into a weighted normalized version, typically facilitated by max normalization. Following this, the method sums up the weighted objective values for each alternative. The one with the highest sum (i.e., performance score) is the recommended choice by the SAW method. Further details, including steps and equations of this method, are available in Wang and Rangaiah [34].

TOPSIS method is one of the most widely used MCDM methods, applied across numerous disciplines. Formulated by Hwang and Yoon [35], this method recommends the alternative, which simultaneously has the greatest Euclidean distance from the negative ideal solution (defined as the aggregation of the worst value for each objective) and the shortest Euclidean distance to the positive ideal solution (i.e., the aggregation of the best value for each objective). For details of the TOPSIS method, including its steps, relevant equations, and visual illustrations, refer to the work by Hwang and Yoon [35].

8.2.2 Weighting methods

In all, three weighting methods are employed in this work; these are chosen for their popularity and ability to capture different aspects of data variability and interdependencies among multiple criteria. Nevertheless, another popular weighting method, analytic hierarchy process (AHP), is not selected for this study, primarily due to its reliance on subjective judgments (i.e., inputs from decision makers) to compute weights. The three chosen weighting methods are outlined below in alphabetical order.

CRITIC weighting method considers the pairwise correlation between objectives within the DOM, along with the standard deviation of each objective, to derive weights. Its underlying principle is to capture both the diversity of objective values and their interdependencies, thereby offering a more comprehensive approach to weight assignment. For a detailed explanation of its rationale, steps, and equations, refer to Diakoulaki et al. [36].

Entropy weighting method is grounded in the probabilistic approach to quantify informational uncertainty. It assigns a higher weight to an objective having a larger variation in its values across the DOM. As highlighted in the review by Hafezalkotob et al. [37], entropy method is frequently cited as the most popular weighting technique in MCDM. Hwang and Yoon [35] present the procedural steps, equations, and illustrations of this method.

StDev weighting method is an intuitive approach for determining the weights of objectives in MCDM. It calculates the standard deviation of values for each objective in DOM to capture the variability or dispersion of values around the mean. The StDev method assigns greater weight to an objective with a higher standard deviation. See Wang et al. [38] for the steps and equations of the method.

Principles of methods such as StDev, GRA, and SAW, and manual calculations for them are relatively simple. However, to avoid manual calculations and potential mistakes in them, use of a well-tested program is better for consistent and extensive studies. From such computational perspective, algorithms of all the weighting and MCDM methods employed in this study are not difficult to understand and implement computer programs for them.

8.2.3 DOM modifications

The three modifications, LTO, ROR, and RA, considered for sensitivity analysis of MCDM methods are outlined in this section. Only one modification at a time is considered in the present study.

LTO: Pamučar and Ćirović [19] utilized the concept of the independence of the value scale (IVS) for analyzing the consistency of MCDM methods. IVS suggests that the rankings produced by an MCDM method should remain unchanged when the units of measurement (or scales) of any objective undergo linear transformations. In this chapter, an LTO in the form of $y = Ax + B$ is applied, where A and B represent positive constants, and x and y denote the original and transformed values of the objectives in DOM, respectively.

ROR: When an objective is equivalently reformulated (e.g., converting maximization of profit to minimization of 1/profit), the ranking of alternatives by an MCDM method should remain unchanged. Essentially, the formulation of objectives, whether as smaller-the-better or larger-the-better, ought not to alter the rankings. Pamučar and Ćirović [19] discussed choosing a forklift in a logistic center, where one objective is the speed of lifting loads (S), measured in meters per minute (larger-the-better). However, this can be converted to a smaller-the-better objective using $1/S$, measured in minutes per meter, signifying the time needed to lift a load by one meter. In this chapter, an ROR transformation, $y = 1/(x + A)$, is applied, where A is a constant (possibly zero), and x and y denote the original and transformed values of the objectives in DOM, respectively.

RA: As mentioned in Section 9.1, in a rational decision-making process, when some solutions are removed from the DOM, the relative rankings of the remaining solutions should not change. However, this modification poses a significant challenge to many MCDM methods, as their calculations often rely on the positive and/or negative ideal solutions, which could include a portion of the solutions that have been removed from the DOM. To assess this sensitivity, this work employs a test where 10% of the solutions from the DOM are randomly removed.

8.2.4 Sensitivity analysis metrics

To measure how the weights by a weighting method change in response to a modification in DOM, the average absolute fractional deviation for weights (AAFD$_w$), defined below, is employed:

$$\text{AAFD}_W = \frac{1}{n} \sum_{j=1}^{n} \frac{\left| WO_j - WM_j \right|}{\left| \frac{WO_j + WM_j}{2} \right|} \tag{8.1}$$

Here, n is the number of objectives, and WO_j and WM_j are the calculated weights for the jth objective in the original and modified DOMs, respectively. Minimal sensitivity requires zero or negligible value of $AAFD_w$.

Analogously, as shown in eq. (8.2), the AAFD for the objective values of the top-ranked alternative is used to quantify the discrepancy between the top-ranked alternatives, derived from the original and the modified DOMs by an MCDM method:

$$\text{AAFD}_o = \frac{1}{n} \sum_{j=1}^{n} \frac{\left| OO_j - OM_j \right|}{\left| \frac{OO_j + OM_j}{2} \right|} \tag{8.2}$$

Here, OO_j and OM_j are the values of the jth objective in the top-ranked alternative based on the original and the modified DOMs, respectively, by an MCDM method. Minimal sensitivity requires zero or negligible value of AAFD$_o$. When needed, the AAFD$_o$ can also be applied to alternatives ranked second, third, and so forth.

Spearman's rank correlation coefficient (r_s), computed using eq. (8.3), is utilized to measure the difference between the ranking of all alternatives found by an MCDM method using the original and the modified DOMs. Subsequently, with eq. (8.4), a t-test is performed to calculate the p-value for r_s. Generally, a r_s is considered statistically significant if its p-value is less than 0.05, providing robust evidence to support its acceptance:

$$r_s = \frac{\sum_{i=1}^{m}(x_i - \bar{x})(y_i - \bar{y})}{\sqrt{\sum_{i=1}^{m}(x_i - \bar{x})^2 \sum_{i=1}^{n}(y_i - \bar{y})^2}} \tag{8.3}$$

$$t = \frac{r_s\sqrt{m-2}}{\sqrt{1-r_s^2}} \tag{8.4}$$

Here, m is the number of all alternatives; x_i and y_i are the ranks of each alternative before and after the DOM modification; and $\bar{x}$ and $\bar{y}$ are the mean ranks of these alternatives. A perfect positive correlation is indicated by a $r_s = 1$.

As stated above, Spearman's rank correlation coefficient quantifies the changes in the ranking of all alternatives due to change(s) in DOM. It is a well-known statistic and has been employed in several studies on sensitivity analysis of MCDM methods [7, 26]. In MCDM, engineers are more likely to be interested in the top-ranked alternatives and changes in them. Hence, besides Spearman's rank correlation coefficient, we employed $AAFD_o$ and frequency of changes in the top-ranked alternative as well as in the top three alternatives in our analysis of sensitivity of MCDM methods, presented in Section 8.4.

8.3 Applications in chemical engineering

The six optimization applications of MCDM in chemical engineering employed in this work are briefly described in this section.

Polyethylene waste co-gasification: The waste gasification involves the decomposition of a solid waste in the presence of a gasifying agent (air or steam) with heating. Its main products are hydrogen and carbon dioxide. Due to the high content of carbon and hydrogen, waste plastics are one of the desirable feeds for this process. The use of these feeds is important from both environmental and economic points of view. Gasification is a thermochemical process, and its efficiency and performance are affected by various variables. The dataset used in this study is for the co-gasification of low- and high-density polyethylene waste with different composition ratios. The MCDM problem has 11 alternatives and 5 energy, environmental and economic criteria of gasification performance; the criteria are hydrogen production rate, energy efficiency and heating value of syngas, all 3 to be maximized, as well as carbon dioxide emission and purchasing cost, both to be minimized [39]. In the LTO, energy efficiency is modified by $y = 2x + 3$; in the ROR, cost is transformed by $y = 1/x$ from minimization to maximization type. The alternative L20H80 (20 wt% low density and 80 wt% high-density polyethylene) is removed in the case of RA.

Synthesis gas production from biomass: In the future, supplying energy from renewable sources will be one of the main solutions to meet the needs of industries and address environmental concerns. The use of biomass as a feedstock for energy production processes has been studied for several decades. Gasification of biomass into valuable intermediates such as synthesis gas will create added value. Synthesis gas is rich in hydrogen and carbon monoxide; it can be used in the combustion process to provide thermal energy and as a feed for the synthesis of basic petrochemicals. Choosing the right type of biomass for the synthesis gas production is important due to the variation in the composition of biomass. In the dataset for MCDM application from Mojaver et al. [40], the original dataset involves 13 alternative biomasses; of these, only seven biomasses (namely, olive refuse, pine sawdust, eucalyptus, straw, legume straw, apricot stone, and pinecone) are non-dominated and are used in the present study. There are seven criteria; of them, the benefit criteria are hydrogen and methane content in the synthesis gas, hydrogen yield, cold gas efficiency, and exergy efficiency, whereas the cost criteria are carbon monoxide and carbon dioxide in the synthesis gas. Both hydrogen and carbon monoxide content are modified by $y = 2x + 3$ in LTO; carbon dioxide content and cold gas efficiency are transformed by $y = 1/x$ from min to max and max to min, respectively. To study RA, legume straw is removed from the dataset of seven biomass alternatives.

Thermal cracking reactor for olefins: The thermal cracking process has been the only industrial process for the production of olefins on a large scale. It is a fundamental and upstream unit in the petrochemical industry. The thermal cracking reactor is the most complex part of this plant. MOO of a thermal cracking reactor with liquefied petroleum gas (LPG) feed was studied by Nabavi et al. [41, 42], considering simultaneous maximization of production of ethylene and propylene, and minimization of heat duty. The dataset used in the present study is from Nabavi et al. [16]; it contains 60 non-dominated solutions with two benefit criteria and one cost criterion. In this dataset, ethylene production is modified by $y = 2x + 3$ for LTO. Propylene production is transformed from max to min by $y = 1/x$ in ROR. For RA, six alternatives are removed randomly from the 60 non-dominated solutions.

Cumene production process: cumene is an important petrochemical intermediate to produce phenol and acetone, which accounts for about 98% of cumene produced in the world. It is produced from the reaction of benzene and propylene in gas phase in catalytic fixed bed reactors. Cumene production process has been optimized for both operation and design, considering operational, economic, and environmental objectives [43, 44]. The dataset of non-dominated/optimal solutions from Amooey et al. [45] for cumene production process is chosen for the present study. It has 120 alternatives and three environmental and economic criteria, namely, damage index, material loss and total capital cost, all of which should be minimized. In this dataset. Total capital cost was modified by $y = 2x + 3$ in LTO and material loss is transformed from min to max by $y = 1/x$ in ROR. For RA, 12 solutions are removed randomly from the original dataset.

Material for cryogenic tank design: Cryogenic tanks are used for storage and transportation of gases like nitrogen, oxygen, hydrogen, helium, and argon. These should be made of special materials to be effective. The number of available materials, together with the complex relationships between the various selection parameters, makes it difficult to choose the most appropriate material. The materials should have good weldability and processability, lower density and specific heat, smaller thermal expansion coefficient and thermal conductivity, and adequate toughness at the operating temperature [46]. The dataset in Mousavi-Nasab et al. [47] is for the material selection for liquid nitrogen tanks. There are seven alternatives, each with seven criteria; of these, toughness index, yield strength, and Young's modulus are the benefit type while density, thermal expansion index, thermal conductivity, and specific heat are the cost type. For LTO in this application, Young's modulus and density are modified by $y = 2x + 3$. For ROR, yield strength is transformed from max to min and thermal conductivity is modified from min to max by $y = 1/x$. The alternative, Al5052-0, is removed from the DOM to study the effect of RA.

Material for high-temperature oxygen-rich environment: This MCDM application is on the selection of a suitable material for an equipment, which needs to be designed for operation in a high-temperature oxygen-rich environment. The material selection matrix involved six alternatives and four criteria [48]; among these, hardness, machinability rating, and corrosion resistance should be maximized, and the cost should be minimized [47]. For LTO in this application, machinability rating is converted by $y = 2x + 3$. For ROR, hardness is transformed from max to min by $y = 1/x$. For RA, material two is removed from DOM.

8.4 Results and discussion

Effect of the three modifications in DOM on weighting and MCDM methods is discussed in the following six subsections: effect on weights, effect on the top-ranked (i.e., rank 1) alternative, effect on the top three (i.e., ranks 1, 2, and 3) alternatives, effect on the ranking of all alternatives, summary of effects of modifications and limitations of the present study.

8.4.1 Effect of modifications on weighting methods

To analyze the effect of the three modifications in DOM on the weights calculated by the entropy, CRITIC, and StDev methods, mean $AAFD_w$ between the weights of criteria based on the original DOM and the weights of criteria based on each of the modified DOMs was computed. As shown in Figure 8.2, weights calculated by the StDev and CRITIC methods for all three modified DOMs are more similar to each other compared to those found by the entropy method. In the case of LTO, the mean AAFD$_w$ is zero for both StDev and CRITIC methods. In other words, the same weights are obtained for both the original and modified DOMs, and the calculated weights are unaffected by LTO. In the case of ROR and RA, the mean AAFD$_w$ is not zero; in these two modifications of DOM, the weights calculated by the StDev and CRITIC methods are different from those values based on the original DOM. The trend of mean AAFD$_w$ is different for the weights calculated by the entropy method; all three modifications of DOM affect the calculated weights for the criteria and the effect of ROR is greater than the effect of LTO and RA.

In summary, the weights calculated by the entropy method are more sensitive to LTO, ROR, and RA modifications compared to StDev and CRITIC methods. This is consistent with the finding of Nabavi et al. [24], who showed that weights calculated by the entropy method are more sensitive to LTO, ROR, and RA compared to those by the CRITIC method, for the 16 applications studied.

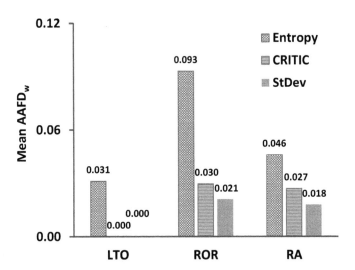

Figure 8.2: Effect (quantified by mean AAFD$_w$) of LTO, ROR, and RA on the weights by entropy, CRITIC, and StDev methods.

8.4.2 Effect of modifications on rank 1 alternative

Figure 8.3 shows the effect of LTO, ROR, and RA on the rank 1 alternative by different MCDM methods with entropy, CRITIC, and StDev weights. Frequency in plots A, C, and E, is the number of changes in the rank 1 alternative due to each modification in the six applications. The maximum frequency is found to be 3, which means rank 1 alternative by the employed weighting and MCDM methods is affected by the modification in three out of six applications tested. Frequency will be zero if there is no change in the rank 1 alternative in all the applications tested; then, $AAFD_O$ in each application and its sum will also be zero. Low/zero frequency and low/zero mean $AAFD_O$ are desirable for a less sensitive (i.e., more robust) MCDM method. GRA is not affected by weight changes because the variant of GRA used in this study does not require weights. Overall, LTO has the least effect on the rank 1 alternative identified by all eight MCDM methods with either of the three weighting methods tested; then, RA has slightly more effect and ROR has the most effect.

Statistics in Table 8.1 show that the LTO and RA have no or little effect on the rank 1 alternative when StDev weights are used with the MCDM methods. However, the effect of ROR is greater when using entropy and StDev weights. Interestingly, the sum of $AAFD_O$ is almost the same for LTO, ROR, and RA when CRITIC weights are used with the MCDM methods. Overall, changes in the rank 1 alternative by the eight methods are less with StDev weights compared to those with CRITIC and entropy weights.

Considering both the frequency and $AAFD_O$ results for finding the rank 1 alternative, less sensitive methods are GRA (without any weights); MABAC, CODAS, and SAW using entropy or CRITIC weights; and CODAS, SAW, and TOPSIS using StDev weights (Figure 8.3). Comparison of total frequency and sum of $AAFD_O$ values in Table 8.1 shows that the effect of LTO, ROR, and RA on the top alternative by all eight MCDM methods is less (i.e., more similar alternatives are selected) when the StDev weights are used, compared to when the entropy and CRITIC weights are used.

Table 8.1: Statistics on the effect of LTO, ROR, and RA on the rank 1 alternative by all eight MCDM Methods with entropy, CRITIC, and StDev weights.

Weighting method	Modification	Frequency	Sum of $AAFD_O$
Entropy	LTO	1	0.03
	ROR	13	5.18
	RA	3	1.32
	Total	**17**	**6.53**
CRITIC	LTO	6	1.96
	ROR	10	1.95
	RA	6	1.96
	Total	**26**	**5.88**
StDev	LTO	0	0
	ROR	11	3.37
	RA	1	0.07
	Total	**12**	**3.44**

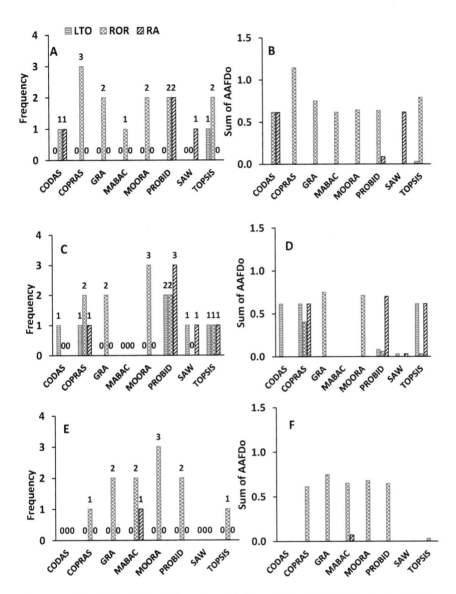

Figure 8.3: Effect of LTO, ROR, and RA on the rank 1 alternative by eight MCDM methods using the weights by entropy (plots A and B), CRITIC (plots C and D), and StDev (plots E and F) methods. In plots A, C, and E, *y*-axis is frequency, whereas it is the sum of AAFD$_O$ in plots B, D, and F.

8.4.3 Effect of modifications on the top three alternatives

Next, the effect of LTO, ROR, and RA modifications on the top three alternatives (i.e., ranks 1, 2, and 3) by each of the eight MCDM methods with entropy, CRITIC, and StDev weights was investigated. Figure 8.4 shows this effect in terms of total frequency of changes (i.e., sum for the top three alternatives) and total AAFD$_O$ (i.e., sum for top three

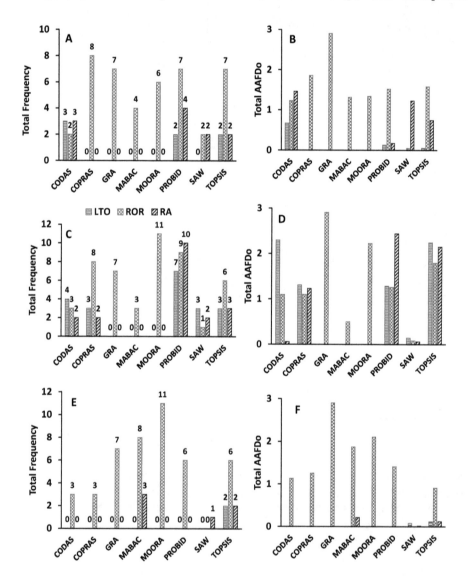

Figure 8.4: Effect of LTO, ROR, and RA on the top three alternatives by eight MCDM methods using weights by the entropy method (plots A and B), CRITIC method (plots C and D), and StDev method (plots E and F). In plots A, C, and E, y-axis is total frequency whereas it is total AAFD$_O$ in plots B, D, and F.

alternatives). As before, ranking by GRA is unaffected by weights and hence its results for total frequency and total $AAFD_O$ in the top, center and bottom plots in Figure 8.4 are identical. Considering both the frequency and $AAFD_O$ results for finding the top three alternatives, the less sensitive methods with the entropy and CRITIC weights are MABAC and SAW; they are SAW, CODAS, and COPRAS with StDev weights.

Statistics in Table 8.2 clearly show the following effects of LTO ROR and RA modifications on the top three alternatives identified by MCDM methods: (a) LTO has the least effect in the case of entropy and StDev weights, (b) compared to both LTO and RA, ROR has more effect, irrespective of which of the three weighting methods is used, and (c) overall, effect of LTO, ROR, and RA is more if CRITIC weights are employed.

Table 8.2: Statistics on the effect of LTO, ROR, and RA on the top three alternatives by all eight MCDM methods with entropy, CRITIC, and StDev weights.

Weighting method	Modification	Frequency	Sum of $AAFD_0$
Entropy	LTO	7	0.86
	ROR	43	11.83
	RA	11	3.61
	Total	**61**	**16.29**
CRITIC	LTO	20	7.29
	ROR	48	10.98
	RA	19	5.94
	Total	**78**	**24.21**
StDev	LTO	2	0.21
	ROR	44	11.60
	RA	6	0.34
	Total	**52**	**12.15**

8.4.4 Effect of modifications on the ranking of all alternatives

The effect of LTO, ROR, and RA on the ranking of all alternatives is quantified by Spearman's rank correlation coefficient between the ranking of the original dataset and that of the modified dataset. The Spearman's rank correlation coefficient of +1 indicates perfect positive relationship; this implies there is no change in the ranking of all alternatives due to the modification in the dataset. Further, Spearman's rank correlation coefficient with p-value less than 0.05 is said to be statistically significant, indicating strong evidence that the calculated correlation coefficient can be accepted.

Table 8.3 summarizes the average correlation coefficient and p-value for the six applications. Each entry in this table is the average of correlation coefficients and of the corresponding p-values for the six applications by an MCDM method with one of

Table 8.3: Effect of LTO, ROR, and RA on ranking of all alternatives: average of Spearman's rank correlation coefficient (with p-value in brackets).

MCDM method	Entropy weights				CRITIC weights				StDev weights			
	LTO	ROR	RA	Global average	LTO	ROR	RA	Global average	LTO	ROR	RA	Global average
CODAS	0.9841 (0.00088)	0.9850 (0.00042)	0.9903 (0.00080)	**0.9865 (0.00070)**	0.9333 (0.01997)	0.9363 (0.01197)	0.9813 (0.00623)	**0.9503 (0.01272)**	0.9986 (4.7E-40)	0.9834 (7.6E-05)	0.9891 (0.00080)	**0.9904 (0.00029)**
COPRAS	0.9994 (4.7E-40)	0.9811 (0.00088)	0.9999 (2.3E-25)	**0.9935 (0.00029)**	0.9814 (0.00050)	0.9570 (0.00700)	0.9802 (0.00314)	**0.9729 (0.00355)**	0.9993 (4.7E-40)	0.9745 (0.00088)	0.9995 (2.3E-25)	**0.9911 (0.00029)**
GRA	1.0000 (4.7E-40)	0.9691 (0.00122)	1.0000 (2.3E-25)	**0.9897 (0.00041)**	1.0000 (4.7E-40)	0.9691 (0.00122)	1.0000 (2.3E-25)	**0.9897 (0.00041)**	1.0000 (4.7E-40)	0.9691 (0.00122)	1.0000 (2.3E-25)	**0.9897 (0.00041)**
MABAC	0.9994 (4.7E-40)	0.9710 (0.00228)	0.9995 (2.3E-25)	**0.9900 (0.00076)**	1.0000 (4.7E-40)	0.9202 (0.00703)	0.9967 (1.5E-8)	**0.9723 (0.00234)**	1.0000 (4.7E-40)	0.9543 (0.00708)	0.9995 (2.3E-25)	**0.9846 (0.00236)**
MOORA	0.9997 (4.7E-40)	0.9744 (0.00228)	0.9999 (2.3E-25)	**0.9913 (0.00076)**	0.9818 (0.00114)	0.9684 (0.00130)	0.9901 (0.00080)	**0.9801 (0.00108)**	0.9997 (4.7E-40)	0.9590 (0.00428)	0.9997 (2.3E-25)	**0.9861 (0.00143)**
PROBID	0.9997 (4.7E-40)	0.9352 (0.00957)	0.9808 (0.00160)	**0.9719 (0.00373)**	0.9752 (0.00121)	0.9422 (0.00814)	0.9680 (0.00394)	**0.9618 (0.00443)**	0.9932 (7.6E-5)	0.9776 (0.00114)	0.9991 (2.3E-25)	**0.9900 (0.00040)**
SAW	0.9994 (4.7E-40)	0.9988 (4.7E-40)	0.9903 (0.00080)	**0.9962 (0.00027)**	0.9936 (7.6E-5)	0.9931 (7.6E-5)	0.9993 (2.3E-25)	**0.9953 (5.1E-5)**	0.9996 (4.7E-40)	0.9991 (4.7E-40)	0.9997 (2.3E-25)	**0.9995 (7.8E-26)**
TOPSIS	0.9997 (4.7E-40)	0.9608 (0.00308)	0.9904 (0.00080)	**0.9836 (0.00130)**	0.9754 (0.00228)	0.9388 (0.00957)	0.9611 (0.01207)	**0.9584 (0.00797)**	0.9991 (4.7E-40)	0.9835 (0.00042)	0.9993 (2.3E-25)	**0.9940 (0.00014)**
Global average	**0.9977**	**0.9719**	**0.9939**		0.9801	0.9531	0.9846		0.9987	0.9751	0.9982	
R (p-value)	(0.00011)	(0.00247)	(0.00050)		(0.00315)	(0.00579)	(0.00327)		(9.5E-6)	(0.00189)	(0.00010)	

the weighting methods. The mean of these average values for the three modifications is the global average for each MCDM method with a weighting method. The global average in the last row of Table 8.3 is the mean of the average values for eight MCDM methods with one of the weighting methods. The maximum of average p-values in Table 8.3 is 0.02, which is well below 0.05; this indicates that the calculated correlation coefficient can be accepted.

In the last row in Table 8.3, the global average of Spearman's rank correlation coefficient for the eight MCDM methods is greater than 0.98 in the case of LTO and RA for all three weighting methods, whereas it is 0.95–0.98 in the case of ROR. These values indicate that ROR has more effect on the ranking of all alternatives, regardless of the weighting methods. This observation coincides with the results in Figure 8.3 and Table 8.1 on the top-ranked alternative, as well as Figure 8.4 and Table 8.2 on the top three alternatives, where ROR has the largest effect, irrespective of the weighting method used.

The global average of Spearman's rank correlation coefficient for the three modifications depends on the MCDM and weighting methods. It is above 0.97, 0.95, and 0.98 in case of entropy, CRITIC, and StDev weights, respectively. This indicates that ranking of all alternatives is more affected if CRITIC weights are used (compared to entropy and StDev weights). This is like the larger effect of the three modifications on the top three alternatives when CRITIC weights are employed (Table 8.2).

8.4.5 Summary of effects of modifications

The results of this study demonstrate that weights calculated by the entropy method are more sensitive to LTO, ROR, and RA compared to those by the StDev and CRITIC methods, for the six applications tested (Figure 8.2). This is consistent with the finding by Nabavi et al. [24], who performed sensitivity analysis considering 16 applications, that the entropy weights are more sensitive to LTO, ROR, and RA compared to CRITIC weights.

Further, Figure 8.2 shows that the effect of ROR on weights is greater than that of LTO and RA. Nabavi et al. [24] made a similar observation for four datasets with a large number of alternatives. However, they found that RA has the greater effect on weights (compared to LTO and ROR) for 12 datasets with small number of alternatives. Recall that four applications in the present study have a small number of alternatives (11 or less) whereas two other applications have a large number of alternatives (60 and 120).

The results in Tables 8.1–8.3 show that ROR has the greatest effect on the ranking of alternatives by MCDM methods. This is consistent with our earlier finding in Nabavi et al. [24] that ROR has the greatest effect on the ranking of alternatives by MCDM methods, for the 16 applications tested, regardless of entropy or CRITIC weights.

Considering both the frequency and $AAFD_O$ values in Figures 8.2 and 8.3, and Spearman's rank correlation coefficients and *p*-values in Table 8.3, GRA (without any weights), MABAC, CODAS, and SAW with entropy or CRITIC weights, and CODAS, SAW, and TOPSIS with StDev weight are found to be less sensitive to LTO, ROR, and RA modifications in DOM, particularly for finding the top alternative. These findings are generally consistent with those reported in Nabavi et al. [24], who concluded that GRA and CODAS with entropy weights, and SAW with entropy or CRITIC weights are less sensitive among the eight MCDM methods tested.

In summary, findings in this work on sensitivity analysis of eight MCDM methods for six applications in chemical engineering are similar to those in Nabavi et al. [24], who performed sensitivity analysis of the same eight MCDM methods for 16 different applications in engineering.

8.4.6 Further studies

The present study is limited to three weighting methods, eight MCDM methods and six applications in chemical engineering. On the other hand, there are many weighting methods, a few hundred MCDM methods and numerous applications. Sensitivity analysis of various combinations of these requires many studies and extensive effort as well as DOMs for a variety of applications. Hence, future studies can classify MCDM applications in chemical engineering into different types such as distillation systems, complete processes, selection of material of construction, energy sources, supply chain systems, etc. Then, sensitivity of popular and/or promising weighting and MCDM methods can be analyzed for each type of application.

Meanwhile, the following strategy can be employed for ranking the alternatives in the application of interest. Consider possible changes in the DOM of this application (e.g., LTO, ROR, RA, and variations in one or more criteria). Choose several weighting and/or MCDM methods based on the literature (including this study and papers cited herein). Then, perform sensitivity analysis of the chosen methods for the possible changes in the DOM. A good program such as our MS Excel program (described in [34, 38]) will be useful for this.

The correct ranking of the alternatives and/or the best alternative for many applications are unknown. Hence, it is not possible to assess the accuracy of ranking results of a MCDM method. However, in some financial applications of MCDM, the best alternative can be found from the subsequent/future performance data. For instance, Baydaş and Elma [49] led the research to leverage the ranking of stocks of companies (listed on a stock exchange) based on their real-life returns, as the benchmark for comparing different MCDM methods.

8.5 Conclusions

In this chapter, we investigated the impact of three types of DOM modifications, namely, LTO, ROR, and RA, on the weights and ranking of alternatives, focusing on both the top-ranked and the top three alternatives, along with the overall rankings. We utilized eight common or recent MCDM methods that are integrated with entropy, CRITIC, or StDev weighting methods across six chemical engineering applications. The analysis of extensive results led to several findings. First, the weights generated using the entropy method displayed higher sensitivity to LTO, ROR, and RA modifications, when compared to those derived from the StDev and CRITIC methods. Second, the analysis identified ROR as having the most significant influence on both the weights and ranking of alternatives by MCDM methods. Finally, (i) GRA without any weights, (ii) MABAC, CODAS and SAW with entropy or CRITIC weights, and (iii) CODAS, SAW and TOPSIS with StDev weights, exhibit lower sensitivity, thereby indicating better robustness to modifications in DOM. These findings help the selection of one or more weighting and MCDM methods for a new optimization application in chemical engineering. If necessary, sensitivity of the chosen methods for the application can be examined utilizing our MS Excel program with many weighting and MCDM methods [38], before finalizing the best alternative (optimal solution).

References

[1]   Görçün, Ö. F., Pamucar, D., Biswas, S. The Blockchain Technology Selection in the Logistics Industry Using a Novel MCDM Framework Based on Fermatean Fuzzy Sets and Dombi Aggregation. In *Information Sciences*. Vol. 635, 2023, pp. 345–374.

[2]   Miškić, S., Stević, Ž., Tadić, S., Alkhayyat, A., Krstić, M. Assessment of the LPI of the EU countries using MCDM model with an emphasis on the importance of criteria. *World Review of Intermodal Transportation Research*, 11(3), 2023. 258–279.

[3]   Chakraborty, S., Raut, R. D., Rofin, T., Chakraborty, S. A comprehensive and systematic review of multi-criteria decision-making methods and applications in healthcare. *Healthcare Analytics*, 2023, 100232.

[4]   Ali, A. M., Abdelhafeez, A., Soliman, T. H., ELMenshawy, K. A probabilistic hesitant fuzzy MCDM approach to selecting treatment policy for COVID-19. *Decision Making: Applications in Management and Engineering*, 7(1), 2024. 131–144.

[5]   Ahmadinejad, B., Bahramian, F., Mousavi, S. A., Jalali, A. Identify the most effective elderly preventive falling interventions using MCDM technique. *Physical & Occupational Therapy In Geriatrics*, 42(1), 2024. 53–69.

[6]   Wang, Z., Baydaş, M., Stević, Ž., Özçil, A., Irfan, S. A., Wu, Z., Rangaiah, G. P. Comparison of fuzzy and crisp decision matrices: An evaluation on PROBID and sPROBID multi-criteria decision-making methods. *Demonstratio Mathematica*, 56(1), 2023. 20230117.

[7]   Elma, O. E., Stević, Ž., Baydaş, M. An alternative sensitivity analysis for the evaluation of MCDA applications: The significance of brand value in the comparative financial performance analysis of BIST high-end companies. *Mathematics*, 12(4), 2024. 520.

[8]   Cicciù, B., Schramm, F., Schramm, V. B. Multi-criteria Decision Making/aid Methods for Assessing Agricultural Sustainability: A Literature Review. In *Environmental Science & Policy*. Vol. 138, 2022, pp. 85–96.

[9]   Kumar, A., Pant, S. Analytical hierarchy process for sustainable agriculture: An overview. *MethodsX*, 2023, 10, 101954.

[10]   Chen, T.-L., Chen, -C.-C., Chuang, Y.-C., Liou, J. J. A hybrid MADM model for product design evaluation and improvement. *Sustainability*, 12(17), 2020. 6743.

[11]   Wang, Z., Nabavi, S. R., Rangaiah, G. P. Selected Multi-criteria Decision-Making Methods and Their Applications to Product and System Design. In *Optimization Methods for Product and System Design*. Springer, Vol. 2023, pp. 107–138.

[12]   Freire, C. A., Ferreira, F. A., Carayannis, E. G., Ferreira, J. J. Artificial intelligence and smart cities: A DEMATEL approach to adaptation challenges and initiatives. *IEEE Transactions on Engineering Management*, 70(5), 2021. 1881–1889.

[13]   Alshahrani, R., Yenugula, M., Algethami, H., Alharbi, F., Goswami, S. S., Naveed, Q. N., Lasisi, A., Islam, S., Khan, N. A., Zahmatkesh, S. Establishing the Fuzzy Integrated Hybrid MCDM Framework to Identify the Key Barriers to Implementing Artificial Intelligence-enabled Sustainable Cloud System in an IT Industry. In *Expert Systems with Applications*. Vol. 238, 2024, pp. 121732.

[14]   AlNouss, A., Alherbawi, M., Parthasarathy, P., Al-Thani, N., McKay, G., Al-Ansari, T. Waste-to-energy Technology Selection: A Multi-criteria Optimisation Approach. In *Computers & Chemical Engineering*. 2024. Vol. 183, 2024, pp. 108595.

[15]   Bhojane, M. S., Murmu, S. C., Chattopadhyay, H., Dutta, A. Application of MCDM Technique for Selection of Fuel in Power Plant. In *Materials Today: Proceedings*. Vol. 2023, 1–10.

[16]   Nabavi, S. R., Jafari, M. J., Wang, Z. Deep learning aided multi-objective optimization and multi-criteria decision making in thermal cracking process for olefines production. *Journal of the Taiwan Institute of Chemical Engineers*, 2023, 152, 105179.

[17]   Wang, Z., Tan, W. G. Y., Rangaiah, G. P., Wu, Z. Machine Learning Aided Model Predictive Control with Multi-objective Optimization and Multi-criteria Decision Making. In *Computers & Chemical Engineering*. Vol. 179, 2023, pp. 108414.

[18]   Rangaiah, G. P., Feng, Z., Hoadley, A. F. Multi-objective optimization applications in chemical process engineering: Tutorial and review. *Processes*, 8(5), 2020. 508–518.

[19]   Pamučar, D., Ćirović, G. The selection of transport and handling resources in logistics centers using multi-attributive border approximation area comparison (MABAC). *Expert Systems with Applications*, 42(6), 2015. 3016–3028.

[20]   Mufazzal, S., Muzakkir, S. A New Multi-criterion Decision Making (MCDM) Method Based on Proximity Indexed Value for Minimizing Rank Reversals. In *Computers & Industrial Engineering*. Vol. 119, 2018, pp. 427–438.

[21]   Wang, Z., Rangaiah, G. P., Wang, X. Preference ranking on the basis of ideal-average distance method for multi-criteria decision-making. *Industrial & Engineering Chemistry Research*, 60(30), 2021. 11216–11230.

[22]   Shih, H.-S. Rank Reversal in TOPSIS. In *TOPSIS and Its Extensions: A Distance-Based MCDM Approach*. Springer, Vol. 2022, 2022, pp. 159–175.

[23]   Dehshiri, S. S. H., Firoozabadi, B. A new multi-criteria decision making approach based on wins in league to avoid rank reversal: A case study on prioritizing environmental deterioration strategies in arid urban areas. *Journal of Cleaner Production*, 2023, 383, 135438.

[24]   Nabavi, S. R., Wang, Z., Rangaiah, G. P. Sensitivity analysis of multi-criteria decision-making methods for engineering applications. *Industrial & Engineering Chemistry Research*, 62(17), 2023. 6707–6722.

[25]   Baydaş, M., Kavacık, M., Wang, Z. Interpreting the determinants of sensitivity in MCDM Methods with a new perspective: an application on E-Scooter selection with the PROBID method. *Spectrum of Engineering and Management Sciences*, 2024, 2, 17–35.

[26] Stević, Ž., Baydaş, M., Kavacık, M., Ayhan, E., Marinković, D. Selection of data conversion technique via sensitivity-performance matching: ranking of Small E-Vans with PROBID method. *Facta Universitatis, Series: Mechanical Engineering*, 2024.

[27] Keshavarz-Ghorabaee, M., Zavadskas, E. K., Turskis, Z., Antucheviciene, J. A new combinative distance-based assessment (CODAS) method for multi-criteria decision-making. *Economic Computation & Economic Cybernetics Studies & Research*, 50(3), 2016. 25–44.

[28] Zavadskas, E., Kaklauskas, A. Determination of an Efficient Contractor by Using the New Method of Multicriteria Assessment. In *The Organization and Management of Construction*. Routledge, 2002, pp. 97–104.

[29] Deng, J.-L. Control problems of grey systems. *Systems & Control Letters*, 1(5), 1982. 288–294.

[30] Martinez-Morales, J. D., Pineda-Rico, U., Stevens-Navarro, E. Performance Comparison between MADM Algorithms for Vertical Handoff in 4G Networks. In *7th International Conference on Electrical Engineering Computing Science and Automatic Control*. 2010.

[31] Brauers, W. K., Zavadskas, E. K. The MOORA method and its application to privatization in a transition economy. *Control and Cybernetics*, 35(2), 2006. 445–469.

[32] Fishburn, P. C. *A Problem-based Selection of Multi-attribute Decision Making Methods*. New Jersey: Blackwell Publishing, 1967.

[33] MacCrimmon, K. R. *Decision-making among Multiple-attribute Alternatives: A Survey and Consolidated Approach*. Rand Corporation Santa Monica, 1968.

[34] Wang, Z., Rangaiah, G. P. Application and analysis of methods for selecting an optimal solution from the Pareto-optimal front obtained by multiobjective optimization. *Industrial & Engineering Chemistry Research*, 56(2), 2017. 560–574.

[35] Hwang, C.-L., Yoon, K. *Multiple Attribute Decision Making: Methods and Applications*. Springer-Verlag, 1981.

[36] Diakoulaki, D., Mavrotas, G., Papayannakis, L. Determining objective weights in multiple criteria problems: The critic method. *Computers & Operations Research*, 22(7), 1995. 763–770.

[37] Hafezalkotob, A., Hafezalkotob, A., Liao, H., Herrera, F. An overview of MULTIMOORA for multi-criteria decision-making: Theory, developments, applications, and challenges. *Information Fusion*, 2019, 51, 145–177.

[38] Wang, Z., Parhi, S. S., Rangaiah, G. P., Jana, A. K. Analysis of weighting and selection methods for Pareto-optimal solutions of multiobjective optimization in chemical engineering applications. *Industrial & Engineering Chemistry Research*, 59(33), 2020. 14850–14867.

[39] Hasanzadeh, R., Mojaver, P., Khalilarya, S., Azdast, T. Air co-gasification process of LDPE/HDPE waste based on thermodynamic modeling: Hybrid multi-criteria decision-making techniques with sensitivity analysis. *International Journal of Hydrogen Energy*, 48(6), 2023. 2145–2160.

[40] Mojaver, P., Jafarmadar, S., Khalilarya, S., Chitsaz, A. Study of synthesis gas composition, exergy assessment, and multi-criteria decision-making analysis of fluidized bed gasifier. *International Journal of Hydrogen Energy*, 44(51), 2019. 27726–27740.

[41] Nabavi, S. R., Rangaiah, G. P., Niaei, A., Salari, D. Multiobjective optimization of an industrial LPG thermal cracker using a first principles model. *Industrial & Engineering Chemistry Research*, 48(21), 2009. 9523–9533.

[42] Nabavi, R., Rangaiah, G. P., Niaei, A., Salari, D. Design optimization of an LPG thermal cracker for multiple objectives. *International Journal of Chemical Reactor Engineering*, 2011, 9(1), 1–8. 2011.

[43] Sharma, S., Chao Lim, Z., Rangaiah, G. P. Process design for economic, environmental and safety objectives with an application to the cumene process. *Multi-Objective Optimization in Chemical Engineering: Developments and Applications*, 2013, 449–477.

[44] Flegiel, F., Sharma, S., Rangaiah, G. P. Development and multiobjective optimization of improved cumene production processes. *Materials and Manufacturing Processes*, 30(4), 2015. 444–457.

[45] Amooey, A. A., Mousapour, M., Nabavi, S. R. Multiobjective particles swarm optimization and multicriteria decision making of improved cumene production process including economic, environmental, and safety criteria. *The Canadian Journal of Chemical Engineering*, 102(3), 2024. 1203–1224.

[46] Dehghan-Manshadi, B., Mahmudi, H., Abedian, A., Mahmudi, R. A novel method for materials selection in mechanical design: Combination of non-linear normalization and a modified digital logic method. *Materials & Design*, 28(1), 2007. 8–15.

[47] Mousavi-Nasab, S. H., Sotoudeh-Anvari, A. A new multi-criteria decision making approach for sustainable material selection problem: A critical study on rank reversal problem. *Journal of Cleaner Production*, 2018, 182, 466–484.

[48] Rao, R. V. A material selection model using graph theory and matrix approach. *Materials Science and Engineering: A*, 431(1–2), 2006. 248–255.

[49] Baydaş, M., Elma, O. E. An objective criteria proposal for the comparison of MCDM and weighting methods in financial performance measurement: An application in Borsa Istanbul. *Decision Making: Applications in Management and Engineering*, 2021, 4, 257–279.

Kai Kruber, Siv Kinau, and Mirko Skiborowski*

Chapter 9
Hybrid optimization methodologies for the design of chemical processes

Abstract: Process synthesis and design problems in chemical engineering usually require a solution to complex nonlinear optimization problems with continuous and discrete decision variables. The resulting mixed-integer nonlinear programming problems are particularly hard to solve, and different strategies are frequently applied for their solution. Gradient-based optimization enables fast computations, exploiting local sensitivity information, but is usually limited to local optima for nonconvex problems, whereas derivative-free optimization methods can be linked to available simulation models, with little effort, but also without any guarantee of optimality. Metaheuristics, especially swarm intelligence and population-based algorithms, are frequently applied for simulation-based process optimization, overcoming the lack of gradient information, at the cost of a considerable number of simulations. Another strategy that is receiving increasing interest builds on surrogate models that are first generated based on an initial sampling of process simulations for systematically varied design variables. Tractable surrogate models do provide the necessary sensitivity information that enables efficient gradient-based optimization, while being only an approximation of the original problem. Each strategy has its advantages and limitations, and no single best option is generally favorable for all kinds of problems. Thoughtful combinations of different strategies have the potential to overcome or at least reduce the individual limitations, while simultaneously combining the strengths of the individual methods. The current chapter provides an introduction and overview of such hybrid optimization methodologies, together with some illustrations of their use for applications in chemical engineering. Several case studies, including utility and entrainer selection, illustrate the performance of hybrid optimization methods and indicate the ability to solve even more complex design problems.

Keywords: hybrid optimization, memetic optimization, global optimization, process design, process optimization

Acknowledgments: The authors would like to acknowledge that the work that led to the writing of this chapter was funded by the Deutsche Forschungsgemeinschaft (DFG, German Research Foundation) – project number 523327609 and SFB 1615 - 503850735: SMART Reactors for Future Process Engineering (subproject B06).

*Corresponding author: **Mirko Skiborowski,** Hamburg University of Technology, Institute of Process Systems Engineering, Am Schwarzenberg-Campus 4, Hamburg 21073, Germany,
e-mail: mirko.skiborowski@tuhh.de
Kai Kruber, Siv Kinau, Hamburg University of Technology, Institute of Process Systems Engineering, Am Schwarzenberg-Campus 4, Hamburg 21073, Germany

https://doi.org/10.1515/9783111383439-009

9.1 Introduction

The need to develop more sustainable and efficient processes fosters the exploitation of optimization strategies as important tools for decision-making over the entire life cycle of chemical processes, ranging from process synthesis and design, and overprocess intensification, to process operations, control, and scheduling [1]. The optimal design and operation of chemical engineering processes require an accurate description of the process performance that usually builds on a combination of first principles models and semi-empirical correlations of different complexity to accurately describe multiphase heat and mass transfer in different kinds of equipment. The respective accuracy of the individual models is usually determined by the model fidelity. Almost all process design models build on mass and energy balances, in combination with thermodynamic models for phase equilibria and enthalpy computations, potentially including reaction, mass, and heat transfer kinetics. Examples of such complex models can be found for the identification of detailed reaction kinetics [2, 3], potentially linked to chemometric models and their parametrization [4], the identification of optimal reaction conditions [5, 6], reactor design [7] and reactor network design [8], or distillation processes, as discussed in Chapter 2, including the design of zeotropic [9, 10] and azeotropic distillation processes [11, 12], intensified distillation processes [13, 14] and hybrid separations processes [15], or even fully integrated reaction-separation [16] and reactive-separation processes [17]. Even the generic synthesis of process flowsheets has been addressed by means of the optimization of general superstructure models [18, 19]. The review paper of Tian et al. [20] provides an elaborate overview of model-based process synthesis and design methods for various applications in chemical engineering, with a focus on process intensification, covering more than 850 references.

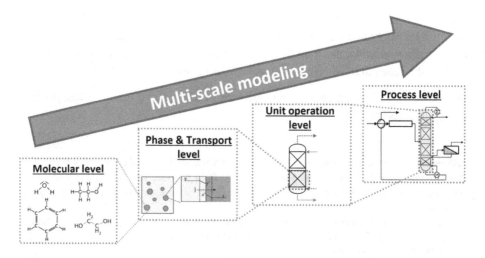

Figure 9.1: Illustration of a multi-scale modeling approach, in accordance with [21, 22].

The aggregation of the different types of models in multi-scale simulation and optimization problems, as illustrated in Figure 9.1, usually results in highly nonlinear and nonconvex large-scale problems, for which the computational effort for the solution increases exponentially with the number of design variables, especially when considering simulation-based optimization methods [23]. In seeking solutions to these optimization problems, engineers need to select proper tools and balance a fundamental trade-off between generality, fidelity, and tractability of the problem formulations [24]. The implementation of hybrid semi-parametric models can help to balance this trade-off and develop accurate and tractable models, tailored for process optimization, by replacing certain parts of the model by means of simpler surrogate models [21]. This has been especially demonstrated for complex thermodynamic models required for phase equilibria computations [25, 26]. The complexity of effectively handling this trade-off is one major reason why, despite the significant potential benefits of large-scale process optimization, there is still limited adoption of optimization-based design methods by practitioners in the chemical industry, as summarized in the study of Tsay et al. [27]. Therefore, it is of fundamental importance that engineers first get acquainted with the prerequisites and opportunities of different optimization methods and tools prior to a dedicated application. The textbooks of Biegler [28] and Grossmann [29] provide an excellent introduction to the different classes of mathematical programming problems and gradient-based optimization with applications in chemical process design and operation. The textbook of Martins and Ning [23], furthermore, provides a very good introduction to simulation-based optimization, especially covering derivative-free optimization (DFO) methods, while the recent article by Agi et al. [30] provides a more general overview of computational toolkits for model-based design and optimization.

While most gradient-based optimization methods operate on a full-space approach, which means that the optimization algorithm solves the minimization/maximization of the objective function simultaneous to the fulfillment of the equality and inequality constraints, DFO methods operate in a reduced space, thereby considering only the design degrees of freedom (DDoF), also referred to as design or independent variables, while utilizing some additional solver to fulfill the equality constraints in each iteration of the optimization, and consider the violation of inequality constraints by means of some additional penalty that is added to the objective function [23]. Each of the different optimization approaches comes with certain advantages and limitations. Deterministic gradient-based optimization can effectively handle large-scale problems with a large number of DDoF, without the need to manually divide the variables in DDoF and calculated variables, and several thousand dependent variables and constraints [31], but is limited to local solutions in case of nonconvex problems [32]. In these cases, the performance of such local optimization algorithms may depend strongly on the initialization, which cannot only result in local optima of low quality but may even impede the determination of a feasible solution. These problems can be effectively overcome by dedicated initialization strategies, problem decomposition, and incremental model refinement, facilitating efficient computation of local optima, even for complex and integrated processes [9, 33]. However, such strategies still leave some degree of uncertainty regarding the quality of the local solution. The requirement of a smooth model formulation further complicates the application of gradient-based optimization, especially in the case of discrete decisions in mixed-integer nonlinear programming (MINLP) problems such as the choice of utilities, auxiliaries or individual unit operations. DFO methods not only operate in a reduced space, spanned solely by the DDoF, but also enable the integration of black-box models that lack accurate derivative information, such as some of the well-established process flowsheet simulators [34]. Therefore, many studies on simulation-based optimization apply DFO methods, which range from deterministic gradient-free methods, such as multi-start heuristics, to all kinds of metaheuristics, such as evolutionary, swarm intelligence, bio-inspired, and physics-based algorithms [35]. Being unable to exploit local gradient information, DFO methods usually require a comparably large number of individual process simulations while lacking any proof of optimality for the original problem. Yet, DFO methods have found considerable application, as they are able to utilize existing simulation models, while overcoming certain restrictions for the application of gradient-based algorithms. Depending on a proper tuning of the respective hyperparameters, these methods can also be quite effective in process optimization. Surrogate-based optimization presents another alternative that builds a bridge between DFO and gradient-based optimization by building a numerically tractable surrogate model that represents the response surface of the original problem, which can be a non-smooth black-box model [23]. Providing a smooth approximation of the original problem, the surrogate model enables the use of both gradient-based and DFO algorithms. The approximation becomes more accurate when more complex is the surrogate model and more samples are provided for training, potentially also requiring an intensive number of simulations. The number of required samples may, however, be reduced by avoiding a space-filling sampling strategy and performing an iterative sampling and optimization strategy [23]. Similar to other DFO methods, the successful application depends strongly on the reliability and convergence of the simulation problems, which form the basis of the surrogate model.

Hybrid optimization approaches effectively integrate different optimization strategies, considering both gradient-based mathematical programming methods as well as metaheuristics and surrogate-based approaches. Consequently, hybrid optimization strategies have the potential to combine the benefits of the different methods while simultaneously overcoming their individual limitations. The main objective of hybrid optimization approaches is therefore the reliable identification of the best possible solution, the global optimum of non-convex problems, at a minimum computational effort. The global solution to process optimization problems is always of interest but may be necessary to provide certain guarantees and can be of fundamental importance in case no information or strategies for good initial solutions are known. The current chapter will provide an overview of different hybrid optimization strategies that have been proposed in the context of process optimization. In order to provide sufficient context for the analysis of the hybrid optimization methods, first, a brief overview of existing strategies for seeking the global optimum solution in process optimization is provided in Section 9.2. Based on the analysis of the limitations of the different approaches, hybrid optimization approaches are introduced in Section 9.3, together with an overview of various chemical engineering applications. Three specific hybrid optimization methods are further illustrated in more detail for specific case studies in Section 9.4, illustrating the possible benefits from the application of the hybrid approach. Finally, Section 9.5 presents conclusions and suggestions for further application and future developments.

9.2 Different strategies for seeking the global optimal solution

In order to identify the best possible solution of a mathematical optimization problem, different optimization strategies have been developed and applied to process optimization problems. Figure 9.2 provides an exemplary overview of the different categories of optimization methods as well as some examples that will further be addressed in the following subsections. Gradient-based global optimization strategies are first discussed in Section 9.2.1 before simulation-based DFO methods are discussed in Section 9.2.2, addressing global search methods and surrogate-based optimization. Both methods are considered as DFO methods in the scope of this chapter as they operate based on a simulation solver and do not require any gradient information in the first place. Surrogate-based optimization enables the use of gradient-based algorithms for the optimization of the surrogate model, including gradient-based global optimization (indicated by the dashed line). The review article of Boukouvala et al. [36] provides an extensive overview of gradient-based and DFO methods with applications in chemical engineering covering more than 500 references.

9.2.1 Gradient-based global optimization methods

Gradient-based methods for NLP and MINLP problems are computationally more efficient than DFO methods due to the exploitation of local sensitivities. This becomes especially evident for problems with a large number of DDoF, as the required number of function evaluations (simulations) in DFO methods scales exponentially with the number of DDoF [23]. Yet, most gradient-based algorithms cannot guarantee convergence to the global optimum in the case of non-convex problems, which is mostly the case for NLP and generally for MINLP problems. However, several rigorous gradient-based global optimization algorithms overcome these limitations by constructing and

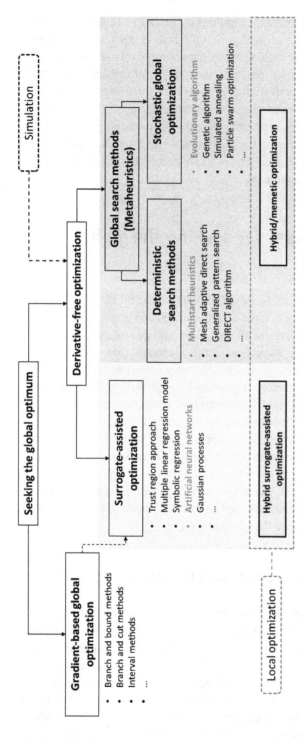

Figure 9.2: Overview of different classes of solution methods for seeking global optima in complex (mixed-integer) nonlinear optimization problems.

iteratively tightening convex envelopes, which allow for quantifying strict bounds on the possible solution [37]. These methods rely on interval methods or branch-and-bound and branch-and-cut algorithms, e.g., in the popular solvers BARON [38] and ANTIGONE [39], or the open-source solvers COUENNE [40] and MAINGO.[1] By repeated partitioning and bound tightening, the optimality gap between the best possible (convex envelope) and currently best (original problem) solution is successively reduced until a solution with a guaranteed optimality, i.e., no feasible solution with, e.g., more than 1% improvement, can be computed. Figure 9.3 provides an illustration of the progression of the lower and upper bounds in the optimization of a distillation column superstructure with BARON based on a simplified model relying on ideal vapor-liquid equilibrium and constant molar overflow assumptions [41].

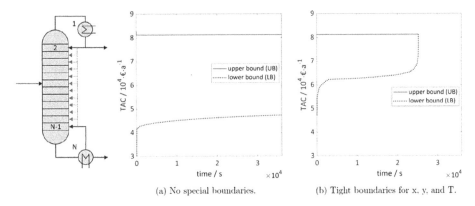

(a) No special boundaries.   (b) Tight boundaries for x, y, and T.

Figure 9.3: Illustration of the superstructure (left) for the design of a single distillation column and the progression of the lower and upper bounds for a gradient-based global optimization with BARON without (a) and with (b) tight bounds on the state variables [41].

While gradient-based global optimization solvers do effectively overcome the need for a good initialization, the specific model formulation and variable bounds have a tremendous effect on the computational effort. This is, e.g., illustrated for the case of simple nonlinear regression problems by Amaran and Sahinidis [42] and for the optimization of a distillation column by Kruber et al. [41]. In both examples, utilizing BARON, the final upper bound is determined after just a few iterations, while the majority of the computational time is required for closing the optimality gap. While the optimization requires just a few minutes for the separation of a binary mixture with ideal thermodynamics (Dalton and Raoult's law), application to a ternary mixture already fails to converge in 10 h, unless tighter bounds for the compositions and the temperature values are provided (see Figure 9.3). Even better performance can be at-

---

1 https://avt-svt.pages.rwth-aachen.de/public/maingo/.

tained with specially tailored problem formulations, as proposed by Ballerstein et al. [43] and Mertens et al. [44], who introduced a tailored bounding strategy that exploits the monotonicity of composition profiles in distillation columns for ideal mixtures and also introduced a linear transformation of the composition variables. While significantly reducing the computational effort, the strategy is limited to ideal mixtures. Other efforts focus on the generation of specific bounding strategies for thermodynamic models [45] or implicit functions in reduced space formulations [46]. Although the considered case studies highlight the significant potential of these strategies, it depends largely on the identification of suitable model formulations on the unit operation and flowsheet level, as pointed out by the authors.

Since problem complexity is a decisive factor for the application of gradient-based global optimization methods, they have primarily been applied for simplified or shortcut models [32]. Gooty et al. [47] apply BARON for the identification of energy-efficient thermally coupled distillation processes on the basis of an MINLP superstructure model, with a special form of the Underwood constraints and variable elimination techniques, while Li et al. [48] apply BARON and ANTIGONE for a building-block based process synthesis. Further two-stage strategies based on problem decomposition have been proposed, which first evaluate rigorous process models to derive fixed performance metrics that are further embedded in a more general superstructure model instead of the rigorous process models. Ulonska et al. [49] apply BARON for the identification of multiproduct biorefinery processes, while Kenkel et al. [50] build a superstructure model for a power-to-methanol plant based on simulation results as performance models, which is finally solved as a mixed-integer linear programming (MILP) problem, building on piecewise linear approximations. These studies exploit the major benefits of gradient-based global optimization, solving complex superstructure models with a large number of DDoF without the need for a good initial solution. In contrast, the simplified performance models limit viability and require significant effort for the formulation, prohibiting a quick and efficient application to new problems.

9.2.2 Derivative-free optimization methods

While gradient-based algorithms generally require an equation-oriented model formulation that provides detailed information on the model structure and sensitivities, DFO methods can operate on so-called black-box models or even experimental setups and plant data without any knowledge of the underlying model structure. Yet, this lack of knowledge has to be compensated by a larger number of function evaluations (simulations) that increases, especially with the number of DDoF [23]. The book of Martins and Ning [23] as well as the review article of Rios and Sahinidis [51] and Kim and Boukouvala [52] provide an overview of algorithms and software implementations for local and global DFO methods. Global DFO methods can overcome local solutions in different ways and the strategies can be broadly clustered in surrogate-based optimization strategies and global search strategies, which are either deterministic or stochastic (cf. Figure 9.2), and further discussed in the subsequent subsections.

9.2.2.1 Surrogate-based optimization strategies

The use of surrogate models in surrogate-based optimization resembles the traditional response surface methodology, known from statistical experimental design but exploits more advanced metamodels from modern machine learning. However, the basic idea of surrogate-based optimization, generating a numerically tractable approximation of the response surface of a complex process model and subsequently determining the optimal choice of DDoF based on the optimization of the surrogate model, has been known for quite some time. Lewandowski et al. [53] applied an artificial neural network (ANN) model for the optimization of a pressure-swing adsorption process in 1998, while Palmer et al. [54] did apply a minimum bias Latin hypercube sampling, in combination with kriging models, to optimize an ammonia synthesis plant in 2002. McBride and Sundmacher [55] provide an overview of surrogate models and applications in surrogate-based and surrogate-assisted design of chemical processes. The latter refers to gray box models for optimization problems in which only the most complex parts are replaced by surrogate models, e.g., the thermodynamic models for gas solubilities or liquid-liquid equilibria [26] or flash computations [56]. Also refer to Chapter 5 for further information on surrogate-based optimization techniques for process systems engineering.

While applicable to non-smooth black-box models that do not provide any gradient information, surrogate-based optimization provides a direct link to gradient-based global optimization algorithms if smooth symbolic surrogate models are derived from the sampled data. Surrogate models can, therefore, be effective in handling non-smooth and even noisy data from experiments or plant data [57], integrating multiple sources of data, as well as replacing expensive evaluations with less complex correlations [23]. However, these benefits do come at the cost of a sufficient amount of sampling data, which scales exponentially with the number of DDoF, also referred to as the course of dimensionality [23]. Another challenge in this approach is the generation of unbiased training data, as space-filling strategies may result in a large number of non-converged and infeasible simulations [58]. Adaptive sampling strategies and classification methods can help to foster convergence [26, 59] and considerably reduce the number of sampling points compared to a single-shot space-filling sampling. As highlighted by Bhosekar and Ierapetritou [60], there is, however, no generally superior surrogate model and sampling strategy. The best combination depends on the specific type of problem, including the tuning of the respective hyperparameters. Yet, there are some available algorithms, such as the EGO solver [61], which builds on kriging models or the MISO solver [62], which builds on radial basis functions, which perform well on a variety of MINLP test problems [63]. Still, most process optimization applications utilize tailored solution approaches.

Ye et al. [64] present a surrogate-based optimization of a distillation column, which first fits an ANN model to data from Aspen Plus® simulations based on a space-filling Latin hypercube sampling. A subsequent genetic algorithm (GA)-based optimization of the surrogate model is used to determine the optimal DDoF, for which the authors demonstrate a great accuracy of the surrogate model. While no information on computational effort is reported, the approach builds on 2,000 simulations for data generation, almost 300 epochs for fitting the ANN, and 1,000 generations of the GA for an optimization of only 4 DDoF. Besides this considerable effort, the authors further highlight an important limitation of such simulation-based optimization, as about 40% of the simulations do not converge and are not considered for building the surrogate model. This is in agreement with the numbers reported by Ibrahim et al. [65] for the surrogate-based optimization of a more complex crude oil distillation unit, for which the 7,000 sampling simulations reportedly took 1.5 h. As a more efficient approach, Carpio et al. [66] apply an adaptive sampling strategy with kriging models for the optimization of a reactor network, a heat exchanger network, and an extractive distillation process, in combination with four different global search strategies. Keßler et al. [67] investigate a similar strategy with explicit and implicit surrogate models for the optimization of a distillation column for the separation of nonideal mixtures using BARON as a gradient-based global optimization method. While they conclude that the approach provides a computationally tractable solution for global optimization of nonideal distillation processes, the optimization of a single distillation column still requires a few hours of computational time. Another interesting approach for surrogate-based global optimization was recently presented by Wang et al. [68] for a simultaneous process and solvent optimization for an extractive distillation process. Instead of training separate surrogate models for solvent properties and process design, an integrated process model is trained on repeated process simulations and subsequently optimized by means of the DFO with NSGA II. While this innovative approach enables simultaneous optimization of the solvent and process, it builds on several hundred thousand individual simulations.

Schweidtmann et al. [69, 70] combine the concept of surrogate-based optimization using ANN and Gaussian process (GP) models, with a specialized approach for gradient-based global optimization on a reduced space problem formulation with special relaxations [71], applying MAINGO. The approach is demonstrated for different applications, including the optimization of a cumene process, modelled as a combination of 14 ANNs trained on Aspen Plus® simulations, as well as the optimization of working fluids for organic Rankine cycles [72] and the simultaneous design of ion separation membrane material and process [73]. The work of Burre et al. [74] further illustrates that a diverse mix of simplified models (Underwood equations for distillation) and ANN and GP models for different unit operations can also be solved efficiently by this approach for a reductive dimethoxymethane synthesis process. As demonstrated for various cases, this advanced combination of effectively trained surrogate models, tailored relaxations, and gradient-based global optimization provides a reliable approach that excels in computational efficiency, but is rather complex to apply.

9.2.2.2 Global search strategies (metaheuristics)

While surrogate-based optimization has received increasing attention in recent years, the direct application of metaheuristics for simulation-based optimization has been applied on a much broader basis as DFO approach for discontinuous and multimodal optimization problems. Besides deterministic global search methods, such as multistart heuristics and mesh adaptive direct search algorithms available in the solvers, OQNLP [75] and NOMAD,[2] especially the use of stochastic population-based and swarm intelligence algorithms have received considerable attention for process optimization applications. The most popular population-based algorithms in this context are evolutionary algorithms (EAs), which have been applied for DFO in different forms for more than 50 years, including GA, evolutionary strategies (ES), and differential evolution (DE), among others [76]. The most popular swarm intelligence methods are particle swarm optimization (PSO) algorithms, which have been applied for DFO for about 30 years [77]. Today, there is a plethora of metaheuristics available. The PlatEmo toolbox[3] already provides implementations of more than 250 different single- and multi-objective metaheuristics, as well as a variety of benchmark problems. The huge variety of metaheuristics is a curse and blessing at the same time, as it provides tremendous flexibility, but causes a complex problem to select and tune a specific approach properly [78]. Also refer to Chapter 4 for more information on the use of metaheuristics for the optimization of chemical processes.

For process optimization, metaheuristics are generally applied as simulation-based optimization methods, operating naturally in the reduced space of the DDoF, but requiring a sufficiently robust process simulation model, as well as a strategy to handle inequality constraints, such as purity specifications. These may either be integrated in the process simulation models by means of design specifications, potentially limiting the convergence of the simulator, or considered as additional penalty term that deteriorates the objective function for the metaheuristic [79]. The latter is the most prominent approach in EA and other population-based methods [80]. As already indicated for the surrogate-based optimization in Section 9.2.2.1, the convergence properties of the process simulator can be the bottleneck of DFO algorithms, since a lack of convergence and an infeasible solution are indistinguishable [81]. While most studies assume that a non-convergent simulation corresponds to an infeasible solution [82, 83], this may limit the identification of the best possible solution, which most

---

2 https://www.gerad.ca/en/software/nomad/.
3 https://github.com/BIMK/PlatEMO.

often is at the limits of validity. This problem was already emphasized by Gross and Roosen [84], who first proposed a simulation-based process optimization with an EA in 1998. This concept was again highlighted in the recent work of Janus et al. [58], who besides a convergence ratio of about 50% of the simulations, stated that less than 0.1% of the simulations actually satisfied the product purities for the heteroazeotropic distillation process under investigation.

Despite the need for a sufficient convergence ratio, stochastic global optimization methods have been applied to various complex process optimization problems. Barakat and Sorensen [85] applied a GA for the optimization of a pervaporation-distillation process, whereas Koch et al. [86] applied a modified DE algorithm for a similar optimization problem with a process model in Aspen Custom Modeler. A similar algorithm was applied by Niesbach et al. [87] for the optimization of a reactive distillation process on the basis of a rate-based reactive distillation model. In order to tackle the convergence problem, Wierschem et al. [88] apply a homotopy approach based on pre-compiled snapshots of converged simulations for the optimization of an enzymatic reactive distillation column.

A huge advantage of simulation-based optimization with metaheuristics is its straightforward extension to multi-objective optimization (MOO), as several complex objectives may be evaluated, subsequent to the simulation of the respective individuals. Gomez-Castro et al. [89] apply the popular NSGA-II solver [90], in combination with flowsheet simulation, in Aspen Plus® for the optimization of multiple dividing wall columns. The same combination is, e.g., also used for the optimization of a hybrid distillation-crystallization process [91], for the optimization of VOC and solvent recovery processes [92], and for the optimization of an industrial styrene reactor [93]. Rangaiah et al. [94] provide an extensive review and tutorial on the use of such combinations for chemical engineering problems.

9.2.3 Summary of the different optimization methods

Each of the discussed methods offers certain advantages but also suffers from limitations that prohibit a widespread application for process optimization. Gradient-based global optimization methods overcome the burden of providing good initial solutions and can effectively handle a large number of decision variables. They are the only methods that provide mathematical proof for global optimality. However, they require sufficiently smooth equation-based model formulations, variable bounds, and proper relaxations, which limit the applicability to general process optimization problems based on rigorous models. Surrogate-based optimization provides a bridge that covers some of these limitations by generating a well-tractable surrogate model that can be effectively optimized by gradient-based global solvers, especially in combination with a reduced-space formulation and special relaxations. Yet, generating these models can be cumbersome, with several thousands of process simulations required as sampling data to train the surrogate models, which scales exponentially with the number of DDoF. Thereby, the success and effort for surrogate-based optimization, similar to other DFO methods, depend strongly on the number of DDoF and the individual process simulations. It is not uncommon that these may fail to converge in many cases and, in even fewer cases, provide results that actually satisfy all constraints, mandating properly tuned process simulation models.

9.3 Hybrid and memetic optimization methods

The fundamental idea behind hybrid algorithms is to advantageously combine global search and local optimization methods by exploiting the benefits of the individual methods. Metaheuristics are effective at identifying promising regions within the search space (exploration) but perform poorly in converging to an optimum (exploitation), even if it is only a local one. In contrast, local optimization and especially gradient-based algorithms can efficiently handle inequality constraints and excel at quickly identifying local optima, when provided a proper initial solution. The purposeful combination of both types of methods can drastically reduce the computational effort while simultaneously increasing the reliability of finding the best possible solution. Liñan et al. [95] provide an overview and classification of hybrid optimization methods in the context of process flowsheet optimization, which is illustrated in Figure 9.4. Sequential hybrid methods represent a straightforward combination of a global search algorithm and a local optimization solver, e.g., applied by Herrera Velazquez et al. [96] or Chia et al. [97, 98] for the optimization of heat-integrated distillation columns and membrane-assisted distillation processes. The sequential combination enables less restrictive termination criteria for the global search to determine a good initial solution for the local optimization that performs a final refinement. However, both algorithms handle the same DDoF and risk convergence to a suboptimal solution if the global search fails to identify the local region with the global optimum, which still depends on the convergence of a simulation model in case of a DFO algorithm. The concept of a parallel hybrid algorithm with parallel execution of a stochastic global search, i.e., a differential evolution with a tabu list algorithm, and a local deterministic optimization, based on a discrete-steepest descent algorithm with variable bounding has only recently been proposed by Liñan et al. [95]. It requires higher-level orchestration that analyzes the results and progress of both algorithms, which, in this case, are even executed on individual computers, and transfers information. The application is shown for the optimization of a thermally coupled side-rectifier and a more complex azeotropic distillation process, for which a comparison with a standalone stochastic global search shows a considerably quicker convergence to the final solution. Refer to the article of Liñan et al. [95] for further details of the implementation and applications.

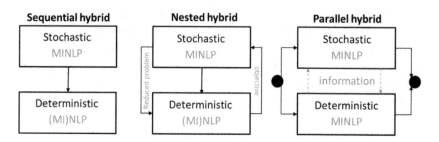

Figure 9.4: Classification of hybrid optimization approaches in agreement with [95].

The most applied class of hybrid algorithms constitutes nested hybrid algorithms, which properly distribute the individual DDoF between the global and local optimization methods, which can considerably improve the performance of the optimization in terms of solution quality and computational time. Athier et al. [99] already proposed such a nested algorithm for the synthesis of a heat exchanger network in 1997. In the outer loop, a metaheuristic is used to determine the structure of the heat exchanger network, while a deterministic algorithm is used to solve the NLP for the exchanged heat duties in the inner loop. Luo et al. [100] present a similar nested hybrid approach based on the combination of a GA and a local deterministic optimization of heat exchanger networks, which was further extended by Rathjens and Fieg [101], applying several heuristics. Such a combination of a population-based EA and a local refinement strategy was introduced before as memetic algorithm (MA) by Moscato [78]. It is evident that the success of a population, particularly in higher organisms, is influenced not only by individual genetic traits but also by the population's collective ability to adapt and learn. As such, the concept of an MA follows the evolutionary theory of Lamarck, mimicking cultural evolution, which depends much less on the aspect of mutation and is focused on a directed transfer of problem-specific knowledge [78, 80]. It has been observed in several studies that extending an EA to an MA significantly reduces the range of population sizes required for success [102]. The application has been shown for various real-world problems, including the design of reactive distillation columns [103, 104], in combination with an additional side-stream reactor [105], the design of distillation-based separation processes [106], and early-stage synthesis of complete processes under uncertainties [107]. Table 9.1 presents an overview of various applications of hybrid optimization methods for process optimization problems, with the respective classification and type of algorithms.

Especially, the combination of a problem-specific EA and an efficient local optimization that handles different sets of the DDoF showed large prospects for solving challenging process design problems. Such an MA was, e.g., proposed by Urselmann et al. [110] and demonstrated for the optimization of reactive distillation columns [104] and additional external reactors [105]. A schematic of the concept is illustrated in Figure 9.5, showcasing the nested structure, in which the evolutionary strategy handles all DDoF,

Table 9.1: Overview of publications on the design and application of hybrid and memetic algorithms in the scope of process optimization problems in chemical engineering.

Problem type	Application	Optimization approach	Type of combination	Reference
MINLP	(Extractive) Distillation processes	Nested stochastic-deterministic hybrid	MA, EA defines superstructure model, local MINLP optimization in GAMS as series of relaxed NLPs [108]	[109]
MINLP	Reactive distillation (with reactor)	Nested stochastic–deterministic hybrid	MA, EA with local solution of NLPs in GAMS	[104, 105, 110, 111]
MINLP	Distillation-based processes	Nested stochastic-deterministic hybrid	MA, EA with local optimization with Nelder-Mead algorithm [112] in ACM	[106]
MINLP	Hetero-azeotropic distillation	Nested stochastic-deterministic hybrid	MA, local solution of NLPs with mesh adaptive direct search algorithm linked to Aspen Plus	[58, 113]
MNILP	Heat exchanger network	Nested stochastic-deterministic hybrid	MA, GA optimizes heat exchanger network structure, which is optimized by Newton-type algorithm	[99–101]
MINLP	Reactive distillation	Nested stochastic-deterministic hybrid	MA, SQP with subsequent SA for continuous DDoF	[103]
MINLP	Heat exchanger network	Nested stochastic/deterministic hybrid	PSO for discrete variables, GA-SQP for continuous variables	[114]
MINLP	Energy-integrated extractive distillation	Nested stochastic/deterministic-deterministic hybrid	MA, EA or multi-start grid search with local MINLP optimization in GAMS as series of relaxed NLPs [108]	[41]
NLP	Flowsheet superstructure	Nested stochastic-deterministic hybrid	Evolutionary cycle, encoded with evolutionary algorithm $(\mu + \lambda, \kappa)$, in FSOpt local search (GAMS, solver: IPOPT)	[107]
MINLP	Building block-based process synthesis	Nested stochastic-deterministic hybrid	MA, Evolutionary strategy (VBA) combined with NLP approach (ACM)	[115, 116]
MINLP	Adsorption desorption process	Nested stochastic-deterministic hybrid	GA and gradient-based optimization (GAMS)	[117]

Table 9.1 (continued)

Problem type	Application	Optimization approach	Type of combination	Reference
MINLP	Distillation process	Sequential stochastic-deterministic hybrid	GA, with subsequent OAERAP (gPROMS)	[118]
MINLP	DWC and hybrid process	Sequential stochastic-deterministic hybrid	GA or PSO, with subsequent OAERAP (gPROMS)	[97, 98]
MINLP	Internally heat-integrated distillation	Sequential deterministic-stochastic optimization	Two-stage optimization: 1. stage problem MILP with CPLEX and 2. stage NLP with simulated annealing with Aspen Plus simulations	[96]
MINLP	Multimodal optimization problem	Nested stochastic-stochastic hybrid	MA, EA and local search covariance Matrix Adaptation Evolution Strategy	[119]
MINLP	Distillation process	Hybrid stochastic deterministic algorithm in parallel	Discrete-steepest descent algorithm with variable bounding and differential evolution with tabu list, with simulations in Aspen Plus	[95]
NLP	Airfoil optimization on the basis of CFD simulation	Nested stochastic surrogate-based hybrid	Multi-objective GA with surrogate-assisted local search	[120]
MINLP	Energy-integrated distillation	Nested surrogate-deterministic hybrid	ANN-based optimization with local MINLP optimization in GAMS [108]	[121]

but fixes only those that define the process structure, which in the case of the reactive distillation column includes the number of stages as well as the number and location of the feed streams. A gradient-based local optimization solves for each individual, an NLP problem to refine the continuous DDoF, minimizing the objective function subject to all equality and inequality constraints. In the case of the reactive distillation, the individual feed flowrates, the reflux and boil-up rate as well as the amount of catalyst on the individual stages are optimized. While the evolutionary strategy still covers the whole set of DDoF, the local refinement has a significant effect on the exploitation efficiency of the MA. Janus et al. [58] modified the concept for simulation-based optimization by replacing the gradient-based optimization with a customized DFO approach, as illustrated in Figure 9.5 as well. They illustrate the simulation-based MA for the optimization of a heteroazeotropic distillation process using the DFO solver NOMAD and Aspen Plus® for process simulation. While the simulation-based MA is capable of generating significantly improved solutions compared to an iterative optimization, it requires a huge number of individual simulations, of which the vast amount (>99%) either fail to converge or provide insufficient product purities. Based on these results, Janus and Engell [59] conclude that the integration of the DFO strategy does not aid significantly beyond an EA-based optimization on top of the simulator, and extend the MA with additional surrogate models that performs a classification of the individuals prior to the simulation, in order to increase the convergence ratio. They illustrate this approach for the optimization of a hydroformylation process, for which flowsheet simulations are performed with recycle streams in Aspen Plus. While the surrogate-assisted optimization approach significantly reduces the number of necessary simulations, the authors indicate that this approach may require additional tuning in the case of a large number of DDoF to avoid an exponential increase in the required training data. Similar concepts for a surrogate-assisted adaptive sampling of flowsheet simulations have also been proposed by Bortz et al. [122, 123] and show considerable potential for the improvement of DFO-based methods in general.

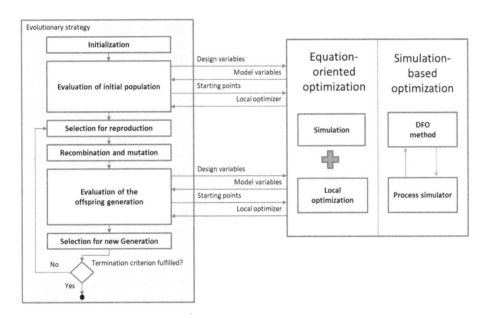

Figure 9.5: Memetic algorithm with nested structure and gradient-based local refinement [111] as well as a DFO-based local refinement [113] on the basis of a process simulator.

Another strategy that improves the convergence ratio and maximizes the synergies in the MA has been proposed by Skiborowski et al. [109], combining a $(\mu + \lambda, \kappa)$ evolutionary strategy with a local optimization approach that solves a series of successively relaxed MINLPs [109]. They demonstrated the application for the optimization of different nonideal and azeotropic distillation processes, for which the continuous DDoF, e.g., operating pressures, heat duties, and solvent-to-feed ratio, are solely handled by the local optimization, which also refines discrete decisions, such as the number of stages and feed-stage locations. Only the initial values of these discrete DDoF are defined by the EA, which further fixes discrete DDoF that would cause larger problems for the gradient-based optimization. Examples of the latter are the choice of utilities that limit applicable pressure ranges or the choice of solvents, which affect the thermodynamic computations.

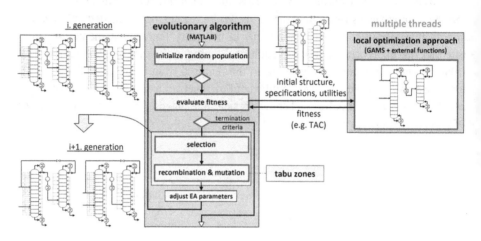

Figure 9.6: Illustration of the hybrid optimization approach, combining an evolutionary algorithm (EA) with a deterministic local optimization. Reprinted (adapted) with permission from Skiborowski et al. [109]. Copyright 2015 American Chemical Society.

The hybrid approach, which is schematically illustrated in Figure 9.6, also integrates the use of tabu zones and other constraints in the EA to further reduce the required number of individuals and generations (see Section 9.4.1). Kruber et al. [41] further showcased the application to a more complex case study for extractive distillation, considering solvent selection and options for energy integration and evaluated an adaptive multi-start grid search as a deterministic alternative to the EA (see Section 9.4.2). Both methods are extremely efficient when compared to other global optimization strategies for the optimization of a single distillation column, for which the best possible solution is identified in almost every run of the algorithm in less than a minute for ideal mixtures [41] and several minutes for strongly nonideal mixtures. The global optimum for the first case is confirmed by BARON, which however requires several hours to confirm global optimality (see Figure 9.3). Kruber et al. [121] further proposed another modification of the hybrid approach toward a surrogate-based optimization, with an ANN replacing the global search algorithm (see Section 9.4.3).

9.4 Case studies

To provide some further insight and to illustrate the potential benefits of hybrid optimization methods, three specific case studies for the optimization of distillation processes on the basis of rigorous models are presented in the following sections. Section 9.4.1 illustrates the implementation and application of the hybrid optimization algorithm proposed by Skiborowski et al. [109] for the design of an entrainer-enhanced pressure-swing distillation process. Section 9.4.2 illustrates the modification of the algorithm with an adaptive multi-start grid search for an energy-integrated extractive distillation process, as presented by Kruber et al. [41]. Finally, Section 9.4.3 illustrates another modification of the hybrid approach with a surrogate-based optimization for an energy-integrated distillation sequence, as presented by Kruber et al. [121].

9.4.1 EA-based hybrid optimization of an entrainer-enhanced pressure-swing distillation process

The first example illustrates the application of the hybrid optimization approach proposed in the work of Skiborowski et al. [109], which is illustrated in Figure 9.6. It combines an EA with an efficient local deterministic optimization of an MINLP superstructure model in a nested structure. The EA employs a $(\mu + \lambda, \kappa)$ evolutionary strategy. In the initial stage, the EA randomly generates a population of λ_{init} individuals, each with a genome containing nominal and ordinal variables of the MINLP problem. The ordinal variables, i.e., the maximum number of stages and initial feed-stage positions of each column in the superstructure model follow a specific order, while the nominal variables, i.e., the selected utilities, do not. In order to build a sufficiently diverse basis for the EA, the size of the initial population is larger than that of subsequent populations ($\lambda_{init} > \lambda$). As illustrated in Figure 9.7, the nominal DDoF are either randomly or uniformly distributed, taking into account different priorities, e.g., the maximum number of stages (high priority) is determined prior to the initial feed-stage location (low priority). The individual is only accepted if it passes further constraints, e.g., demanding that the feed location is below the maximum number of stages, and the check on existing tabu zones, which avoids the evaluation of individuals in close proximity. If the DdoF values do not meet the constraints or fall within the tabu zones, the algorithm reassigns random values until a valid new individual is generated or a maximum number of trials is reached. By carefully managing the assignment of DdoF and rigorously checking the constraints, the method ensures that the initial population is diverse while minimizing the risk of infeasible solutions.

Each of the individuals defines a superstructure model, which is further optimized with an efficient local optimization approach that builds on the solution of a series of successively relaxed MINLP problems with gradient-based algorithms. The

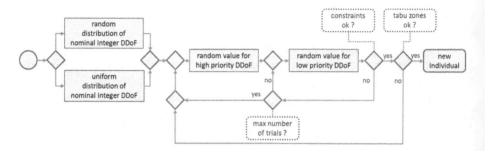

Figure 9.7: Procedure for initializing an individual within the initial population. Reprinted (adapted) with permission from Skiborowski et al. [109]. Copyright 2015 American Chemical Society.

approach successively builds the final superstructure model and determines a good initial solution by means of a polylithic modeling and solution approach. In the specific case of the pressure-swing distillation process, both column models are first initialized individually by solving the respective NLP problems for a fixed column structure, with purity constraints starting from the state variable values of an initial flash calculation. After integrating the individual column models and solving an NLP problem for minimizing the total heat duty, the structural constraints are relaxed and the sizing and costing equations are added to finally solve the MINLP problem for minimizing the TAC. The interested reader is referred to the work of Skiborowski et al. [11] for further details on the specific local optimization strategy and the work of Kallrath [124] for a more general introduction to polylithic modeling and solution approaches. Therefore, The EA operates only on a subset of the overall DDoF, whereas some of the DDoF are refined by the local optimization, and others are handled solely by the local optimization. As illustrated in Figure 9.8 for the specific case of the entrainer-enhanced pressure-swing distillation column, the optimization problem has overall 12 DDoF, of which only 7 are defined or initialized by the EA, while 5 are refined and 5 more are handled entirely by the local MINLP solver. The EA only determines the two utility choices and specifies the maximum number of stages, as well as the initial feed and recycle stages for each column. The local optimization determines the final stage numbers, operating pressure of the second column, and heat duties for a cost optimal design. The integration of the algorithms thereby drastically reduces the search space for the EA, which reduces the complexity of the local optimization by avoiding singularities for the utility choices and varying the initial superstructure.

The resulting cost-optimal local solution with a minimized total annualized cost (TAC) defines the fitness of the respective individual. The local optimization is deterministic, and it is expected that the same local optimum will be determined for closely related superstructure models, which are therefore disclosed from further evaluation by means of the tabu zones. For the creation of the next generation, a (μ, λ, κ)-selection is applied, picking the best μ individuals from the λ individuals of the current generation and previously considered parents that have not exceeded a maxi-

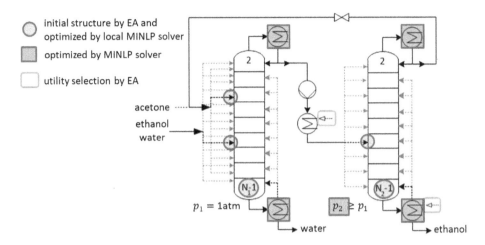

○ initial structure by EA and
optimized by local MINLP solver

▢ optimized by MINLP solver

▢ utility selection by EA

Figure 9.8: Superstructure for the entrainer-enhanced pressure-swing distillation. Reprinted (adapted) with permission from Skiborowski et al. [109]. Copyright 2015 American Chemical Society.

mum age of κ generations. This approach factors in some elitism for preserving good solutions while avoiding early convergence to sub-optimal local solutions. As illustrated in Figure 9.9, a similar two-stage process with additional constraint and tabu-zone testing, as applied for the creation of the initial population, is applied when creating the offspring for the next generation by applying recombination and mutation. Discrete and intermediate recombination as well as subsequent mutation with specific operators and probabilities are applied, depending on the specific DDoF. To foster a transition from an early exploration to a later exploitation phase, the mutation parameters are further adjusted after a certain number of generations.

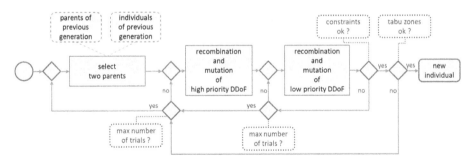

Figure 9.9: Procedure for initializing an individual within a new generation. Reprinted (adapted) with permission from Skiborowski et al. [109]. Copyright 2015 American Chemical Society.

Refer to Skiborowski et al. [109] for further details. The hybrid algorithm is supposed to converge and terminate in case of a lack of improvement over consecutive generations, or if the maximum number of trials was reached for the generation of new individuals. Due to the population-based EA, the hybrid optimization approach, by default, generates a multitude of solutions, which due to the local optimization, all reflect locally optimal solutions. It, therefore, is not only able to escape local optima and seek the best possible solution but also provides an excess of information that fosters an improved understanding of the optimization problem. Figure 9.10 illustrates the progression of the resulting TAC of the respective individuals over the number of generations (left) and the individual cost distribution in terms of annual capital and operating costs (right).

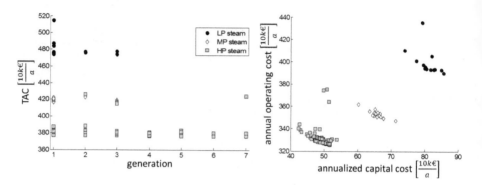

Figure 9.10: Evolution of the total annualized cost (TAC) across generations (left) and the cost distribution (right) for a single optimization run. Reprinted with permission from Skiborowski et al. [109]. Copyright 2015 American Chemical Society.

The hybrid algorithm maintains the initial diversity regarding the utility choices only over the first few generations while subsequently focusing on the most effective choice. While this is obviously high-pressure (HP) steam, the cost distribution highlights that for each utility, an individual Pareto front is approached by the individual local optima. While there is some flexibility in the distribution of capital and investment cost regarding the overall least expensive design, there are also many suboptimal local solutions, which originate from an improper initial superstructure model. Due to the stochastic nature of the EA, it is elementary to analyze the reproducibility of the results based on repetitive evaluations. Table 9.2 summarizes the results of 15 repetitive runs of the algorithm for the case study, showcasing a variation of less than 0.1% concerning the TAC of the best solution, which, however, shows some flexibility with respect to the DDoF, as was expected from the form of the Pareto front in Figure 9.10.

Besides the added information offered by the multitude of locally optimal solutions, the high convergence ratio of more than 75% underscores the practical advantages of the hybrid optimization approach, in comparison to the simulation-based DFO methods described in Section 9.2.2, while the application of gradient-based global optimization methods, as described in Section 9.2.1, has not yet been demonstrated for rigorous process models of similar complexity.

Table 9.2: Performance evaluation of 15 consecutive runs for optimizing the entrainer-enhanced pressure-swing distillation process.

	Overall			Best individual per run		
	n_g	t_{tot} (s)	Convergence (%)	TAC (mio€·a^{-1})		t_{opt} (s)
min	5.00	7,379	75.56	3.7638		239
Ø	5.87	8,887	78.96	3.7643		370
max	8.00	10,800	85.56	3.7653		536

	Best individual per run						
	$N_{trays}^{Col\ 1}$	$N_{feed}^{Col\ 1}$	$N_{recycle}^{Col\ 1}$	$N_{trays}^{Col\ 2}$	$N_{feed}^{Col\ 2}$	$\dot{Q}_B^{Col\ 1}$ / kW	$\dot{Q}_B^{Col\ 2}$ / kW
min	79	75	70	127	66	8.81	7.81
Ø	82	77	72	132	70	8.84	7.83
max	85	81	76	135	74	8.85	7.84

Reprinted (adapted) with permission from Skiborowski et al. [109]. Copyright 2015 American Chemical Society.

9.4.2 Hybrid optimization of an energy-integrated extractive distillation with entrainer selection

The hybrid approach is modified by replacing the EA with a systematic multi-start grid search (MSGS) in the work of Kruber et al. [41]. In this way, a fully deterministic hybrid method is generated, which overcomes the need for repeated evaluations. A schematic of the MSGS-based hybrid approach is depicted in Figure 9.11 (left), together with an illustration of the grid refinement (right). It starts with an initial population represented by a systematic grid, for which a central representation of each grid segment (individual) is evaluated by the local deterministic optimization, similar to the individuals in the EA-based hybrid method (cf. Figure 9.6). The grid dimensions are equivalent to the DDoF handled by the EA. The best μ individuals are selected, and the grid is refined around these candidates.

The MSGS-based hybrid and the EA-based hybrid were applied for the optimal design of an extractive distillation process, including the choice of solvent and possible options for heat integration. For both metaheuristics, a uniform distribution of the respective combinations is enforced in the initial population. The metaheuristic fixes the solvent choice and the type of process configuration, for which the maximum stage number and initial feed and recycle stages for the superstructure are defined. Thus, the local optimization of the individual process configurations, which was developed in the work of Waltermann et al. [33], focuses on the superstructure of the specific configurations, which are illustrated in Figure 9.12. Thus, each individual local optimization is a confined MINLP problem.

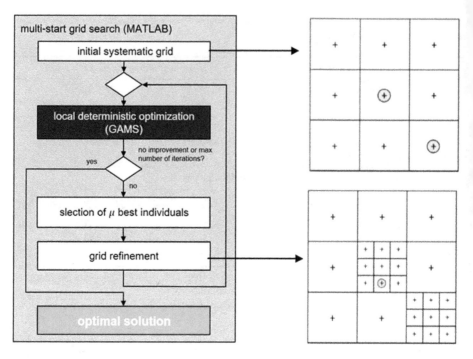

Figure 9.11: General procedure of the multi-start grid search (MSGS) hybrid optimization approach. Reprinted from Kruber et al. [41], with permission from Elsevier.

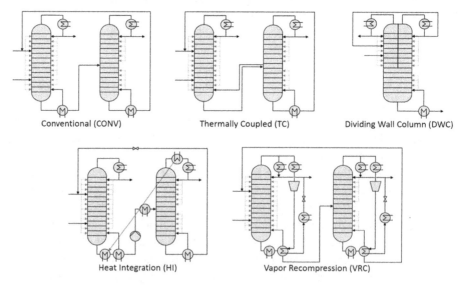

Figure 9.12: Superstructure models for a conventional (CONV), thermally coupled (TC), dividing wall column (DWC), heat-integrated (HI), and vapor recompression (VRC) process. Reprinted from Kruber et al. [41], with permission from Elsevier.

Figure 9.13 illustrates the results for the separation of an equimolar acetone–methanol mixture with six solvents, determined by a single run of the two hybrid optimization approaches. While the initial generation covers the full range of process-solvent combinations, the EA-based hybrid approach quickly focuses on the most promising combination, which is, in this case, using water as solvent. However, the first generations maintain a certain diversity, and some other solvents are occasionally considered in later generations due to random mutation. Due to the strict selection of a few best configurations, the MSGS-based approach basically focuses solely on the water-based processes after the initial population has been evaluated. Both approaches identify the same process-solvent combination as most promising, whereas the MSGS converges after just four generations, after having started with a larger number of individuals in the initial generation.

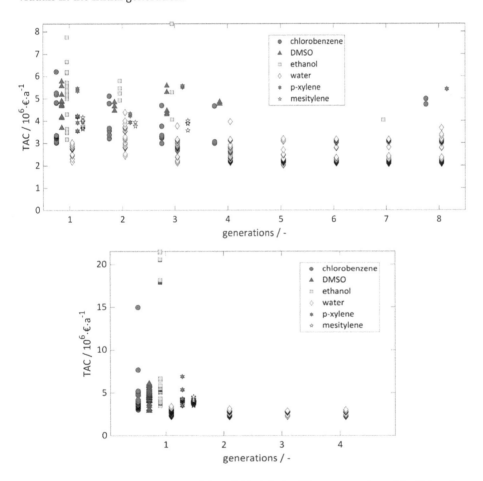

Figure 9.13: Total annualized costs (TAC) of the individuals in the different generations of the EA-based hybrid (top) and MSGS-based hybrid (bottom) for a single run. Reprinted from Kruber et al. [41], with permission from Elsevier.

While the MSGS-based hybrid is fully deterministic, the reproducibility of the results of the EA-based hybrid needs to be evaluated based on repeated runs. Table 9.3 provides an analysis of five repeated runs, showcasing high reproducibility, with a variation in the objective of the best individual of about 1.5%. Depending on the number of generations, the optimization takes between 5 and 15 h, which is still acceptable, considering the complexity of the problem, allowing for a quick adaption to other separation problems.

Table 9.3: Performance evaluation of five consecutive runs for optimizing the extractive distillation (ED) process.

	Overall			Best individual per run				
	n_g	t_{tot} (s)	λ_{tot}	Convergence (%)	TAC (mio€·a⁻¹)	t_{opt} (s)	$\dot{Q}_{heat}^{total}$ (kW)	$\dot{Q}_{cool}^{total}$ (kW)
min	4.0	17,672	300	77.0	2.0241	743.49	7,882.6	7,668.1
Ø	7.2	31,680	492	80.2	2.0345	1,020.20	7,926.6	7,706.3
max	10.0	52,578	660	82.4	2.0547	1,127.16	7,970.4	7,759.2

	Best individual per run						
	$N_{trays}^{Col\ 1}$	$N_{feed}^{Col\ 1}$	$N_{recycle}^{Col\ 1}$	$N_{trays}^{Col\ 2}$	$N_{feed}^{Col\ 2}$	$N_{solvent}$	$N_{process}$
min	71.0	60.0	37.0	31.0	22.0	4	2
Ø	74.4	63.4	38.6	32.0	23.6	4	2
max	77.0	67.0	40.0	33.0	24.0	4	2

Reprinted from Kruber et al. [41], with permission from Elsevier.

While both methods enable the identification of the best process–solvent combination, the EA-based hybrid determines overall better solutions, as indicated by the analysis of the cost distribution in Figure 9.14. Despite the single-objective optimization, the results of both methods approach a similar approximation of the Pareto front for each solvent. On closer inspection of the best solutions in the bottom diagrams, it becomes apparent that the EA-based hybrid identifies a set of water-based processes, which are not covered by the MSGS-based hybrid, resulting in a more than 5% more expensive process design, which exceeds the variation of the individual runs of the EA-based hybrid.

The results illustrate that although the MSGS-based hybrid has the advantage of representing a fully deterministic approach, it may miss the best possible solution. This limitation of the MSGS-based approach may likely be resolved by a modification of the refinement approach, increasing the initial resolution and number of grid points for refinement. This will obviously increase the computational cost and require a case-dependent calibration. Thus, the evaluation of a proper parameterization that balances computational efficiency and reliability will require a comparable effort to the repeated runs for the EA-based hybrid.

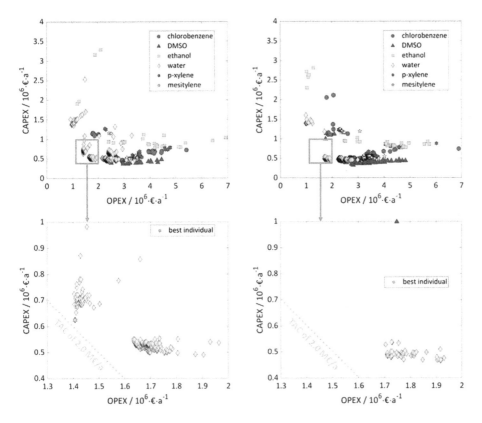

Figure 9.14: Distribution of capital expenditures (CAPEX) and operational expenditures (OPEX) of all (top) and the best-performing individuals (bottom) for a single run of EA-based hybrid (left) and MSGS-based hybrid (right). Reprinted from Kruber et al. [41], with permission from Elsevier.

9.4.3 Surrogate-assisted optimization of heat-integrated distillation sequence

As another modification of the hybrid algorithm, a surrogate-based optimization with a top-level ANN trained on the results of local optimization runs was proposed by Kruber et al. [121]. The modified hybrid approach is illustrated in Figure 9.15. It implements an adaptive sampling on the results of the local optimization, which again reduces the design space for the surrogate model's training, reflecting the results of the individual local MINLP optimizations rather than the response surface of the simulations. The approach starts with an initial Latin hypercube sampling that spans the space of the DDoF, which are considered as features of the ANN. For the energy-integrated distillation sequence considered as a case study, the type of process configuration is reflected by four binary variables, following the principle of one-hot encoding, whereas the maximum number of trays and the initial feed-stage location for the definition of the superstructure model are the respective DDoF for the ANN. After local optimization of each superstructure model, training of the ANN is performed and a mixed adaptive sampling algorithm is used to identify a new set of candidates in each generation, based on a combination of a space-filling approach, an approach to reduce the local variance of the ANN, and an approach to reduce the variance in accordance with the current best solution.

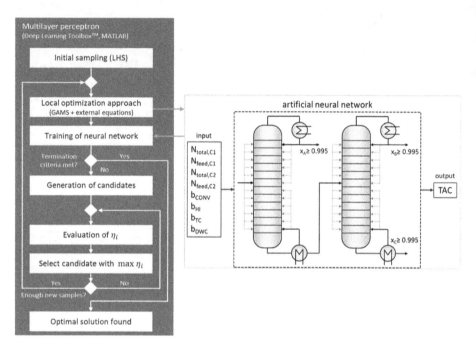

Figure 9.15: General procedure of the surrogate-assisted hybrid optimization approach. Reprinted from Kruber et al. [121], with permission from Elsevier.

As a case study, the ANN-based hybrid was evaluated for the design of an energy-integrated distillation sequence for the separation of a benzene–toluene–xylene mixture with a molar feed composition of (0.3, 0.3, 0.4) and product purities of at least 99.5 mol%. The employed ANN was represented by a multilayer perceptron with two hidden layers of five and three neurons, each utilizing a hyperbolic tangent sigmoid transfer function. The initial sampling comprises 100 data points with a subsequent sampling rate of 25% with respect to the current number of samples. Figure 9.16 represents a reduced illustration of the response surface of the ANN concerning the total number of stages for each column (left) as well as the best process design identified from the optimization of the ANN (right). The reduced response surface of the ANN indicates the existence of a flat optimum with respect to the TAC, as already indicated

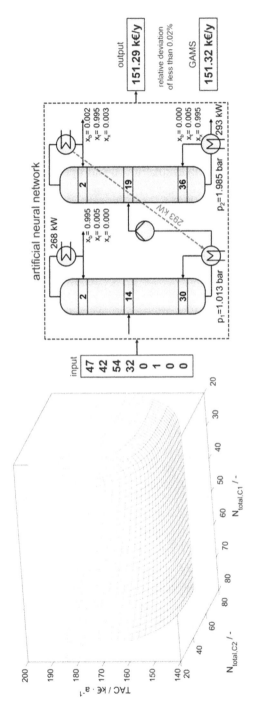

Figure 9.16: Reduced response surface of the ANN for the heat-integrated sequence (left) and the corresponding process design with objective values for the ANN and rigorous model (right). Reprinted from Kruber et al. [121], with permission from Elsevier.

by the results of the other two hybrid optimization algorithms (cf. Figures 9.10 and 9.14). The result of the surrogate-based optimization is in excellent agreement with the local optimization of the rigorous model in GAMS, showing a deviation of less than 0.1%.

9.5 Conclusion and outlook

This chapter introduces hybrid and memetic algorithms for global optimization, building on a detailed analysis of the benefits and limitations of gradient-based global optimization and global search methods. Apart from a general overview of existing applications of hybrid optimization methods in the context of chemical process optimization, three specific examples of a hybrid DFO–gradient-based optimization approach are illustrated for the conceptual design of complex distillation processes on the basis of rigorous process models. While the generation of hybrid optimization methods requires a thoughtful combination of the individual global and local optimization methods, it can effectively exploit the synergies between both methods and enable computationally efficient global optimization of real-world problems with complex rigorous process models and a large number of discrete and continuous DDoF. The examples discussed in Section 9.4 highlight the reliability and computational efficiency of such hybrid optimization methods and imply that the solution to even more complex process optimization problems should be feasible with similar strategies. When building such hybrid optimization methods, it is generally recommended to maximize the use of the local optimization by handling as many continuous and (relaxed) discrete DDoF as possible to limit the search space of the global search. For any hybrid algorithm that integrates a stochastic optimization method, a validation of the results by repeated runs is mandatory and the convergence rate of the local optimization should always be critically assessed.

References

[1]   Biegler, L. T. Multi-level optimization strategies for large-scale nonlinear process systems. *Computational Chemical Engineering*, 185, 2024, 108657.
[2]   Waldron, C., Pankajakshan, A., Quaglio, M., Cao, E., Galvanin, F., Gavriilidis, A. Model-based design of transient flow experiments for the identification of kinetic parameters. *Reaction Chemistry & Engineering*, 5, 2020, 112–123.
[3]   Blanquero, R., Carrizosa, E., Jiménez-Cordero, A., Rodríguez, J. F. A global optimization method for model selection in chemical reactions networks. *Computational Chemical Engineering*, 93, 2016, 52–62.

[4]   Schenk, C., Short, M., Rodriguez, J. S., Thierry, D., Biegler, L. T., García-Muñoz, S., Chen, W. Introducing KIPET: A novel open-source software package for kinetic parameter estimation from experimental datasets including spectra. *Computational Chemical Engineering*, 134, 2020, 106716.

[5]   Peschel, A., Freund, H., Sundmacher, K. Methodology for the design of optimal chemical reactors based on the concept of elementary process functions. *Industrial & Engineering Chemistry Research*, 49, 2010, 10535–10548.

[6]   Freund, H., Maußner, J., Kaiser, M., Xie, M. Process intensification by model-based design of tailor-made reactors. *Current Opinion In Chemical Engineering*, 26, 2019, 46–57.

[7]   Shi, J., Biegler, L. T., Hamdan, I., Wassick, J. Optimization of grade transitions in polyethylene solution polymerization process under uncertainty. *Computational Chemical Engineering*, 95, 2016, 260–279.

[8]   Zhou, C., Deng, J., Gao, F., Zhu, Y., Jia, C., Wang, J. Integration of reactor network synthesis: A literature review and opportunities for future research. *Asia-Pacific Journal of Chemical Engineering*, 2023, 18.

[9]   Kraemer, K., Kossack, S., Marquardt, W. Efficient optimization-based design of distillation processes for homogeneous azeotropic mixtures. *Industrial & Engineering Chemistry Research*, 48, 2009, 6749–6764.

[10]   Caballero, J. A., Grossmann, I. E. Optimization of distillation processes. *Distillation: Fundamentals and Principles*, 2014, 437–496.

[11]   Skiborowski, M., Harwardt, A., Marquardt, W. Efficient optimization-based design for the separation of heterogeneous azeotropic mixtures. *Computational Chemical Engineering*, 72, 2015, 34–51.

[12]   Kossack, S., Kraemer, K., Gani, R., Marquardt, W. A systematic synthesis framework for extractive distillation processes. *Chemical Engineering Research and Design*, 86, 2008, 781–792.

[13]   Waltermann, T., Skiborowski, M. Conceptual design of highly integrated processes – optimization of dividing wall columns. *Chemie Ingenieur Technik*, 89, 2017, 562–581.

[14]   Waltermann, T., Skiborowski, M. Efficient optimization-based design of energy-integrated distillation processes. *Computational Chemical Engineering*, 129, 2019, 106520.

[15]   Scharzec, B., Waltermann, T., Skiborowski, M. A systematic approach towards synthesis and design of pervaporation-assisted separation processes. *Chemie Ingenieur Technik*, 89, 2017, 1534–1549.

[16]   Recker, S., Skiborowski, M., Redepenning, C., Marquardt, W. A unifying framework for optimization-based design of integrated reaction–separation processes. *Computational Chemical Engineering*, 81, 2015, 260–271.

[17]   Kiss, A. A., Jobson, M., Gao, X. Reactive distillation: Stepping up to the next level of process intensification. *Industrial & Engineering Chemistry Research*, 58, 2019, 5909–5918.

[18]   Kuhlmann, H., Skiborowski, M. Optimization-based approach to process synthesis for process intensification: General approach and application to ethanol dehydration. *Industrial & Engineering Chemistry Research*, 56, 2017, 13461–13481.

[19]   Demirel, S. E., Li, J., Hasan, M. M. F. A general framework for process synthesis, integration, and intensification. *Industrial & Engineering Chemistry Research*, 58, 2019, 5950–5967.

[20]   Tian, Y., Demirel, S. E., Hasan, M. F., Pistikopoulos, E. N. An overview of process systems engineering approaches for process intensification: State of the art. *Chemical Engineering and Processing*, 133, 2018, 160–210.

[21]   McBride, K., Sanchez Medina, E. I., Sundmacher, K. Hybrid semi-parametric modeling in separation processes: A review. *Chemie Ingenieur Technik*, 92, 2020, 842–855.

[22]   Skiborowski, M., Sudhoff, D. 1 Introduction to Process Intensification and Synthesis Methods. In Skiborowski, M., Górak, A. (Eds.). *Process Intensification*. De Gruyter, 2022, pp. 1–48.

[23]   Martins, J. R. R. A., Ning, A. *Engineering Design Optimization*. Cambridge University Press, 2022.

[24]   Chen, Q., Liu, Y., Seastream, G., Siirola, J. D., Grossmann, I. E. Pyosyn: A new framework for
       conceptual design modeling and optimization. *Computational Chemical Engineering*, 153, 2021,
       107414.

[25]   Kunde, C., Keßler, T., Linke, S., McBride, K., Sundmacher, K., Kienle, A. Surrogate modeling for
       liquid–liquid equilibria using a parameterization of the Binodal Curve. *Processes*, 7, 2019, 753.

[26]   Nentwich, C., Engell, S. Surrogate modeling of phase equilibrium calculations using adaptive
       sampling. *Computational Chemical Engineering*, 126, 2019, 204–217.

[27]   Tsay, C., Pattison, R. C., Piana, M. R., Baldea, M. A survey of optimal process design capabilities and
       practices in the chemical and petrochemical industries. *Computational Chemical Engineering*, 112,
       2018, 180–189.

[28]   Biegler, L. T. *Nonlinear programming: Concepts, algorithms, and applications to chemical processes.
       Society for Industrial and Applied Mathematics; Mathematical Optimization Society: Philadelphia*, 2010.

[29]   Grossmann, I. E. *Advanced Optimization for Process Systems Engineering*. Cambridge, New York, NY,
       Port Melbourne, VIC, New Delhi: Cambridge University Press, 2021.

[30]   Agi, D. T., Jones, K. D., Watson, M. J., Lynch, H. G., Dougher, M., Chen, X., Carlozo, M. N., Dowling,
       A. W. Computational toolkits for model-based design and optimization. *Current Opinion in Chemical
       Engineering*, 43, 2024, 100994.

[31]   Biegler, L. New directions for nonlinear process optimization. *Current Opinion in Chemical
       Engineering*, 21, 2018, 32–40.

[32]   Chen, Q., Grossmann, I. E. Recent developments and challenges in optimization-based process
       synthesis. *Annual Review of Chemical and Biomolecular Engineering*, 8, 2017, 249–283.

[33]   Waltermann, T., Grueters, T., Muenchrath, D., Skiborowski, M. Efficient optimization-based design of
       energy-integrated azeotropic distillation processes. *Computational Chemical Engineering*, 133, 2020,
       106676.

[34]   Caballero, J. A., Grossmann, I. E. Rigorous Design of Complex Liquid-Liquid Multi-Staged Extractors
       Combining Mathematical Programming and Process Simulators. In Brito Alves, R. M., De Oller Do
       Nascimento, C. A., Biscaia, E. C. (Eds.). *10th International Symposium on Process Systems Engineering:
       Part A*. Elsevier, 2009, pp. 981–986.

[35]   Dhiman, G. ESA: A hybrid bio-inspired metaheuristic optimization approach for engineering
       problems. *Engineering with Computers*, 37, 2021, 323–353.

[36]   Boukouvala, F., Misener, R., Floudas, C. A. Global optimization advances in mixed-integer nonlinear
       programming, MINLP, and constrained derivative-free optimization, CDFO. *European Journal of
       Operational Research*, 252, 2016, 701–727.

[37]   Adjiman, C. S., Androulakis, I. P., Floudas, C. A. Global optimization of mixed-integer nonlinear
       problems. *AIChE Journal*, 46, 2000, 1769–1797.

[38]   Tawarmalani, M., Sahinidis, N. V. A polyhedral branch-and-cut approach to global optimization.
       Mathematics Program, 103, 2005, 225–249.

[39]   Misener, R., Floudas, C. A. ANTIGONE: Algorithms for coNTinuous / integer global optimization of
       nonlinear equations. *Journal of Global Optimization*, 59, 2014, 503–526.

[40]   Belotti, P., Lee, J., Liberti, L., Margot, F., Wächter, A. Branching and bounds tightening techniques
       for non-convex MINLP. *Optimization Methods and Software*, 24, 2009, 597–634.

[41]   Kruber, K. F., Grueters, T., Skiborowski, M. Advanced hybrid optimization methods for the design of
       complex separation processes. *Computational Chemical Engineering*, 147, 2021, 107257.

[42]   Amaran, S., Sahinidis, N. V. Global optimization of nonlinear least-squares problems by branch-and-
       bound and optimality constraints. *TOP*, 20, 2012, 154–172.

[43]   Ballerstein, M., Kienle, A., Kunde, C., Michaels, D., Weismantel, R. Deterministic global optimization
       of binary hybrid distillation/melt-crystallization processes based on relaxed MINLP formulations.
       Optimization and Engineering, 16, 2015, 409–440.

[44] Mertens, N., Kunde, C., Kienle, A., Michaels, D. Monotonic reformulation and bound tightening for global optimization of ideal multi-component distillation columns. *Optimization and Engineering*, 19, 2018, 479–514.

[45] Najman, J., Bongartz, D., Mitsos, A. Relaxations of thermodynamic property and costing models in process engineering. *Computational Chemical Engineering*, 130, 2019, 106571.

[46] Bongartz, D., Mitsos, A. Deterministic global flowsheet optimization: Between equation-oriented and sequential-modular methods. *AIChE Journal*, 65, 2019, 1022–1034.

[47] Tumbalam Gooty, R., Mathew, T. J., Tawarmalani, M., Agrawal, R. An MINLP formulation to identify thermodynamically efficient distillation configurations. *Computational Chemical Engineering*, 178, 2023, 108369.

[48] Li, J., Demirel, S. E., Hasan, M. M. F. Process synthesis using block superstructure with automated flowsheet generation and optimization. *AIChE Journal*, 64, 2018, 3082–3100.

[49] Ulonska, K., König, A., Klatt, M., Mitsos, A., Viell, J. Optimization of multiproduct biorefinery processes under consideration of biomass supply chain management and market developments. *Industrial & Engineering Chemistry Research*, 57, 2018, 6980–6991.

[50] Kenkel, P., Wassermann, T., Rose, C., Zondervan, E. A generic superstructure modeling and optimization framework on the example of bi-criteria Power-to-Methanol process design. *Computational Chemical Engineering*, 150, 2021, 107327.

[51] Rios, L. M., Sahinidis, N. V. Derivative-free optimization: A review of algorithms and comparison of software implementations. *Journal of Engineering and Optimization*, 56, 2013, 1247–1293.

[52] Kim, S. H., Boukouvala, F. Surrogate-based optimization for mixed-integer nonlinear problems. *Computational Chemical Engineering*, 140, 2020, 106847.

[53] Lewandowski, J., Lemcoff, N. O., Palosaari, S. Use of neural networks in the simulation and optimization of pressure swing adsorption processes. *Chemical Engineering Technology*, 21, 1998, 593–597.

[54] Palmer, K., Realff, M. Optimization and validation of steady-state flowsheet simulation metamodels. *Chemical Engineering Research and Design*, 80, 2002, 773–782.

[55] McBride, K., Sundmacher, K. Overview of surrogate modeling in chemical process engineering. *Chemie Ingenieur Technik*, 39, 2019, 2233.

[56] Schweidtmann, A. M., Bongartz, D., Huster, W. R., Mitsos, A. Deterministic Global Process Optimization: Flash Calculations via Artificial Neural Networks. In *29th European Symposium on Computer Aided Process Engineering* . Elsevier, 2019, pp. 937–942.

[57] Clayton, A. D., Schweidtmann, A. M., Clemens, G., Manson, J. A., Taylor, C. J., Niño, C. G., Chamberlain, T. W., Kapur, N., Blacker, A. J., Lapkin, A. A. et al. 2020. Automated self-optimisation of multi-step reaction and separation processes using machine learning. *Chemical Engineering Journal*, 384, 123340.

[58] Janus, T., Foussette, C., Urselmann, M., Tlatlik, S., Gottschalk, A., Emmerich, M., Bäck, T., Engell, S. Optimization-based process synthesis based on a commercial flowsheet simulator. *Chemie Ingenieur Technik*, 89, 2017, 655–664.

[59] Janus, T., Engell, S. Iterative process design with surrogate-assisted global flowsheet optimization. *Chemie Ingenieur Technik*, 93, 2021, 2019–2028.

[60] Bhosekar, A., Ierapetritou, M. Advances in surrogate based modeling, feasibility analysis, and optimization: A review. *Computational Chemical Engineering*, 108, 2018, 250–267.

[61] Jones, D. R., Schonlau, M., Welch, W. J. Efficient global optimization of expensive black-box functions. *Journal of Global Optimization*, 13, 1998, 455–492.

[62] Müller, J. MISO: Mixed-integer surrogate optimization framework. *Optimization and Engineering*, 17, 2016, 177–203.

[63] Ploskas, N., Sahinidis, N. V. Review and comparison of algorithms and software for mixed-integer derivative-free optimization. *Journal of Global Optimization*, 82, 2022, 433–462.

[64] Ye, L., Zhang, N., Li, G., Gu, D., Lu, J., Lou, Y. Intelligent optimization design of distillation columns using surrogate models based on GA-BP. *Processes*, 11, 2023, 2386.

[65] Ibrahim, D., Jobson, M., Li, J., Guillén-Gosálbez, G. Optimization-based design of crude oil distillation units using surrogate column models and a support vector machine. *Chemical Engineering Research and Design*, 134, 2018, 212–225.

[66] Carpio, R. R., Giordano, R. C., Secchi, A. R. Enhanced surrogate assisted framework for constrained global optimization of expensive black-box functions. *Computational Chemical Engineering*, 118, 2018, 91–102.

[67] Keßler, T., Kunde, C., McBride, K., Mertens, N., Michaels, D., Sundmacher, K., Kienle, A. Global optimization of distillation columns using explicit and implicit surrogate models. *Chemical Engineering Science*, 197, 2019, 235–245.

[68] Wang, Z., Zhou, T., Sundmacher, K. Data-driven integrated design of solvents and extractive distillation processes. *AIChE Journal*, 2023, 69.

[69] Schweidtmann, A. M., Mitsos, A. Deterministic global optimization with artificial neural networks embedded. *Journal of Optimization Theory and Applications*, 180, 2019, 925–948.

[70] Schweidtmann, A. M., Bongartz, D., Grothe, D., Kerkenhoff, T., Lin, X., Najman, J., Mitsos, A. Deterministic global optimization with Gaussian processes embedded. *Mathematical Programming Computation*, 13, 2021, 553–581.

[71] Bongartz, D., Mitsos, A. Deterministic global optimization of process flowsheets in a reduced space using McCormick relaxations. *Journal of Global Optimization*, 69, 2017, 761–796.

[72] Huster, W. R., Schweidtmann, A. M., Mitsos, A. Working fluid selection for organic rankine cycles via deterministic global optimization of design and operation. *Optimization and Engineering*, 21, 2020, 517–536.

[73] Rall, D., Schweidtmann, A. M., Aumeier, B. M., Kamp, J., Karwe, J., Ostendorf, K., Mitsos, A., Wessling, M. Simultaneous rational design of ion separation membranes and processes. *Journal of Membrane Science*, 600, 2020, 117860.

[74] Burre, J., Kabatnik, C., Al-Khatib, M., Bongartz, D., Jupke, A., Mitsos, A. Global flowsheet optimization for reductive dimethoxymethane production using data-driven thermodynamic models. *Computational Chemical Engineering*, 162, 2022, 107806.

[75] Ugray, Z., Lasdon, L., Plummer, J. C., Glover, F., Kelly, J., Martí, R. A Multistart Scatter Search Heuristic for Smooth NLP and MINLP Problems. In Sharda, R., Voß, S., Rego, C., Alidaee, B. (Eds.). *Metaheuristic Optimization via Memory and Evolution*. Boston: Kluwer Academic Publishers, 2005, pp. 25–57.

[76] Beyer, H.-G., Schwefel, H.-P. Evolution strategies – A comprehensive introduction. *Natural Computing*, 1, 2002, 3–52.

[77] Jain, N. K., Nangia, U., Jain, J. A review of particle swarm optimization. *Journal of the Institution of Engineers (India): Series B*, 99, 2018, 407–411.

[78] Moscato, P. On evolution, search, optimization, genetic algorithms and martial arts: Towards memetic algorithms. *Caltech Concurrent Computation Program, C3P Report*, 826, 1989, 1–67.

[79] Leboreiro, J., Acevedo, J. Processes synthesis and design of distillation sequences using modular simulators: A genetic algorithm framework. *Computational Chemical Engineering*, 28, 2004, 1223–1236.

[80] Weicker, K. *Evolutionäre Algorithmen*. Wiesbaden: B.G. Teubner Verlag / GWV Fachverlage, 2007.

[81] Silva, H. G., Salcedo, R. L. R. A coupled strategy for the solution of NLP and MINLP optimization problems: Benefits and pitfalls. *Industrial & Engineering Chemistry Research*, 48, 2009, 9611–9621.

[82] Linke, P., Kokossis, A. On the robust application of stochastic optimisation technology for the synthesis of reaction/separation systems. *Computational Chemical Engineering*, 27, 2003, 733–758.

[83] Modla, G. Energy saving methods for the separation of a minimum boiling point azeotrope using an intermediate entrainer. *Energy*, 50, 2013, 103–109.

[84] Gross, B., Roosen, P. Total process optimization in chemical engineering with evolutionary algorithms. *Computational Chemical Engineering*, 22, 1998, 229–236.

[85] Barakat, T. M. M., Sørensen, E. Simultaneous optimal synthesis, design and operation of batch and continuous hybrid separation processes. *Chemical Engineering Research and Design*, 86, 2008, 279–298.

[86] Koch, K., Sudhoff, D., Kreiß, S., Górak, A., Kreis, P. Optimisation-based design method for membrane-assisted separation processes. *Chemical Engineering and Processing*, 67, 2013, 2–15.

[87] Niesbach, A., Kuhlmann, H., Keller, T., Lutze, P., Górak, A. Optimisation of industrial-scale n-butyl acrylate production using reactive distillation. *Chemical Engineering Science*, 100, 2013, 360–372.

[88] Wierschem, M., Skiborowski, M., Górak, A., Schmuhl, R., Kiss, A. A. Techno-economic evaluation of an ultrasound-assisted Enzymatic Reactive Distillation process. *Computational Chemical Engineering*, 2017.

[89] Gómez-Castro, F. I., Rodríguez-Ángeles, M. A., Segovia-Hernández, J. G., Gutiérrez-Antonio, C., Briones-Ramírez, A. Optimal designs of multiple dividing wall columns. *Chemical Engineering Technology*, 34, 2011, 2051–2058.

[90] Deb, K., Agrawal, S., Pratap, A., Meyarivan, T. A Fast Elitist Non-dominated Sorting Genetic Algorithm for Multi-objective Optimization: NSGA-II. In Goos, G., Hartmanis, J., Van Leeuwen, J., Schoenauer, M., Deb, K., Rudolph, G., Yao, X., Lutton, E., Merelo, J. J., Schwefel, H.-P. (Eds.). *Parallel Problem Solving from Nature PPSN VI*. Berlin Heidelberg: Berlin, Heidelberg: Springer, 2000, pp. 849–858.

[91] Bravo-Bravo, C., Segovia-Hernández, J. G., Gutiérrez-Antonio, C., Durán, A. L., Bonilla-Petriciolet, A., Briones-Ramírez, A. Extractive dividing wall column: Design and optimization. *Industrial & Engineering Chemistry Research*, 49, 2010, 3672–3688.

[92] Lee, E. S.-Q., Rangaiah, G. P. Optimization of recovery processes for multiple economic and environmental objectives. *Industrial & Engineering Chemistry Research*, 48, 2009, 7662–7681.

[93] Yee, A. K., Ray, A. K., Rangaiah, G. P. Multiobjective optimization of an industrial styrene reactor. *Computational Chemical Engineering*, 27, 2003, 111–130.

[94] Rangaiah, G. P., Feng, Z., Hoadley, A. F. Multi-objective optimization applications in chemical process engineering: Tutorial and review. *Processes*, 8, 2020, 508.

[95] Liñán, D. A., Contreras-Zarazúa, G., Sánchez-Ramírez, E., Segovia-Hernández, J. G., Ricardez-Sandoval, L. A. A hybrid deterministic-stochastic algorithm for the optimal design of process flowsheets with ordered discrete decisions. *Computational Chemical Engineering*, 180, 2024, 108501.

[96] Herrera Velázquez, J. J., Zavala Durán, F. M., Chávez Díaz, L. A., Cabrera Ruiz, J., Alcántara Avila, J. R. Hybrid two-step optimization of internally heat-integrated distillation columns. *Journal of the Taiwan Institute of Chemical Engineers*, 130, 2022, 103967.

[97] Chia, D. N., Duanmu, F., Sorensen, E. Single- and multi-objective optimisation of hybrid distillation-pervaporation and dividing wall column structures. *Chemical Engineering Research and Design*, 194, 2023, 280–305.

[98] Chia, D. N., Sorensen, E. Optimal design of hybrid distillation-membrane processes based on a superstructure approach. *Chemical Engineering Research and Design*, 194, 2023, 256–279.

[99] Athier, G., Floquet, P., Pibouleau, L., Domenech, S. Process optimization by simulated annealing and NLP procedures. Application to heat exchanger network synthesis. *Computational Chemical Engineering*, 21, 1997, 475–480.

[100] Luo, X., Wen, Q.-Y., Fieg, G. A hybrid genetic algorithm for synthesis of heat exchanger networks. *Computational Chemical Engineering*, 33, 2009, 1169–1181.

[101] Rathjens, M., Fieg, G. A novel hybrid strategy for cost-optimal heat exchanger network synthesis suited for large-scale problems. *Applied Thermal Engineering*, 167, 2020, 114771.

[102] Molina, D., Lozano, M., García-Martínez, C., Herrera, F. Memetic algorithms for continuous optimisation based on local search chains. *Evolutionary Computation*, 18, 2010, 27–63.

[103] Gómez, J. M., Reneaume, J. M., Roques, M., Meyer, M., Meyer, X. A mixed integer nonlinear programming formulation for optimal design of a catalytic distillation column based on a generic nonequilibrium model. *Industrial & Engineering Chemistry Research*, 45, 2006, 1373–1388.

[104] Urselmann, M., Barkmann, S., Sand, G., Engell, S. Optimization-based design of reactive distillation columns using a memetic algorithm. *Computational Chemical Engineering*, 35, 2011, 787–805.

[105] Urselmann, M., Engell, S. Design of memetic algorithms for the efficient optimization of chemical process synthesis problems with structural restrictions. *Computational Chemical Engineering*, 72, 2015, 87–108.

[106] Holtbruegge, J., Kuhlmann, H., Lutze, P. Process analysis and economic optimization of intensified process alternatives for simultaneous industrial scale production of dimethyl carbonate and propylene glycol. *Chemical Engineering Research and Design*, 2015, 411–431.

[107] Steimel, J., Engell, S. Conceptual design and optimization of chemical processes under uncertainty by two-stage programming. *Computational Chemical Engineering*, 81, 2015, 200–217.

[108] Skiborowski, M., Harwardt, A., Marquardt, W. Conceptual Design of Azeotropic Distillation Processes. In *Distillation* . Elsevier, 2014, pp. 305–355.

[109] Skiborowski, M., Rautenberg, M., Marquardt, W. A Hybrid Evolutionary–Deterministic Optimization Approach for Conceptual Design. *Industrial & Engineering Chemistry Research*, 54, 2015, 10054–10072.

[110] Urselmann, M., Barkmann, S., Sand, G., Engell, S. A memetic algorithm for global optimization in chemical process synthesis problems. *IEEE Transactions on Evolutionary Computation*, 15, 2011, 659–683.

[111] Urselmann, M., Engell, S. Optimization-based Design of Reactive Distillation Columns Using a Memetic Algorithm. In *20th European Symposium on Computer Aided Process Engineering* . Elsevier, 2010, pp. 1243–1248.

[112] Nelder, J. A., Mead, R. A simplex method for function minimization. *Computer Journal*, 7, 1965, 308–313.

[113] Urselmann, M., Janus, T., Foussette, C., Tlatlik, S., Gottschalk, A., Emmerich, M. T., Bäck, T., Engell, S. Derivative-Free Chemical Process Synthesis by Memetic Algorithms Coupled to Aspen Plus Process Models. In Kravanja, Z., Bogataj, M. (Eds.). *26th European Symposium on Computer Aided Process Engineering*. Amsterdam, Netherlands: Elsevier, 2016, Vol. 38, pp. 187–192.

[114] Chanthasuwannasin, M., Kottitutum, B., Srinophakun, T. A mixed coding scheme of a particle swarm optimization and a hybrid genetic algorithm with sequential quadratic programming for mixed integer nonlinear programming in common chemical engineering practice. *Chemical Engineering Communications*, 204, 2017, 840–851.

[115] Kuhlmann, H., Möller, M., Skiborowski, M. Analysis of TBA-based ETBE production by means of an optimization-based process-synthesis approach. *Chemie Ingenieur Technik*, 32, 2018, 3.

[116] Kuhlmann, H., Veith, H., Möller, M., Nguyen, K.-P., Górak, A., Skiborowski, M. Optimization-based approach to process synthesis for process intensification: Synthesis of reaction-separation processes. *Industrial & Engineering Chemistry Research*, 57, 2018, 3639–3655.

[117] Zhou, T., Zhou, Y., Sundmacher, K. A hybrid stochastic–deterministic optimization approach for integrated solvent and process design. *Chemical Engineering Science*, 159, 2017, 207–216.

[118] Chia, D. N., Duanmu, F., Sorensen, E. Optimal Design of Distillation Columns Using a Combined Optimisation Approach. In *31st European Symposium on Computer Aided Process Engineering* . Elsevier, 2021, pp. 153–158.

[119] Lacroix, B., Molina, D., Herrera, F. Region-based memetic algorithm with archive for multimodal optimisation. *Information Science*, 367–368, 2016, 719–746.

[120] Palar, P. S., Tsuchiya, T., Parks, G. T. A comparative study of local search within a surrogate-assisted multi-objective memetic algorithm framework for expensive problems. *Applied Soft Computing*, 43, 2016, 1–19.

[121] Kruber, K. F., Miroschnitschenko, A., Skiborowski, M. ANN-assisted Optimization-based Design of Energy-integrated Distillation Columns. In *32nd European Symposium on Computer Aided Process Engineering*. Elsevier, 2022, pp. 1261–1266.

[122] Ludl, P. O., Heese, R., Höller, J., Asprion, N., Bortz, M. Using machine learning models to explore the solution space of large nonlinear systems underlying flowsheet simulations with constraints. *Frontiers of Chemical Science and Engineering*, 16, 2022, 183–197.

[123] Höller, J., Bubel, M., Heese, R., Ludl, P. O., Schwartz, P., Schwientek, J., Asprion, N., Wlotzka, M., Bortz, M. Adaptively exploring the feature space of flowsheets. *AIChE Journal*, 2024, 70.

[124] Kallrath, J. Polylithic modeling and solution approaches using algebraic modeling systems. *Optimization Letters*, 5, 2011, 453–466.

Urmila Diwekar*

Chapter 10
Optimization under uncertainty in process systems engineering

Abstract: Process systems engineering (PSE) covers a broad range of applications involving optimization amid uncertainty. These applications range from product development to chemical synthesis, process design, energy optimization, environmental management, and operational tasks such as scheduling, supply chain management, and control. Dealing with optimization under uncertainty poses significant challenges, requiring both optimization techniques and uncertainty analysis. This chapter introduces methods for quantifying uncertainty, sampling techniques for uncertainty analysis, and algorithms designed to tackle optimization problems in uncertain environments. Additionally, it explores the practical applications of these algorithms across various PSE areas.

Keywords: optimization under uncertainty, stochastic programming, stochastic optimal control, Ito processes, sampling techniques

10.1 Introduction

Within the deterministic optimization literature, problems are categorized as linear programming (LP), nonlinear programming (NLP), integer programming (IP), mixed-integer LP (MILP), mixed-integer NLP (MINLP), disjunctive programming, and optimal control, based on the characteristics of decision variables, objectives, and constraints. However, the dynamic nature of the future precludes perfect forecasting, necessitating consideration of randomness or uncertainty. Optimization under uncertainty comprises a significant branch of optimization wherein uncertainties exist within the data or model. This field is commonly referred to as stochastic programming or stochastic optimization problems, where "stochastic" denotes randomness and "programming" encompasses mathematical programming techniques like LP, NLP, IP, MILP, disjunctive programming, and MINLP.

In discrete optimization (IP, MILP, and MINLP), probabilistic techniques such as simulated annealing and genetic algorithms are employed, sometimes labeled as sto-

*Corresponding author: Urmila Diwekar, Vishwamitra Research Institute, Clarendon Hills, IL 60514, USA; Richard and Loan Hill Department of Biomedical Engineering, University of Illinois at Chicago, Crystal Lake, IL 60012, USA, e-mail: urmila@vri-custom.org

https://doi.org/10.1515/9783111383439-010

chastic optimization techniques due to their probabilistic nature. However, stochastic programming and stochastic optimization entail making optimal decisions amid uncertainties. Parametric programming, a method rooted in sensitivity analysis, also accounts for uncertainties, illustrating how optimal decisions and designs fluctuate with varying degrees of uncertainty.

The recognition of the necessity to incorporate uncertainty into complex decision models emerged early in the history of mathematical programming. Early models, incorporating action followed by observation and reaction (recourse), were introduced in seminal works [1, 2]. An illustrative example of such a recourse problem is the news vendor or newsboy problem, which is elaborated below [3]. In this problem, the vendor must decide on the quantity of papers to purchase at the current cost and selling price per paper, considering uncertain demand. The lineage of the news vendor problem can be traced back to the economist Edgeworth [4], who applied variance to a bank cash flow problem. However, it was not until the 1950s, spurred by the demands of wartime operations that this problem, among others in the field of operations research and management science, became the subject of comprehensive academic study [5].

Example 10.1: Consider the simplest form of a stochastic program, exemplified by the news vendor (known as the newsboy) problem. In this scenario, the vendor decides how many papers (denoted as x) to purchase at a cost of c cents amid uncertain demand. Each paper is sold at a price of s_p cents. For a specific instance where the weekly demand is outlined below, the cost per paper is $c = 20$ cents, and the selling price is $s_p = 25$ cents. Let us tackle this problem assuming that the news vendor is aware of demand uncertainties (as shown in Table 10.1) but needs to gain prior knowledge of the demand curve for the upcoming week (as shown in Table 10.2). Moreover, let us assume no salvage value ($s = 0$), meaning any excess papers beyond demand are simply discarded without any return.

Table 10.1: Weekly demand probabilities.

Demand, d	Probability
50	0.714
100	0.143
140	0.143

Table 10.2: Weekly demand.

Day	Demand (u)
Monday	50
Tuesday	50
Wednesday	50
Thursday	50
Friday	50
Saturday	100
Sunday	140

Solution: To solve this problem, we aim to determine the optimal number of papers (x) the vendor should procure to maximize profit. Let r represent effective sales and w denote the excess inventory destined for disposal. This scenario falls within the realm of stochastic programming with recourse, involving action (x), followed by observation (profit), and subsequent reaction (or recourse) (r and w). Given that excess papers are discarded, the objective is to minimize waste while maximizing sales.

Initially, our inclination might be to ascertain the average demand and set the supply (x) accordingly. From Table 10.1, the average demand is 70 papers, suggesting a solution of $x = 70$. However, with this approach, wherein the daily supply is 70 papers, the vendor incurs a weekly loss of 50 cents (Table 10.3) as calculated using the demand values given in Table 10.2. This indicates that such a solution might not be optimal. Can we improve upon this? To find out, we need to account for the uncertainty in demand and evaluate its impact on the objective function, thereby determining the optimal value of x. This information can be reformulated to analyze daily profit accordingly.

Table 10.3: Supply and profit.

Day	Supply, x	Profit
Monday	70	−150
Tuesday	70	−150
Wednesday	70	−150
Thursday	70	−150
Friday	70	−150
Saturday	70	350
Sunday	70	350
Avg. (weekly)	70	−50

There are three demands shown in Table 10.1, so we can write the profit function as follows:

$$\text{Profit} = -cx + s_p x$$

$$\text{if } x \leq 50 \tag{10.1}$$

$$\text{Profit} = -cx + 0.714 \ s_p \times 50 + 0.143 \ s_p \ x + 0.143 \ s_p \ x$$

$$\text{if } 50 \leq x \leq 100 \tag{10.2}$$

$$\text{Profit} = -cx + 0.714 \ s_p \times 50 + 0.143 \ s_p \times 100 + 0.143 \ s_p \ x$$

$$\text{if } 100 \leq x \leq 140 \tag{10.3}$$

Notice that the problem represents three equations for the objective function, resulting in a discontinuous function that is no longer an LP. The optimal solution to this problem is $x = d_1 = 50$, which will give the news vendor an optimum profit of 1750 cents per week.

The difference between taking the average value of the uncertain variable as the solution and using stochastic analysis (propagating the uncertainties through the model and finding the effect on the objective function as above) is defined as the value of stochastic solution, VSS. If we take the average value of the demand, i.e., $x = 70$ as the solution, we obtain a loss of 50 cents per week. Therefore, the VSS is $1,750 - (-50) = 1,800$ cents per week.

Now consider the case where the vendor knows the exact demand (Table 10.2) a priori. This is the perfect information problem, where we want to find the solution x_i for each day i. This can be directly calculated as a daily profit with different supply values (Table 10.4).

Table 10.4: Supply and profit with perfect information.

Day	Supply, x	Profit
Monday	50	250
Tuesday	50	250
Wednesday	50	250
Thursday	50	250
Friday	50	250
Saturday	100	500
Sunday	140	700
Avg. (weekly)	70	2,450

One can see the difference between the two values: (1) when the news vendor has the perfect information ($2,450 cents per week) and (2) when he does not have the perfect information ($1,750 cents per week) but can represent it using probabilistic functions. This is the expected value of perfect information, EVPI. Therefore, the EVPI for this problem is 700 cents.

The newsboy problem shown in Example 10.1 falls under the category of multi-stage stochastic programming with recourse. Recourse problems with multiple stages involve decisions that are made before uncertainty realization (e.g., x and r in Example 10.1), and recourse action is taken when information is disclosed (w in Example 10.1).

Optimization under uncertainty allows us to optimize systems in the face of uncertainties. This requires that the objective function and constraints be represented in probabilistic representations like expected value, variance, and fractiles (Figure 10.1). For example, in chance-constrained programming (CCP), the objective function is expressed in terms of expected value and the constraints are expressed in terms of fractiles (probability of constraint violation). In Taguchi's offline quality control method, the objective is to minimize variance.

Unlike deterministic optimization problems, stochastic optimization problems require evaluating the probabilistic function of the objective function and constraints. This depends upon modeling and propagating uncertainties in the formulation, which is discussed in the next section.

10.2 Uncertainty analysis and propagation

As far as uncertainty modeling is concerned, the uncertainties commonly encountered in chemical systems can be divided into two groups [6].
- Static uncertainties
- Dynamic uncertainties

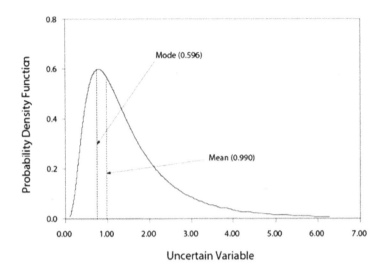

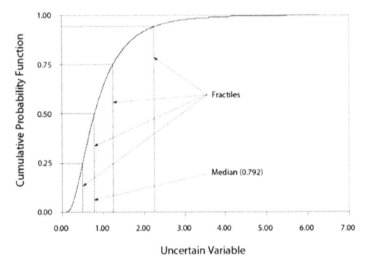

Figure 10.1: Different probabilistic performance measures.

10.2.1 Static uncertainties

Probability distributions normally represent static uncertainties. In the first step of the stochastic modeling framework, uncertainties in key input variables are represented by probability distribution functions. An example of uncertainty characterization and quantification by probability distributions was presented by Kim and Diwekar [7] in a computer-aided molecular design (CAMD) problem. Discrepancies between the experimental data for predicting a thermodynamic property and the models are commonly

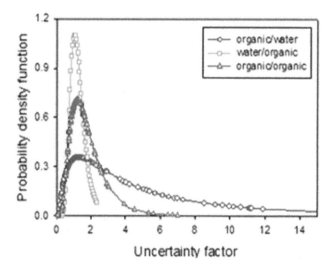

Figure 10.2: UNIFAC uncertainty factor uncertainty: organic/water (lognormal distribution), water/organic (normal distribution), and organic/organic(lognormal distribution.

encountered in CAMD. For example, Figure 10.2 shows the uncertainties in more than 1,800 interaction parameters present in the UNIFAC activity coefficient model to predict solvent selection objectives for acetic acid separations. Uncertainty factors (UFs) were established as the ratio between the experimental and the calculated values of activity coefficients at infinite dilution, as defined in eq. (10.4). Furthermore, uncertainty factors were divided into three categories based on the type of family: organic/water (lognormal distribution), water/organic (normal distribution), and organic/organic (lognormal distribution):

$$UF = \frac{\gamma_{exp}^{\infty}}{\gamma_{cal}^{\infty}} \tag{10.4}$$

The type of distribution for an uncertain variable is a function of the amount of data available and the characteristic of the distribution function. The simplest distribution for an uncertain variable is a uniform distribution, which has a constant probability. This means that the uncertain variable can take any value within an interval $[a,b]$ with equal probability. On the other hand, if the uncertain variable is represented by a normal (Gaussian) distribution, there is a symmetric but equal probability that the value of the uncertain variable will be above or below a mean value. In lognormal or some triangular distributions, there is a higher probability that the value of uncertain variable will be on one side of the median, resulting in a skewed shape. A beta distribution provides a wide range of shapes and is a very flexible means of representing variability over a fixed range. In some special cases, user-supplied distributions are used such as chance distribution.

It is easier to assume the upper and lower bound of uncertain variables and hence uniform distribution is the first step toward uncertainty quantification. Most of the papers in chemical engineering use this simplistic approach and use upper and lower bounds of uncertain variables. Kim and Diwekar [8–11] identified most likely values and use triangular distributions. They used extensive data obtained from DECHEMA and obtained realistic quantification of uncertainties related to UNIFAC parameters.

Interval methods are also proposed [12–14] for handling uncertainties where there is no information about uncertainties. For interval methods, interval mathematics is used to propagate the uncertainty through the model. Recently, Stadtherr et al. [13, 14] proposed the use of probability boxes (P-box) instead of just intervals for this purpose. However, sampling methods modified for P-boxes are necessary to solve these problems.

10.2.2 Dynamic uncertainties

Due to their dynamic nature, dynamic uncertainties are prevalent in chemical systems, particularly in batch processes. Even static uncertainties can manifest as dynamic uncertainties in such processes. An illustrative example can be observed in Figure 10.3, where the uncertainties in activity coefficients, predicted by the UNIFAC method (Figure 10.3a), impact the time-varying relative volatility profile in a batch distillation column (Figure 10.3b) [6]. Despite this, a comprehensive approach to addressing dynamic uncertainties in chemical systems was only recently introduced [3, 6, 15]. This approach draws upon Ito processes and real options theory from the finance literature.

The Wiener process exhibits three crucial properties. Firstly, it adheres to the Markov property, where the probability distribution of future values depends on the present value. Secondly, it features independent increments, meaning the probability distribution of changes in the process over nonoverlapping time intervals is unrelated to other intervals. Thirdly, changes in the process over any finite time interval follow a normal distribution, with variance linearly dependent on the interval's length, dt. The general equation of an Ito process is provided below:

$$dx = a(x, t)dt + b(x, t)dz \tag{10.5}$$

where

$$dz = \varepsilon_t \sqrt{dt}$$

In this equation, dz is the increment of a Wiener process, and $b(x, t)$ are known functions. ε_t is a random number drawn from a unit normal distribution.

Different forms of functions a and b define various Ito processes. The simplest Ito process is Brownian motion with drift. Other examples of Ito processes are geometric Brownian motion, mean-reverting processes, etc.

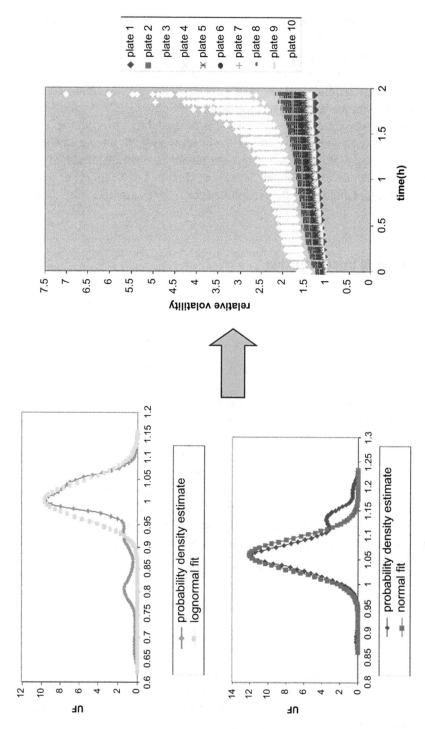

Figure 10.3: (a) The uncertainties in activity coefficients predicted by UNIFAC and (b) the effect of static uncertainties on the relative volatility profile on each plate in a batch distillation column [6].

Depending on the character of uncertainty, the dynamic uncertainties in chemical processes (batch processes) could be represented by these Ito processes. For example, the dynamic uncertainty shown in Figure 10.3b can be modeled as a geometric mean reverting process. For details, please see Diwekar [3].

10.2.3 Stochastic modeling and sampling techniques

The inclusion of uncertainties in a deterministic model result in a stochastic model. Stochastic modeling is an iterative procedure that consists of the next three steps [16], as shown in Figure 10.4:
1. Sampling (or generating scenarios) distribution of the specified parameter in an iterative fashion
2. Propagating the effects of uncertainties through the model
3. Applying statistical techniques to obtain the probabilistic representation of objective function and constraints

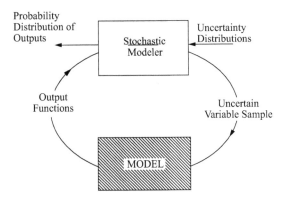

Figure 10.4: Stochastic modeling for uncertainty propagation.

Sampling techniques can be categorized into two main groups: probability sampling and non-probability sampling. Probability sampling involves selecting samples based on the principles of probability theory, wherein each potential unit has an assigned probability of selection. These samples are chosen randomly, with known confidence intervals for estimates. In contrast, non-probability sampling does not entail random selection. For instance, quota sampling divides the population into subgroups, and individuals are selected based on judgment or convenience, making it challenging to determine sampling error through probabilistic means.

A robust sampling technique ensures that all plausible values of input and output variables have a chance of occurrence without excluding any segment of the population.

Additionally, estimates should closely approximate the actual values being estimated while allowing for an assessment of the relative importance of each input variable.

Probabilistic sampling techniques, particularly those based on Monte Carlo methods, are pertinent to this discussion. This section delves into three subsections concerning Monte Carlo sampling, variance reduction techniques to enhance the efficiency of Monte Carlo methods, and Bayesian and adaptive sampling techniques, utilized when complete probability information is lacking.

10.2.3.1 Monte Carlo sampling

The Monte Carlo method is among the most fundamental and widely employed sampling methods. Monte Carlo methods are numerical techniques that offer approximate solutions to diverse physical and mathematical problems through random sampling. Coined by Nicholas Metropolis, the term "Monte Carlo" derives from Monaco, famed for its casinos, drawing an analogy between statistical experiments and the stochastic nature of games like roulette.

Initially developed during World War II for the Manhattan Project to simulate probabilistic scenarios concerning random neutron diffusion in fissile material, Monte Carlo methods gained prominence across various scientific domains, following the advent of electronic computers in 1945. Metropolis and Ulam's seminal work [17] in 1949 is among the earliest publications outlining the Monte Carlo algorithm.

The fundamental concept underlying Monte Carlo simulation is the random generation of input samples to model random outputs. In a basic Monte Carlo approach, a value is randomly drawn from the probability distribution for each input, and the corresponding output is computed. This process is reiterated n times, yielding n corresponding output values, constituting a random sample from the probability distribution over the output induced by the input probability distributions.

Notably, this approach allows for estimating output distribution precision using standard statistical techniques, with an average approximation error ε of the order $O(N - 1/2)$. Remarkably, the error bound is independent of the dimension k, though it remains probabilistic, offering no concrete assurance of achieving the expected accuracy in a specific calculation.

The efficacy of a Monte Carlo calculation hinges on selecting an appropriate random sample. Computers generate the requisite random numbers and vectors through deterministic algorithms, earning the designation "pseudorandom numbers" or "pseudorandom vectors." Among the earliest and most renowned methods for generating pseudorandom numbers for Monte Carlo sampling is the linear congruential generator (LCG), introduced by D.H. Lehmer in 1949.

Pseudorandom numbers of different sample sizes on a unit square generated using this method are given in Figure 10.5. This figure shows that the pseudorandom number generator produces samples that may be clustered in certain regions of the unit square and does not produce uniform samples. Therefore, larger sample sizes are needed to reach high accuracy, which adversely affects this method's efficiency. Here, I define the efficiency of the method by measuring the number of samples required to attain a specified accuracy in measurements like mean, variance, and fractiles. Variance reduction techniques address this problem of increasing the efficiency of Monte Carlo methods and are described in the following subsection.

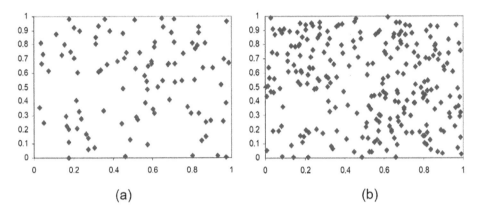

Figure 10.5: (a) One hundred pseudorandom numbers on a unit square and (b) 250 pseudorandom numbers on a unit square obtained by linear congruential generator developed by Wichmann and Hill (1982).

10.2.3.2 Variance reduction techniques

Variance reduction techniques have been devised to enhance the efficiency of Monte Carlo simulations and mitigate drawbacks such as probabilistic error bounds. James [19] categorized variance reduction techniques.

The first category of sampling techniques extracts more information from a run than what is initially apparent regarding parameter values. An example is the control variate sampling, one of the most adaptable variance reduction techniques. Control variates utilize information about errors in estimates of known quantities to diminish the error in estimating unknown quantities [20]. They are particularly effective in estimating the mean of the outcome distribution and can also aid in variance estimation. Control variates are configured to utilize a simplified version of a model. However, one challenge with control variates is selecting effective controls. Additionally, control variates assume a specific probabilistic structure for the simulation output process, typically joint normality of the response and control variates, which may not always hold true [20]. These limitations constrain the widespread adoption of control variates, explaining why this method has yet to permeate extensively in PSE.

Sampling techniques in the second category ensure that each run is unbiased, concerning the estimated mean outcome measure. For instance, in antithetic sampling, a negative correlation is introduced between two unbiased estimators of a variable X [21]. This technique is applied when only one significant variable within the model is sampled once during a run. Similar to control variate sampling, this method hasn't been widely used in chemical engineering literature.

The sampling approaches commonly used for variance reduction in chemical engineering applications include importance sampling, Latin hypercube sampling (LHS) [22, 23], descriptive sampling [24], Hammersley sequence sampling (HSS) [25, 26], and LHS-HSS [27]. The latter technique belongs to the quasi-Monte Carlo methods group, introduced to enhance Monte Carlo methods' efficiency by utilizing quasi-random sequences with better statistical properties and deterministic error bounds.

10.2.3.3 Importance sampling

Importance sampling, also known as biased sampling, is a variance reduction technique that improves the efficiency of Monte Carlo algorithms. Monte Carlo methods are commonly employed to integrate a function F over the domain D:

$$\int_D F(x)dx \tag{10.6}$$

If random numbers are drawn from a normal distribution, information is distributed over the sampling interval. However, using a nonuniform (biased) distribution $G(x)$, which draws more samples from areas making substantial contributions to the integral, results in a more accurate approximation of the integral and enhances the process's efficiency. This forms the basic concept of importance sampling, where the approximated integral is given by

$$I = \frac{1}{N}\sum_{i=1}^{N} \frac{F(x_i)}{G(x_i)} \tag{10.7}$$

Importance sampling is particularly crucial for sampling low-probability events. The critical aspect of implementing importance sampling is selecting the biased distribution that emphasizes the essential regions of the input variables. An example of importance sampling is the Metropolis criterion used in molecular simulations [28]. In molecular simulations, the configurational phase space is explored, involving the evaluation of a multidimensional integral over $3N$ degrees of freedom. The Metropolis approach generates a Markov chain of states and biases the generation of configurations toward those making the most significant contribution to the integral. Specifically, it generates states with a probability $= e - \Delta E/kT$, where ΔE is the change in energy, k is the Boltzmann factor, and T is the temperature. This algorithm enables more efficient sampling of low-energy configurations where the Boltzmann factor has a significant value, resulting in more accurate calculations of the thermodynamic properties of fluids.

Example 10.2: Let us consider estimating the integral:

$$\int_0^\infty x^2 \exp(-x^2)dx \qquad (10.8)$$

This function cannot be integrated analytically, but its value is known to be approximately 0.44311328.

As observed from Figure 10.6, the function value rapidly decreases when x is greater than about 3.5, indicating that the integral has a significant value only within a specific interval.

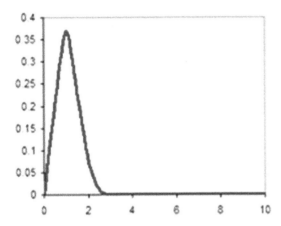

Figure 10.6: The function $x^2 \exp(-x^2)$.

Using a Monte Carlo integration with uniform sampling between 0 and 1,000, most points will be from areas where the integral has a minimal value, resulting in inefficient estimation. However, by utilizing a nonuniform distribution function like the lognormal distribution, which samples more from the important areas contributing significantly to the integral, the accuracy of the estimation can be improved with fewer samples. For instance, considering a lognormal distribution with mean $\mu = 1$ and standard deviation $\sigma = 1.7$ (Figure 10.7), importance sampling requires far fewer samples to accurately estimate the function than crude Monte Carlo methods using uniform distribution.

Table 10.5: Estimation of the integral $\int_0^\infty x^2 \exp(-x^2)dx$
by using uniform random sampling and importance sampling.

N	Uniform random sampling	Importance sampling
10	0	0.11054
100	0.00095	0.44363
1,000	0.07585	0.44312
10,000	0.44131	0.44311

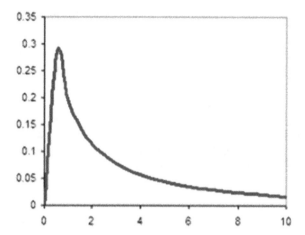

Figure 10.7: Lognormal distribution with mean $\mu = 1$ and standard deviation $\sigma = 1.7$.

This comparison demonstrates the effectiveness of importance sampling in efficiently estimating integrals with fewer samples, particularly when the integral has significant values only within specific intervals.

10.2.3.4 Stratified sampling

Stratification involves categorizing individuals in a population into distinct groups (strata) before sampling takes place. These strata should be nonoverlapping and collectively cover the entire population, ensuring that each individual belongs to only one stratum. Furthermore, every stratum should be represented in the sample in proportion to its representation in the population. Latin hypercube sampling belongs to stratified sampling.

Latin hypercube sampling (LHS)
Latin hypercube sampling [22, 23] is a stratified sampling technique to enhance the accuracy of estimating distribution functions. By dividing each uncertain parameter's range into equal probability intervals, LHS ensures thorough coverage of the sampled distribution. Within each interval, values are randomly selected according to the probability distribution. These assigned values are then combined systematically by randomly selecting ranks of each variable sample to form sets, ensuring random and representative samples across all parameters.

To consider how intervals are formed, consider the first random variable X_1, which has a normal distribution with a mean value of $\mu = 8$ and a standard deviation of $\sigma = 1$ shown in probability density function (PDF) (Figure 10.8). If I want to generate five LHS points, I can divide the PDF into equal probability zones, as shown in Figure 10.8 (top). This gets translated into CDF, shown in Figure 10.8. For each interval, I can generate a random number if I denote these values by U_m, where $m = 1, 2, 3, 4, 5$. Each random number U_m is scaled to obtain a cumulative probability P_m, so that each P_m lies within mth interval:

$$P_m = \frac{1}{5}U_m + \frac{m-1}{5} \tag{10.9}$$

I can then use the CDF, as shown in Figure 10.8, to get the samples for this variable. However, this will generate a monotonic sample, so I need to randomize it. This can be done by randomizing ranks 1 to 5 for the sample and arranging the new sample according to this rank, as shown in Table 10.6.

Table 10.6: Rank arrangement for the sample for randomization.

Randomizing ranks	X_1
2	7.648
5	9.213
3	7.899
1	6.808
4	8.737

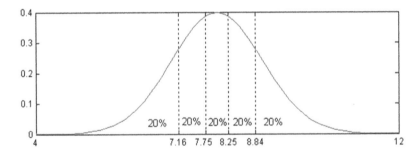

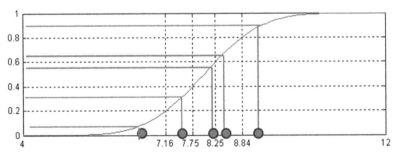

Figure 10.8: Intervals used with a Latin hypercube sample of size $n = 5$ in terms of the density function and cumulative distribution function for a normal random variable X_1. Sample values (monotonic) are shown by blue circles.

The primary limitation of the stratification scheme employed in LHS is its uniformity, which is confined to one dimension and lacks uniform properties across multiple dimensions. Similarly, techniques such as sampling based on quadrature [29], cubature techniques [30] or collocation techniques [31] encounter comparable constraints. These sampling methods exhibit superior performance when dealing with lower-dimensional uncertainties. Consequently, many of these techniques incorporate correlations to simplify the integration process into one or two dimensions. However, such transformations are only viable for certain distribution functions, mainly when the uncertain variables are strongly correlated. In instances where samples exhibit high correlation, akin to observations in thermodynamic phase equilibria, employing a sampling technique based on confidence region estimates becomes feasible [32].

10.2.3.5 Quasi-Monte Carlo methods

Quasi-Monte Carlo methods aim to generate point sequences that outperform traditional Monte Carlo methods, typically exhibiting an average complexity, approximately given by the following equation:

$$\text{average complexity} = \frac{1}{\epsilon^2} \tag{10.10}$$

By selecting an appropriate set of samples, the quasi-Monte Carlo method offers a deterministic error bound (eq. (10.11)) without imposing stringent assumptions on the integrand:

$$\epsilon = \frac{(\log(N))^k}{N} \tag{10.11}$$

Notable quasi-Monte Carlo sequences include Halton, Hammersley, Sobol, Faure, Korobov, and Neiderreiter [33]. The selection of a suitable quasi-Monte Carlo sequence depends on its discrepancy. The upper and lower deterministic error bounds of any sequence for integration are articulated in terms of the discrepancy measure, which quantifies the deviation of the sequence from a uniform distribution. Hence, the preference lies in low-discrepancy sequences like Halton, Hammersley, and Sobol.

Hammersley sequence sampling (HSS) is an efficient sampling technique pioneered by Diwekar and colleagues [25, 26], leveraging quasi-random numbers. HSS employs Hammersley points to sample a unit hypercube uniformly, and then inversely maps these points over the joint cumulative probability distribution, yielding a sample set for the variables of interest.

Sobol sampling also enjoys considerable popularity, particularly within financial literature. Figure 10.9 illustrates the two-dimensional uniformity of Hammersley and Sobol, compared to Monte Carlo and LHS. It can be seen from this figure that Hammersley has the best two-dimensional uniformity among these four. However, HSS incorporates spurious correlations beyond 40 dimensions and SOBOL beyond 60.

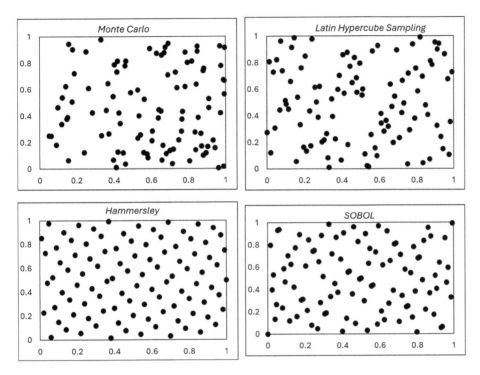

Figure 10.9: Two-dimensional uniformity of HSS and Sobol versus MCS and LHS.

I know that uniformity plays an important role in the design of sampling. LHS provides good one-dimensional uniformity but not k-dimensional uniformity like HSS and Sobol sampling (Sobol) and vice versa. New variants of HSS and Sobol have been proposed by the Diwekar group [27] called LHS-HSS and LHS-Sobol sampling. These sampling techniques generate the monotonic LHS sample for each variable and change the ranks of these variables using random numbers generated by HSS in LHS-HSS and Sobol in LHS-Sobol sampling. A general guideline for the use of these sampling techniques is provided in Figure 10.10.

10.2.3.6 Bayesian and adaptive methods

Bayesian probability theory was originally developed by Bayes [34]. Bayesian and adaptive methods are used when the probability functions are inaccurate. Bayesian method uses two steps. The first step is to identify the conceptual models and the distribution of model parameters. The second step compares the model results with existing observa-

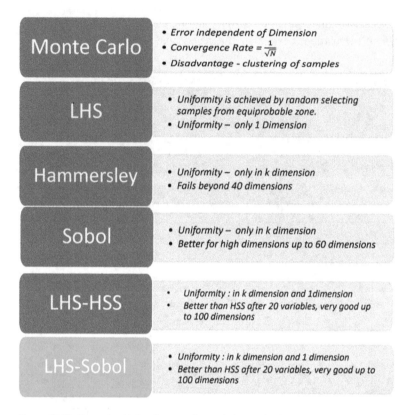

Figure 10.10: General guideline for sampling technique selection.

tions through a structured probabilistic methodology. In the Bayesian context, probability represents a degree of belief based on all the relevant information. Classical statistical approaches are ineffective in predicting low-frequency, rare but consequential (e.g., accidents or chemical spills) events. Bayesian theory could be applied to these cases [35]. A Bayesian approach is used for sensor fault detection [36] and experimental design [37].

10.3 Optimization algorithms

Despite the inevitability of uncertainties in real-world systems, the literature on optimization under uncertainty remains relatively sparse compared to other optimization domains. This scarcity can be attributed to the additional complexity of simultaneously addressing uncertainty analysis and optimization objectives. The preceding section detailed algorithms and methods for uncertainty analysis, while this section focuses on algorithms for addressing stochastic linear, nonlinear, discrete optimization, parametric programming, and stochastic optimal control problems.

10.3.1 Chance-constrained programming

In the earlier section, we looked at the scenario or sample generation for uncertainty analysis. Can uncertainty be effectively managed by utilizing input uncertainty moments like mean and variance to derive a deterministic representation of the problem? This concept forms the foundation of CCP [38], pioneered early on in the history of optimization under uncertainty, notably by Charnes and Cooper. In CCP, certain constraints may not need to be strictly upheld, as previously assumed. These problems can be expressed as follows:

$$\text{Opt}_x \quad E(z(x, u)) \tag{10.12}$$

subject to

$$P(g(x) \le u) \le \alpha \tag{10.13}$$

In the above equations, eq. (10.13) represents the chance constraint with probability α and the objective function is the expected value of Z. In chance-constraint formulation, this constraint (or constraints) is transformed into deterministic equivalents, assuming the distribution of uncertain variables, u, which follows a stable distribution. Stable distributions such as normal, Cauchy, uniform, and chi-square facilitate the conversion of probabilistic constraints into deterministic ones. These deterministic constraints are formulated in terms of moments of the uncertain variable u (input uncertainties), so the problem gets transformed into a deterministic equivalent given below:

$$\text{Opt}_x \quad z(x, u) \tag{10.14}$$

subject to

$$g(x) \le F^{-1}(\alpha) \tag{10.15}$$

where F^{-1} represents the inverse of cumulative distribution function F.

Maranas et al. [39] have applied chance-constraint programming to address (1) a chemical synthesis problem concerning polymer design and (2) a metabolic pathway synthesis problem in biochemical reaction engineering. A recent review [40] presents applications of chance constraints in PSE.

10.3.2 Decomposition techniques

Even seemingly linear optimization problems under uncertainty, such as the newsboy problem, introduce nonlinearities due to probabilistic functions. Decomposition methods offer a common strategy for addressing such complexities. This approach involves breaking down the problem into a master problem and subproblems. The master problem provides a simplified, often linear, approximation derived from multiple subproblems. Initially proposed for mixed-integer problems, Bender's decomposition is the foundation for algorithms tackling deterministic mixed IP and stochastic linear or mixed IP problems.

In stochastic programming, two prominent algorithms are commonly employed for problems with fixed recourse: the L-shaped method [41–43] and stochastic decomposition methods [44, 45]. Recourse is the ability to take corrective action after a random event has taken place. The L-shaped method suits scenarios where discrete distributions define uncertainties. Conversely, stochastic decomposition methods utilize sampling for random variables, represented by continuous distribution functions.

Despite the prevalence of stochastic LP and stochastic MILP in chemical engineering applications, researchers in the field have yet to fully explore generalized algorithms until recently [46]. Instead, they have primarily focused on leveraging the specific structure of individual problems [47–50], such as flexibility, to develop unique decomposition schemes and bounds. However, the problem's domain-specific nature often limits the broader applicability of these methods. A recent review [46] presents stochastic linear and stochastic MILP applications in PSE.

10.3.3 Sample approximation methods

As previously mentioned, stochastic programming formulations often involve approximations of the underlying probability distribution. However, fully solving the yth approximation through sampling approaches can lead to wasted effort when the approximation is inaccurate [43]. In cases where the L-shaped method is applicable, two approaches address this issue by incorporating sampling within another algorithm without complete optimization. These approaches include the method proposed by Dantzig and Infanger [51], which utilizes the importance sampling to reduce variance in each cut, based on a large sample and the stochastic decomposition method introduced by Higle and Sen [44, 45]. However, these methods have not been seen used in process engineering literature.

Sample average approximation methods have also been employed to enhance computational efficiency and accuracy in stochastic process design problems. Wei and Realff [52] presented a method comprising two algorithms: the optimality gap method (OGM) and the confidence level method (CLM) for solving convex stochastic mixed-integer nonlinear problems. In this approach, decisions are made using a smaller sample size with several replications, while a larger sample size is used to reevaluate the objective value with fixed decision variables. The sample sizes and replication numbers are iteratively increased until a stopping criterion is met. In the OGM algorithm, sample sizes are augmented until the optimality gap of each upper and lower bounds is sufficiently small, whereas in the CLM algorithm, sample sizes are increased until an overall accuracy probability falls within a certain tolerance.

Given that engineering models typically exhibit nonlinearity and uncertainties characterized by nonuniform or non-normal distributions, the aforementioned algorithms and methodologies are constrained in their applicability to large-scale problems across various domains of chemical engineering. To address this limitation, Sahin and Diwekar [53–55] introduced the Better Optimization of Nonlinear Uncertain Systems (BONUS) algorithm. BONUS employs a sampling-based approach and replaces the conventional inner model evaluation loop with a reweighting scheme. This scheme provides values of the probabilistic functions associated with objective functions and constraints, obviating the need for iterative sampling in each optimization iteration. Additionally, NLP problems rely on quasi-Newton methods and entail derivative calculations for each decision variable in every optimization iteration. BONUS circumvents these iterations as well. The BONUS algorithm has been successfully applied to a wide array of real-world problems in the fields of Energy and Environment. Please refer to Diwekar and David [55].

10.3.4 Sampling accuracy and probabilistic methods

Many chemical engineering applications like chemical synthesis, process synthesis under uncertainty and planning under uncertainty involve discrete decision variables (mixed-integer problems). Although decomposition methods have been used to solve these problems, the applicability of these algorithms to solve problems of real-world scale is very limited, as also indicated by Birge in his review on stochastic programming [56]. This is mainly because most of these methods rely heavily on convexity conditions and simplified approximations of the probabilistic functions containing uncertainties. In this regard, variants of probabilistic methods like the stochastic annealing algorithm [69] show great promise.

In almost all stochastic optimization problems, the major bottleneck is the computational time involved in generating and evaluating probabilistic functions, which represent the objective function and constraints. The number of samples required for a given accuracy in a stochastic optimization problem depends on factors such as the uncertainty type and the decision variables' point values [57]. For optimization problems, the number of samples required depends on the location of the trial point solution in the optimization space. Figure 10.11 shows how the shape of the surface over a range of uncertain parameter values changes at different optimization iterations. Therefore, selecting the number of samples is an important step that ultimately decides the accuracy of the optimal solution in stochastic programming. This is the basis of the Stochastic Annealing algorithm.

In the stochastic annealing algorithm, the optimizer obtains the decision variables and the number of samples required for the stochastic model. Furthermore, it provides the trade-off between accuracy and efficiency by selecting more samples as one approaches the optimum. Stochastic annealing and its recent variants have been used

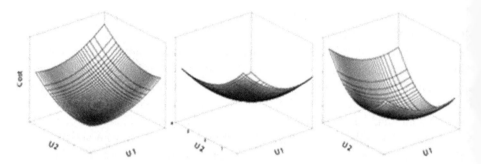

Figure 10.11: Uncertainty space at different optimization iterations.

to solve real-world problems such as (1) nuclear waste management problem [58], (2) computer-aided molecular design polymer design under uncertainty [59], (3) greener solvent selection [7], and (4) methylene chloride process synthesis [60]. New variants of genetic algorithms based on the theory used in stochastic annealing were developed and applied to solvent selection and recycling problems [61–63].

10.3.5 Parametric programming

Parametric programming is based on the sensitivity analysis theory, distinguishing from the latter in the targets. Sensitivity analysis provides solutions in the neighborhood of the nominal value of the varying parameters. In contrast, parametric programming offers a complete map of the optimal solution in the space of the varying parameters. Theory and algorithms for the solution of a wide range of parametric programming problems have been reported in the literature [64].

10.3.6 Robust optimization

In robust optimization, the aim is to ensure feasibility within a specified uncertainty set. At the same time, stochastic programming involves making decisions based on anticipating recourse actions, once uncertainties are revealed through a predefined scenario tree with discrete probabilities. Generally, robust optimization tends to yield more conservative outcomes compared to stochastic programming. Chance-constrained optimization can be seen as an extension of robust optimization, where distribution functions are defined for uncertainties along with a specified probability level for constraint feasibility [40]. Another approach involves incorporating variance into the objective function [65].

In the realm of PSE, robust optimization (primarily the "traditional" approach without recourse) has mainly been employed in production scheduling problems [66].

A recent paper presents a machine learning-based approach to robust optimization [67].

10.3.7 Stochastic optimal control

In essence, control involves a closed-loop system wherein the desired operating point is compared to the actual fact, with any difference (error) being fed back to adjust the system toward the desired state. Traditional frequency domain techniques are typically employed for controller design. However, optimal control scenarios deviate from this framework. Optimal control problems are open-loop, where a decision-variable profile (e.g., time-dependent profile) is crafted to optimize a performance index. The dynamic nature of these decision variables renders these problems more complex than standard optimization challenges involving scalar variables. Introducing stochastic optimal control introduces dynamic uncertainties and Ito processes, compounding the difficulty of solving such problems.

It's important to recognize that conventional mathematical techniques used for solving optimal control problems (e.g., calculus of variations and NLP) cannot be directly applied to scenarios involving dynamic uncertainties. Within financial literature and real options theory, Ito's lemma and dynamic programming are leveraged to tackle stochastic optimal control challenges. However, equations stemming from stochastic dynamic programming tend to be cumbersome and computationally inefficient. In 2004, Rico-Ramirez and Diwekar [68] introduced the stochastic maximum principle as a solution method for these problems. In recent years, this approach has been found to be applicable to solving optimal control problems within batch and bioprocessing domains and environmental management.

10.4 Applications

The life cycle of a chemical manufacturing process spans from the discovery of a product, selection of raw materials (chemical synthesis), and development of processes (process synthesis and design) to operations, planning, and management. The latter tasks encompass scheduling, supply chain management, and optimal control. Modern process design methodologies necessitate engineers to consider not only cost-effectiveness but also various other factors such as reliability, flexibility, operability, controllability, environmental and ecological impacts, safety, and quality throughout different analysis and design stages. This introduces additional complexities and uncertainties. This section will explore the utilization of optimization under uncertainty in product discovery, chemical synthesis, process synthesis, energy and environmental problems, and process operations, including scheduling, supply chain management, and control.

10.4.1 Chemical synthesis and computer-aided molecular design (CAMD)

The process design initiates with chemical synthesis in a laboratory where a chemical pathway from reactants to products is delineated. This entails searching for molecules possessing desired physical, chemical, and biological properties. Computer-aided chemical synthesis relies on group contribution methods [69, 70], which assign numerical values to functional groups forming each molecule through experimental data and theoretical approaches. Combining these functional groups can calculate a wide range of characteristics for any given chemical.

A fundamental diagram of computer-aided molecular design (CAMD) is depicted in Figure 10.12. In CAMD, the process begins with a set of functional groups. All conceivable combinations of these functional groups are explored to generate molecules that meet feasibility constraints. The properties of each group and the interaction parameters between groups can be theoretically obtained, experimentally determined or derived through statistical regression techniques. Once molecules are generated, their properties are inferred from the properties of the functional groups comprising them. If a generated molecule meets specific criteria, it is added to the list of candidate molecules. This method can produce a list of candidate molecules for any purpose with reasonable accuracy within a moderate timeframe. CAMD is the reverse use of group contribution models (GCMs). There are three primary CAMD approaches: generation-and-test, mathematical optimization, and combinatorial optimization approaches [71]. However, all methodologies for CAMD are subject to uncertainties arising from experimental errors, imperfect theoretical models/parameters, and incomplete knowledge of the systems. Additionally, group parameters may not always be available, and current GCMs may be unable to estimate all necessary properties.

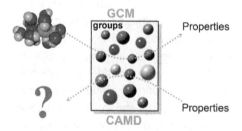

Figure 10.12: CAMD for chemical synthesis.

Various publications have addressed uncertainties in CAMD. For instance, Maranas [39] investigated polymer design with optimal thermophysical and mechanical properties. To model uncertainties in group contribution parameters, probability distribution functions are utilized, leading to a chance-constrained formulation. These chance constraints represent the probability of meeting target property values. Solutions to the optimal molecular design problem under uncertainty ensure at least an α chance of meeting performance objectives and a β chance of maintaining property values within their designated bounds. However, integrating multivariate probability density distributions poses computational challenges. To overcome this, stochastic constraints are transformed into equivalent deterministic ones, enabling the exact solution of resulting convex MINLP formulations.

Tayal and Diwekar [59] also tackled property prediction uncertainty in polymer design and presented a generalized stochastic framework based on Hammersley sequence sampling (HSS) and stochastic annealing. This framework is applicable to nonlinear or even black-box property prediction models, nonlinear objective functions and constraints, and stable and non-stable distributions for uncertain variables. Moreover, it provides a set of solutions, offering flexibility to the designer.

Kim and Diwekar [7, 11, 72] applied the CAMD approach to select environmentally benign solvents under uncertainty for extractive distillation. Property prediction model uncertainty was quantified using available experimental data (Figure 10.2). The Hammersley stochastic annealing (HSTA) algorithm was employed to solve this combinatorial optimization problem. This algorithm utilizes the efficient HSS technique to update discrete combinations, reduce Markov chain length, and automatically determine the number of samples.

Genetic algorithms were also used to address the problem of selecting environmentally benign solvents [62, 63]. The Hammersley stochastic genetic algorithm (HSGA) was developed, which outperforms the HSTA algorithm by selecting solvents with better targeted properties in less computational time.

10.4.2 Process synthesis and design

Process synthesis and design involves translating chemical synthesis concepts into practical chemical processes. This entails selecting appropriate unit operations, determining their interconnections, and optimizing the proposed plant layout. Process design activities commence at this stage, culminating in the creation of a plant flowsheet based on these decisions, with process simulators utilized to forecast mass and energy flows. Various methodologies are commonly employed to select optimal process flowsheet configurations, which can be categorized into four groups:

-   Optimization-based approach: This involves formulating a comprehensive flowsheet, incorporating all alternative process configurations, known as superstructures, and identifying an optimal design configuration to meet specific performance and cost objectives. Combinatorial optimization methods, such as MINLP algorithms, are then applied to solve the synthesis problem.

-   Hierarchical heuristic approach
-   Thermodynamic phenomena-driven approach
-   Evolutionary methods

Early work in optimization under uncertainty in chemical engineering utilized a "here and now" formulation, where the expected cost value served as the objective function for a given risk level [73–76]. Subsequently, researchers recognized the importance of incorporating operability considerations such as feasibility, flexibility, reliability, safety, and controllability into the optimal design problem [75, 77–79]. Flexibility ensures feasible steady-state operation over various conditions, while reliability addresses the probability of normal operation, given potential failures, safety concerns, and hazards resulting from failures. Controllability pertains to the quality and stability of the system's dynamic response.

Early approaches to achieving flexible designs involved determining overdesign factors [74]. Swaney [79] introduced the concept of flexibility index, presenting a well-structured optimization problem that could be easily transformed into a mini-max deterministic problem. Literature addresses two main problems: the "optimal flexibility" and "flexibility index" problems, with the latter extended to the "stochastic flexibility index" [29]. Flexibility has garnered significant attention in chemical engineering literature due to its structured nature. It allows solutions using deterministic approaches with decomposition [47, 79–81] or new approximations to derive probabilistic functions describing uncertainties' effects [82]. For further details, refer to a review paper by Grossmann et al. [46].

Earlier literature on process synthesis and design under uncertainty also focused on energy and environmental issues, which will be discussed separately below.

10.4.3 Energy and environmental problems

The earlier papers in the synthesis under uncertainty with pollution prevention focused on integrated environmental control systems for coal-based power systems. The work continued and extended to address synthesis problems in this area [16, 83, 84]. Nuclear waste management posed a very hard process synthesis problem. The combinatorial, nonconvex nature of the problem took a lot of work to solve, even with deterministic optimization methods. Uncertainties associated with waste tank contents and models caused further problems and demanded new algorithms. The new stochastic annealing algorithm provided an optimal and robust solution to this problem in the face of uncertainties with reasonable computational time [58]. A multiobjective extension of this problem to include policy aspects was possible due to these new algorithms [85, 86]. Dantus and High [60] also used this new algorithm for waste minimization in methylene chloride process synthesis.

In early 2000, applications in this area, including waste reduction in pharmaceutical industries [87, 88], hybrid fuel cell power plant design under uncertainty [89], industrial pollution management [90, 91], flexible carbon capture process plant scheduling [92], and environmentally benign heterogeneous azeotropic distillation system design [63], have been presented. Kheawhom and Hirao [93] also presented a two-layer algorithm for environmentally benign process synthesis under uncertainty. In the outer layer, the synthesis problem is represented by a multi-objective optimization problem, considering the performances associated with design parameters. In the inner layer, the problem is expressed as a single-objective optimization problem, taking into account the operating performances in the presence of uncertainty. This algorithm was applied to a membrane-based toluene recovery process. Identifying the environmental impacts of a process earlier is essential because the opportunity to overcome environmental problems in later stages of process development diminishes. However, in the early design stages, there is high uncertainty in various economic, ecological, and technical process parameters. For this purpose, Hoffman et al. [94] proposed a new approach to select promising process alternatives in the early stages of design. The method is based on approximating flowsheets by polynomial response surfaces with a lower complexity. A multi-objective optimization problem was solved for selecting a production process for hydrocyanic acid with 400 uncertain variables. Latin hypercube sampling technique was performed on the substituted response surface to obtain Pareto optimal solutions.

Contemporary environmental advocacy has shifted attention toward renewable energy and the intricate relationship between energy and water resources. Energy systems facing optimization challenges amid uncertain conditions encompass solar, wind, bioenergy, and biofuels. This optimization spans various applications such as designing plants optimally, considering weather and cost uncertainties [95], addressing the energy-water nexus [55, 96–99], expanding capacities [55, 100, 101], and scheduling and planning operations [102, 103].

Sensors are pivotal in environmental stewardship. Optimizing sensor placement in settings like chemical or power plants [104–106], water distribution networks [107, 108], or urban pollution monitoring [109] entails grappling with uncertainties. These uncertainties may stem from demand fluctuations, traffic patterns, measurement inaccuracies or meteorological variations.

10.4.4 Planning, managing, and scheduling processes

Most challenges in management, scheduling, and planning involve combinatorial problems and uncertainties. Pekny [110], and more recently, Li and Ierapetritou [66] and Ye et al. [111] have extensively reviewed this field. These issues are prevalent in batch processing due to the time-dependent nature of chemical processes commonly utilized in high-value, low-volume specialty chemicals and pharmaceuticals. Uncertainties often arise in batch plant operations, with much of the literature focusing on demand uncertainties.

Scheduling problems encompass sequencing, task allocation to equipment, maintenance planning over a specified timeframe, and inventory management. Typically, scheduling problems are encountered in batch processing scenarios. Addressing the uncertain nature of processes is crucial for establishing realistic production plans, with uncertainties stemming from processing time variations, equipment reliability, availability, and demand fluctuations.

Two primary approaches are employed to tackle scheduling uncertainties: reactive scheduling and stochastic scheduling. Reactive scheduling adjusts schedules in response to uncertain parameters or unforeseen events using heuristic methods. Conversely, stochastic scheduling incorporates uncertainties at the initial scheduling stage to develop optimal and reliable schedules in the face of uncertainty.

Recent applications [112, 113] include electricity planning and scheduling and oil and gas planning [114].

10.4.5 Supply chain management

Supply chain management extends scheduling challenges by managing multiple facilities and material shipments to customers through associated transportation networks. It encompasses activities related to storing and transporting raw materials and products from production facilities to consumption points. Given the dynamic market conditions and evolving customer demands, efficient and adaptable supply chains are crucial for businesses.

Short-term uncertainties in these systems include variations in processing parameters such as processing times, yields, or equipment availability. Long-term uncertainties involve fluctuations in raw material and final product prices and seasonal demand variations occurring over extended periods. Despite their critical importance, research on supply chain management under uncertainty in chemical engineering remains relatively scarce and has only recently begun to be explored [115, 116].

10.4.6 Reliability

Reliability-based design optimization (RBDO) emerged in its early stages to address the inherent uncertainties stemming from equipment failures while determining the structure and parameters of a system. Kuo and Wan [117] provided a comprehensive overview of research on reliability optimization problems and solution methodologies, emphasizing the significance of discrete decisions concerning parallel redundancies in RBDO. Aguilar et al. [118] focused on optimizing the design and operation of flexible utility plants while considering reliability and availability factors. Ye et al.

[119] introduced a rigorous nonconvex MINLP model to select redundant units in serial systems, aiming to optimize availability and cost. Thomaidis et al. [120] developed an MILP model for designing an integrated site vulnerable to random failures. Design choices impacting availability include increasing process capacity, incorporating parallel units, and integrating intermediate storage facilities. Recently, Chen et al. [121] proposed a two-stage stochastic programming GDP (generalized disjunctive programming) model with reliability constraints to deal with uncertainties in process synthesis, where the reliability model is incorporated into the flowsheet superstructure optimization.

10.4.7 Quality control

The parameter design methodology is widely employed to ensure robustness, known as an offline quality control approach for developing products and manufacturing processes resistant to uncontrollable variations, as popularized by Taguchi [122]. The factors influencing a product's performance are categorized into two groups: (1) design parameters with specified nominal settings and (2) noise parameters representing uncontrollable variations across a product's lifespan and different units. Two approaches are commonly utilized to establish the relationship between noisy input parameters and process output: (1) conducting physical experiments by varying input parameters across the noise space to generate a response surface or (2) developing computational models. Monte Carlo methods are employed to propagate the effects of input variability through a model and analyze output variability. A representative sample of input vectors is generated, reflecting the uncertainty distribution, and outputs are evaluated for each sample.

The significance of sampling efficiency in generating samples from the multivariate space for formulating a stochastic optimization problem for parameter design was highlighted in an earlier study by Diwekar and Rubin [123]. They presented the problem of offline quality control of a continuous-stirred tank reactor. Latin hypercube sampling (LHS) was employed instead of the Monte Carlo technique to reduce the required number of samples. Subsequently, Kalagnanam and Diwekar [25] applied the HSS technique to further improve efficiency by exploiting the k-dimensional uniformity properties of this technique. Another study on robust batch distillation column design using the HSS technique also demonstrated its effectiveness. Sahin and Diwekar [54] demonstrated the efficiency of a new algorithm called Better Optimization of Nonlinear Uncertain Systems (BONUS) for the same problem. Terwiesch and Agarwal [124] presented an optimization procedure to achieve robustness in batch reactor optimal control under parametric uncertainties using probability distributions for uncertain process parameters and optimizing the expectation of the cost function for the entire parameter space.

10.4.8 Optimal control

An extensively studied control problem in the literature is optimal control, where an optimal trajectory for a control variable is computed through dynamic optimization to maximize/minimize a performance index such as cost, product yield, or time. To compute optimal control trajectories in the presence of time-dependent uncertainties, a stochastic maximum principle formulation was developed based on real options theory and utilizing Ito processes [15, 68]. Using this methodology, thermodynamic model uncertainties in batch distillation were characterized, and stochastic optimal control profiles were computed [6, 15]. This approach was extended to address problems related to crystallization [125], biodiesel production [126], dynamic mercury management in lakes using pH control [127], and, recently, for addressing the problem of missing components in the specialty chemical industry [128].

In recent years, considering simultaneous process design and control has become important. Sakizlis et al. [129] provided a review of recent advances toward the integration of process design, process control and process operability, where time-varying disturbances and uncertainties are considered. More recently, Pajula and Ritala [130] discussed the effects of measurement uncertainty on process performance and its incorporation in the design of the control structure. Dynamic scenarios were utilized; each assigned a probability of occurrence based on knowledge gained from prior experiences. This method was applied to a papermaking process, studying the effect of measurement uncertainty of fiber and filler consistency (concentration) on controller performance. For batch separations and solvent recycling in the pharmaceutical industry, Ulas and Diwekar [131] presented a framework that couples product design, process design, and optimal control in the face of time-dependent uncertainties.

10.5 Conclusions

Optimization problems under uncertainty present formidable challenges within the field of optimization. This type of optimization involves two main loops: the optimization loop and the uncertainty analysis loop. Probabilistic representation can represent uncertainties, and sampling is used to generate uncertainty scenarios. Depending on the number of uncertain variables, efficient sampling techniques can be selected, and the efficiency of optimization can be increased by managing the interactions of optimization and sampling loops. PSE encompasses numerous applications of optimization under uncertainty, from product discovery to chemical synthesis, process synthesis, energy and environmental issues, and process operations such as scheduling, supply chain management, and control. The recent focus on renewable energy systems, with their inherent weather, cost, and demand uncertainties, has significantly heightened interest in optimization under uncertainty within PSE.

References

[1]   Bealet, E. M. L. On minimizing a convex function subject to linear inequalities. *Journal of the Royal Statistical Society*, 1955(17B), 173.

[2]   Dantzig, G. B. Linear programming under uncertainty. *Management Science*, 1955, 1, 197.

[3]   Diwekar, U. M. *Introduction to Applied Optimization*. 3rd Edition, Springer, 2020.

[4]   Edgeworth, E. The mathematical theory of banking. *Journal of the Royal Statistical Society*, 1888, 51, 113.

[5]   Petruzzi, N. C., M. Dada, M. Pricing and the newsvendor problem: A review with extensions. *Operations Research*, 1999, 47(2), 183.

[6]   Ulas, S., Diwekar, U. M. Thermodynamic uncertainties in batch processing and optimal control. *Computers & Chemical Engineering*, 2004, 28(11), 2245–2258.

[7]   Kim, K.-J., Diwekar, U. M. Efficient combinatorial optimization under uncertainty. 2. Application to stochastic solvent selection. *Industrial & Engineering Chemistry Research*, 2002, 41(5), 1285–1296.

[8]   Diwekar, U. M., Rubin, E. S., Frey, H. C. Optimal design of advanced power systems under uncertainty. *Energy Conversion and Management Journal*, 1997, 38, 1725.

[9]   Diwekar, U. M. A process analysis approach to pollution prevention. *AIChE Symposium Series on Pollution Prevention through Process and Product Modifications*, 1995, 90, 168.

[10] Subramanian, D., Pekny, J. F., Reklaitis, G. V. A simulation optimization framework for research and development pipeline management. *AIChE Journal*, 2001, 47(10), 2226.

[11] Kim, K.-J., Diwekar, U. M., Joback, K. G. Greener Solvent Selection under Uncertainty. In Abraham, M., Moens, L. (Eds.). *ACS Symposium Series No. 819 Clean Solvents: Alternative Media for Chemical Reactions and Processing*. Washington, DC: American Chemical Society, 2001.

[12] Gwaltney, C. R., Lin, Y., Simoni, L. D., Stadtherr, M. A. Interval Methods for Nonlinear Equation Solving Applications. In Pedrycz, W., Skowron, A., Kreinovich, V. (Eds.). *Handbook of Granular Computing*. John Wiley & Sons Ltd, 2008.

[13] Enszer, J. A., Lin, Y., Ferson, S., Corliss, G. F., Stadtherr, M. A. Probability bounds analysis for nonlinear dynamic process models. *AIChE Journal*, 2010.

[14] Stadtherr, M. A. Interval Analysis: Application to Chemical Engineering Design Problems. In Floudas, C. A., Pardalos, P. M. (Eds.). *Encyclopedia of Optimization*. Kluwer Academic Publishers, 2001.

[15] Rico-Ramirez, V., Diwekar, U. M., Morel, B. Real option theory from finance to batch distillation. *Computers & Chemical Engineering*, 2003, 27(12), 1867–1882.

[16] Diwekar, U. M., Rubin, E. S. Stochastic modeling of chemical processes. *Computers & Chemical Engineering*, 1991, 15(2), 105–114.

[17] Metropolis, N., Ulam, S. The Monte Carlo method. *Journal of American Statistical Association*, 1949, 44(247), 335–341.

[18] Lehmer, D. H. Mathematical methods in large-scale computing units. *Proceedings of the 2nd Symposium on Large Scale Digital Calculating Machinery*, 1949, 141–146.

[19] James, B. A. P. Variance reduction techniques. *The Journal of the Operational Research Society*, 1985, 36(6), 525–530.

[20] Nelson, B. L. Control variate remedies. *Operations Research*, 1990, 38(6), 974–992.

[21] Wilson, J. R. Antithetic sampling with multivariate inputs. *American Journal of Mathematical and Management Sciences*, 1983, 3, 121–144.

[22] McKay, M. D., Beckman, R. J., Conover, W. J. A comparison of three methods of selecting values of input variables in the analysis of output from a computer code. *Technometrics*, 1979, 21(2), 239–245.

[23] Iman, R. L., Conover, W. J. Small sample sensitivity analysis techniques for computer models, with an application to risk assessment. *Communications in Statistics*, 1982, A17, 1749–1842.

[24] Saliby, E. Descriptive sampling: A better approach to Monte Carlo simulations. *Journal of Operations Research Society*, 1990, 41(12), 1133–1142.

[25] Kalagnanam, J. R., Diwekar, U. M. An efficient sampling technique for off-line quality control. *Technometrics*, 1997, 39(3), 308–319.

[26] Diwekar, U. M., Kalagnanam, J. R. Efficient sampling technique for optimization under uncertainty. *AIChE Journal*, 1997, 43(2), 440–447.

[27] Wang, R. Y., Diwekar, U. M., Padro, C. E. G. Efficient sampling techniques for uncertainties in risk analysis. *Environmental Progress*, 2004, 23(2), 141–157.

[28] Frenkel, D., Smit, B. *Understanding Molecular Simulation: From Algorithms to Applications*. San Diego: Academic Press, 1996.

[29] Straub, D. A., Grossmann, I. E. Integrated stochastic metric of flexibility for systems with discrete state and continuous parameter uncertainties. *Computers & Chemical Engineering*, 1990, 14(9), 967–985.

[30] Bernardo, F. P., Pistikopoulos, E. N., Saraiva, P. M. Quality costs and robustness criteria in chemical process design optimization. *Computers & Chemical Engineering*, 2001, 25(1), 27–40.

[31] Wendt, M., Li, P., Wozny, G. Nonlinear chance-constrained process optimization under uncertainty. *Industrial & Engineering Chemistry Research*, 2002, 41(15), 3621–3629.

[32] Vasquez, V. R., Whiting, W. B. Uncertainty and sensitivity analysis of thermodynamic models using equal probability sampling (EPS. *Computers & Chemical Engineering*, 2000, 23(11–12), 1825–1838.

[33] Niederreiter, H. *Random Number Generation and Quasi-Monte Carlo Methods*. Philadelphia, PA: SIAM Publications, 1992.

[34] Bayes, T. An essay towards solving a problem in the doctrine of chances. *Philosophical Transactions of the Royal Society of London*, 1763, 53, 370–418.

[35] Meel, A., O'Neill, L. M., Seider, W. D., Oktem, U., Keren, N. Frequency and Consequence Modeling of Rare Events Using Accidents Databases. In *AIChE Spring National Meeting*. Paper 106d, 2006.

[36] Mehranbod, N., Soroush, M., Panjapornpon, C. A method of sensor fault detection and identification. *Journal of Process Control*, 2005, 15(3), 321–339.

[37] Lainez-Aguirre, J., Mockus, L., Reklaitis, G. V. A stochastic programming approach for the Bayesian experimental design of nonlinear system. *Computers and Chemical Engineering*, 2015, 72, 312–324.

[38] Charnes, A., Cooper, W. W. Chance-constrained programming. *Management Science*, 1959, 5, 73.

[39] Maranas, C. D. Optimal molecular design under property prediction uncertainty. *AIChE Journal*, 1997, 43(5).

[40] P., L., Arellano-Garcia, H., Wozny, G. Chance constrained programming approach to process optimization under uncertainty. *Computers and Chemical Engineering*, 2008, 32, 25–45.

[41] VanSlyke, R., Wets, R. J. B. L-shaped linear programs with application to optimal control and Stochastic Programming. *SIAM Journal on Applied Mathematics*, 1969, 17, 638.

[42] Birge, J. R., Louveaux, F. A multicut algorithm for two-stage stochastic linear programming. *European Journal of Operations Research*, 1988, 34, 384.

[43] Birge, J. R., Louveaux, F. *Introduction to Stochastic Programming*. New York: Springer-Verlag, 1997.

[44] Higle, J. L., Sen, S. Stochastic decomposition: An algorithm for two-stage linear programs with recourse. *Mathematics of Operations Research*, 1991, 16(3), 650–669.

[45] Higle, J. L., Sen, S. *Stochastic Decomposition*. Kluwer Academic Publisher, 1996.

[46] Grossmann, I., . R., Calfa, B., García-Herreros, P., Zhang, Q. Recent advances in mathematical programming techniques for the optimization of process systems under uncertainty. *Computers and Chemical Engineering*, 2016, 91, 3–14.

[47] Ahmed, S., Sahinidis, N. V., Pistikopoulos, E. N. An improved decomposition algorithm for optimization under uncertainty. *Computational Chemical Engineering*, 2000, 23, 1589.

[48] Floudas, C. A., Guemues, Z. H., Ierapetritou, M. G. Global optimization in design under uncertainty: Feasibility test and flexibility index problems. *Industrial Engineering Chemistry Research*, 2001, 40, 4267.

[49] Ierapetritou, M. G., Pistikopoulos, E. N. Novel optimization approach of stochastic planning models, Ind. *Engineering Chemistry Research*, 1994, 33, 1930.

[50] Ostrovskii, G. M., Volin, Y. M. On new problems in the theory of flexibility and optimization of chemical process under uncertainty. *Theoretical Foundations of Chemical Engineering*, 1999, 33(5), 524.

[51] Dantzig, G. B., Infanger, G. Large Scale Stochastic Linear Programs – Importance Sampling and Bender Decomposition. In Brezinski, C., Kulisch, U. (Eds.). *Computational and Applied Mathematics I-Algorithms and Theory. Proceedings of the 13th IMCAS World Congress*. Dublin, Ireland, 1991, pp. 111–120.

[52] Wei, J., Realff, M. J. Sample average approximation methods for stochastic MINLPs. *Computers & Chemical Engineering*, 2004, 28(3).

[53] Sahin, K., Diwekar, U. Better optimization of nonlinear uncertain systems (BONUS): A new algorithm for stochastic programming using reweighting through kernel density estimation. *Annals of OR*, 2004, 132, 47.

[54] Sahin, K., Diwekar, U. Stochastic nonlinear optimization -The BONUS algorithm. *SIAM Newsletter*, 2002, 13, 1.

[55] Diwekar, U. M., David, A. *BONUS Algorithm for Large Scale Stochastic Nonlinear Programming Problems*. Springer, 2015.

[56] Birge, J. R. Stochastic Programming computation and applications. *INFORMS Journal on Computing*, 1997, 9(2), 111.

[57] Painton, L. A., Diwekar, U. M. Stochastic annealing under uncertainty. *European Journal of Operations Research*, 1995, 83, 489.

[58] Chaudhuri, P., Diwekar, U. Synthesis approach to optimal waste blend under uncertainty. *AIChE Journal*, 1999, 45, 1671.

[59] Tayal, M., Diwekar, U. Novel sampling approach to optimal molecular design under uncertainty: A polymer design case study. *AIChE Journal*, 2001, 47(3), 609.

[60] Dantus, M. M., High, K. A. Evaluation of waste minimization alternatives under un- certainty: A multiobjective optimization approach. *Computational Chemical Engineering*, 1999, 23, 1493.

[61] Xu, W., Diwekar, U. Improved genetic algorithms for deterministic optimization and optimization under uncertainty-Part I: Algorithms development. *Industrial Engineering Chemistry Research*, 2005, 44, 7132.

[62] Xu, W., Diwekar, U.. Improved genetic algorithms for deterministic optimization and optimization under uncertainty-Part II: Solvent selection under uncertainty, Ind. *Engineering Chemistry Research*, 2005, 44, 7138.

[63] Xu, W., Diwekar, U. Multiobjective integrated Solvent selection and solvent recycling under uncertainty using new genetic algorithms. *International Journal of Environment and Pollution*, 2006, 29, 70.

[64] Pistikopoulos, E. N., Georgiadis, M. C., Dua, V. *Editors Multi-parametric Programming: Theory, Algorithms, and Applications*. Wiley-VCH, 2007.

[65] Sahinidis, N. Optimization under uncertainty: State-of-the-art and opportunities. *Computers and Chemical Engineering*, 28(2004), 971–983.

[66] Li, Z., Ierapetritou, M. Robust optimization for process scheduling under uncertainty. *Industrial & Engineering Chemistry Research*, 2008, 47(12), 4148–4157. Li Z, Ierapetritou M. Capacity expansion planning.

[67] Ning, C., You, F. Data-driven stochastic robust optimization: General computational framework and algorithm leveraging machine learning for optimization under uncertainty in the big data era. *Computers and Chemical Engineering*, 111(2018), 115–13.

[68] Rico-Ramirez, V., Diwekar, U. M. Stochastic maximum principle for optimal control under uncertainty. *Computers & Chemical Engineering*, 2004, 28(12), 2845–2849.

[69] Joback, K. G., Reid, R. C. Estimation of pure-component properties from group-contributions. *Chemical Engineering Communications*, 1987, 57(1-6), 233–243.

[70] Hansen, H. K., Rasmussen, P., Fredenslund, A., Schiller, M., Gmehling, J.. Vapor-Liquid-Equilibria by Unifac Group Contribution. 5. Revision and Extension. *Industrial & Engineering Chemistry Research*, 1991, 30(10), 2352–2355.

[71] Harper, P. M., Rafiqul, G., Kolar, P., Ishikawa, T. Computer-aided molecular design with combined molecular modeling and group contribution. *Fluid Phase Equilibria*, 1999, 158-160, 337–347.

[72] Kim, K.-J., Diwekar, U. M. Integrated solvent selection and recycling for continuous processes. *Industrial & Engineering Chemistry Research*, 2002d, 41(18), 4479–4488.

[73] Kittrel, J. R., Watson, C. C. Don't overdesign equipment. *Chemical Engineering Progress*, 1966, 63, 79.

[74] Watanbe, N., Nishimura, Y., Matsubara, M. Optimal design of chemical processes involving parameter uncertainty. *Chemical Engineering Science*, 1973, 28, 905.

[75] Wen, C. Y., Chang, T. M. Optimal design of systems involving parameter uncertainty. *Industrial & Engineering Chemistry Process Design and Development*, 1973(7), 49.

[76] Weisman, J., Holzman, A. G. Optimal process system design under conditions of risk. *Industrial & Engineering Chemistry Process Design and Development*, 1972, 11, 386.

[77] Malik, R. K., Hughes, R. R. Optimal design of flexible chemical processes. *Computational Chemical Engineering*, 1983, 7, 423.

[78] Nishida, N., Ichikawa, A., Tazaki, E. Synthesis of optimal process systems with uncertainty. *Industrial Engineering Chemistry Research*, 1974, 13, 209.

[79] Swaney, R. S. *Analysis of Operational Flexibility in Chemical Process Design*. Ph.D. Thesis. Pittsburgh, PA, USA: Department of Chemical Engineering, Carnegie Mellon University, 1983.

[80] Floudas, C. A., Guemues, Z. H., Ierapetritou, M. G. Global optimization in design under uncertainty: Feasibility test and flexibility index problems, *Industrial Engineering Chemistry Research*, 2001, 40, 4267.

[81] Ierapetritou, M. G., Pistikopoulos, E. N. Novel optimization approach of stochastic planning models, Ind. *Engineering Chemistry Research*, 1994, 33, 1930.

[82] Raspanti, C. G., Bandoni, J. A., Biegler, L. T. New strategies for flexibility analysis and design under uncertainty. *Computational Chemical Engineering*, 2000, 24, 2193.

[83] Diwekar, U. M. A process analysis approach to pollution prevention. *AIChE Symposium Series on Pollution Prevention through Process and Product Modifications*, 1995, 90(303), 168–179.

[84] Diwekar, U. M., Rubin, E. S., Frey, H. C. Optimal design of advanced power systems under uncertainty. *Energy Convers Management*, 1997, 38, 1725–1735.

[85] Johnson, T. L., Diwekar, U. M. Hanford waste blending and the value of research: Stochastic optimization as a policy tool. *Journal of Multi-Criteria Decision Analysis*, 2001, 10, 87–99.

[86] Johnson, T. L., Diwekar, U. M. The Value of Design Research: Stochastic Optimization as a Policy Tool. In Malone, M. F., Tainham, J. A., Carnahan, B. (Eds.). *Foundations of Computer-Aided Design*, AIChE Symposium Series. Vol. 96, pp. 454–461.

[87] Linninger, A. A., Chakraborty, A. Pharmaceutical waste management under uncertainty. *Computers & Chemical Engineering*, 2001, 25(4–6), 675–681.

[88] Linninger, A. A., Chakraborty, A., Colberg, R. D. Planning of waste reduction strategies under uncertainty. *Computers & Chemical Engineering*, 2000, 24(2-7), 1043–1048.

[89] Subramanyan, K., Diwekar, U. M. The "value of research" methodology and hybrid power plant design. *Industrial & Engineering Chemistry Research*, 2006a, 45(2), 681–695.

[90] Diwekar, U. M. Greener by Design. *Environmental Science and Technology*, 2003, 37(23), 5432–5444.

[91] Shastri, Y., Diwekar, U. Industrial Pollution Management: A Sustainability Perspective. In *Invited Chapter in New Research in Chemical Engineering*. Nova Science Publishers, 2009.

[92] Zantve, M., Arora, A., Hasan, M. Operational power plant scheduling with flexible carbon capture: A multistage stochastic optimization approach. *Computers and Chemical Engineering*, 2019, 130, 106544.

[93] Kheawhom, S., Hirao, M. Decision support tools for environmentally benign process design under uncertainty. *Computers & Chemical Engineering*, 2004, 28(9), 1715–1723.

[94] Hoffmann, V. H., McRae, G. J., Hungerbuhler, K. Methodology for early-stage technology assessment and decision making under uncertainty: application to the selection of chemical processes. *Industrial & Engineering Chemistry Research*, 2004, 43(15), 4337–4349.

[95] Vaderobli, A., Parikh, D., Diwekar, U. Optimization under uncertainty to reduce the cost of energy for parabolic trough solar power plants for different weather conditions. *Energies*, 2020, 13, 3131.

[96] Salazar, J., Zitney, S., Diwekar, U. Minimization of water consumption under uncertainty for a Pulverized Coal power plant. *Environmental Science and Technology, Water-Energy Nexus Virtual Issue*, 2011, 45, 4545.

[97] Salazar, J., Diwekar, U. An efficient stochastic optimization framework for studying the impact of seasonal variation on the minimum water consumption of Pulverized coal (PC) power plants. *Energy Systems*, 2011, 2, 263.

[98] Salazar, J., Diwekar, U., Constantinescu, E., Zavala, V. Stochastic optimization approach to water management in cooling-constrained power plants. *Applied Energy*, 2013, 112, 12.

[99] Li, Y., Wei, J., Yuan, Z., Chen, B., Gani, R. Sustainable synthesis of integrated process, water treatment, energy supply, and CCUS networks under uncertainty. *Computers and Chemical Engineering*, 2022, 157, 107636.

[100] Pulsipher, J. L., Zavala, V. M. A scalable stochastic programming approach for the design of flexible systems. *Computers and Chemical Engineering*, 2019, 128, 69–76.

[101] Ikonen, T. J., Han, D., Lee, J. H., Harjunkoski, I. Stochastic programming of energy system operations considering terminal energy storage levels. *Computers and Chemical Engineering*, 2023, 179, 108449.

[102] Cramer, E., Paeleke, L., Mitsos, A., Dahmen, M. Normalizing flow-based day-ahead wind power scenario generation for profitable and reliable delivery commitments by wind farm operators. *Computers and Chemical Engineering*, 2022, 166, 107923.

[103] Arabi, M., Yaghoubi, S., Tajik, J. Algal biofuel supply chain network design with variable demand under alternative fuel price uncertainty: A case study. *Computers and Chemical Engineering*, 2019, 130, 106528.

[104] Lee, A., Diwekar, U. Optimal sensor placement in integrated gasification combined cycle power systems. *Applied Energy*, 2012, 99, 255.

[105] Sen, P., Sen, K., Diwekar, U. Multiobjective optimization approach to sensor placement in IGCC power plant. *Applied Energy*, 2016, 181, 527.

[106] Sen, P., Diwekar, U., Bhattacharyya, D. Stochastic programming approach versus estimator-based approach for sensor network design for maximizing efficiency. *Smart and Sustainable Manufacturing*, 2018, 2(2), 44.

[107] Mukherjee, R., Diwekar, U. Optimal sensor placement with mitigation strategy for water network system under uncertainty. *Computers and Chemical Engineering*, 2017, 103, 91.

[108] Rico-Ramirez, S. F.-H., Diwekar, U., Hernandez-Castro, S. Water networks security: A two-stage mixed-integer stochastic program for sensor placement under uncertainty. *Computers and Chemical Engineering*, 2007, 31, 565.

[109] Mukherjee, R., Diwekar, U., Kumar, N. Real-time optimal spatiotemporal sensor placement for monitoring air pollutants. *Clean Technologies and Environmental Policy*, 2020, 22, 2091.

[110] Pekny, J. F. Algorithm architectures to support large-scale process systems engineering applications involving combinatorics, uncertainty, and risk management. *Computers & Chemical Engineering*, 2002, 26(2), 239–267.

[111] Ye, Y., Li, J., Li, Z., Tang, Q., Xiao, X., Floudas, C. Robust optimization and stochastic programming approaches for medium-term production scheduling of a large-scale steelmaking continuous casting process under demand uncertainty. *Computers and Chemical Engineering*, 2014, 66, 165–185.

[112] Shabazbegian, V., Ameli, H., Ameli, M. T., Strbac, G. Stochastic optimization model for coordinated operation of natural gas and electricity networks. *Computers and Chemical Engineering*, 2020, 142, 107060.

[113] Leo, E., Ave, G. D., Harjunkoski, I., Engell, S. Stochastic short-term integrated electricity procurement and production scheduling for a large consumer. *Computers and Chemical Engineering*, 2021, 145, 107191.

[114] Li, F., Qian, F., Minglei Yang, W. D., Long, J., Mahalec, V. Refinery production planning optimization under crude oil quality uncertainty. *Computers and Chemical Engineering*, 2021, 151, 107361.

[115] Lima, C., Relvas, S., Barbosa-Póvoa, A. Stochastic programming approach for the optimal tactical planning of the downstream oil supply chain. *Computers and Chemical Engineering*, 2018, 108, 314–336.

[116] Mohamadpour Tosarkani, B., Amin, S. An environmental optimization model to configure a hybrid forward and reverse supply chain network under uncertainty. *Computers and Chemical Engineering*, 2019, 121, 540–555.

[117] Kuo, W., Wan, R. Recent advances in optimal reliability allocation. *IEEE Transactions on Systems, Man, and Cybernetics - Part A*, 2007, 37(2), 143–156.

[118] Aguilar, O., Kim, J.-K., Perry, S., Smith, R. Availability and reliability considerations in the design and optimisation of flexible utility systems. *Chemical Engineering Science*, 2008, 63(14), 3569–3584.

[119] Ye, Y., Grossmann, I. E., Pinto, J. M. Mixed-integer nonlinear programming models for optimal design of reliable chemical plants. *Computers & Chemical Engineering*, 2018, 116, 3–16.

[120] Thomaidis, T. V., Terrazas-Moreno, S., Grossmann, I. E., Wassick, J. M., Bury, S. J. Optimal design of reliable integrated chemical production sites. *Computers & Chemical Engineering*, 2010, 34(12), 1919–1936.

[121] Chen, Y., Ye, Y., Yuan, Z., Grossmann, I. E., Chen, B. Integrating stochastic programming and reliability in the optimal synthesis of chemical processes. *Computers & Chemical Engineering*, 2022, 157, 107616.

[122] Taguchi, G. *Introduction to Quality Engineering*. Tokyo, Japan: Asian Productivity Center, 1986.

[123] Diwekar, U. M., Rubin, E. S. Parameter design methodology for chemical processes using a simulator. *Industrial & Engineering Chemistry Research*, 1994, 33(2), 292–298.

[124] Terwiesch, P., Agarwal, M. Robust input policies for batch reactors under parametric uncertainty. *Chemical Engineering Communications*, 1995, 131, 33–52.

[125] Yenkie, K., Diwekar, U. Stochastic optimal control of batch crystallization with Ito processes. *Industrial & Engineering Chemistry Research*, 2013, 52(1), 108–122.

[126] Benavides, P., Diwekar, U. Studying various optimal control problems in biodiesel production in a batch reactor under uncertainty. *Fuel*, 2013, 103, 585–592.

[127] Shastri, Y., Diwekar, U. Optimal control of lake pH for mercury bioaccumulation control. *Ecological Modelling*, 2008, 216(1).

[128] Shastri, Y., Diwekar, U. Optimal control of lake pH for mercury bioaccumulation control. *Ecological Modelling*, 2008, 216(1).

[129] Sakizlis, V., Perkins, J. D., Pistikopoulos, E. N. Recent advances in optimization-based simultaneous process and control design. *Computers & Chemical Engineering*, 2004, 28(10), 2069–2086.

[130] Pajula, E., Ritala, R. Measurement uncertainty in integrated control and process design–A case study. *Chemical Engineering and Processing*, 2006, 45(4), 312–322.

[131] Ulas, S., Diwekar, U. M. Integrating product and process design with optimal control: A case study of solvent recycling. *Chemical Engineering Science*, 2006a, 61(6), 2001–2009.

Riju De* and Yogendra Shastri

Chapter 11
Optimal control of batch processes in the continuous time domain

Abstract: Batch processes are vital in manufacturing specialty and fine chemicals, which are commonly employed in small-scale industries targeting a lower production volume. Optimal control problems (OCPs) are paramount to optimizing the batch processes owing to their nonlinear dynamical behavior and a wider range of operating conditions. This chapter briefly reviews various OCPs applied to batch processes along with their mathematical formulations while considering case-specific constraints defined on the control and state variables. The key challenges and commonly appeared decision variables pertaining to chemical or biochemical batch operations are summarized. Numerical techniques to solve the OCPs using an indirect method, i.e., Pontryagin's minimum principle and a direct method involving sequential and simultaneous collocation-based approaches are provided. This study further explores the quadratic regulator problem and optimal tracking controller design for the batch processes using linearized dynamical models by illustrating a case study. Selected applications of the OCPs on diverse batch processes from chemical, biochemical, and ecosystem engineering domains, viz., transesterification, acid pre-treatment, hydro-thermal liquefaction, enzymatic hydrolysis, and predator-prey system are reviewed from the literature. Current trends and progress toward developing new OCPs algorithms are also discussed.

Keywords: optimal control, batch processes, direct methods, indirect methods

11.1 Introduction

Over the years, chemical process industries involved in the production of various polymers, dyes, paints, lubricants, foods, as well as specialty and fine chemicals have been relying on batch-process manufacturing, where usually a low production volume is targeted with a requirement to attain a desirable product quality [1]. Batch processes are primarily employed in small-scale process industries. In contrast to continuous processes, batch processes can be operated at much lower investment costs

---

*Corresponding author: **Riju De**, Department of Chemical Engineering, Birla Institute of Technology and Science Pilani, K.K. Birla Goa Campus, Zuarinagar, Goa 403726, India
Yogendra Shastri, Department of Chemical Engineering, Indian Institute of Technology, Bombay, Mumbai 400076, India, e-mail: yshastri@che.iitb.ac.in

https://doi.org/10.1515/9783111383439-011

with greater flexibility that allows for the manipulation of key process variables such as temperature, flow rate, and time to obtain the desired products [2–3]. Moreover, batch processes, owing to greater flexibility in operations, are also better placed to handle uncertainties and disturbances, such as changes in the feedstock composition [2].

The optimization of batch processes has been an important topic of research, from both design and operations perspective. Batch processes are generally optimized to meet specific goals, such as maximization of the desired product yield, maximization of productivity, minimization of batch time for specific yield, and maximization of profit. Other goals such as minimization of total energy consumption during the batch have also been commonly used [4–6]. These various performance indicators can also be translated into equivalent economic indicators, and cost minimization problems can also be solved. Many reactions occurring in a batch process involve multiple reaction pathways leading to the generation of undesired products along with the formation of the desired product. Therefore, minimization of the undesired product is another crucial objective to maximize the desired product's selectivity [3]. These goals need to be achieved while simultaneously meeting process safety and product quality constraints.

Notwithstanding the aforementioned benefits of batch processing, it also has its own challenges from operations and control standpoint. All batch processes are dynamic in nature, which is inherently more challenging. The batch process control and optimization problems also are significantly challenging to solve [2, 3]. Due to the transient nature of a batch process, the decision variables also need to be time-dependent. This significantly expands the feasible space of the control and optimization problems. However, it also provides certain benefits as one has greater degree of freedom to steer the batch process to its desired objective.

Optimal control theory has been a popular approach to optimize batch processes in a systematic manner [7, 8]. The theory has evolved over the years to address several complexities such as uncertainties and disturbances. The objective of this chapter is to briefly review various optimal control problems (OCPs) that are more frequently encountered in batch processes while discussing their general formulations along with the proposed solution methodologies. Subsequently, selected applications of optimal control theory using these approaches are discussed.

11.2 Batch process control: challenges and formulations

A typical batch process is highly nonlinear and transient, does not attain a steady state, and usually runs over a broader range of operating conditions. Such characteristics impose challenges in the optimization and control tasks. For instance, due to the absence of a steady state, a nonlinear process model cannot be linearized around a fixed operating point that is required for the controller design. Owing to the dynamic nature of a batch process, the process variables such as temperature, pressure, concentration, and flow rates of the coolant in the reactor jacket vary with time. Hence, a trajectory-based optimization approach is essential to optimize the time-varying profiles of the control variables. However, to devise such an optimization policy, knowledge of a well-defined and fairly accurate process model of the batch process is essential. If a first principles-based mechanistic batch model is available, model-based control can be implemented to generate the optimized profiles of the manipulated variables. Optimal control, also termed as dynamic optimization, is a method for identifying optimal time-varying trajectories of the decision variables to attain a specific objective while satisfying a set of constraints for any given batch process [9]. The OCPs can be classified as unconstrained and constrained optimization problems, which are discussed next.

11.2.1 Unconstrained optimal control problem

Equation (11.1) represents a typical unconstrained optimal control problem. $x(t)$ represents the state variables, while $u(t)$ represents the decision variables, both of which could be more than one. Let us consider a nonlinear batch process whose dynamics can be represented by a time-varying ordinary differential equation, as expressed in eq. (11.1). Here, $u(t) \epsilon R^{Nu}$ denotes the control input variable, and $x(t) \epsilon R^{Nx}$ indicates the state variables. The main goal of an unconstrained OCP is to deduce an open loop time-varying profile of the control variable $u^*(t)$ such that the states of the batch plant $(x(t))$ can be driven from initial time (t_0) to the final time (t_f) along an optimal profile $x^*(t)$ by either minimizing or maximizing a performance index $J(x)$ while satisfying the model constraints expressed as follows.

$$\min J(x)$$

$$\text{s.t. } \dot{x} = f(x(t), u(t), t); \ x(t=0) = x_0 \tag{11.1}$$

$$u(t) \in U(t)$$

Here, $U(t)$ belongs to the subset of all Nu admissible piecewise continuous control variables defined over a fixed batch horizon of $[t_0, t_f]$. The performance index $J(x)$ may comprise of two components: a terminal cost or fixed cost, represented by $\varphi(t_f, x(t_f))$, and a running or stage cost denoted by a Lagrange function, i.e., $L(x(t), u(t), t)$, defined inside the integral as shown in eq. (11.2). Such a representation of the objective function is known as the Bolza form, given as follows:

$$J(x) = \varphi(t_f, x(t_f)) + \int_{t_0}^{t_f} L(x(t), u(t), t)dt \tag{11.2}$$

The Bolza form can be simplified to Mayer's form type objective function when the desired performance index is to optimize only the terminal cost $\varphi(t_f, x(t_f))$. However, for certain problems, it is required to deal only with the optimization of the stage cost, which reduces the Bolza form to a Lagrange form by setting the final cost to zero.

11.2.2 Constrained optimal control problem

In batch processes, several equality and inequality constraints may exist on the state and control variables. Such a constrained OCP in the continuous time domain can be represented by eqs. (11.3)–(11.7) as follows:

$$\min_{u(t), x(t), t_f} \varphi(t_f, x(t_f)) + \int_{t_0}^{t_f} L(x(t), u(t), t)dt$$

$$\text{s.t.} \ \dot{x} = f(x(t), u(t), t); \ x(t=0) = x_0 \tag{11.3}$$

$$g(x(t), u(t), t) \le 0 \tag{11.4}$$

$$a(x(t), t) \le 0 \tag{11.5}$$

$$m(x(t_f), t_f) \le 0 \tag{11.6}$$

$$h(x(t_f), t_f) = 0 \tag{11.7}$$

Here, $g \in R^{Ng}$ belongs to the mixed control and state inequality constraints; $a \in R^{Na}$ signifies the inequality state constraints, $m \in R^{Nm}$ corresponds to the terminal state inequality constraints, and $h \in R^{Nh}$ indicates the terminal state equality constraints. Several batch processes emerging from diverse applied engineering areas, e.g., manufacturing of specialty chemicals [3], biochemicals [10], pharmaceuticals [11], polymers [12], and semiconductors [13] have been successfully optimized through the solution of the OCPs presented in eqs. (11.3)–(11.7). In addition, certain electricity-driven batch processes, such as the operation of batch electric arc furnaces used in steel making [14], hydrochloric acid recovery from batch electrodialysis [15], and fed-batch microbial electrolysis cells used for hydrogen production, have also been optimized through such OCPs to enhance the system performance [16]. These OCPs usually seek to meet specific objectives such as maximum yield, maximum profit, maximum degree of separation, minimum time and minimum energy cost while determining the time-varying profiles of the decision variables.

11.3 Methods to solve optimal control problems

There are two popular methods to solve an OCP, which can be classified into indirect and direct methods. The former is based on the classical Hamiltonian formulation, which can determine an accurate and exact solution of the optimal control profile. On the other hand, the latter approach falls under the category of direct collocation-based methods, which produces an approximate solution of the optimal control input. Here, we will mainly discuss the indirect method that is applied to a generic nonlinear batch process in the continuous time domain. The Hamiltonian function for the OCP of the batch process described in eqs. (11.3)–(11.7) can be expressed using eq. (11.8) as follows:

$$H(x(t), u(t), \lambda(t), t) = L(x(t), u(t), \lambda(t), t) + \lambda^T(t)f(x(t), u(t), t) \tag{11.8}$$

Here, $\lambda(t) \in R^n$ denotes the adjoint or the co-state variable associated with the Nx state variables. The first-order necessary conditions of optimality given by the Pontryagin's minimum/maximum principle [17] require the integration of the state and the co-state variables given by eqs. (11.9) and (11.10) in the forward and backward directions, respectively:

$$\dot{x}(t) = \frac{\partial H}{\partial \lambda}; \; x(t=0) = x0 \tag{11.9}$$

$$\dot{\lambda}(t) = -\frac{\partial H}{\partial x} \tag{11.10}$$

The solution of the co-state variable, as in eq. (11.10), requires the knowledge of the transversality boundary condition [18], as given by eq. (11.11):

$$\left[\frac{\partial \varphi}{\partial x(t_f)} + m_x^T(x(t_f), t_f)\mu - \lambda \right]^T \bigg|_{t_f} dx(t_f)$$

$$+ \left[\frac{\partial \varphi}{\partial t} + m_t^T(x(t_f), t_f)\mu + H \right]^T \bigg|_{t_f} dt_f = 0 \tag{11.11}$$

The necessary conditions of optimality require that eqs. (11.9) and (11.10) must be satisfied with the specified boundary conditions, and the partial derivative of the Hamiltonian evaluated with respect to the control variable must vanish at a stationary trajectory, as defined by the following equation:

$$\frac{\partial H(x^*(t), u^*(t), \lambda^*(t), t)}{\partial u} = 0 \tag{11.12}$$

Specific variations of the OCPs may arise, depending on the type of constraints enforced on the state and control variables for a given batch process. These can be categorized as (1) fixed final time problem with final state fixed or (2) free, and (3) free final time OCP. Therefore, depending on the class of the OCP, the transversality boundary conditions, as in eq. (11.11), may vary, which should be carefully included along with the optimality conditions discussed.

11.3.1 OCP-1: Fixed final time problem with final state fixed

In this OCP, the terminal state and time are fixed to a finite value. The fixed final state boundary condition can be expressed as follows:

$$x(t_f) = x_f \tag{11.13}$$

Certain batch processes may require strict bounds on the terminal states, as defined in eqs. (11.6) and (11.7), where a particular state variable cannot be permitted to exceed a specific threshold value, to ensure the requirements of process safety and economics. Only a few batch processes, where such strict bounds become necessary on the final state variable, can be optimized using this OCP structure.

11.3.2 OCP-2: Fixed final time problem with free final state

This OCP formulation can determine the transient optimal trajectories of the decision variables for many batch processes involving the manufacturing of specialty chemicals, polymers, biofuels such as biodiesel, bio-ethanol, and a plethora of pharmaceutical products. This OCP is concerned with determining an open-loop trajectory of the control variable such that the concentration or yield of the desired product can be maximized in a pre-specified batch time while satisfying the dynamic model constraints. Additionally, specific bounds can also be enforced on the state and control variables, depending on the thermodynamics or physical boundaries of the problem, as mentioned in eqs. (11.3)–(11.7). Since the final state is free, $dx(t_f) \neq 0$, the transversality boundary condition, as in eq. (11.11), simplifies to eq. (11.14) as follows:

$$\left[\frac{\partial \varphi}{\partial x(t_f)} + m_x^T(x(t_f), t_f)\mu - \lambda\right]^T \bigg|_{t_f} = 0 \tag{11.14}$$

The control variable selected in such applications is preferably either the reactor temperature or the flow rate of the coolant contained in the reactor jacket. For instance, Benavides et al. [19] postulated an OCP to maximize the batch endpoint biodiesel concentration at 100 min, where the control variable, i.e., the reactor temperature, was constrained between 298 and 365 K. The process dynamics of the state variables, i.e., the concentrations of each species, were described by the rate constraints modelled using first-order ordinary differential equations. In another similar study, De et al.

[20] proposed an OCP for the same system, considering the flow rate of hot water flowing in the reactor jacket as the control variable, to maximize the concentration of fatty acid methyl esters at a fixed final time. Few other studies concerning the OCPs of batch processes considered control variables other than temperature and flow rates of the utilities. For instance, reflux policy was chosen as the control variable in the OCP for batch distillation with fixed batch time, which involved maximization of thermodynamic efficiency, provided a target product concentration [21]. In a recent study involving a seeded and unseeded batch crystallization undergoing primary and secondary nucleation, the supersaturation trajectories were treated as the decision variable to maximize the final weight mean size of the crystals at 200 min [22].

11.3.3 OCP-3: Free final time problem

This class of OCP considers that the final time is free, and it appears as a decision variable that needs to be optimized. In the case of batch processes, the most common OCP is the final batch time minimization problem. This OCP relies on identifying an open-loop control variable profile $(u^*(t))$ that minimizes the final batch time (t_f) while attaining a target constraint, which is defined on the desired product, as given in eq. (11.16), subject to the process model constraints. The target constraints can be the optimal concentration of the biodiesel [5, 20], final monomer conversion in a batch polymerization reactor [23], or the optimal glucose concentration obtained from batch enzymatic hydrolysis [24]. For such time minimization OCPs, the temperature and coolant flow rates passing through the batch reactor jacket are widely selected control variables. Additionally, terminal or path constraints on certain state variables can also be enforced to meet the product quality requirements. A general representation of the minimum time OCP without any bounds on the control variable can be given by eqs. (11.15), (11.16), and (11.3) as follows:

$$J[u(t)] = \min_{u(t)} \int_{t0}^{t_f} dt \qquad (11.15)$$

$$\text{s.t. Plant model eq. (11.3) holds}$$

$$m(x(t_f), t_f) \geq 0 \qquad (11.16)$$

11.3.4 Numerical solution

Most of the commonly encountered batch processes in chemical engineering are highly nonlinear. For instance, the reaction rates are nonlinearly dependent on the temperature, as modelled by the Arrhenius equation, and the reaction kinetics for a series reaction involving multiple species is nonlinear due to the dependency of the reaction rates on the concentrations of several reactants and intermediate products. Moreover, the mass transfer rates between the liquid-solid or gas-liquid interfaces vary nonlinearly as a function of the interfacial area and the concentration difference. Therefore, analytical solutions to OCPs are not possible. In such cases, a numerical solution is necessary to solve such problems. As discussed in Section 11.3, the numerical techniques widely applied to solve the OCPs include indirect and direct methods, which are discussed as follows.

11.3.4.1 Indirect methods

The indirect method is based on the Pontryagin's maximum principle, which requires the solving of necessary conditions of optimality provided by eqs. (11.9), (11.10), and (11.12). Here, usually, a two-point boundary value problem is formulated and solved through the forward and backward integration of the state and co-state variables with the knowledge of the boundary conditions, as in eq. (11.11). The two most popular approaches implemented under the indirect methods are, namely, single shooting and multiple shooting, which are summarized as follows:

- **Single shooting:** In this method, an initial guess of the co-state variable is made at the initial time step and forward integration is performed from time t_0 to t_f. The co-state profile generated through this integration is then forced to track the known boundary condition at the final time by employing a Newton's method. If a deviation exists between the final boundary condition resulting from the integration and the known terminal conditions by more than a specified tolerance, the unknown initial guess is updated [25, 26]. This process is iterated until the difference between the integrated final conditions and the desired terminal conditions become less than the prespecified tolerance value. However, this method suffers from a major pitfall that if the accurate initial guess is not selected, the Hamiltonian dynamics would become ill-conditioned, which, in turn, results in divergent trajectories along the integration direction.
- **Multiple shooting:** To overcome the numerical challenges associated with the single shooting method, a multiple shooting approach is adopted, which involves the discretization of the time interval $[t_0, t_f]$ into smaller subintervals. The initial conditions of the state and co-state variables at each of these internal segments become the unknowns, which are computed by performing integration over each subinterval. The continuity of the states across each subinterval is ensured by adding equality constraints [25, 26].

11.3.4.2 Direct methods

Direct methods find an advantage over indirect methods in solving the OCPs due to their flexibility to tackle path constraints and ability to handle the differential-algebraic equation (DAE) systems, which essentially describe batch process models. Under the direct methods, the collocation-based approaches have gained extensive popularity in solving the practical OCPs arising from various engineering domains, which can be categorized as follows [25, 26]:

–   **Sequential approach:** This method is often referred to as control vector parameterization, where the control variable trajectory is discretized into a finite number of parameters over time intervals with the help of a piecewise constant basis function. Such an approximation of the control leads to an optimization problem where the control parameters become the decision variables, which are optimized using a nonlinear programming (NLP) solver at the parameterized time step. This approach is sequential, in the sense that first, the state variable trajectory is computed from the model integration performed at an inner loop, which evaluates the objective function. Subsequently, the gradient information is supplied to the outer loop, i.e., the NLP solver, where the decision variables are iteratively adjusted until the necessary conditions of optimality are satisfied.

–   **Simultaneous approach:** Unlike the sequential approach, the simultaneous method approximates both the control as well as the state variable trajectories at the collocation points using piecewise continuous functions chosen from the orthogonal family, e.g., Chebyshev or Lagrange interpolating polynomials. This discretization strategy is known as orthogonal collocation on finite elements (OCFE), which partitions the time horizon into equally spaced finite elements, where the collocation points appear inside each finite element. Thus, this approach transforms the system of ordinary differential equations (ODES) into a set of algebraic equations. Parameterization yields the coefficients of the basis functions as the decision variables, which are optimized at each iteration with the help of an NLP solver. Because of the simultaneous discretization of control and states, any number of path constraints can be easily handled using this numerical scheme. Recently, some direct collocation-based nonlinear programming solvers, e.g., *FMINCON* and *IPOPT*, embedded with modelling frameworks, namely, DYNOPT [27], PYOMO.DAE [28], and GEKKO [29] have gained significant attention to solve the OCPs, transformed as dynamic optimization problems in MATLAB® or Python platforms.

11.4 Optimal control problems for linear systems

This category of OCPs deals with linear systems where the control and state variables of the plant model appear in linear forms, and a quadratic cost function is optimized. Therefore, the OCPs for linear systems are expressed as quadratic programming problems. Such OCPs can be broadly classified into (1) linear quadratic regulator problems and (2) linear quadratic tracking problems, which are discussed next.

11.4.1 Linear quadratic regulator problem

Let us recall the nonlinear time-dynamical batch plant, as in eq. (11.3), which describes the nonlinear mapping of states (x) with the inputs (u). The first step in control-relevant modelling toward implementing a model-based optimization is to develop a linear approximation of the plant model around a steady-state operating point. However, due to the time-varying characteristics of a batch process, instead of a fixed-point approximation, a trajectory-based linearization should be performed around the model eq. (11.1), which gives rise to a linear time-varying perturbation model (LTVP) as follows:

$$\dot{x}(t) = A(t)x(t) + B(t)u(t) \tag{11.17}$$

We assume a finite batch time horizon as $[t_0, t_f]$, where the final time is fixed, and the terminal state is free. Hence, the goal here is to determine an optimal control trajectory $u^*(t)$ that will maintain the states of the LTVP model near to the origin with minimum control effort by minimizing a quadratic cost function (J), defined by eq. (11.18) as follows:

$$J[x(t), u(t)] = \frac{1}{2}x^T(t_f)Mx(t_f) + \frac{1}{2}\int_{t0}^{t_f}[x^T(t)Q(t)x(t) + u^T(t)R(t)u(t)]dt \tag{11.18}$$

Here, M and $Q(t)$ are real symmetric and positive-semidefinite matrices, each having dimensions of $N_x \times N_x$. $R(t)$ is a real symmetric positive-definite matrix with dimensions of $N_u \times N_u$. Such an OCP is termed as a linear quadratic regulator (LQR) problem. The Hamiltonian function for the LQR problem is written as follows:

$$H(t) = \frac{1}{2}x^T(t)Q(t)x(t) + \frac{1}{2}u^T(t)R(t)u(t) + \lambda^T(t)[A(t)x(t) + B(t)u(t)] \tag{11.19}$$

The first-order necessary conditions of optimality for the LQR are given by eqs. (11.17), (11.20), and (11.21), which can be derived similarly from the definitions of eqs. (11.9), (11.10), and (11.12), as follows:

$$\dot{\lambda}(t) = \frac{-\partial H}{\partial x} = -Q(t)x(t) - A^T(t)\lambda(t)$$ (11.20)

$$\frac{\partial H}{\partial u} = R(t)u(t) + B^T(t)\lambda(t) = 0$$ (11.21)

Further simplification of eq. (11.21) generates an explicit expression of the optimal control profile $u(t)$ as a function of $\lambda(t)$, defined as follows:

$$u(t) = -R(t)^{-1}B^T(t)\lambda(t)$$ (11.22)

Substituting the above expression of $u(t)$ in eq. (11.17) has resulted into a set of $2N_x$ ODES, represented as follows [30]:

$$\begin{bmatrix} \dot{x}(t) \\ \dot{\lambda}(t) \end{bmatrix} = \begin{bmatrix} A(t) & -B(t)R(t)^{-1}B^T(t) \\ -Q(t) & -A^T(t) \end{bmatrix} \begin{bmatrix} x(t) \\ \lambda(t) \end{bmatrix}$$ (11.23)

The solution of $\lambda(t)$ requires information about the boundary condition on the co-state variable at the final time, given by eq. (11.24), which can be obtained from eq. (11.11):

$$\lambda(t_f) = Mx(t_f)$$ (11.24)

The optimal control can also be expressed as a function of the state, usually given in terms of a linear state feedback law as defined as follows:

$$u(t) = K(t)x(t)$$ (11.25)

Here, $K(t)$ is referred to the optimal state feedback gain matrix, given as follows:

$$K(t) = -R(t)^{-1} B^T(t)G(t)$$ (11.26)

Here, $G(t)$ is a $N_x \times N_x$ positive-definite symmetric matrix, often termed as the Riccati coefficient matrix, which satisfies the Riccati differential equation given by eq. (11.27), subject to the boundary condition $G(t_f) = M$, where $G(t_f)$ is a positive-semidefinite matrix [30]:

$$\dot{G}(t) = -G(t)A(t) - A^T(t)G(t) - Q(t) + G(t)B(t)R(t)^{-1}B^T(t)G(t)$$ (11.27)

It is noteworthy that eq. (11.27) must be integrated backward in time from $[t_f, t_0]$ to obtain the solution for $G(t)$. It should be noted that the computation of $G(t)$ does not require the knowledge of the optimal state trajectory. It is dependent only on the linearized system matrices, i.e., $A(t)$ $B(t)$, and the weighting matrices M, $Q(t)$ and $R(t)$ of the quadratic objective function, as in eq. (11.18).

11.4.2 Linear quadratic tracking problem

Let us reconsider the linear time-varying batch plant given by eq. (11.17). Let $Y(t) = Cx(t)$, where $Y(t)$ indicates the measurement vector containing $x(t)$ measurable states, denoted by o outputs. Here, C is a matrix with dimension $N_o \times N_x$. In this problem, the state variables are guided to track a desired reference trajectory $r(t)$ by the minimization of the quadratic objective function J as follows [31]:

$$J = \frac{1}{2} \left(Cx(t_f) - r(t_f) \right)^T M \left(Cx(t_f) - r(t_f) \right)$$

$$+ \frac{1}{2} \int_{t_0}^{t_f} \left[\left(Cx(t) - r(t) \right)^T Q(t) \left(Cx(t) - r(t) \right) + u(t)^T R(t) u(t) \right] dt \tag{11.28}$$

Therefore, this OCP is termed as a linear quadratic tracking (LQT) problem. Here, the matrices M, $Q(t)$ and $R(t)$ have the equivalent definitions as given by the eq. (11.18). For the same LTVP model as in eq. (11.17), the Hamiltonian formulation can be represented for this LQT problem by eq. (11.29) as follows:

$$H(t) = \frac{1}{2} \left[\left(Cx(t) - r(t) \right)^T Q(t) \left(Cx(t) - r(t) \right) + u(t)^T R(t) u(t) \right]$$

$$+ \lambda^T(t) A(t) x(t) + \lambda^T(t) B(t) u(t) \tag{11.29}$$

The necessary conditions of optimality can be determined by integrating the states as in eq. (11.17) and co-states given by eq. (11.30) in the forward and backward directions, respectively, and setting the gradient of the Hamiltonian with respect to the control variable as zero. The set of equations developed for the LQR problem, as in eqs. (11.20), (11.21), and (11.22), can be transformed into eqs. (11.30), (11.31), and (11.32) to generate the necessary conditions of optimality for the LQT problem as follows:

$$\dot{\lambda}(t) = -A^T(t)\lambda(t) - C^T Q(t) Cx(t) + C^T Q(t) r(t) \tag{11.30}$$

$$R(t)u(t) + B^T(t)\lambda(t) = 0 \tag{11.31}$$

$$u(t) = -R(t)^{-1} B^T(t)\lambda(t) \tag{11.32}$$

Substituting the expression of $u(t)$ from eq. (11.32) into eq. (11.17) again gives rise to a set of $2N_x$ ODES provided by eq. (11.33), as follows:

$$\begin{bmatrix} \dot{x}(t) \\ \dot{\lambda}(t) \end{bmatrix} = \begin{bmatrix} A(t) & -B(t)R(t)^{-1}B^T(t) \\ -C^T Q(t) C & -A^T(t) \end{bmatrix} \begin{bmatrix} x(t) \\ \lambda(t) \end{bmatrix} + \begin{bmatrix} 0 \\ C^T Q(t) r(t) \end{bmatrix} \tag{11.33}$$

The backward integration of the co-state variable $\lambda(t)$ requires the boundary condition marked by eq. (11.34) as follows:

$$\lambda(t_f) = C^T M [Cx(t_f) - r(t_f)] \tag{11.34}$$

After substituting the solution of $\lambda(t)$ in eq. (11.32) and further performing algebraic manipulations, the optimal control law $u(t)$ can again be expressed as a linear state feedback law, given by eq. (11.35) as follows:

$$u(t) = K(t)x(t) + R(t)^{-1}B^T(t)S(t) \tag{11.35}$$

Here, $K(t)$ is the feedback gain matrix, with the same definition as of eq. (11.26). However, $S(t)$ comprises N_x vector equations, which can be obtained by solving a nonhomogeneous vector differential equation, represented by eq. (11.36) with the boundary condition, $S(t_f) = C^T Mr(t_f)$:

$$\dot{S}(t) = -\left[A^T(t) - G(t)B(t)R(t)^{-1}B^T(t)\right]S(t) - C^T Q(t)r(t) \tag{11.36}$$

The solution of $G(t)$ can be computed similarly by integrating the Riccati eq. (11.27) backward with the boundary condition $G(t_f) = C^T MC$. Recalling the fact that $G(t)$ is a symmetric matrix having a dimension $N_x \times N_x$ and $S(t)$ consists of N_x vectors, both the differential equations, i.e., eq. (11.27) and eq. (11.36), together represent a set of $N_x(N_x + 1)/2 + N_x$ first-order ODES, which must be integrated backward in time using the boundary conditions $G(t_f)$ and $S(t_f)$, as specified earlier [31]. On substituting $u(t)$ from eq. (11.35) into the LTVP model, i.e., in eq. (11.17), the optimal state (x^*) equation takes the form of eq. (11.37), as follows:

$$\dot{x}^*(t) = \left[A(t) - B(t)R(t)^{-1}B^T(t)G(t)\right]x^*(t) + B(t)R(t)^{-1}B^T(t)S(t) \tag{11.37}$$

It can be noted that either the $G(t)$ or its boundary condition is not affected by the desired reference trajectory $r(t)$. This implies that once the LTVP model is formulated with the time-varying matrices $A(t)$, $B(t)$, and C, and the weighting matrices associated with the performance index, i.e., M, $Q(t)$, and $R(t)$ are specified, the Riccati coefficient matrix can be easily computed [31]. On comparing the LQR and LQT controllers, it can be inferred that the presence of the desired reference trajectory in the latter acts as a forcing function, which produces $S(t)$. This brings out the major difference between the optimal regulatory and tracking control modes. Another interesting observation from the LQT controller reveals that the matrix $[A(t) - B(t)R(t)^{-1}B^T(t)G(t)]$, as in eq. (11.37) does not depend on the desired reference profile $r(t)$. Therefore, for the closed-loop optimal tracking system, the eigenvalues are also independent of $r(t)$.

To demonstrate the LQT problem, let us consider a single input-single output LTVP batch plant having $u(t)$ and $x(t)$ as the control and state variables, respectively. The dynamics of the LTVP model can be described as follows:

$$\dot{x}(t) = 0.13x(t) + 0.1u(t) \tag{11.38}$$

$$y(t) = x(t) \tag{11.39}$$

Let the batch time be fixed and defined over a finite horizon between [0,2] min and the $A(t)$ and $B(t)$ matrices obtained at a particular time instance are given by 0.13 and 0.1, respectively. Since $x(t)$ is the only output variable, $C = 1$. Let $x_{ref}(t)$ denote the desired reference trajectory that is maintained at a constant value over the entire batch time of 2 mins. Now, we intend to solve an LQT problem aiming to find a closed-loop optimal control profile $(u^*(t))$ for which $x(t)$ can be tracked to the desired reference profile, i.e., $x_{ref}(t)$, with minimum control effort by the minimization of the quadratic cost function, represented in eq. (11.28). The algorithm for solving the LQT problem can be summarized as follows:

- **Step 1.** Set $x_{ref}(t) = 8$. Select M, $Q(t)$ and $R(t)$ matrices as 10, 10 and 0.01, respectively. A higher weight is assigned on M and $Q(t)$ compared to $R(t)$ in order to ensure that the final state can be tracked to the desired reference value with minimum control effort.
- **Step 2.** Integrate eqs. (11.27) and (11.36) in the backward direction from [2,0] min to solve for the trajectories of $G(t)$ and $S(t)$ by applying the boundary conditions $G(t) = 10$ and $S(t_f) = 80$.
- **Step 3.** Determine the optimal state trajectory $x^*(t)$ by integrating eq. (11.37) from [0,2] min with the initial condition $x(0) = 0$.
- **Step 4.** Compute the closed-loop optimal control law $u^*(t)$ from eq. (11.35).

The closed-loop optimal control profile is plotted in Figure 11.1(a). The profiles of $G(t)$ and $S(t)$ are plotted in Figure 11.1(b), and the optimal state profile is depicted in Figure 11.1(c). It can be seen from Figure 11.1(c) that the designed LQT controller has forced the state to track the desired reference profile at the final batch time.

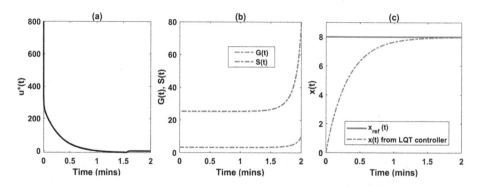

Figure 11.1: (a) Closed-loop optimal control profile resulting from the LQT controller. (b) Profiles of $G(t)$ and $S(t)$ computed from the proposed LQT problem. (c) Evolution of the state profile to attain the desired reference trajectory.

11.5 Applications of optimal control theory to batch processes

Numerous research works have been carried out highlighting the application of OCPs on various batch processes. It is not possible to review the applications in an exhaustive manner. Instead, some examples in the emerging areas are discussed here as follows.

11.5.1 Batch transesterification control

De et al. [6] proposed certain OCPs for a commercial-grade batch transesterification reactor of rapeseed oil to optimize biodiesel production. Here, the hot utility profile circulating in the reactor jacket was treated as the control variable to optimize certain performance indices, viz., maximum concentration of biodiesel at final time, minimum reactor energy cost and maximum profit. The OCPs were converted into dynamic optimization problems using orthogonal collocation on finite elements and subsequently solved using *fmincon* in MATLAB®. Furthermore, the batch reactor system was integrated with continuous distillation columns to perform downstream separation of biodiesel. Here, the separation cost of biodiesel was modelled as another economic cost function, which aimed to minimize the reboiler heat duties of the distillation columns. Finally, specific OCPs were defined, translated into multi-objective dynamic optimization problems, and solved to capture the internal trade-offs among the competing cost functions arising from the integrated reactor-separator system. Another study by the same group of authors demonstrated the application of OCP on a laboratory-scale batch reactor of 1 L undergoing transesterification of soybean oil to optimize the biodiesel yield in a final time of 70 min [20]. Here, the kinetic parameters of the batch model were estimated first from the time-series measurements of the state profiles using nonlinear least squares regression. The OCP was then solved with the optimized kinetic parameters to identify the optimal hot water flow rate profile, which maximized the biodiesel yield compared to the base case.

Kern and Shastri [32] proposed an advanced model-based optimization and control strategy for a batch transesterification reactor with the objective of productivity maximization. The batch model for biodiesel production was developed based on the reaction kinetics integrated with material and energy balances. Since the system was identified as unobservable, the control problem was solved in two tiers instead of employing a model predictive controller. First, the uncertain parameters of the batch model were optimized using the online measurements of alcohol concentration and temperatures of the reactor and the jacket. In the top tier, a productivity maximization problem was defined to find an open-loop optimal reactor temperature profile by solving an OCP subject to the model constraints with the updated parameters. The OCP was solved on ACADO toolkit based on the multiple shooting method. In the second tier, the set-point reactor temperature profile was tracked using a nonlinear model predictive controller, which produced optimal flow rate trajectories of the hot and cold utilities. Such a closed-loop control approach was capable of rejecting online disturbances while handling uncertainties, besides generating higher revenues.

11.5.2 Lignocellulosic biofuels and biochemicals

With increasing emphasis on the sustainability of the energy and chemicals sector, the transition to a biomass-based economy has generated interest. Conversion of biomass to energy and chemicals is expected to be more sustainable than that from fossil sources, since biomass is a renewable source that is produced locally. However, most of the biomass processing steps are slow and involve solid–liquid multiphase mixtures. In such cases, batch or fed-batch processing becomes essential. Ethanol from lignocellulosic biofuels is one such example. The production involves pretreatment, hydrolysis and fermentation as the three main steps, all of which are typically conducted in batch or fed-batch mode.

Vegi and Shastri [33] investigated an OCP for the batch dilute acid pretreatment of lignocellulosic biomass, followed by enzymatic hydrolysis, to determine the optimal mass flow rate profiles of the hot and cold utilities flowing through the reactor jacket. For the upstream pretreatment process, the OCPs formulated with objectives, e.g., maximum concentration and maximum profit, improved the hemicellulose solubilization by enhancing the xylose concentration and profit by 6.83% and 124%, respectively, compared to the chosen base cases. Another OCP posed a time minimization problem, resulting in a reduction in the batch time by 43%, where the optimal xylose concentration was provided as the target constraint. The OCPs considered for the downstream enzymatic hydrolysis yielded similar optimized results in terms of maximum glucose concentration, maximum profit, and minimum hydrolysis time through the identification of the optimal flow rate profiles of the coolants.

Verma and Shastri [34, 35] applied optimal control to the batch acid pretreatment reactor. The pretreatment involves a series of reactions in which the middle product is of interest. Therefore, optimal control becomes very important to ensure greater conversion and lower production of undesired degradation products. Verma and Shastri [34] used temperature as the decision variable. The maximum principle, as described previously, was used to formulate the problem, and the steepest ascent/descent method was used to solve it. The optimal profiles typically had high temperatures at the beginning of the batch to enhance the reaction kinetics. The optimal temperatures are reduced with time to restrict the formation of degradation products.

Fenila and Shastri [36] used optimal control for the enzymatic hydrolysis process, which was operated in a batch mode. Temperature was used as a control variable. Here, higher temperature enhanced reaction kinetics but also degraded the enzyme.

Therefore, a balance is essential. Optimal control application was shown to increase glucose yield by 3.2% and reduce batch time by 5.8% for two different objective functions.

In both processes, i.e., the dilute acid pretreatment and enzymatic hydrolysis, uncertainty plays an important role. Feedstock variability is a major concern since biomass is a natural feedstock. Similarly, the solid-liquid nature of the system leads to nonideality in mixing [37]. Therefore, both problems were extended to solve stochastic optimal control problems [34, 38]. The stochastic optimal control theory has been developed based on the concepts of Ito lemma and random processes [38]. Although not discussed here, these ideas are becoming quite important as most practical problems will have several uncertainties. Therefore, solving a deterministic optimization problem may not give an acceptable solution.

11.5.3 Hydrothermal liquefaction control

Over the past decade, a significant upsurge in studies has been noticed in decarbonizing the transportation sector by replacing fossil-based diesel fuels with carbon-neutral green fuels. In this regard, hydrothermal liquefaction (HTL) has gained considerable attention, which utilizes either wet biomass such as microalgae, sewage sludge, animal manure, or municipal solid wastes, or the lignocellulosic biomass to produce a net CO_2-neutral fuel, referred to as bio-oil. HTL is a thermochemical process in which the biomass is mixed with water or preferably a solvent, e.g., methanol, in a batch reactor and allowed to react for 30–60 min under a high-pressure environment of 20–25 MPa at a moderate temperature, ranging between 200 and 375 °C [39]. The solid biomass undergoes a sequence of hydrolysis, depolymerization, deamination, decarboxylation and repolymerization reactions, which eventually yields the liquid fuel called bio-oil, along with the formation of non-condensable gases, e.g., methane (CH_4), CO_2, carbon-monoxide (CO), and hydrogen (H_2).

Very recently, De [40] demonstrated the first-ever application of an OCP on a batch HTL reactor to enhance bio-oil production and improve reactor performance. Here, the OCPs were posed as dynamic optimization problems (DOP), which were solved for the HTL of two microalgal strains, namely, *Nannochloropsis* sp. and *Aurantiochytrium* sp. KRS101 to optimize the economic performance indices viz. bio-oil yield maximization, batch time minimization and reactor thermal energy minimization. The direct collocation scheme based on orthogonal collocation on finite elements was employed to solve the resulting NLPS using the *fmincon* solver in MATLAB®. The DOP results revealed optimal temperature trajectories, which contributed to a significant increase in the bio-oil yields by 6.18% and 11% for the proteins and lipid-rich algae, respectively, compared to the base cases. Furthermore, the DOP implementation successfully decreased the batch times by 61.66% and 78.2%, respectively, and also reduced the reactor thermal energies by 28% and 26%, respectively, to produce the targeted optimal bio-oil yields, in contrast to the base cases. This study also showed its potential to benefit the existing algal biorefineries undergoing bio-oil production from batch HTL, through the proposed DOP strategy by generating higher biocrude yields with reduced thermal energy consumption, thereby improving the process economics.

11.5.4 Novel applications in sustainability

The earlier applications of optimal control theory have only covered batch processes, which are directly coming from the chemical engineering field. However, optimal control of systems that are dynamic is nature has much wider applications. Increasingly, systematic methods are being used to solve problems in sustainability in which the ecosystem is the "process" of interest. Since ecosystems are continuously evolving in time, they are akin to batch processes. Therefore, the theory discussed for batch processes is directly applicable to such systems. In fact, optimization and optimal control of natural resources has been a topic of interest for a long time, and systematic methods have been used for systems such as fisheries and forests.

Shastri et al. [41, 42] demonstrated such an application for a simple three-species predator-prey model and later extended that approach to more complex multi-compartment ecosystem models [43]. In all cases, the application of optimal control theory was shown to provide definite benefits. An important challenge while solving such problems is the identification of control variables. For traditional chemical processes, control variables are generally obvious. However, for interdisciplinary problems related to ecosystems, control variables need to be carefully selected. The success of the control strategy may depend more on this selection than the exactness of the OCP solution.

11.6 Conclusion and emerging trends

OCPs belong to a branch of algorithms that are specifically designed to optimize batch processes. This chapter presents a brief review of challenges faced during batch operations and discusses methods to solve OCPs for batch processes, with applications emerging from diverse engineering sectors to enable effective design, smooth decision-making and control. The problem formulations and solution methods of unconstrained and constrained OCPs based on Pontryagin's minimum principle are covered while mentioning specific cases differing in the constraint's formulation, depending on the terminal state and final time appearing as fixed or free. A section is also presented on the approach to solving the OCPs applied to transient linear systems, which reviews the linear quadratic regulator and tracking problems. Furthermore, research works carried out by Shastri et al. discussing the applications of OCPs on batch chemical/bio-chemical processes, e.g., biodiesel production, biomass pretreatment using dilute acid, and enzymatic hydrolysis, are also summarized.

Although the optimal control theory has been applied to optimize the performance of several batch processes, recent advancements have been made toward developing new algorithms for further improvements. Very recently, efforts have been made to device optimal control methods that can directly handle partial differential equations (PDES) as constraints to describe the spatial and temporal variation of the states and controls. For instance, the space-time orthogonal collocation on finite elements (ST-OCFE) approach is explored to solve the PDE-constrained dynamic optimization problem involving single, multiple, and simultaneous shooting methods [44]. The emerging trends in this field focus on identifying model reduction techniques, developing numerical methods, and solving inverse problems pertaining to data assimilation and parameter estimation.

Over the years, model predictive control (MPC) has been a widespread and popular approach to solving closed-loop OCPs, which uses a process model to predict the future evolution of the system states by computing the optimal control inputs. The emerging trends in MPC research include the development of learning-based [45], stochastic [46], and distributed MPC [47] frameworks in order to enhance their stability, performance, and robustness under uncertainties.

With the advent of artificial intelligence and data science, machine learning methods have garnered significant attention. To this end, a recent surge has been observed in optimal control research, which emphasizes data-driven-based learning approaches [48]. For instance, reinforcement learning (RL) has gained significant popularity among batch processes for the identification of optimal control policies [49]. RL is a model-free neural network-based learning scheme, which can solve the dynamic optimization problem directly using data. Here, the trained RL agent can learn from its own actions and rewards while accounting for process uncertainties.

Furthermore, current data-driven optimal control methods have a limitation, in that for a newly established batch process, the data is available for a shorter period or may be insufficient. To overcome this shortcoming, a transfer learning-based approach built on latent variable process transfer models (LV-PTM) has been proposed to transfer data from similar other existing batches to the recent ones [50]. Next, a batch-to-batch adaptive control was implemented through model updating, data cleaning, and modifier adaptation while accounting for the mismatch of necessary conditions of optimality arising from differences in the batch data.

References

[1]   Tomazi, K. G., Linninger, A. A., Daniel, J. R. Batch Processing Industries. In *Batch Processes*. CRC Press, 2005, pp. 19–54.

[2]   Aziz, N., Mujtaba, I. M. Optimal operation policies in batch reactors. *Chemical Engineering Journal*, 2002, 85(2–3), 313–325.

[3]   Bonvin, D. Control and Optimization of Batch Processes. In *Encyclopedia of Systems and Control*. Cham: Springer International Publishing, 2021, pp. 255–260.

[4]   Lee, J., Lee, K. S., Lee, J. H., Park, S. An on-line batch span minimization and quality control strategy for batch and semi-batch processes. *Control Engineering Practice*, 2001, 9(8), 901–909.

[5]   Benavides, P. T., Diwekar, U. Studying various optimal control problems in biodiesel production in a batch reactor under uncertainty. *Fuel*, 2013, 103, 585–592.

[6]   De, R., Bhartiya, S., Shastri, Y. Multi-objective optimization of integrated biodiesel production and separation system. *Fuel*, 2019, 243, 519–532.

[7]   Luus, R., Okongwu, O. N. Towards practical optimal control of batch reactors. *Chemical Engineering Journal*, 1999, 75(1), 1–9.

[8]   Swartz, C. L., Kawajiri, Y. Design for dynamic operation-A review and new perspectives for an increasingly dynamic plant operating environment. *Computers and Chemical Engineering*, 2019, 128, 329–339.

[9]   Biegler, L. T. Advanced optimization strategies for integrated dynamic process operations. *Computers and Chemical Engineering*, 2018, 114, 3–13.

[10]   Yuan, J., Liu, C., Zhang, X., Xie, J., Feng, E., Yin, H., Xiu, Z. (2016). Optimal control of a batch fermentation process with nonlinear time-delay and free terminal time and cost sensitivity constraint. Journal of Process Control, 44, 41–52.

[11]   Cezerac, J., Garcia, V., Cabassud, M., Le Lann, M. V., Casamatta, G. Optimal control under environmental considerations application to a pharmaceutical synthesis reaction. *Computers and Chemical Engineering*, 1995, 19, 415–420.

[12]   Tian, Y., Zhang, J., Morris, J. Modeling and optimal control of a batch polymerization reactor using a hybrid stacked recurrent neural network model. *Industrial & Engineering Chemistry Research*, 2001, 40(21), 4525–4535.

[13]   Koo, P. H., Moon, D. H. A review on control strategies of batch processing machines in semiconductor manufacturing. *IFAC Proceedings Volumes*, 2013, 46(9), 1690–1695.

[14]   Shyamal, S., Swartz, C. L. Real-time dynamic optimization-based advisory system for electric arc furnace operation. *Industrial & Engineering Chemistry Research*, 2018, 57(39), 13177–13190.

[15]   Rohman, F. S., Aziz, N. Optimization of batch electrodialysis for hydrochloric acid recovery using orthogonal collocation method. *Desalination*, 2011, 275(1–3), 37–49.

[16]   Rahman, M. Z. U., Rizwan, M., Liaquat, R., Leiva, V., Muddasar, M. Model-based optimal and robust control of renewable hydrogen gas production in a fed-batch microbial electrolysis cell. *International Journal of Hydrogen Energy*, 2023, Zhang, G. P., & Rohani, S. (2003). On-line optimal control of a seeded batch cooling crystallizer. Chemical Engineering Science, 58(9), 1887-1896.

[17]   Kirk, D. E. *Optimal Control Theory: An Introduction*. Courier Corporation, 2004.

[18]   Lewis, F. L., Vrabie, D., Syrmos, V. L. *Optimal Control*. John Wiley & Sons, 2012.

[19]   Benavides, P. T., Diwekar, U. Optimal control of biodiesel production in a batch reactor: Part I: Deterministic control. *Fuel*, 2012, 94, 211–217.

[20]   De, R., Bhartiya, S., Shastri, Y. Parameter estimation and optimal control of a batch transesterification reactor: An experimental study. *Chemical Engineering Research and Design*, 2019, 157, 1–12.

[21]   Zavala, J. C., Coronado, C. Optimal control problem in batch distillation using thermodynamic efficiency. *Industrial & Engineering Chemistry Research*, 2008, 47(8), 2788–2793.

[22] Chien, W. T., Pan, H. J., Ward, J. D. Optimal control of batch crystallization processes with primary nucleation. *Industrial & Engineering Chemistry Research*, 2023.

[23] Ekpo, E. E., Mujtaba, I. M. Optimal control trajectories for a batch polymerisation reactor. *International Journal of Chemical Reactor Engineering*, 2007, 5(1).

[24] Fenila, F., Shastri, Y. Optimal control of enzymatic hydrolysis of lignocellulosic biomass. *Resource Efficient Technologies*, 2016, 2, S96–S104.

[25] Rao, A. V. A survey of numerical methods for optimal control. *Advances in the Astronautical Sciences*, 2009, 135(1), 497–528.

[26] Nolasco, E., Vassiliadis, V. S., Kähm, W., Adloor, S. D., Al Ismaili, R., Conejeros, R., . . . Zhang, Q. Optimal control in chemical engineering: Past, present and future. *Computers and Chemical Engineering*, 2021, 155(107528).

[27] Cizniar, M., Salhi, D., Fikar, M., Latifi, M. A. A MATLAB package for orthogonal collocations on finite elements in dynamic optimisation. *Proceedings of the 15th International Conference Process Control*, 2005 June, 5, 058f.

[28] Nicholson, B., Siirola, J. D., Watson, J. P., Zavala, V. M., Biegler, L. T. pyomo. dae: A modeling and automatic discretization framework for optimization with differential and algebraic equations. *Mathematical Programming Computation*, 2018, 10, 187–223.

[29] Beal, L. D., Hill, D. C., Martin, R. A., Hedengren, J. D. Gekko optimization suite. *Processes*, 2018, 6(8), 106.

[30] Andrés-Martínez, O., Palma-Flores, O., Ricardez-Sandoval, L. A. Optimal control and the Pontryagin's principle in chemical engineering: History, theory, and challenges. *AIChE Journal*, 2022, 68(8), e17777.

[31] Naidu, D. S. Constrained Optimal Control Systems. In *Optimal Control Systems*. CRC press, 2018, pp. 293–364.

[32] Kern, R., Shastri, Y. Advanced control with parameter estimation of batch transesterification reactor. *Journal of Process Control*, 2015, 33, 127–139.

[33] Vegi, S., Shastri, Y. Optimal control of dilute acid pretreatment and enzymatic hydrolysis for processing lignocellulosic feedstock. *Journal of Process Control*, 2017, 56, 100–111.

[34] Verma, S. K., Shastri, Y. Deterministic and stochastic optimization of dilute acid pretreatment of sugarcane bagasse. *Biofuels*, 2021, 12(8), 987–998.

[35] Verma, S. K., Shastri, Y. Economic optimization of acid pretreatment: Structural changes and impact on enzymatic hydrolysis. *Industrial Crops and Products*, 2020, 147, 112236.

[36] Fenila, F., Shastri, Y. Optimal control of enzymatic hydrolysis of lignocellulosic biomass. *Resource-Efficient Technologies*, 2016, 2(1), S96–S104.

[37] Fenila, F., Shastri, Y. Optimization of cellulose hydrolysis in a non-ideally mixed batch reactor. *Computers and Chemical Engineering*, 2019, 128, 340–351.

[38] Fenila, F., Shastri, Y. Stochastic Optimization of enzymatic hydrolysis of lignocellulosic biomass. *Computers and Chemical Engineering*, 2020, 135, 106776.

[39] Bassoli, S. C., Da Fonseca, Y. A., Wandurraga, H. J. L., Baeta, B. E. L., De Souza Amaral, M. Research progress, trends, and future prospects on hydrothermal liquefaction of algae for biocrude production: A bibliometric analysis. *Biomass Convers Biorefin*, 2023, 1–16.

[40] De, R. Comparative dynamic optimization study of batch hydrothermal liquefaction of two microalgal strains for economic bio-oil production. *Bioresource Technology*, 2024, 398, 130523.

[41] Shastri, Y., Diwekar, U. Sustainable ecosystem management using optimal control theory: Part 1 (Deterministic systems). *Journal of Theoretical Biology*, 2006, 241(3), 506–521.

[42] Shastri, Y., Diwekar, U. Sustainable ecosystem management using optimal control theory: Part 2 (Stochastic systems). *Journal of Theoretical Biology*, 2006, 241(3), 522–531.

[43] Shastri, Y., Diwekar, U., Cabezas, H. Optimal control theory for sustainable environmental management. *Environmental Science & Technology*, 2008, 42(14), 5322–5328.

[44] Jie, H., Zhu, G., Hong, W. Direct approaches for PDE-constrained dynamic optimization based on space–time orthogonal collocation on finite elements. *Journal of Process Control*, 2023, 124, 187–198.

[45] Hewing, L., Wabersich, K. P., Menner, M., Zeilinger, M. N. Learning-based model predictive control: Toward safe learning in control. *Annual Review of Control, Robotics, and Autonomous Systems*, 2020, 3, 269–296.

[46] Shang, C., You, F. A data-driven robust optimization approach to scenario-based stochastic model predictive control. *Journal of Process Control*, 2019, 75, 24–39.

[47] Kohler, M., Berberich, J., Müller, M. A., Allgower, F. Data-driven distributed MPC of dynamically coupled linear systems. *IFAC-PapersOnLine*, 2022, 55(30), 365–370.

[48] Prag, K., Woolway, M., Celik, T. Toward data-driven optimal control: A systematic review of the landscape. *IEEE Access*, 2022, 10, 32190–32212.

[49] Joshi, T., Makker, S., Kodamana, H., Kandath, H. Twin actor twin delayed deep deterministic policy gradient (TATD3) learning for batch process control. *Computers and Chemical Engineering*, 2021, 155 (107527).

[50] Chu, F., Zhao, X., Yao, Y., Chen, T., Wang, F. Transfer learning for batch process optimal control using LV-PTM and adaptive control strategy. *Journal of Process Control*, 2019, 81, 197–208.

Yulissa Mercedes Espinoza-Vázquez, Nereyda Vanessa
Hernández-Camacho, Jahaziel Alberto Sánchez-Gómez,
José Ezequiel Santibañez-Aguilar, and Fernando Israel Gómez-Castro*

Chapter 12
Supply chain optimization for chemical and biochemical processes

Abstract: The optimal design of conversion processes is highly relevant for the chemical and biochemical industries, besides the determination of optimal operation policies. However, the selection of an adequate strategy to distribute the raw materials to the facilities and the products to the markets, *i.e.*, the supply chain, is indispensable to ensure the economic feasibility of the macroscopic production scheme. In this chapter, the importance of the optimization of supply chains for chemical and biochemical processes is described. The use of logic-based and mixed-integer models to represent and optimize supply chains is discussed. Guidelines to properly select the objective function are presented, grouping the objectives as economic, environmental, and social functions. Finally, an example of the modeling and optimization of a supply chain for the conversion of glycerol into valuable derivatives is discussed.

Keywords: supply chain, mathematical modeling, optimization

12.1 Introduction

The optimization of chemical and biochemical processes is of major importance nowadays, where global interactions demand the highest efficiency to the producers in order to remain competitive. The production processes must be designed to obtain the highest yield of the desired products, while requiring the lowest energy and external mass agents inputs, leading to reduced production costs. However, economic aspects are not the only important in the optimization of chemical and biochemical processes in the current scenario. Reducing the environmental impact is a consideration

---

*Corresponding author: Fernando Israel Gómez-Castro, Universidad de Guanajuato, Campus Guanajuato, División de Ciencias Naturales y Exactas, Departamento de Ingeniería Química, Noria Alta S/N Col. Noria Alta, Guanajuato, Guanajuato 36050, Mexico, e-mail: fgomez@ugto.mx
Yulissa Mercedes Espinoza-Vázquez, Nereyda Vanessa Hernández-Camacho, Jahaziel Alberto Sánchez-Gómez, Universidad de Guanajuato, Campus Guanajuato, División de Ciencias Naturales y Exactas, Departamento de Ingeniería Química, Noria Alta S/N Col. Noria Alta, Guanajuato, Guanajuato 36050, Mexico
José Ezequiel Santibañez-Aguilar, Instituto Tecnológico y de Estudios Superiores de Monterrey, Escuela de Ingeniería y Ciencias, Av. Eugenio Garza Sada 2501, Monterrey, Nuevo León 64849, México

https://doi.org/10.1515/9783111383439-012

that must be taken into account in the design or revamping of production schemes. At a process scale, this may imply reducing the greenhouse gases emissions, the concentration of a given pollutant in the effluents, and the use of nonrenewable materials, among other indicators. Moreover, in the last few years, additional objectives have been highlighted as important in the optimization of chemical and biochemical processes, as those related with the controllability of the processes [1, 2] and safety [3, 4].

Although the optimization of the production processes is of great importance, it is not enough to ensure the feasibility of the macroscopic production scheme if the supply chain is not properly designed. The negative effects of an inadequate supply chain are even greater when the raw materials come from different locations, and there are various markets where the products are demanded. The concept of a unique facility used to transform the raw materials and distribute the products is easy to visualize, and can be the best alternative for a local, small-scale production. However, for a large-scale production where raw materials are distributed in various locations, it has been demonstrated that a decentralized production is the best alternative, which proposes distributing the facilities in strategic locations, satisfying the markets in the nearby areas. This leads to reducing the transportation costs [5, 6] by avoiding the transportation of raw materials and products over large distances. Additionally, the distribution of smaller chemical or biochemical facilities allows reducing the risk associated to the production processes [3]. Moreover, the decentralized scheme has lower negative effects when supply disruptions occur [7]. Nevertheless, this leads to making a great number of potential decisions, including the source of raw materials, the best location of the facilities, and the logistics for the distribution of the products. Thus, formal optimization strategies are required to design the most adequate supply chain. Due to the discrete nature of the decisions, supply chain modeling usually implies a mixed-integer approach, or a logic-based model. Various studies have been developed to perform the optimization of supply chains for biochemical processes, as those involved in the conversion of biomass to biofuels and bioproducts [8–10], and for chemical processes [11]. Other related areas where supply chain analysis have been reported are the generation of renewable energy [12], the extraction of fossil energy sources as shale gas [13] and the capture of carbon dioxide [14]. This indicates the great importance of this topic for the modern industry.

In this chapter, the use of logic-based models to represent supply chains is described. Once the mathematical model has been developed, it is important to properly define the criterion for decision-making, *i.e.*, the objective function. For supply chain optimization, economic functions are fundamental criteria. However, environmental and social functions have taken importance over the last decade. Thus, the chapter describes various objective functions that can be used in supply chain optimization. Also, a discussion on the generation of data required for these kinds of models is presented. Optimization strategies for the determination of the best supply chain design are discussed, and an application example is presented.

12.2 Supply chain modeling

The design of supply chains can be performed through mathematical modeling, allowing a detailed analysis of the effect of the potential decisions on the performance of
the supply chain. However, before developing the detailed model, a superstructure
can be useful to better understand the interactions between the elements of the supply chain. A superstructure is a graphical representation in the form of a network
composed of arrows, nodes, and columns that help in dividing a process into different
stages and tasks, taking into consideration all the feasible routes. In general, the columns represent the processing steps, the nodes can be the techniques to be used and
the arrows indicate the flow of material or information [15]. Figure 12.1 shows several
examples of superstructure representations that can be applied to the area of chemi-

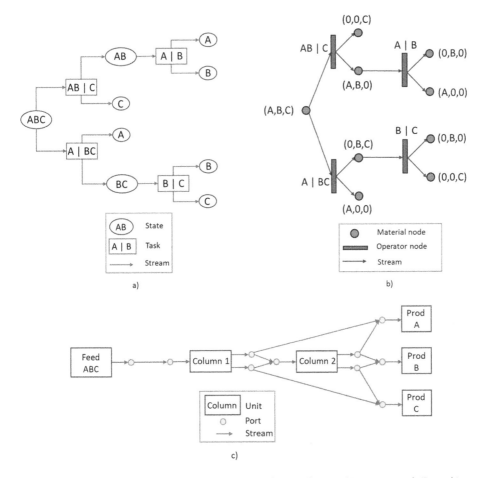

Figure 12.1: Different types of superstructures: (a) state-task network (STN), (b) process graph (P-graph),
and (c) R-graph.

cal engineering. Figure 12.1a represents the state-task network (STN), which is used for multiproduct/multipurpose programming problems, where it is applied to the synthesis of distillation sequences. It is made up of nodes that indicate states and tasks. Figure 12.1b shows an example of process graph (P-graph)-type superstructure, very similar to the STN, made up of material nodes and operating unit nodes, where an operating unit accepts one or more input materials and produces one or more output products. Finally, Figure 12.1c shows the representation of the R-graph, where the nodes correspond to the input and output ports of each considered process unit. This type of graph also represents the source and sink units, corresponding to raw materials and products. Within this type of graph, each node must be connected to another node. It is considered an improvement to the STN, since it has an accommodation that allows generating a better solution strategy [16].

Although these are some of the main types of superstructures, they are not the only ones. Over time, improvements have been made to the main types and coupled to other case studies. For example, for the specific case of supply chains, Figure 12.2 shows a basic scheme for a superstructure. This example is a superstructure of processing step-interval network (PSIN) type. Here, the columns represent processing steps, and the arrows represent connections. This type of superstructure takes into consideration everything from raw materials to the sales market, but it may include more or fewer stages, depending on the case study. Other configurations may include the interaction between intermediate and initial decision levels. To exemplify, if at the end of the processing step N, the possibility of recirculating some of the by-products exists, a line connecting such stage with the beginning of the processing steps could be added to include the possibility of reusing one or more by-products of waste materials. This leads to circular supply chains.

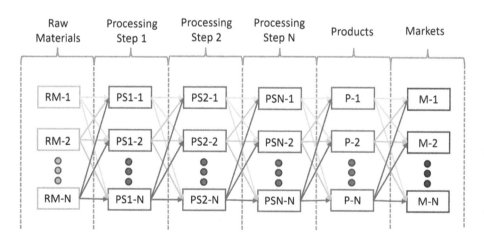

Figure 12.2: Superstructure of processing step-interval network (PSIN) type [17].

Superstructures represent decision-making at various levels. Thus, continuous and discrete variables are used to represent the relationships among their elements as linear and nonlinear mixed-integer programming models [17]. There are various techniques that allow solving the optimization of supply chains where many aspects can be included for the solution. Two of the most used strategies are hierarchical decomposition and superstructure synthesis. The first involves a sequential procedure that is developed by levels, and at each level, specific decisions are made. So, this approach represents a solution in stages and not a simultaneous solution. On the contrary, for the superstructure synthesis technique, an attempt is made to solve the supply chain design as a mathematical programming problem. This second technique is preferred because of its more precise results. The superstructure synthesis technique is composed of three steps:

1. Postulation of a superstructure that encompasses all the feasible options to be considered.
2. Translation of the superstructure to a mathematical programming model.
3. Calculation of the optimal structure by solving the mathematical optimization model [18].

The blocks within a superstructure can be encoded using logic-based models to translate them into a mathematical form through generalized disjunctive programming (GDP). This kind of models can intuitively express the relationship between different proposed alternatives for process paths, while capturing the connection between logical clauses and algebraic logic [16]. This type of programming makes use of disjunctions, equations that represent the balances or characteristics of the superstructure as well as logical propositions. Below is the generalized form of a GDP problem [19]:

$$\min Z = \sum_i c_i + f(x)$$

$$\text{s.t.}$$

$$g(x) \le 0$$

$$\begin{bmatrix} Y_i \\ h_i(x) \le 0 \\ c_i = \gamma_i \end{bmatrix} \vee \begin{bmatrix} \neg Y_i \\ B^i x = 0 \\ c_i = 0 \end{bmatrix}, \quad i \epsilon D \tag{12.1}$$

$$\Omega(Y) = \mathit{True}$$

$$x \epsilon R^n$$

$$c_i \epsilon R$$

$$Y_i \epsilon \{\mathit{True}, \mathit{False}\}$$

x is a vector of continuous variables, while c_i are variables associated with fixed costs. Y_i is a Boolean variable associated to the disjunction. $g(x)$ and $h_i(x)$ are functions. B^i ensures that the proper subset of the continuous variables is zero. Finally, y_i represents either a function or a constant, related to the fixed costs c_i. Strategies to solve a GDP problem will be discussed in Section 12.5. However, a discussion related to the selection of the objective function is first presented.

12.3 Defining the objective function

Objective function is a crucial part of any optimization model since it is the criterion to assess in order to obtain a solution. The purpose of an optimization problem is searching for the combination of decision variables that lead to the minimum or maximum value of the objective function. Thus, the election of an adequate function objective is of great importance to obtain the solution that best fits to the needs of the decision-maker. Moreover, each kind of objective function would lead to different solutions. An optimization problem can be classified into two main groups: a) single-objective optimization problem and b) multi-objective optimization problem, regarding the number of objective functions to be evaluated. In multi-objective optimization, the objective functions are commonly competing among themselves. This implies that a good solution in terms of one objective leads to inadequate values for the other objectives. Thus, the common approach is to find solutions with a good balance between all the objectives.

Concerning chemical engineering, most of relevant problems are multi-objective formulations; some typical examples are listed below:

- Number of trays in distillation columns versus purity of products. If the number of trays in the distillation column is low, the cost of the column is reduced. However, to reach high purities, more trays are necessary.
- Conversion versus reactor size. If the volume of the reactor is small, its cost is low. However, to increase the conversion and, consequently, the profit, the volume of the reactor must be increased.
- Heat exchanger network cost versus minimum temperature approach. When the temperature approach is lower, the need for further cooling or heating is reduced, but the area of the exchanger is increased, and the cost of the network is high.

Furthermore, supply chain optimization problems are very versatile in terms of objective functions. These can be related to supply chain production aspects as well as sustainability dimensions. Table 12.1 presents some common objective functions in supply chain optimization problems.

Table 12.1: Summary of the main objective functions used in supply chain optimization problems (case: biomass, bioenergy, and biofuels supply chains).

Supply chain type	Economic objective	Environmental objective	Social objective	Supply chain production	Source
Bioenergy and biofuel supply chain	Net present value				[20]
Biogas supply chain	Profit	Emissions			[21]
Biomass supply chain	Profit			Product demand	[22]
Cellulosic biofuels supply chain	Net present value	Greenhouse gas emissions	Jobs		[23]
Biofuels supply chain	Net profit	Economic indicator 99	Jobs		[24]
Biomass supply chain	Net profit	Greenhouse gas emissions	Marginalization level		[25]
Bioethanol supply chain	Cost	Economic indicator	Social responsibility		[26]

As seen, there is a large variety of objective functions in supply chain optimization. These objective functions are classified into four main groups: a) Supply chain production aspects, b) economic aspects, c) environmental aspects, and d) social aspects. A brief description of these groups as well as representative examples is provided.

Regarding supply chain production aspects, these are directly related to production and performance of a supply chain. In this respect, production level could be defined as the amount of product generated in a supply chain, which can be local production and international production, depending on the specifications of the supply chain designer in terms of considering or not exportations and/or importations [27]. It must be noted that production level is directly related to the satisfied demand. In fact, the satisfied demand is another supply chain objective function that can be used in an individual form, or combined with other objectives as shown in [28], where product demand is included through a penalty for unsatisfied demand in an economic objective function. Also, the satisfied demand can be used to evaluate the performance of a supply chain to illustrate if a supply chain configuration can be used to satisfy the demand requirements.

Another classification is given by the economic objective functions. This is probably the most used kind in supply chain optimization problems, since economic performance can be evaluated by a large number of criteria such as net annual profit, net present value, internal rate of return, total annual cost, expected profit, minimum sustainable price and investment cost. In fact, most of approaches presented in Table 12.1 take into account at least one economic objective function. For a supply chain to be feasible, it must have positive economic indicators. Otherwise, the production scheme will not endure.

Net annual profit considers the total revenue from products minus the total capital and operation cost for a supply chain; it is important to mention that capital cost must be annualized for a fair comparison. Moreover, net present value corresponds to the summation of all cash flow at the analysis year for an investment project. In this case, it could include the total revenue from products, total raw material cost and all involved building cost for processing plants. Similarly, the internal rate of return is used to evaluate investment projects; it is computed using the cash flow and a discount interest rate to obtain a net present value equal to zero. Also, a similar metric is the minimum sustainable price, which is an economic metric to obtain the minimum price of a product that allows obtaining a net present value equal to zero [27]. Finally, total annual cost considers the annualized capital and operating cost for a production system.

Environmental objective functions have also been addressed in supply chain optimization problems. One of the environmental objectives is given by the total CO_2 emissions, which are caused by transportation, and processing and production of raw materials in a manufacturing system. In addition, another environmental objective function is given by the environmental impact measured through methodologies based on life cycle analysis such as EcoIndicator99 [29] and tools as OpenLCA [30]. Such strategies generate numerical values representing the whole impact of the production scheme in terms of different impact categories. For example, the EcoIndicator99 includes the impacts of emissions to the atmosphere and the components released to the water and earth over the human health and the ecosystems. It also evaluates the impacts associated with the consumption of nonrenewable materials, such as fossil fuels and minerals. In other approaches, water consumption is used as a metric for environmental impact [31].

In contrast to environmental and economic objective functions, there is a lack of well-defined mathematical formulations that take into account social aspects. In this respect, one of the differences is that economic and environmental objectives are established and there are metrics that can be used to evaluate the performance of a production system. However, it has been hard to find an adequate quantitative criterion to assess the social impact. Even though one of the most used criterion is the number of generated jobs in the supply chain [23, 31], the number of generated jobs do not permit to identify clear differences between the different types of jobs in a manufacturing system, or the location where the jobs take place. Social impact also has been related to the inherent safety through functions for expected fatalities or solvent toxicity [32].

Recently, social objective functions based on social indexes have been proposed to include multiple factors, as well as the social impact in a specific geographic region. Some of the social indices used in objective functions are the marginalization index and the human development index, since these indices can be computed for different geographical regions, being indicators of the current social conditions in the studies zones [25, 33].

As discussed, there are several objective functions that can be used in supply chain optimization approaches that can consider economic, environmental, and social criteria. In addition, most of methodologies involve at least two objectives. Decision concerning the set of objective functions depends on factors such as feasibility to measure, interest of decision makers and data availability, for computing the objectives. This last topic is discussed in detail in the next section.

12.4 Data generation

Data generation is a crucial stage when optimization problems are solved, since the quality and realism of the solution is directly related to the quality and data feasibility. In this respect, some of main data for any supply chain optimization problems are:

- Unitary cost for raw materials and products
- Manufacturing cost
- Raw material availability and product demand
- Transportation cost
- Distance between supply chain nodes or entities
- Emission factors for pollutants
- Raw material-to-product ratio under operating conditions
- Physical and thermodynamic constraints

Unitary cost for raw material and products can be estimated through market analysis and reports, which can be accessed via companies such as Intratec [34]. Also, data for raw material price are reported in public databases by government, such as the Information System for Agriculture and Fishing database (SIAP) for Mexico [35] as well as data from The United States Department of Agriculture (USDA) [36].

Manufacturing cost for processes can be initially estimated from reported data in scientific papers, reports or data from industry and then the data could be scaled, since in most times, the data is available for a different production capacity than that required for a supply chain. It is worth noting that manufacturing cost might be affected by economies of scale and therefore estimation of additional parameters for prediction of costs is needed. Some examples or industrial data are reported in [37, 38].

Data for raw material availability and product demand could be obtained from reports from industry, forecast of government institutions based on development goals, and correlation between similar products such as gasoline, diesel, bioethanol, and biodiesel. In addition, some commodities have direct relation between population and product demand via per capita consumption. It is important to highlight that the product demand must be limited to a geographic region because of population lifestyle.

Whereas raw material availability, manufacturing cost, and raw material price are reported in a general way, transportation cost is not so, because it depends on several factors such as departure and destination sites, distance, geographic region, fuel cost, transport capacity, and material transported. Thus, in most cases, transportation cost is computed from other parameters or via direct freight quotes requested for hypothetical case studies. Concerning distance between supply chain sites, it is necessary to consider the real distance, and the use of distance by straight line should be avoided.

In case environmental objectives are considered, required parameters like emission factors and unitary environmental impact need to be estimated. These parameters could involve carrying out life cycle analysis to determine the quantity of a pollutant with respect to a functional unit for industrial processes. Nevertheless, these factors are reported in specialized databases in governmental institutions (e.g., Environmental Protection Agency [39]).

Additionally, processes in supply chain and product-to-raw material ratio should consider rigorous modeling. In this respect, product-to-raw material ratio can be obtained from experimental reports and scientific papers. Also, process simulators could be combined with mathematical programming and other strategies to model the process and thermodynamic properties into a supply chain, which allows obtaining more realistic modeling [25].

12.5 Supply chain optimization

Once the whole GDP has been formulated, a proper strategy must be used to solve the resulting optimization problem. Figure 12.3 shows a diagram of the existing alternatives to solve a GDP model. Because superstructures, by their nature, mostly feature MINLP and MILP models, they are more likely to use the convex hull and Big-M relaxation methods. Both methods lead to good approximations when the disjunctions include linear terms. However, the convex hull approach leads to tighter approximations for nonlinear terms.

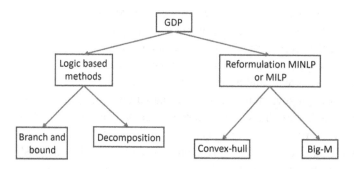

Figure 12.3: Alternatives to solve GDP problems [40].

Any of the relaxation methods allow replacing the disjunctions in eq. (13.1) by algebraic terms. This leads to MILP or MINLP formulations that can be solved using an adequate algorithm for such kinds of optimization problems.

To exemplify the relaxation techniques, a generic linear constraint in a disjunction will be represented as follows:

$$V_{i\epsilon D}[A_i x - b_i \leq 0] \tag{12.2}$$

Such logical constraint is then transformed into the following inequality through the Big-M approach:

$$A_i x - b_i \leq M_i(1-y_i), \quad i\epsilon D \tag{12.3}$$

where

$$\sum_{i\epsilon D} y_i = 1 \tag{12.4}$$

y_i represents binary variables associated with the logical propositions and M_i is a parameter specific to the Big-M methodology. If the logical proposition is true, then $y_i = 1$, otherwise $y_i = 0$. The constraint given by eq. (13.4) ensures that only one of the propositions in the disjunction is true. M_i must be large enough for the inequality to be redundant if y_i is zero.

On the other hand, if the convex hull strategy is applied, the relaxation of eq. (13.2) leads to the following set of equations:

$$x = \sum_{i\epsilon D} v_i \tag{12.5}$$

$$A_i v_i - b_i y_i \leq 0, \quad i\epsilon D \tag{12.6}$$

$$\sum_{i\epsilon D} y_i = 1 \tag{12.7}$$

$$0 \leq v_i \leq v_i^U y_i, \quad i\epsilon D \tag{12.8}$$

$$y_i \epsilon \{0,1\} \tag{12.9}$$

Where the superscript U represents an upper bound, y_i represent the binary variables and v^i are disaggregated variables. This methodology requires the disaggregation of variables to help the selection method to decide whether the proposition is true or not [41]. Although the GDP relaxation using a convex hull involves the generation of a greater number of variables, this methodology allows working in a more defined field to find the optimal solution. On the other hand, the Big-M method is simpler to execute but the searching field is usually more dispersed.

When modeling through GDP, in addition to disjunctions, the use of logical propositions aids in defining the relationship between the elements of the superstructure. To define these propositions, the use of logical structures, as those shown in Table 12.2, is necessary. A representation of the propositions in terms of binary variables is also shown. Such equivalences are incorporated to the relaxed GDP to complete the model.

Table 12.2: Logical propositions and their equivalences [42].

Logical relation	Logical expression	Equivalence
OR	$Y_1 \vee Y_2 \vee \cdots \vee Y_i$	$y_1 + y_2 + \cdots + y_i \geq 1$
AND	$Y_1 \wedge Y_2 \wedge \cdots \wedge Y_i$	$y_1 \geq 1; y_2 \geq 1; \ldots; y_i \geq 1$
Implication	$\neg Y_1 \vee Y_2$	$y_1 - y_2 \leq 0$
Equivalence (double implication)	$(\neg Y_1 \vee Y_2) \wedge (\neg Y_2 \vee Y_1)$	$y_1 = y_2$
Exclusive OR	$Y_1 \oplus Y_2 \oplus \cdots \oplus Y_i$	$y_1 + y_2 + \cdots + y_i = 1$

These expressions are closely related to the Boolean variables and help to better restrict what happens in the supply chain, ensuring that the solution is attached to the logic that is sought to be represented. All the elements of the model for the superstructure, i.e., the relaxed GDP model, the binary form of the logical propositions as well as the additional constraints and boundaries, generate MILP or MINLP problems. Due to the high number of variables and constraints commonly involved in a supply chain model, optimization software is commonly used to solve the resulting mixed-integer model. Among the most common tools, GAMS, Python, and MATLAB can be mentioned.

12.6 Case study: optimization of the supply chain for the valorization of glycerol derivatives

In this section, a case study related to the determination of the optimal supply chain for the valorization of glycerol is developed. The potential for the use of glycerol as a raw material to produce high value-added derivatives is analyzed. Glycerol is a by-product obtained from biodiesel production, with a generation of approximately 10 kg of crude glycerol for every 100 kg of biodiesel produced [43]. Crude glycerol finds applications as livestock food, or it is directly combusted for heat and power generation [44, 45]. However, these applications do not yield significant economic benefits. On the other hand, since glycerol is a saturated chemical component with high hydrogen and carbon content, it can be used as a platform molecule to obtain high value-added chemical products [46].

Chemical products that can be produced from glycerol include solvents such as lactic acid, acetone, and propylene glycol, and precursors for the synthesis of polymers such as acrylic acid, epichlorohydrin, and glycerol carbonate [47, 48]. In this study, two potential sources of crude glycerol from the biodiesel industry in Mexico are considered. These are two facilities with the highest biodiesel production: Cooperativa Agrícola Luz Michell S.C. de R.L de C.V., located in the state of Durango (DGO) and CEDA in Mexico City (CDMX) [49].

Additionally, the state of Querétaro (QRO) has been identified as a promising location for establishing the glycerol biorefinery. This choice is based on its proximity to feedstock sources and its robust industrial infrastructure capacity [50]. Each facility can produce one or more products. Four possible high value-added products were selected due to their demand in the domestic market: acrylic acid (AA), epichlorohydrin (ECH), lactic acid (LA), and propylene glycol (PG). These products are distributed to the states with the highest international purchases for each respective product. Mathematical optimization is used to represent the supply chain, taking a GDP model as basis. In Figure 12.4, the superstructure for the case study is presented. Figure 12.5 shows a map of Mexico where the location of the facility, the potential sources of glycerol, and the potential markets are indicated.

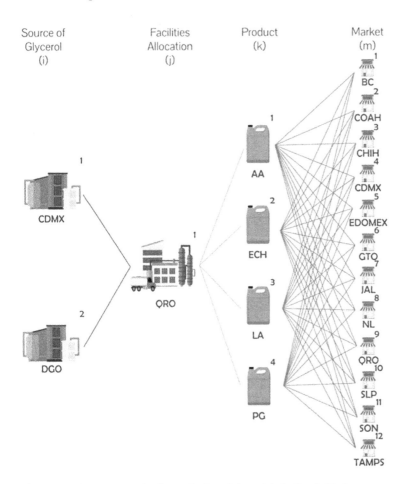

Figure 12.4: Superstructure for the production of glycerol derivatives in Mexico.

Source

Facility allocation

Market

Figure 12.5: Location of the glycerol biorefinery, the potential glycerol sources, and the potential markets.

The selection of the best supply chain to produce one or more high-added value products is carried out, aiming to maximize the profit. The profit is estimated as follows:

$$\text{Profit} = \sum_k \sum_m \text{HVAPP}(k,\text{QRO},m) \cdot \text{SP}(k) - \sum_k \text{CB}(k) - \sum_k \text{CF}(k)$$

$$- \frac{\sum_k \text{LCB}(k)}{\text{PP}} - \sum_i \text{C2TGF}(i,\text{QRO}) - \sum_m \text{C2TP2M}(\text{QRO},m) \qquad (12.10)$$

where HVAPP(k, QRO,m) is the amount of each high value-added product k produced in QRO and sent to market m. SP(k) is the selling price for the product k. CB(k) and CF(k) are the operational costs of the biorefinery and the cost of feedstock to produce each bioproduct k, respectively. LCB(k) is the land cost of the biorefinery devoted to produce k, and PP is the payment period, established as 5 years [51]. Finally, C2TGF(i,QRO) is the cost to transport glycerol from the source i to the facility in QRO, and C2TP2M(QRO,m) is the cost to transport any product from QRO to the market m. This optimization problem is subject to:

$$Y(AA) \vee Y(ECH) \vee Y(LA) \vee Y(PG) \tag{12.11}$$

This equation implies that at least one of the products k must be obtained in the facility j. Each product k can either be produced or not. A logical variable $Y(k)$ is assigned to each product. These considerations can be presented in terms of a disjunction for each product, as follows:

$$
\begin{bmatrix}
Y(AA) \\
\sum_m HVAPP(AA,QRO,m) = \sum_i GAU(i,QRO,AA) \cdot Yld(AA) \\
CB(AA) = \sum_m HVAPP(AA,QRO,m) \cdot CP(AA) \\
CF(AA) = \sum_i GAU(i,QRO,AA) \cdot CG \\
LCB(AA) = \sum_m HVAPP(AA,QRO,m) \cdot RAB(AA) \cdot LC(QRO)
\end{bmatrix}
$$

$$
V
\begin{bmatrix}
\neg Y(AA) \\
\sum_i GAU(i,QRO,AA) = 0 \\
\sum_m HVAPP(AA,QRO,m) = 0 \\
CB(AA) = 0 \\
CF(AA) = 0 \\
LCB(AA) = 0
\end{bmatrix}
\tag{12.12}
$$

$$
\begin{bmatrix}
Y(ECH) \\
\sum_m HVAPP(ECH,QRO,m) = \sum_i GAU(i,QRO,ECH) \cdot Yld(ECH) \\
CB(ECH) = \sum_m HVAPP(ECH,QROm) \cdot CP(ECH) \\
CF(ECH) = \sum_i GAU(i,QRO,ECH) \cdot CG \\
LCB(ECH) = \sum_m HVAPP(ECH,QRO,m) \cdot RAB(ECH) \cdot LC(QRO)
\end{bmatrix}
$$

$$
V
\begin{bmatrix}
\neg Y(ECH) \\
\sum_i GAU(i,QRO,ECH) = 0 \\
\sum_m HVAPP(ECH,QRO,m) = 0 \\
CB(ECH) = 0 \\
CF(ECH) = 0 \\
LCB(ECH) = 0
\end{bmatrix}
\tag{12.13}
$$

$$
\begin{bmatrix}
Y(\text{LA}) \\
\sum_m \text{HVAPP}(\text{LA,QRO},m) = \sum_i \text{GAU}(i,\text{QRO,LA}) \cdot \text{Yld}(\text{LA}) \\
\text{CB}(\text{LA}) = \sum_m \text{HVAPP}(\text{LA,QRO},m) \cdot \text{CP}(\text{LA}) \\
\text{CF}(\text{LA}) = \sum_i \text{GAU}(i,\text{QRO,LA}) \cdot \text{CG} \\
\text{LCB}(\text{LA}) = \sum_m \text{HVAPP}(\text{LA,QRO},m) \cdot \text{RAB}(\text{LA}) \cdot \text{LC}(\text{QRO})
\end{bmatrix}
$$

$$
\vee
\begin{bmatrix}
\neg Y(\text{ECH}) \\
\sum_i \text{GAU}(i,\text{QRO,LA}) = 0 \\
\sum_m \text{HVAPP}(\text{LA,QRO},m) = 0 \\
\text{CB}(\text{LA}) = 0 \\
\text{CF}(\text{LA}) = 0 \\
\text{LCB}(\text{LA}) = 0
\end{bmatrix}
\tag{12.14}
$$

$$
\begin{bmatrix}
Y(\text{PG}) \\
\sum_m \text{HVAPP}(\text{PG,QRO},m) = \sum_i \text{GAU}(i,\text{QRO,PG}) \cdot \text{Yld}(\text{PG}) \\
\text{CB}(\text{PG}) = \sum_m \text{HVAPP}(\text{PG,QRO},m) \cdot \text{CP}(\text{PG}) \\
\text{CF}(\text{PG}) = \sum_i \text{GAU}(i,\text{QRO,PG}) \cdot \text{CG} \\
\text{LCB}(\text{PG}) = \sum_m \text{HVAPP}(\text{PG,QRO},m) \cdot \text{RAB}(\text{PG}) \cdot \text{LC}(\text{QRO})
\end{bmatrix}
$$

$$
\vee
\begin{bmatrix}
\neg Y(\text{PG}) \\
\sum_i \text{GAU}(i,\text{QRO,PG}) = 0 \\
\sum_m \text{HVAPP}(\text{PG,QRO},m) = 0 \\
\text{CB}(\text{PG}) = 0 \\
\text{CF}(\text{PG}) = 0 \\
\text{LCB}(\text{PG}) = 0
\end{bmatrix}
\tag{12.15}
$$

Where GAU(i,QRO,k) is the glycerol available sent from source i to the facility in QRO to produce k, and Yld(k) is the yield to the product k. On the other side, CB(k) is the total production cost for each product, given by the production of k HVAPP(k,QRO,m) multiplied by the unitary production cost, CP(k). CF(k) is the total feedstock cost, given by GAU(i,QRO,k), multiplied by the unitary cost of the glycerol CG, established as 0.11 USD/kg [52]. Finally, LCB(k) is the land cost, given by the production of k, HVAPP(k,QRO,m), multiplied by the required area for each biorefinery k, RAB(k), and the land cost for the facility, LC(QRO). This last cost has been determined as 90.59 USD/m^2, according to the reported price for land in the industrial zone, Parque Industrial de Querétaro [53]. The values of the parameters used in the model are shown in Table 12.3.

Table 12.3: Parameters used in the supply chain model [54, 55].

Product	Production cost (USD/kg)	Yield (kg$_{glycerol}$/kg$_{product}$)	Sales price (USD/kg)	Required area (m^2/kg)
Acrylic acid (AA)	1.4	2	1.5	0.00098
Epichlorohydrin (ECH)	1.1	2.7	1.8	0.00309
Lactic acid (LA)	0.8	1.56	1.8	0.00112
Propylene glycol (PG)	0.66	1.28	1.4	0.00098

Other constraints are described next. The amount of each high value-added product HVAPP(k,QRO,m) must not be higher than the total demand for that product k in that market m, TPD(k,m). This is represented as follows:

$$HVAPP(k,QRO,m) \leq TPD(k,m), \quad \forall k,m \tag{12.16}$$

In the same way, the glycerol used from source i, GAU(i,j,k) cannot be higher than the availability of glycerol in each source i, GA(i):

$$\sum_k GAU(i,QRO,k) \leq GA(i), \quad \forall i \tag{12.17}$$

The total amount of each product k, HVAPP(k,QRO,m) is given by the total amount of used glycerol, GAU(i,QRO,k) times the yield of each product Yld(k):

$$\sum_m HVAPP(k,QRO,m) = \sum_i GAU(i,QRO,k) \cdot Yld(k), \quad \forall k \tag{12.18}$$

Finally, the cost to transport glycerol from source i to the facility in QRO, C2TGF(i,QRO) is given by

$$C2TGF(i,QRO) = \sum_k GAU(i,QRO,k) \cdot D1(i,QRO) \cdot TG, \quad \forall i \tag{12.19}$$

Where D1(i,QRO) is the distance from source i to QRO and TG is the transportation cost of glycerol, taken as 0.000168 USD/km·kg. On the other side, the cost to transport each product from the facility in QRO to the market m, C2TP2M(QRO,m) is

$$C2TP2M(QRO,m) = \sum_k HVAPP(k,QRO,m) \cdot D2(QRO,m) \cdot TUnit, \quad \forall m \tag{12.20}$$

Where D2(QRO,m) is the distance from the facility in QRO to the market m, and TUnit is the transportation cost for the product, taken as 0.000298 USD/km·kg. TG and TUnit are computed according to data reported by the Instituto Mexicano del Transporte (Mexican Institute of Transportation) [56].

This problem is solved in Python v. 3.11.5. The Big M strategy is used to relax the disjunctions associated with each product k, through the command Transformation-Factory('gdp.bigm').apply_to(model).

The resulting MILP problem consists of 91 variables. The optimization problem is solved with the solver *Bonmin* in an Acer equipment with an Intel Core i7-11,800 H CPU, 16.00 GB of RAM.

According to the results, the maximum profit is 193,978.1 USD/y, implying the use of all the glycerol available in both sources to produce epichlorohydrin. This production scheme allows satisfying 16.9% of the demand of this product. Table 12.4 shows the amount of ECH received by the selected markets. The total demand in EDOMEX is completely satisfied since that location is close to the biorefinery. On the other hand, only 12.8% of the demand of CDMX is satisfied. Thus, the solution implies satisfying the whole demand of the state with the lowest distance to the biorefinery and then sending the remaining production to the second market. It is important to mention that only these two states have demand for ECH in the country, hence their selection as markets in the supply chain. The final supply chain is shown in Figure 12.6. It is important to notice that, to present a simplified example of supply chain optimization, a single location for the facility has been defined. However, the implementation of glycerol conversion facilities in other states with industrial infrastructure would lead to different supply chains, given the distance between the glycerol source in Durango and the location of the proposed facility in Queretaro.

Table 12.4: Distribution of epichlorohydrin (ECH) to each market.

Market	Amount of ECH (kg/year)
CDMX	274,142.98
EDOMEX	105,316.10

According to the results, the highest contribution to the cost in the ECH supply chain is given by the price of the production process, representing 85.35% of the total outflows, as shown in Figure 12.7. This could be reduced by considering the integration of production processes within the framework of a biorefinery, i.e., producing two or even all four high value-added products simultaneously, which would allow the simultaneous generation of multiple products within a unified facility. Another approach is a detailed analysis of the production processes to implement strategies to reduce the production costs, as energy integration and process intensification, together with further optimization of the production schemes. Additionally, the existing model can be improved by incorporating the other biodiesel industries in Mexico,

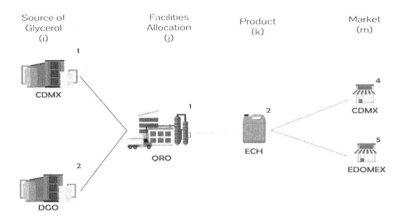

Figure 12.6: Optimized supply chain for the production of high value-added glycerol derivatives.

that is, not only the two with the largest biodiesel generation, but also the seven existing in the country or even projecting the installation of new biodiesel facilities. Moreover, exploring additional locations for the glycerol biorefineries may help to obtain more profitable solutions for the entire supply chain. Furthermore, a multi-objective analysis, including functions addressing environmental and social aspects, can contribute to the development of a sustainable scheme.

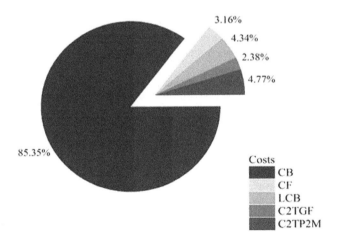

Figure 12.7: Contribution of each cost component to the total cost of the supply chain.

12.7 Conclusions

Process optimization is of great importance for the transformation industry, since it allows determining the best conversion routes, the best design of the process and the best operational conditions. However, if the supply chain is not properly selected, the economic feasibility of the complete conversion scheme could be at risk. In this chapter, the most important factors associated with the optimization of supply chains for chemical and biochemical processes have been discussed. Different potential objective functions have been described, which include not only economical criteria but also environmental and social aspects, necessary to ensure the sustainability of the conversion schemes. Generalized disjunctive programming has been presented as a powerful tool to develop mathematical models to represent all the potential decisions in a supply chain. Finally, the optimization of the supply chain has been explained with a case study related to the conversion of renewable glycerol into high value-added products, allowing the determination of areas with opportunity to increase the economic potential of such a transformation scheme.

References

[1]   Cabrera-Ruiz, J., Santaella, M. A., Alcántara-Ávila, J. R., Segovia-Hernández, J. G., Hernández, S. Open loop based controllability criterion applied to stochastic global optimization for intensified distillation sequences. *Chemical Engineering Research and Design*, 2017, 123, 165–179.

[2]   Contreras-Zarazúa, G., Vázquez-Castillo, J. A., Ramírez-Márquez, C., Segovia-Hernández, J. G., Alcántara-Ávila, J. R. Multi-objective optimization involving cost and control properties in reactive distillation processes to produce diphenyl carbonate. *Computers & Chemical Engineering*, 2017, 105, 185–196.

[3]   Vega-Guerrero, D. B., Gómez-Castro, F. I., López-Molina, A. Production of biodiesel with supercritical ethanol: Compromise between safety and costs. *Chemical Engineering Research and Design*, 2022, 184, 79–89.

[4]   López-Villarreal, F., López-Molina, A., Gómez-Castro, F. I., Conde-Mejía, C., Valenzuela, L. M. Multiobjective optimization of process equipment distribution: Effect of pumping cost. *Chemical Engineering Research and Design*, 2022, 187, 461–469.

[5]   Güler, M. G., Geçici, E., Erdogan, A. Design of a future hydrogen supply chain: A multi period model for Turkey. *International Journal of Hydrogen Energy*, 2021, 46(30), 16279–16298.

[6]   Kang, K., Klinghoffer, N. B., ElGhamrawy, I., Berruti, F. Thermochemical conversion of agroforestry biomass and solid waste using decentralized and mobile systems for renewable energy and products. *Renewable and Sustainable Energy Reviews*, 2021, 149, 111372.

[7]   Schmitt, A. J., Sun, S. A., Snyder, L. V., Shen, Z.-J. M. Centralization versus decentralization: Risk pooling, risk diversification, and supply chain disruptions. *Omega*, 2015, 52, 201–212.

[8]   Mohammadi, M., Martín-Hernández, E., Martín, M., Harjunkoski, I. Modeling and analysis of organic waste management systems in centralized and decentralized supply chains using Generalized Disjunctive Programming. *Industrial & Engineering Chemistry Research*, 2021, 60(4), 1719–1745.

[9]   Contreras-Zarazúa, G., Martín, M., Ponce-Ortega, J. M., Segovia-Hernández, J. G. Sustainable Design of an Optimal Supply Chain for Furfural Production from Agricultural Wastes. *Industrial & Engineering Chemistry Research*, 2021, 60(40), 14495–14510.

[10] Espinoza-Vázquez, Y. M., Gómez-Castro, F. I., Ponce-Ortega, J. M. Multiobjective optimization of the supply chain for the production of biomass-based fuels and high-value added products in Mexico. *Computers & Chemical Engineering*, 2022, 157, 107598.

[11] Hoseinzade, L., Adams, T. A. II. Supply chain optimization of flare-gas-to-butanol processes in Alberta. *The Canadian Journal of Chemical Engineering*, 2016, 94(12), 2336–2354.

[12] Martínez-Guido, S. I., García-Trejo, J. F., Gutiérrez-Antonio, C., Domínguez-González, A., Gómez-Castro, F. I., Ponce-Ortega, J. M. The integration of pelletized agricultural residues into electricity grid: Perspectives from the human, environmental and economic aspects. *Journal of Cleaner Production*, 2021, 321, 128932.

[13] Gao, J., You, F. Shale gas supply chain design and operations toward better economic and life cycle environmental performance: MINLP model and global optimization algorithm. *ACS Sustainable Chemistry & Engineering*, 2015, 3(7), 1282–1291.

[14] Ravi, N. K., Van Sint Annaland, M., Fransoo, J. C., Grievink, J., Zondervan, E. Development and implementation of supply chain optimization framework for CO_2 capture and storage in the Netherlands. *Computers & Chemical Engineering*, 2017, 102, 40–51.

[15] Ding, X., Li, J., Chen, H., Zhou, T. Superstructure-based carbon capture and utilization process design. *Current Opinion in Chemical Engineering*, 2024, 43, 100995.

[16] Mercarelli, L., Chen, Q., Pagot, A., Grossmann, I. A review on superstructure optimization approaches in process system engineering. *Computers and Chemical Engineering*, 2020, 1–15.

[17] Bertran, M., Frauzem, R., Sanchez-Arcilla, A., Zhang, L., Woodley, J. M., Gani, R. A generic methodology for processing route synthesis and design based on superstructure optimization. *Computers and Chemical Engineering*, 2017, 106, 892–910.

[18] Umeda, T., Hirai, A., Ichikawa, A. Synthesis of optimal processing system by an integrated approach. *Chemical Engineering Science*, 1972, 27, 795–804.

[19] Türkay, M., Grossmann, I. E. Logic-Based MINLP algorithms for the optimal synthesis of process networks. *Computers and Chemical Engineering*, 1996, 20(8), 959–978.

[20] Akhtari, S., Sowlati, T. Hybrid optimization-simulation for integrated planning of bioenergy and biofuel supply chains. *Applied Energy*, 2020, 259, 114124.

[21] Durmaz, Y. G., Bilgen, B. Multi-objective optimization of sustainable biomass supply chain network design. *Applied Energy*, 2020, 272, 115259.

[22] Espinoza-Vázquez, Y. M., Gómez-Castro, F. I., Ponce-Ortega, J. M. Optimization of the supply chain for the production of bio-mass-based fuels and high-added value products in Mexico. *Computers & Chemical Engineering*, 2021, 145, 107181.

[23] You, F., Tao, L., Graziano, D. J., Snyder, S. W. Optimal design of sustainable cellulosic biofuel supply chains: Multiobjective optimization coupled with life cycle assessment and input–output analysis. *AIChE Journal*, 2012, 58(4), 1157–1180.

[24] Malladi, K. T., Sowlati, T. Bi-objective optimization of biomass supply chains, considering carbon pricing policies. *Applied Energy*, 2020, 264, 114719.

[25] Santibañez-Aguilar, J. E., Quiroz-Ramírez, J. J., Sánchez-Ramírez, E., Segovia-Hernández, J. G., Flores-Tlacuahuac, A., Ponce-Ortega, J. M. Marginalization index as social measure for Acetone-Butanol-Ethanol supply chain planning. *Renewable and Sustainable Energy Reviews*, 2022, 154, 111816.

[26] Ghaderi, H., Moini, A., Pishvaee, M. S. A multi-objective robust possibilistic programming approach to sustainable switchgrass-based bioethanol supply chain network design. *Journal of Cleaner Production*, 2018, 179, 368–406.

[27] Castellanos, S., Santibañez-Aguilar, J. E., Shapiro, B. B., Powell, D. M., Peters, I. M., Buonassisi, T., Kammen, D. M., Flores-Tlacuahuac, A. Sustainable silicon photovoltaics manufacturing in a global market: A techno-economic, tariff and transportation framework. *Applied Energy*, 2018, 212, 704–719.

[28] Santibañez-Aguilar, J. E., González-Campos, J. B., Ponce-Ortega, J. M., Serna-González, M., El-Halwagi, M. M. Optimal planning and site selection for distributed multi-product biorefineries involving economic, environmental and social objectives. *Journal of Cleaner Production*, 2014, 65, 270–294.

[29] Rivas-Interian, R. M., Sanchez-Ramirez, E., Quiroz-Ramírez, J. J., Segovia-Hernandez, J. G. Feedstock planning and optimization of a sustainable distributed configuration biorefinery for biojet fuel production via ATJ process. *Biofuels, Bioproducts and Biorefining*, 2023, 17(1), 71–96.

[30] Acevedo-García, B., Santibañez-Aguilar, J. E., Alvarez, A. J. Integrated multiproduct biorefinery from Ricinus communis in Mexico: Conceptual design, evaluation, and optimization, based on environmental and economic aspects. *Bioresource Technology Reports*, 2022, 19, 101201.

[31] Fuentes-Cortés, L. F., González-Bravo, R., Flores-Tlacuahuac, A., Ponce-Ortega, J. M. Optimal sustainable water-Energy storage strategies for off-grid systems in low-income communities. *Computers & Chemical Engineering*, 2019, 123, 87–109.

[32] Santibañez-Aguilar, J. E., Martinez-Gomez, J., Ponce-Ortega, J. M., Nápoles-Rivera, F., Serna-González, M., González-Campos, J. B., El-Halwagi, M. M. Optimal planning for the reuse of municipal solid waste, considering economic, environmental, and safety objectives. *AIChE Journal*, 2015, 61(6), 1881–1889.

[33] Pulido-Ocegueda, J. C., Santibañez-Aguilar, J. E., Ponce-Ortega, J. M. Strategic Planning of Biorefineries for the Use of Residual Bio-mass for the Benefit of Regions with Low Human Development Index. *Waste and Biomass Valorization*, 2023, 14(9), 2825–2841.

[34] IntraTec, n.d. https://www.intratec.us/ (accessed April 10, 2024).

[35] Servicio de Información Agroalimentaria y Pesquera (SIAP), n.d. https://www.gob.mx/siap/ (accessed April 10, 2024). Spanish

[36] U.S. Department of Agriculture, n.d. https://www.usda.gov/ (accessed April 10, 2024).

[37] Incitec Pivot Limited, Louisiana Ammonia Plant, 2013. https://www.asx.com.au/asxpdf/20130417/pdf/42f9brx9b4xscx.pdf (accessed April 10, 2024).

[38] BASF, Yara and BASF break ground on new ammonia plant in Freeport, Texas, 2015. https://www.basf.com/global/en/media/news-releases/2015/07/p-15-300.html (accessed April 10, 2024).

[39] U.S. Environmental Protection Agency (EPA), n.d. https://www.epa.gov/ (accessed April 10, 2024).

[40] Grossmann, I. E., Ruiz, J. P. Generalized Disjunctive Programming: A Framework for Formulation and Alternative Algorithms for MINLP Optimization. In Lee, J., Leyffer, S. (eds.). *Mixed Integer Nonlinear Programming*. Springer, 2012, pp. 93–115.

[41] Lee, S., Grossmann, I. E. New algorithms for nonlinear generalized disjunctive programming. *Computers and Chemical Engineering*, 2000, 24, 2125–2141.

[42] Raman, R., Grossmann, I. E. Relation between MILP modelling and logical inference for chemical process synthesis. *Computers and Chemical Engineering*, 1991, 15(2), 73–84.

[43] Bansod, Y., Crabbe, B., Forster, L., Ghasemzadeh, K., D'Agostino, C. Evaluating the environmental impact of crude glycerol purification derived from biodiesel production: A comparative life cycle assessment study. *Journal of Cleaner Production*, 2024, 437, 140485.

[44] Kholif, A. E. Glycerol use in dairy diets: A systemic review. *Animal Nutrition*, 2019, 5(3), 209–216.

[45] McCann, T., Marek, E. J., Zheng, Y., Davidson, J. F., Hayhurst, A. N. The combustion of waste, industrial glycerol in a fluidised bed. *Fuel*, 2022, 322, 124169.

[46] Zhang, J., Wang, Y., Muldoon, V. L., Deng, S. Crude glycerol and glycerol as fuels and fuel additives in combustion applications. *Renewable and Sustainable Energy Reviews*, 2022, 159, 112206.

[47] Inrirai, P., Keogh, J., Centeno-Pedrazo, A., Artioli, N. Recent advances in processes and catalysts for glycerol carbonate production via direct and indirect use of CO_2. *Journal of CO_2 Utilization*, 2024, 80, 102693.

[48] Kaur, J., Sarma, A. K., Jha, M. K., Gera, P. Valorisation of crude glycerol to value-added products: Perspectives of process technology, economics and environmental issues. *Biotechnology Reports*, 2020, 27, e00487.

[49] Cabrera-Munguia, D. A., Romero, A., López, R. A., Rios, L. J., Leyva, Z. C. Producción de Biodiésel en México: Estado Actual y perspectivas. *CienciAcierta*, 2022, 72, 334–383. Spanish.

[50] The Logistic World, 2024. Panorama industrial en Querétaro: Cuáles son los giros de las principales empresas locales. https://thelogisticsworld.com/logistica-y-distribucion/panorama-industrial-en-queretaro-cuales-son-los-giros-de-las-principales-empresas-locales/ (accessed February 28, 2024). (Spanish).

[51] Susmozas, A., Moreno, A. D., Romero-García, J. M., Manzanares, P., Ballesteros, M. Designing an olive tree pruning biorefinery for the production of bioethanol, xylitol and antioxidants: A techno-economic assessment. *Holzforschung*, 2018, 73(1), 15–23.

[52] Jitjamnong, J., Khongprom, P., Ratanawilai, T., Ratanawilai, S. Techno-economic analysis of glycerol carbonate production by glycerolysis of crude glycerol and urea with multi-functional reactive distillation. *Case Studies in Chemical and Environmental Engineering*, 2023, 8, 100465.

[53] La Sombra de Arteaga, 2023. Tabla de valores unitarios de construcción 2024. https://lasombradearteaga.segobqueretaro.gob.mx/getfile.php?p1=20231296-01.pdf (accessed February 28, 2024). (Spanish).

[54] Lari, G. M., Pastore, G., Haus, M., Ding, Y., Papadokonstantakis, S., Mondelli, C., Pérez-Ramírez, J. Environmental and economical perspectives of a glycerol biorefinery. *Energy & Environmental Science*, 2018, 11, 1012–1029.

[55] European Investment Bank, n.d. Non-Technical Summary. https://www.eib.org/attachments/regis ters/137575361.pdf (accessed 28 February 2024).

[56] Instituto Mexicano del Transporte, 2022. Estudio exploratorio sobre el movimiento de hidrocarburos por autotransporte de carga en México. https://imt.mx/descarga-archivo.html?l=YXJ jaGl2b3MvUHVibGljYWNpb25lcy9QdWJsaWNhY2lvblRlY25pY2EvcHQ2ODkucGRm (accessed 28 February 2024). (Spanish).

David Esteban Bernal Neira, Fernando Israel Gómez-Castro*,
Ehecatl Antonio del-Río-Chanona, and Vicente Rico-Ramírez

Chapter 13
Future insights for optimization in chemical engineering

Abstract: Although there have been several advances in the development of optimization strategies over the last few decades, there are still challenges in the area. This chapter discusses the main areas of development identified by the authors for the following years, where a detailed analysis of the complex interactions in real-world systems is mandatory to make better decisions in fields where chemical engineering has influence. The chapter is focused on three main areas of study: the development of efficient optimization algorithms, the use of data-based approaches, and the development of powerful computational hardware.

Keywords: optimization, chemical engineering, algorithms, data-based optimization, hardware

13.1 Introduction

Optimization is an area of constant evolution. From the first developments of the mathematical fundamentals of optimization that followed Johann Bernoulli's brachistochrone challenge in 1696 [1], researchers in mathematics and computational sciences have focused their efforts on the enhancement of traditional optimization techniques and the development of new strategies. Although the fundamental concepts of mathematical optimization, such as stationary solutions, optimality conditions, feasible regions, and solution algorithms, among others, are well-known nowadays, the identification of global solutions for real-world optimization problems remains a considerable challenge. This

---

*Corresponding author: Fernando Israel Gómez-Castro**, Departamento de Ingeniería Química, División de Ciencias Naturales y Exactas, Universidad de Guanajuato, Campus Guanajuato, Noria Alta S/N Col. Noria Alta, Guanajuato 36050, Mexico, e-mail: fgomez@ugto.mx
David Esteban Bernal-Neira, Davidson School of Chemical Engineering, Purdue University, West Lafayette, IN, USA; Research Institute of Advanced Computer Science, Universities Space Research Association, Mountain View, CA, USA; Quantum Artificial Intelligence Laboratory, NASA Ames Research Center, Moffett Field, CA, USA
Ehecatl Antonio del-Río-Chanona, Department of Chemical Engineering, Sargent Centre for Process Systems Engineering, Imperial College London, London SW72AZ, UK
Vicente Rico-Ramírez, Departamento de Ingeniería Química, Tecnológico Nacional de México en Celaya, Av. Tecnológico y García Cubas S/N, Celaya, Guanajuato 38010, Mexico

https://doi.org/10.1515/9783111383439-013

occurs because of the large number of equations and variables involved in the modeling of practical optimization problems, the nonlinearities and non-convexities commonly found in engineering applications, and the need for simultaneously dealing with integer and continuous variables, among other factors [2].

Chemical engineering has long been both a source of challenging problems and a catalyst for developing innovative optimization methods. The field presents complex applications that test the limits of optimization techniques while simultaneously advancing algorithmic development. These contributions have not only addressed the specific needs of chemical engineering but have also influenced the broader landscape of optimization across various disciplines [3, 4].

There are several systems related to chemical engineering that show these characteristics. There are diverse examples of the application of optimization techniques to determine the optimal design of processing equipment, such as reactors (e.g., [5]), separators (e.g., [6, 7]), and heat exchangers (e.g., [8]), among others. The operation of processing units also leads to typical optimization problems related to reaction and separation systems [9, 10]. Other examples include the optimization of parameters for thermodynamic models to represent phase equilibrium [11]. These case studies are commonly nonlinear, occasionally involving functions with bilinear terms that may cause non-convexity or even functions that may lead to discontinuities or non-differentiable functions. The need to handle both integer and continuous variables is a common task as that occurs in stage-based separations [12]. Even for cases where the models are commonly linear, such as the optimization of supply chains, the consideration of the effects of the economy of scale implies nonlinearities, requiring assumptions to maintain the linearity of the representation [13].

Mathematical programming has become a foundational framework in this field, facilitating both the modeling and solution of optimization problems. This approach is flexible enough to incorporate a wide range of constraints, including linear, nonlinear, black-box, and surrogate data-trained models, and can handle both continuous and discrete variables. The versatility of mathematical programming allows it to address complex and diverse optimization scenarios encountered in chemical engineering, ensuring its ongoing relevance and applicability.

The strong tradition of mathematical programming in chemical engineering has motivated significant contributions in various areas of optimization. These areas include mixed-integer nonlinear programming (MINLP) [14–16], global optimization [2, 17–22], generalized disjunctive programming [23, 24], dynamic optimization [25–27], and optimization under uncertainty [28–30]. These are only examples of the crucial contributions made to the area by the chemical engineering community, evidencing the great importance of optimization to the discipline. Contributions in software, models, algorithms, frameworks, and many other required tools for optimization are provided in a recent review article by the current and former editors of the journal *Computers and Chemical Engineering* [31].

Optimization problems in chemical engineering commonly imply large, nonlinear models where the integrations across timescales and length scales are involved. Moreover, tighter coordination between industrially relevant problems and academically studied instances is required to address them [32]. Problems relevant to chemical engineering include situations ranging from a molecular scale to macroscopic systems, such as global supply chains. Applications in chemical engineering require representing an ever-evolving complex world, implying models that consider aspects such as real-time optimization, optimization under uncertainty, multi-scale modeling, and the integration of process dynamics and control. However, the more detailed the model, the higher are the complexities that might arise in its solution. Therefore, current developments in our field point toward a sustained development of algorithms and methods, as well as innovative use of tools from other areas such as artificial intelligence, machine learning, and data science. Additionally, advancements in computing technology, including quantum computing and parallel processing architectures, are expected to play a pivotal role in addressing the growing complexity of optimization problems. This chapter is devoted to highlighting some of the areas of development for optimization applied to chemical engineering in the following years.

13.2 Optimization algorithms

The future of algorithmic development in optimization for chemical engineering scenarios is expected to evolve toward approaches that are increasingly tailored to the specific characteristics of the problems we face. As our understanding of the underlying structure and the nature of these optimization problems deepens, we have the opportunity to design algorithms that are not only more efficient but also better suited to address the unique challenges posed by these problems [33]. While general-purpose algorithms have broad applicability, their effectiveness can be limited when dealing with the computational complexity and scale of real-world chemical engineering problems, many of which are NP-hard [2].

In this context, custom-made, application-specific methods become invaluable. The "no-free-lunch" theorem [34] suggests that no single optimization approach will be universally applicable for all types of problems. Instead, the development of specialized algorithms that exploit the particular structure of a given problem class can provide significant advantages in terms of performance and scalability. This approach involves crafting algorithms that capitalize on known properties, such as problem convexity, sparsity patterns, or the presence of specific types of constraints, to enhance computational efficiency and solution quality [18, 19].

Application-dependent algorithms should also aim to achieve Pareto-optimal solutions, balancing the trade-off between optimality guarantees and computational efficiency. In practice, this often involves selecting or designing algorithms that provide good approximations of the global optimum within acceptable time frames, particularly for large-scale or real-time applications [29]. For example, the use of metaheuristic approaches, such as genetic algorithms or simulated annealing, can be effective when exact solutions are computationally prohibitive [35, 36]. However, incorporating elements of mathematical programming into these heuristics can help guide the search process more effectively, providing a hybrid approach that leverages the strengths of both rigorous mathematical formulations and flexible heuristic techniques [37, 38].

The integration of mathematical programming and metaheuristic strategies represents a significant opportunity for advancing optimization in chemical engineering. By combining the structured search capabilities of mathematical programming with the exploration capabilities of metaheuristics, algorithms that are not only more robust but also capable of solving increasingly complex problems can be developed [39, 40]. These hybrid approaches can help navigate the trade-off between finding high-quality solutions and maintaining computational feasibility, especially in scenarios where uncertainty or real-time decision-making are critical factors.

13.3 Data-based optimization

Chemical and manufacturing industries are experiencing a shift toward automation and digitalization; as a result, processing activities are currently generating large amounts of information that can be applied for improving process performance. Gathering data is becoming a strategic tool to process sustainability and improve productivity. Hence, it is increasingly common to find applications involving data mining, big data, artificial intelligence and machine learning. In the most simplistic sense, these approaches basically involve the use of experimental/practical data along with statistical analysis to model reality, and such models can be used in multiple engineering applications. In particular, process systems engineering (PSE) is a discipline that involves mathematical programming methods and tools for solving complex problems related to simulation, control, design, and optimization. Modeling is a fundamental task that provides the basis for PSE applications. Therefore, PSE in general, and in particular, the area of optimization, are inherently related to data science and artificial intelligence approaches. A schematic illustration of that idea is shown in Figure 13.1.

The role of data-driven optimization in chemical engineering is becoming increasingly prominent as the field moves toward integrating large-scale data analysis with traditional modeling techniques. Contrary to the perception that data-driven modeling is entirely distinct from first-principles modeling, there is growing recognition that the most effective strategies often combine these approaches. The integration of data-driven models with first-principles models allows for leveraging the strengths of both: the empirical accuracy of data-driven methods and the theoretical rigor of first-

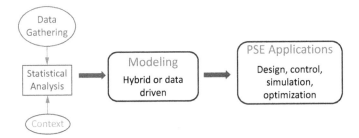

Figure 13.1: Different types of superstructures: (a) state-task network (STN), (b) process graph (P-graph), and (c) R-graph.

principles approaches [41,42]. This hybrid modeling paradigm provides a more robust framework for optimizing complex systems that are otherwise difficult to model with a purely mechanistic or purely data-driven approach. Several instances of such combined approaches have already been reported in the literature. These include, for instance, applications to chemical reaction engineering [43], manufacturing [44], process safety [45], stochastic optimization [46], real-time optimization [47], scheduling [48], and optimal control [49]. Data-driven techniques are bounded to become essential and represent useful tools in the field of optimization in chemical engineering. These tools more certainly will help achieve scientific breakthroughs and will contribute to an efficient and sustainable industry in the near future.

One of the emerging tools in this integrated approach is the use of physics-informed neural networks (PINNs). PINNs embed physical laws and constraints into the architecture of neural networks, ensuring that the predictions adhere to known principles of physics and chemistry [41]. By incorporating constraints directly into data-driven models, PINNs can enhance model accuracy and reliability, particularly in scenarios where data might be sparse or noisy. This capability is crucial for optimizing chemical processes that are governed by complex sets of equations; in some cases, even nonlinear partial differential equations (PDEs) and other physical constraints. PINNs represent a step forward in the fusion of data-driven and first-principles modeling, ensuring that optimization solutions are both data-consistent and physically plausible. Moreover, new developments in integrating physical constraints into chemical engineering systems for optimization have already been published [50].

Another essential aspect of data-driven optimization is the shift in terminology from "surrogate models" to "varying fidelity" models. This change reflects a more profound understanding that all models, whether empirical or theoretical, serve as surrogates for the actual physical systems they represent. George Box's famous saying, "All models are wrong, but some are useful," captures this idea succinctly. By adopting the term "varying fidelity," we acknowledge that models can be ranked based on their accuracy and reliability rather than being viewed as perfect or flawed representations [51]. This perspective is particularly valuable in chemical engineering, where models often need to balance complexity and computational efficiency.

Varying fidelity models enable a multi-scale approach to optimization, where high-fidelity models (based on first principles) are used to capture detailed behaviors of small-scale systems. In contrast, lower fidelity models (data-driven or simplified mechanistic models) are applied to larger-scale or less critical parts of the process [52]. This hierarchical modeling strategy allows for more efficient optimization, as it can dynamically allocate computational resources to the most critical aspects of the problem.

The future of data-driven optimization in chemical engineering lies in this seamless integration of models with varying levels of fidelity, utilizing advanced machine learning techniques like PINNs to bridge the gap between empirical data and theoretical knowledge. As the availability of process data continues to increase, so too will the potential for refining these models, leading to more accurate predictions and more efficient optimization of chemical processes.

13.4 Optimization using novel hardware

As the complexity and scale of optimization problems in chemical engineering increase, traditional computing methods are often insufficient to meet the demands of real-time decision-making and large-scale simulations. To overcome these limitations, researchers are exploring novel hardware platforms that provide significant computational power and efficiency. These emerging technologies promise to transform optimization capabilities, enabling the solution of problems previously deemed computationally intractable. This section explores how emerging and advanced computing technologies can enhance optimization strategies, focusing mainly on parallel computing, GPUs, and quantum computing.

13.4.1 Parallel computing

Parallel computing has been essential in the evolution of computational systems, contributing to the continual growth of computational power, a trend often associated with Moore's law. Moore's law, initially observed by Gordon Moore, states that the number of transistors on a microchip doubles approximately every two years, leading to an increase in computational power and a decrease in relative cost [53]. While the physical limitations of transistor scaling are increasingly becoming a challenge, parallel computing has provided a means to maintain performance improvements by allowing multiple processing cores to work simultaneously on different parts of a problem [54]. This capability offers significant potential for algorithmic development in various fields [55].

By distributing computational tasks across multiple processors, researchers can achieve improved performance and scalability [55]. In chemical engineering, the adoption of parallel computing techniques has facilitated more efficient handling of certain large-scale optimization problems and simulations. However, while parallel computing has enhanced computational efficiency, there is still much room for further advancement in the development and application of parallel algorithms, specifically tailored for chemical engineering optimization and simulation.

Modern high-performance computing (HPC) architectures, such as multi-core CPUs and clusters with distributed memory, provide a scalable solution for such tasks by enabling the concurrent execution of multiple operations [56]. This approach is particularly beneficial for decomposing large-scale optimization problems into smaller, more manageable subproblems that can be solved concurrently.

Critical techniques for implementing parallel computing include the use of message-passing interfaces (MPI) and parallel linear algebra libraries, such as BLAS and LAPACK. These tools facilitate communication between processing units and optimizing memory usage, which are crucial for handling large datasets common in chemical engineering applications [57]. By efficiently distributing the workload across multiple cores or nodes, parallel computing reduces the time required to achieve solutions, making real-time optimization and control more feasible. This capability is essential for applications like real-time monitoring and optimization of chemical processes, where timely decision-making can significantly impact operational efficiency and safety. An example of MPI's application is the PyNumero and ParaPint framework [58], which is designed to solve dynamic optimization problems using scalable parallel nonlinear optimization techniques.

Despite its advantages, parallel computing also poses challenges, such as the need for specialized knowledge to develop parallel algorithms and manage data dependencies effectively. Additionally, the communication overhead between processing units can become a bottleneck if not handled correctly, particularly in distributed memory systems, where data must be transferred between nodes [59].

In addition, recent developments in the use of parallel computing have shown integration with graphical processing unit (GPU) architectures to solve complex optimization problems efficiently, providing substantial speedups and scalability for block-structured nonlinear programs [60]. This highlights the potential of combining parallel computing strategies with advanced hardware to address the increasing demands of chemical engineering optimization [56].

13.4.2 Graphical processing units

Graphics processing units (GPUs) have become a significant tool in modern computational optimization, especially in scenarios that benefit from high levels of parallelism. Initially designed for rendering graphics, GPUs have been adapted for general-purpose computing because of their ability to manage thousands of simultaneous threads efficiently. This capability makes them well-suited for tasks involving large-scale data processing, matrix operations, and iterative methods. Their architecture allows for massively parallel execution, which is particularly beneficial for algorithms such as those used in deep learning, molecular dynamics simulations and other computationally intensive tasks [61, 62].

In particular, GPUs have been used for dense linear algebra tasks. Libraries such as the basic linear algebra subprograms (BLAS) [63] have seen the implementation of linear algebra matrix and vector operations (e.g., matrix multiplication), optimized for particular hardware, including GPUs [64]. Libraries built on top of BLAS, such as LAPACK [65], implement higher-level linear algebra operations such as eigenvalue algorithms and linear equation solvers, and have also been able to harness the benefits on GPUs [66]. Contrary to deep learning and computer graphics, only a few optimization algorithms benefit from dense linear algebra computations, avoiding them harnessing the speedups provided by GPUs directly.

Regardless of these limitations, by specializing optimization algorithms, GPUs have been employed to accelerate various tasks. An example is nonlinear programming (NLP) where, by compressing the usually sparse linear algebra operations involved in solving systems of optimality conditions and performing interior-point iterations on this reduced space, orders of magnitude improvements have been achieved when using GPUs [67]. Specialized algorithms have been designed for specifically structured NLP problems, such as those arising from dynamic optimization or stochastic programming [60, 68] or defined over graphs [69], showing promise of harvesting the acceleration provided by this novel hardware. Applications of these optimization algorithms in chemical engineering have covered integrated networks of gas and electricity, as well as optimal control.

For linear programming (LP) problems, methods that rely on first-order methods have been adapted to be efficiently implemented in GPUs, making them competitive with highly engineered methods running on (multi-threaded) CPUs [70]. These methods seem to be most effective when handling really large-scale LP problems, usually beyond the scope of chemical engineering applications, both from a problem perspective, as the problems in chemical engineering involve nonlinearities, and from a size point of view, as the problems that concern chemical engineers that are linear tend not to reach scales that becomes challenging for existing solution methods.

Generalizing these gains for the mixed-integer case has been more evasive. Parallel implementations of branch-and-bound algorithms, on which most of the mixed-integer LP (MILP) problems rely, need to tackle issues mainly derived from the imbalance in the various subproblems that arise when using these algorithms. Such an imbalance results in inefficient parallel implementations, which have been tackled in the multi-core CPU systems available nowadays [71], although not yet for GPU systems.

For MINLP, the main algorithmic framework has been problem decomposition. In this approach, the original problem is broken into smaller subproblems, which should be easily solvable, given the suitable algorithm-subproblem matching. Coordinating the solution of the different subproblems can lead to a feasible and even globally optimal solution to these complex optimization problems. As a challenge, one could design decomposition approaches such that the resulting subproblems are both relevant to the solution of the original problem and efficiently solvable using GPUs. An example of this approach is given in [72], where linear relaxations of MINLP problems are generated and evaluated parallelly in GPUs, efficiently providing a dual bound for these problems, which, combined with feasible solutions, leads to guarantees on the global optimality of a solution. These decomposition approaches are relevant not only for GPU algorithms but also for all other approaches mentioned herein and elsewhere. By designing algorithms with the hardware in mind, potential performance gains can be exploited [56].

13.4.3 Quantum computing

Quantum computing represents a transformative approach to solving computational problems, leveraging principles such as superposition and entanglement, to process information potentially more efficiently than traditional or *classical* computational methods. Unlike classical computers, which process information using bits that can be either 0 or 1, quantum computers use qubits that can exist in superposition states, defined by linear combinations of states using complex coefficients. These properties result in the possibility of performing computational tasks exponentially faster than by classical computers. These advantages can be proved theoretically, relying on the fact that the coefficients, which define the probability of outcomes in a computation, can interact either constructively or destructively, and that the states that a system of N qubits requires 2^N coefficient to be expressed, resulting in an apparent parallelism [73].

Such a computational advantage from quantum computers has been observed experimentally in a few cases related to the sampling of multi-body systems [74–76]. Moreover, various algorithms with practical applications have been proposed to be run in quantum computers with provable asymptotic advantage with respect to any known classical algorithm, albeit these methods require quantum computers of sizes and characteristics not yet available [56].

One of the applications that quantum algorithms can prove to yield asymptotic advantages is in search and optimization, with the Grover's method of unstructured search [77], where finding a particular element in a dataset of n elements takes quadratically fewer steps than the best classical algorithm, which, on average, would require in the order of n queries to an oracle to verify the output. This search procedure would not consider the structure of the database at all, aiming to find the "marked" best solution without any other information than its value compared to the alternatives. This procedure can be generalized for optimization in the sense that the corresponding search value is the minimizer of a given objective function.

In classical optimization, exploiting the structure of the problem leads to algorithms that can potentially turn search procedures from impractical to practical. This has allowed, in the case that the problems to be solved have a worst-case complexity that grows non-polynomially or exponentially (also known as NP-hard problems), the tackling of practical applications. Many of the developments mentioned in this book make explicit usage of the problem structure, e.g., the type of constraints a problem has or the information available on the variables and functions, leading to significant progress in the state-of-the-art classical optimization. New proposals to accelerate these algorithms, including those incorporating quantum computation, need to overcome the best available alternatives in the literature.

Even with the sustained progress over the previous years, there is still plenty of room to improve our optimization solutions. Mainly motivated by the increasingly complicated and larger problems that we aim to address, we must explore algorithmic alternatives to tackle these instances.

Algorithms to be executed in quantum computers, or quantum algorithms, have been proposed to tackle different types of optimization problems. A broad classification includes those algorithms derived for continuous optimization problems and those for discrete optimization problems. We refer the interested reader to the recent review on quantum optimization that covers the topic in depth [78], yet we provide an overview of these methods herein.

Quantum algorithms for continuous optimization have been mainly proposed for convex problems. Classical efficient algorithms exist for these problems, in the sense that the worst-case complexity to address them grows polynomially, usually known as belonging to complexity class P. Even in this case, the use of quantum subroutines can lead to provable speedups, mainly by relying on the acceleration of linear algebra operations such as the solution of linear systems of equations or the factorization of matrices using quantum algorithms [79, 80]. Such algorithms have been proposed to tackle linear and semidefinite optimization problems. These algorithms are of first or second order, relying on either first- or second-order derivative information, and provide speedups with respect to the best-known algorithms in certain parameter regimes [81, 82]. These algorithms are to be implemented in fault-tolerant quantum computers and, at the point of the publication of this book, are beyond the reach of existing quantum computational devices. Moreover, their acceleration results depend on the usage of quantum random access memory, which is currently undeveloped [78].

When considering discrete optimization problems, an alternative beyond classical or von Neumann computers is the use of neural networks, implemented using analog electronic circuits, providing an alternative to solving these problems and the energy minimization of a physical system. Examples of these are the acclaimed Hopfield Networks, awarded with the Nobel Prize for Physics in 2024 [83, 84]. Another example relies on quantum techniques, where the underlying algorithm is the quantum adiabatic algorithm [85], which proves that by performing an adiabatic evolution of a quantum system, one can guarantee that its initial and final states, described by discrete levels, each particle in the system sitting at a discrete level of energy will have total energies with the same ranking at each time step of the system, at each time-step of the adiabatic evolution. Using this algorithm, one can aim to address a start at a state that is easy to set up and whose minimum energy is known, and evolve it adiabatically to a final system. That final system can be a representation of a discrete optimization problem of interest, where the objective function corresponds to the function that describes the energy of that final system. Once the adiabatic evolution is performed, the final state is guaranteed to be sitting at the lowest energy state, which, by virtue of the mapping, corresponds to the optimal solution of the discrete optimization problem. This algorithm alone can be proved to be asymptotically more efficient for certain families of discrete optimization problems [86]. Yet, it involves the implementation of an adiabatic evolution, namely one that consists of no information/heat exchange with the environment. This is practically infeasible to achieve, as it would require, e.g., execution at zero kelvin. Yet, practical implementations of this algorithm have motivated the development of specialized hardware operating in extreme conditions, running algorithms aiming to replicate the pathway described by the adiabatic evolution. Because of the equivalence of the energy function describing the systems in question to the well-known Ising model of spin systems in statistical mechanics, hardware implementations of the quantum adiabatic algorithm are an example of Ising machines. Those relying on quantum phenomena can be classified as "quantum annealers" [87, 88] and coherent Ising machines [89, 90]. Since guaranteeing an adiabatic evolution might be beyond the practical implementation of these Ising machines, the convergence to optimal solutions of these methods is no longer granted, leading to these becoming heuristic methods for discrete optimization. Another limitation is that the Ising model is equivalent to a quadratic unconstrained binary optimization (QUBO) problem, which, as its name mentions, has no constraints, besides the nature of its variables. Although nonlinear, the lack of constraint severely limits the scope of QUBO.

Quantum annealing receives its name from the classical heuristic for discrete optimization, simulated annealing [35], as the evolution aiming to approximate the adiabatic pathway can be interpreted as an annealing process allowing for quantum phenomena [91]. One of the best-known implementations of quantum annealing is given by the company D-Wave, which, in their latest version, uses up to 5,000 superconducting qubits connected in a particular topology that allows for the compilation of sparse problems with as many variables as qubits [92]. Given the conditions under which these systems need to operate, accessing them is done through cloud resources, making them an accessible resource for those aiming to integrate quantum computation on optimization workloads. One of the main limitations of these systems in addressing arbitrary optimization problems is that not every pair of qubits in their chips is connected, leading to a pre-compilation routine known as embedding, which is a challenging optimization problem on its own [93]. Considering these limitations, experiments dealing with problems that natively fit in the hardware have proved that these devices can perform more efficiently than any other available optimization heuristic [94].

Coherent Ising machines make use of light to perform computations and arise as an alternative to superconducting qubits and electronic networks as computational models for solving the Ising problem. These machines use optical pulses within a laser system or optical circuit to simulate the spin interactions, leveraging quantum-like effects such as coherence and superposition. As the system evolves, the optical pulses adjust to reach a stable, low-energy state, ideally corresponding to the optimal or near-optimal solution to the problem. In particular, the system uses degenerate optical parametric oscillators (DOPO) to encode the problem and make use of coherent light, where photons that are in-phase oscillate at the same frequency and have the same wavelengths. The computation with light makes the construction of these devices less restrictive than quantum annealers, allowing for room temperature usage and providing the possibility of enabling full connectivity among its spins, avoiding the embedding problem. Performance comparisons have shown that depending on the instance to be solved, coherent Ising machines or quantum annealers exhibit better performance than others [95].

Another alternative to solving Ising models is to make use of the most popular alternative of quantum computers currently available. These are known as gate-based computers and allow implementing algorithms by enabling the modification of systems of qubits by quantum operators, also known as quantum gates [73]. A number of different implementations currently exist to physically realize these qubit systems and their gates, including superconducting qubits, ion-based quantum computers, and neutral atoms. At the moment of publishing this book, none of these devices have reached the threshold of implementing scalable schemes to counteract the decoherence of quantum states, limiting the number of qubits and gates that can be implemented. These limitations must be overcome to be able to use quantum computers for continuous optimization, as mentioned above. Yet, there is a potential of using current noisy intermediate-scale quantum (NISQ) devices [96] for discrete optimization.

Inspired by the quantum adiabatic algorithm, one can define a discrete approximation of the adiabatic pathway involving quantum gates. This approximation uses the time evolution of the system in two directions, one related to a mixing of possible solutions and another one toward the energy function encoding the optimization problem. This approximation then depends on the number of discretization steps to be taken and on parameters that define for how long each direction is evolved. This approach is known as the quantum approximation optimization algorithms (QAOAs) [97], and it can potentially be implemented in NISQ devices, as the sequence of gates required for each evolution is comparatively small, to the point that it could be implemented within the windows that current devices can maintain a coherent state. The number of discretization steps is defined a priori, and the length of each time evolution is solved using a classical computer, which involves solving a classical nonconvex optimization problem, similar to how neural networks are trained. This algorithm has also received attention as specific performance metrics can be guaranteed, achieving in polynomial time a solution always within a smaller optimality gap than classical alternatives [98]. There exist empirical studies that compare the performance of different quantum optimization techniques for QUBO, highlighting the current and evolving landscape of this technology [99].

However, the practical implementation of quantum computing in chemical engineering optimization is still in its infancy, primarily due to the current limitations of quantum hardware. Noise, error rates, and the need for quantum error correction are significant challenges that must be addressed before quantum computers can reliably outperform classical systems in practical applications [56].

As mentioned above, the problem tackled by quantum methods in discrete optimization limits QUBO. This places a barrier to use these methods for practical applications in optimization and, consequently, in chemical engineering. As chemical engineers, we are accustomed to model optimization problems as mixed-integer nonlinear programs (MINLPs) that incorporate the physical rules that our system is subjected to by the constraints. Although one can approximate and, in certain cases, exactly reformulate MINLPs into QUBOs, and there are even open-source implementations that automate this process [100], these reformulations lead to an increase in the size of the instances that might limit the usability of these quantum devices.

An exciting avenue for future research is the study of hybrid approaches that combine quantum algorithms with classical optimization techniques. These can leverage the strengths of both paradigms, potentially offering a path to scalable quantum-enhanced optimization. Possible ideas that have served the purpose of making classical optimization practical, such as decomposition techniques, can allow the integration of QUBO solution methods with other subroutines to address more general optimization problems. Initial proposals have appeared in the literature. Yet, a common observation that needs to be assessed is whether the resulting decomposition method becomes performant because of the use of potentially quantum subroutines or despite it. The observation of reliance on quantum subroutines for performant use would lead to practical quantum advantage, a goal sought by the quantum computing community. Yet, the development of new classical solution methods, inspired by the use of novel hardware, would still be beneficial for the optimization community and, by extension, for chemical engineering practitioners.

Moreover, the use of quantum computational techniques for problems beyond optimization is of interest to the chemical engineering community, with the potential to tackle problems in computational chemistry and machine learning [101].

13.5 Conclusions

Chemical engineering and optimization have become deeply connected and have been cross-pollinating each other for several decades. This book gathers both state-of-the-art developments and future perspectives in optimization within chemical engineering, with this chapter especially offering a forward-looking view. The training in chemical engineering, guided by the mission of understanding and designing processes that transform matter and energy into valuable products for society, is inherently an optimization challenge. Chemical engineers operate in complex systems with many degrees of freedom and near-infinite solution alternatives, from which we must select those that optimize, minimize, or maximize specific objectives. These selections are governed by limited resources and are bound by the unyielding constraints of physics and chemistry.

This chapter has highlighted what we anticipate will be exciting avenues of research and development at the intersection of optimization and chemical engineering. While no predictions can fully capture the future, we intend to expose readers to ideas and potential advancements that we believe will inspire new pathways in these interconnected fields. The ever-evolving needs of society drive both fields forward, and the dialogue between them is fundamental in solving real-world problems that grow increasingly complex.

Learning optimization in chemical engineering means learning to apply the entirety of our discipline's knowledge in real-world contexts. This application is not only transformative but also stretches the boundaries of what optimization can achieve on a continual basis. By approaching optimization through the lens of chemical engineering, we help shape the field's frontiers, transforming theoretical tools into impactful solutions that meet the technological and industrial challenges of today and tomorrow. We hope that this book serves as both a guide and a catalyst for engineers and optimization experts, fostering a collaborative space where the next generation of research and innovation can flourish at the vibrant intersection of these fields.

References

[1]   Shafer, D. S. The brachistochrone: Historical gateway to the Calculus of Variations. *Materials Matemàtics*, 2007, 1–14.
[2]   Floudas, C. A., Pardalos, P. M., Adjiman, C. S., Esposito, W. R., Gümüş, Z. H., Harding, S. T., Schweiger, C. A. *Handbook of Test Problems in Local and Global Optimization*. Springer, 2010.
[3]   Biegler, L. T., Grossmann, I. E., Westerberg, A. W. *Systematic Methods of Chemical Process Design*. Prentice Hall, 1997.
[4]   Sargent, R. W. H. Introduction: 25 years of progress in process systems engineering. *Computers & Chemical Engineering*, 2004, 28(4), 437–439.

[5]   Živković, L. A., Pohar, A., Likozar, B., Nikačević, N. M. Reactor conceptual design by optimization for hydrogen production through intensified sorption- and membrane-enhanced water-gas shift reaction. *Chemical Engineering Science*, 2020, 211, 115174.

[6]   Sargent, R. W. H. Advances in modelling and analysis of chemical process systems. *Computers & Chemical Engineering*. 1983, 7(4), 219–237.

[7]   Hu, Y., Li, F., Wei, S., Jin, S., Shen, W. Design and optimization of the efficient extractive distillation process for separating the binary azeotropic mixture methanol-acetone based on the quantum chemistry and conceptual design. *Separation and Purification Technology*, 2020, 242, 116829.

[8]   Lara-Montaño, O. D., Gómez-Castro, F. I., Gutiérrez-Antonio, C. Comparison of the performance of different metaheuristic methods for the optimization of shell-and-tube heat exchangers. *Computers & Chemical Engineering*, 2021, 152, 107403.

[9]   Sharikov, Y. V., Sharikov, F. Y., Krylov, K. A. Mathematical model of optimum control for petroleum coke production in a rotary tube kiln. *Theoretical Foundations of Chemical Engineering*, 2021, 55, 711–719.

[10] Wu, X., Hou, Y., Zhang, K. Optimal control approach for nonlinear chemical processes with uncertainty and application to a continuous stirred-tank reactor problem. *Arabian Journal of Chemistry*, 2022, 15(11), 104257.

[11] Sabrina, H. Regression of NRTL parameters from liquid–liquid equilibria for water + ethanol + solvent (dichloromethane, diethyl ether and chloroform) using particle swarm optimization and discussions at T =293.15 K. *Algerian Journal of Engineering Research*, 2022, 5(2), 12–27.

[12] Lee, H.-Y., Chen, -Y.-Y., Eiamsuttitam, P., Alcántara-Ávila, J. R. Comparison of optimization methods for the design and control of reactive distillation with inter condensers. *Computers & Chemical Engineering*, 2022, 164, 107871.

[13] Méndez-Vázquez, M. A., Gómez-Castro, F. I., Ponce-Ortega, J. M., Serafín-Muñoz, A. H., Santibañez-Aguilar, E., El-Halwagi, M. M. Mathematical optimization of a supply chain for the production of fuel pellets from residual biomass. *Clean Technologies and Environmental Policy*, 2017, 19, 721–734.

[14] Duran, M. A., Grossmann, I. E. An outer-approximation algorithm for a class of mixed-integer nonlinear programs. *Mathematical Programming*, 1986, 36, 307–339.

[15] Varvarezos, D. K., Grossmann, I. E., Biegler, L. T. An outer-approximation method for multiperiod design optimization. *Industrial & Engineering Chemistry Research*, 1992, 31(6), 1466–1477.

[16] Kronqvist, J., Bernal, D. E., Grossmann, I. E. Using regularization and second order information in outer approximation for convex MINLP. *Mathematical Programming*, 2020, 180, 285–310.

[17] Sahinidis, N. V. BARON: A general purpose global optimization software package. *Journal of Global Optimization*, 1996, 8, 201–205.

[18] Maranas, C. D., Floudas, C. A. Global optimization in generalized geometric programming. *Computers & Chemical Engineering*, 1997, 21(4), 351–369.

[19] Adjiman, C. S., Androulakis, I. P., Floudas, C. A. A global optimization method, αBB, for general twice-differentiable constrained NLPs – II. Implementation and computational results. *Computers & Chemical Engineering*, 1998, 22(9), 1159–1179.

[20] Tawarmalani, M., Sahinidis, N. V. Global optimization of mixed-integer nonlinear programs: A theoretical and computational study. *Mathematical Programming*, 2004, 99, 563–591.

[21] Misener, R., Floudas, C. A. ANTIGONE: Algorithms for continuous/integer global optimization of nonlinear equations. *Journal of Global Optimization*, 2014, 59(2), 503–526.

[22] Boukouvala, F., Misener, R., Floudas, C. A. Global optimization advances in mixed-integer nonlinear programming, MINLP, and constrained derivative-free optimization, CDFO. *European Journal of Operational Research*, 2016, 252(3), 701–727.

[23] Raman, R., Grossmann, I. E. Modelling and computational techniques for logic based integer programming. *Computers & Chemical Engineering*, 1994, 18(7), 563–578.

[24] Lee, S., Grossmann, I. E. New algorithms for nonlinear generalized disjunctive programming. *Computers & Chemical Engineering*, 2000, 24(9–10), 2125–2141.

[25] Logsdon, J. S., Biegler, L. T. Decomposition strategies for large-scale dynamic optimization problems. *Chemical Engineering Science*, 1992, 47(4), 851–864.

[26] Tanartkit, P., Biegler, L. T. Stable decomposition for dynamic optimization. *Industrial & Engineering Chemistry Research*, 1995, 34(4), 1253–1266.

[27] Daoutidis, P., Lee, J. H., Harjunkoski, I., Skogestad, S., Baldea, M., Georgakis, C. Integrating operations and control: A perspective and roadmap for future research. *Computers & Chemical Engineering*, 2018, 115, 179–184.

[28] Diwekar, U. M., Kalagnanam, J. R. Efficient sampling technique for optimization under uncertainty. *AIChE Journal*, 1997, 43(2), 440–447.

[29] Sahinidis, N. V. Optimization under uncertainty: State-of-the-art and opportunities. *Computers & Chemical Engineering*, 2004, 28(6–7), 971–983.

[30] Dige, N., Diwekar, U. M. Efficient sampling algorithm for large-scale optimization under uncertainty problems. *Computers & Chemical Engineering*, 2018, 115, 431–454.

[31] Pistikopoulos, E. N., Barbosa-Povoa, A., Lee, J. H., Misener, R., Mitsos, A., Reklaitis, G. V., Venkatasubramanian, V., You, F., Gani, R. Process systems engineering–the generation next? *Computers &Chemical Engineering*, 2021, 147, 107252.

[32] Grossmann, I. E., Harjunkoski, I. Process systems engineering: Academic and industrial perspectives. *Computers & Chemical Engineering*, 2019, 126, 474–484.

[33] Biegler, L. T. Tailoring optimization algorithms to process applications. *Computers & Chemical Engineering*, 1992, 16, S81–S95.

[34] Wolpert, D. H., Macready, W. G. No free lunch theorems for optimization. *IEEE Transactions on Evolutionary Computation*, 1997, 1, 67–82.

[35] Kirkpatrick, S., Gelatt, C. D., Vecchi, M. P. Optimization by simulated annealing. *Science*, 1983, 220(4598), 671–680.

[36] Deb, K., Pratap, A., Agarwal, S., Meyarivan, T. A fast and elitist multiobjective genetic algorithm: NSGA-II. *IEEE Transactions on Evolutionary Computation*, 2002, 6(2), 182–197.

[37] Talbi, E. G. *Metaheuristics: From Design to Implementation*. Wiley, 2009.

[38] Blum, C., Puchinger, J., Raidl, G. R., Roli, A. Hybrid metaheuristics in combinatorial optimization: A survey. *Applied Soft Computing*, 2011, 11(6), 4135–4151.

[39] Biegler, L. T., Grossmann, I. E. Retrospective on optimization. *Computers & Chemical Engineering*, 2004, 28(8), 1169–1192.

[40] Glover, F., Kochenberger, G. A. *Handbook of Metaheuristics*. Kluwer Academic Publishers, 2003.

[41] Raissi, M., Perdikaris, P., Karniadakis, G. E. Physics-informed neural networks: A deep learning framework for solving forward and inverse problems involving nonlinear partial differential equations. *Journal of Computational Physics*, 2019, 378, 686–707.

[42] Karpatne, A., Atluri, G., Faghmous, J. H., Steinbach, M., Banerjee, A., Ganguly, A., Shekhar, S., Samatova, N., Vipin, K. Theory-guided data science: A new paradigm for scientific discovery from data. *IEEE Transactions on Knowledge and Data Engineering*, 2017, 29(10), 2318–2331.

[43] Xu, W., Wang, Y., Zhang, D., Yang, Z., Yuan, Z., Lin, Y., Yan, H., Zhou, X., Yang, C. Transparent AI-assisted chemical engineering process: Machine learning modeling and multi-objective optimization for integrating process data and molecular-level reaction mechanisms. *Journal of Cleaner Production*, 2024, 448, 141412.

[44] Jin, L. X. Z., Wang, K., Zhang, K., Wu, D., Nazir, A., Jiang, J., Liao, W. Big data, machine learning, and digital twin assisted additive manufacturing: A review. *Materials and Design*, 2024, 244, 113086.

[45] Belaud, J. P., Negny, S., Dupros, F., Michéa, D., Vautrin, B. Collaborative simulation and scientific big data analysis: Illustration for sustainability in natural hazards management and chemical process engineering. *Computers in Industry*, 2014, 65(3), 521–535.

[46] Ning, C., You, F. Optimization under uncertainty in the era of big data and deep learning: When machine learning meets mathematical programming. *Computers & Chemical Engineering*, 2019, 125, 434–448.

[47] Powell, B. K. M., Machalek, D., Quah, T. Real-time optimization using reinforcement learning. *Computers & Chemical Engineering*, 2020, 143, 107077.

[48] Hubbs, C. D., Li, C., Sahinidis, N. V., Grossmann, I. E., Wassick, J. M. A deep reinforcement learning approach for chemical production scheduling. *Computers & Chemical Engineering*, 2020, 141, 106982.

[49] Jiang, Y., Fan, J., Chai, T., Li, J., Lewis, F. L. Data-Driven Flotation Industrial Process Operational Optimal Control Based on Reinforcement Learning. *IEEE Transactions on Industrial Informatics*, 2018, 14, 1974–1989.

[50] Ryu, Y., Shin, S., Liu, J. J., Lee, W., Na, J. Physics-informed neural networks for optimization of polymer reactor design. *Computer Aided Chemical Engineering*, 2023, 52, 493–498.

[51] Box, G. E. P., Draper, N. R. *Empirical Model-Building and Response Surfaces*. Wiley, 1987.

[52] Peherstorfer, B., Willcox, K., Gunzburger, M. Survey of multifidelity methods in uncertainty propagation, inference, and optimization. *SIAM Review*, 2016, 60(3), 550–591.

[53] Moore, G. E. Cramming more components onto integrated circuits. *Electronics*, 1965, 38(8), 114–117.

[54] Waldrop, M. M. The chips are down for Moore's law. *Nature*, 2016, 530(7589), 144–147.

[55] Grama, A., Gupta, A., Karypis, G., Kumar, V. *Introduction to Parallel Computing*. Pearson Education, 2003.

[56] Bernal Neira, D. E., Laird, C. D., Lueg, L. R., Harwood, S. M., Trenev, D., Venturelli, D. Utilizing modern computer architectures to solve mathematical optimization problems: A survey. *Computers & Chemical Engineering*, 2024, 184, 108627.

[57] Bernal Neira, D. E., Laird, C. D., Harwood, S. M., Trenev, D., Venturelli, D. Impact of emerging computing architectures and opportunities for process systems engineering applications. *Foundations of Computer-Aided Process Operations (Focapo)/chemical Process Control (CPC) 2023*, 2023.

[58] Rodriguez, J. S., Parker, R. B., Laird, C. D., Nicholson, B. L., Siirola, J. D., Bynum, M. L. Scalable parallel nonlinear optimization with PyNumero and Parapint. *INFORMS Journal on Computing*, 2023, 35(2), 509–517.

[59] Dongarra, J. An Overview of High Performance Computing and Challenges for the Future. In Laginha, J. M., Palma, M., Amestoy, P. R., Daydé, M., Mattoso, M., Lopes, J. C. (Eds.). *High Performance Computing for Computational Science – VECPAR*. Springer, 2008. pp. 1.

[60] Pacaud, F., Schanen, M., Shin, S., Maldonado, D. A., Anitescu, M. Parallel interior-point solver for block-structured nonlinear programs on SIMD/GPU architectures. *Optimization Methods & Software*. In Press, https://doi.org/10.1080/10556788.2024.2329646.

[61] Owens, J. D., Houston, M., Luebke, D., Green, S., Stone, J. E., Phillips, J. C. GPU computing. *Proceedings of the IEEE*, 2008, 96(5), 879–899.

[62] Dongarra, J. J. The evolution of mathematical software. *Communications of the ACM*, 2022, 65(12), 66–72.

[63] Blackford, L. S., Demmel, J., Dongarra, J., Duff, I., Hammarling, S., Henry, G., Heroux, M., Kaufman, L., Lumsdaine, A., Petitet, A., Pozo, R., Remington, K., Whaley, R. C. An updated set of basic linear algebra subprograms (BLAS). *ACM Trans. Math. Software*, 2002, 28(2), 135–151.

[64] NVIDIA Developer, cuBLAS, 2013. At https://developer.nvidia.com/cublas

[65] Anderson, E., Bai, Z., Bischof, C., Blackford, L. S., Demmel, J., Dongarra, J., Du Croz, J., Greenbaum, A., Hammarling, S., McKenney, A., Sorensen, D. LAPACK Users' Guide. *SIAM*, 1999.

[66] Dongarra, J., Gates, M., Haidar, A., Kurzak, J., Luszczek, P., Tomov, S., Yamazaki, I. Accelerating Numerical Dense Linear Algebra Calculations with GPUs. In At Kindratenko, V. (Ed.). *Numerical Computations with GPUs*. Cham: Springer, 2014, pp. 3–28.

[67] Pacaud, F., Shin, S., Schanen, M., Maldonado, D. A., Anitescu, M. Accelerating condensed interior-point methods on SIMD/GPU architectures. *Journal of Optimization Theory and Applications*, 2024(202), 184–203.

[68] Kang, J., Cao, Y., Word, D. P., Laird, C. D. An interior-point method for efficient solution of block-structured NLP problems using an implicit Schur-complement decomposition. *Computers and Chemical Engingeering*, 2014(71), 563–573.

[69] Shin, S., Zavala, V. M., Anitescu, M. Decentralized schemes with overlap for solving graph-structured optimization problems. *IEEE Transactions on Control of Network Systems*, 2020, 7(3), 1225–1236.

[70] Applegate, D., Díaz, M., Hinder, O., Lu, H., Lubin, M., O'Donoghue, B., Schudy, W. Practical large-scale linear programming using primal-dual hybrid gradient. *Advances in Neural Information Processing Systems*, 2021, 34, 20243–20257.

[71] Ralphs, T., Shinano, Y., Berthold, T., Koch, T. Parallel Solvers for Mixed Integer Linear Optimization. In Hamadi, Y., Sais, L. (Eds.). *Handbook of Parallel Constraint Reasoning*. Cham: Springer, Vol. 2018.

[72] Gottlieb, R. X., Xu, P., Stuber, M. D. Automatic source code generation of complicated models for deterministic global optimization with parallel architectures. *Optimization Methods & Software*. In Press, https://doi.org/10.1080/10556788.2024.2396297.

[73] Nielsen, M. A., Chuang, I. L. *Quantum Computation and Quantum Information*. Cambridge University Press, 2010.

[74] Arute, F., Arya, K., Babbush, R., Bacon, D., Bardin, J. C., Barends, R., Biswas, R., Boixo, S., Brandao, F. G. S. L., Buell, D. A., Burkett, B., Chen, Y., Chen, Z., Chiaro, B., Colins, R., Courtney, W., Dunsworth, A., Farhi, E., Foxen, B., Fowler, A., Gidney, C., Giustina, M., Graff, R., Guerin, K., Habegger, S., Harrigan, M. P., Hartmann, M. J., Ho, A., Hoffmann, M., Huang, T., Humble, T. S., Isakov, S. V., Jeffrey, E., Jiang, Z., Kafri, D., Kechedzhi, K., Kelly, J., Klimov, P. V., Knysh, S., Korotkov, A., Kostritsa, F., Landhuis, D., Lindmark, M., Lucero, E., Lyakh, D., Mandrà, S., McClean, J. R., McEwen, M., Megrant, A., Mi, X., Michielsen, K., Mohseni, M., Mutus, J., Naaman, O., Neeley, M., Neill, C., Niu, M. Y., Ostby, E., Petukhov, A., Platt, J. C., Quintana, C., Rieffel, E. G., Roushan, P., Rubin, N. C., Sank, D., Satzinger, K. J., Smelyansjiy, V., Sung, K. J., Trevithick, M. D., Vainsencher, A., Villalonga, B., White, T., Yao, Z. J., Yeh, P., Zalcman, A., Neven, H., Martinis, J. M. Quantum supremacy using a programmable superconducting processor. *Nature*, 2019, 574, 505–510.

[75] Wu, Y., Bao, W.-S., Cao, S., Chen, F., Chen, M.-C., Chen, X., Chung, T.-H., Deng, H., Du, Y., Fan, D., Gong, M., Guo, C., Guo, C., Guo, S., Han, L., Hong, L., Huang, H.-L., Huo, Y.-H., Li, L., Li, N., Li, S., Li, Y., Liang, F., Lin, C., Lin, J., Qian, H., Qiao, D., Rong, H., Su, H., Sun, L., Wang, L., Wang, S., Wu, D., Xu, Y., Yan, K., Yang, W., Yang, Y., Ye, Y., Yin, J., Ying, C., Yu, J., Zha, C., Zhang, C., Zhang, H., Zhang, K., Zhang, Y., Zhao, H., Zhao, Y., Zhou, L., Zhu, Q., Lu, C.-Y., Peng, C.-Z., Zhu, X., Pan, J.-W. Strong quantum computational advantage using a superconducting quantum processor. *Physical Review Letters*, 2021, 127(18), 180501.

[76] Madsen, L. S., Laudenbach, F., Askarani, M. F., Rortais, F., Vincent, T., Bulmer, J. F. F., Miatto, F. M., Neuhaus, L., Helt, L. G., Collins, M. J., Lita, A. E., Gerrits, T., Nam, S. W., Vaidya, V. D., Menotti, M., Dhand, I., Vernon, Z., Quesada, N., Lavoie, J. Quantum computational advantage with a programmable photonic processor. *Nature*, 2022, 606, 75–81.

[77] Grove, L. K. Quantum mechanics helps in searching for a needle in a haystack. *Physical Review Letters*, 1997, 79(2), 325.

[78] Abbas, A., Ambainis, A., Augustino, B., Bärtschi, A., Buhrman, H., Coffrin, C., Cortiana, G., Dunjko, V., Egger, D. J., Elmegreen, B. G., Franco, N., Fratini, F., Fuller, B., Gacon, J., Gonciulea, C., Gribling, S., Gupta, S., Hadfield, S., Heese, R., Kircher, G., Kleinert, T., Koch, T., Korpas, G., Lenk, S., Marecek, J., Markov, V., Mazzola, G., Mensa, S., Mohseni, N., Nannicini, G., O'Meara, C., Peña Tapia, E., Pokutta, S., Proissl, M., Rebentrost, P., Sahin, E., Symons, B. C. B., Tornow, S., Valls, V., Woerner, S., Wolf-Bauwens, M. L., Yard, J., Yarkoni, S., Zechiel, D., Zhuk, S., Zoufal, C. Quantum optimization: Potential,

challenges, and the path forward. 2023, *arXiv preprint arXiv:2312.02279*. https://arxiv.org/abs/2312. 02279

[79] Harrow, A. W., Hassidim, A., Lloyd, S. Quantum algorithm for linear systems of equations. *Physical Review Letters*, 2009, 103, 150502.

[80] Childs, A. M., Kothari, R., Somma, R. D. Quantum algorithm for systems of linear equations with exponentially improved dependence on precision. *SIAM Journal on Computing*, 2017, 46, 1920–1950.

[81] Van Apeldoorn, J., Gilyén, A. Improvements in Quantum Sdp-solving with Applications. In *46th International Colloquium on Automata, Languages, and Programming (ICALP 2019). Leibniz International Proceedings in Informatics (Lipics)*. Vol. 132, 99:1–99:15, Schloss Dagstuhl – Leibniz-Zentru für Informatik, 2019.

[82] Augustino, B., Nannicini, G., Terlaky, T., Zuluaga, L. Solving the semidefinite relaxation of QUBOS in matrix multiplication time, and faster with a quantum computer. 2023, *arXiv preprint arXiv:2301.04237*. https://arxiv.org/abs/2301.04237

[83] Hopfield, J. J., Tank, D. W. Computing with neural circuits: A model. *Science*, 1986, 233, 625–633.

[84] Tank, D., Hopfield, J. Simple 'neural' optimization networks: An A/D converter, signal decision circuit, and a linear programming circuit. *IEEE Transactions on Circuits and Systems*, 1986, 33, 533–541.

[85] Farhi, E., Goldstone, J., Gutmann, S., Sipser, M. Quantum computation by adiabatic evolution. 2000, *arXiv preprint* arXiv:quant-ph/0001106. https://arxiv.org/abs/quant-ph/0001106.

[86] Albash, T., Lidar, D. A. Adiabatic quantum computation. *Reviews of Modern Physics*, 2018, 90, 015002.

[87] King, A. D., Suzuki, S., Raymond, J., Zucca, A., Lanting, T., Altomare, F., Berkley, A. J., Ejtemaee, S., Hoskinson, E., Huang, S., Ladizinsky, E., MacDonald, A. J. R., Marsden, G., Oh, T., Poulin-Lamarre, G., Reis, M., Rich, C., Sato, Y., Whittaker, J. D., Yao, J., Harris, R., Lidar, D. A., Nishimori, H., Amin, M. H. Coherent quantum annealing in a programmable 2,000 qubit Ising chain. *Nature Physics*, 2022, 18, 1324–1328.

[88] Johnson, M. W., Amin, M. H. S., Gildert, S., Lanting, T., Hamze, F., Dickson, N., Harris, R., Berkley, A. J., Johansson, J., Bunyk, P., Chapple, E. M., Enderud, C., Hilton, J. P., Karimi, K., Ladizinsky, N., Oh, T., Perminov, I., Rich, C., Thom, M. C., Tolkacheva, E., Truncik, C. J. S., Uchaikin, S., Wang, J., Wilson, B., Rose, G. Quantum annealing with manufactured spins. *Nature*, 2011, 473, 194–198.

[89] Takesue, H., Inagaki, T., Inaba, K., Ikuta, T., Honjo, T. Large-scale coherent Ising machine. *Journal of the Physical Society of Japan*, 2019, 88(6), 061014.

[90] Yamamoto, Y., Aihara, K., Leleu, T., Kawarabayashi, K.-I., Kako, S., Fejer, M., Inoue, K., Takesue, H. Coherent Ising machines – Optical neural networks operating at the quantum limit. *Npj Quantum Information*, 2017, 3, 49.

[91] Rajak, A., Suzuki, S., Dutta, A., Chakrabarti, B. K. Quantum annealing: An overview. *Philosophical Transactions of the Royal Society A*, 2023, 381(2241), 20210417.

[92] McGeoch, C., Farré, P. The advantage system: Performance update. Technical Report. https://www. dwavesys.com/media/kjtlcemb/14-1054a-a_advantage_system_performance_update.pdf

[93] Bernal, D. E., Booth, K. E. C., Dridi, R., Alghassi, H., Tayur, S., Venturelli, D. Integer Programming Techniques for Minor-embedding in Quantum Annealers. In Hebrard, E., Musliu, N. (eds.). *Integration of Constraint Programming, Artificial Intelligence, and Operations Research: 17th International Conference, CPAIOR 2020*. Vienna, Austria September 21–24 2020Proceedings Springer 2020 112–129.

[94] Tasseff, B., Albash, T., Morrell, Z., Vuffray, M., Lokhov, A. Y., Misra, S., Coffrin, C. On the Emerging Potential of Quantum Annealing Hardware for Combinatorial Optimization. *Journal of Heuristics*, 2024, 30, 325–358.

[95] Hamerly, R., Inagaki, T., McMahon, P. L., Venturelli, D., Marandi, A., Onodera, T., Ng, E., Langrock, C., Inaba, K., Honjo, T., Enbutsu, K., Umeki, T., Kasahara, R., Utsunomiya, S., Kako, S., Kawarabayashi, K.-I., Byer, R. L., Fejer, M. M., Mabuchi, H., Englund, D., Rieffel, E., Takesue, H., Yamamoto,

Y. Experimental investigation of performance differences between coherent Ising machines and a quantum annealer. *Science Advances*, 2019, 5(5), eaau0823.

[96]   Preskill, J. Quantum computing in the NISQ era and beyond. *Quantum*, 2018, 2, 79.

[97]   Farhi, E., Goldstone, J., Gutmann, S. A quantum approximate optimization algorithm. 2014, *arXiv preprint arXiv:1411.4028*. https://arxiv.org/abs/1411.4028.

[98]   Farhi, E., Goldstone, J., Gutmann, S., Zhou, L. The quantum approximate optimization algorithm and the Sherrington-Kirkpatrick model at infinite size. *Quantum*, 2022, 6, 759.

[99]   Lubinski, T., Coffrin, C., McGeoch, C., Sathe, P., Apanavicius, J., Bernal Neira, D. Optimization applications as quantum performance benchmarks. *ACM Transactions on Quantum Computing*, 2024, 5(3), 1–44.

[100] Maciel Xavier, P., Ripper, P., Andrade, T., Dias Garcia, J., Maculan, N., Bernal Neira, D. E. QUBO. jl: A Julia ecosystem for quadratic unconstrained binary optimization. 2023, *arXiv preprint arXiv:2307.02577*. https://arxiv.org/abs/2307.02577.

[101] Bernal, D. E., Ajagekar, A., Harwood, S. M., Stober, S. T., Trenev, D., You, F. Perspectives of quantum computing for chemical engineering. *AIChE Journal*, 2022, 68(6), e17651.

Index

https://doi.org/10.1515/9783111383439-014

www.ingramcontent.com/pod-product-compliance
Lightning Source LLC
Jackson TN
JSHW052053270325
81542JS00005B/47